PLANNING IN THE USA

This extensively revised and updated fourth edition of *Planning in the USA* continues to provide a comprehensive introduction to the policies, theory, and practice of planning. Outlining land use, urban planning, and environmental protection policies, this fully illustrated book explains the nature of the planning process and the way in which policy issues are identified, defined, and approached.

This full color edition incorporates new planning legislation and regulations at the state and federal layers of government, updated discussion on current economic issues, and examples of local ordinances in

- a new chapter on planning and sustainability;
- a new discussion on the role of foundations and giving to communities;
- a discussion regarding the aftermath of Katrina in New Orleans;
- a discussion on deindustrialization and shrinking cities;
- a discussion on digital billboards;
- a discussion on recent comprehensive planning efforts;
- a discussion on land banking;
- a discussion on unfunded mandates;
- a discussion on community character;
- a companion website with multiple choice questions, and downloadable tables and figures from the book.

This book gives a detailed account of urbanization in the United States and reveals the problematic nature and limitations of the planning process, the fallibility of experts, and the difficulties facing policy-makers in their search for solutions. *Planning in the USA* is an essential book for students, planners, and all who are concerned with the nature of contemporary urban and environmental problems.

Barry Cullingworth was Emeritus Professor of Urban Affairs and Public Policy at the University of Delaware, and held posts at the universities of Manchester, Durham, Glasgow, Birmingham, Toronto, and Cambridge.

Roger W. Caves is a Professor of City Planning at San Diego State University (SDSU).

PLANNING IN THE USA

"This updated edition improves upon an already great planning textbook. Cullingworth and Caves, once again, provide a thorough, unbiased and high-quality overview of the most important debates in the field with sufficient depth and detailed information to give students entree to the actual practice of planning. The newly added sections on the impacts of Hurricane Katrina, shrinking cities, and sustainability expand the appeal and utility of this text in a variety of classroom settings."

Harley F. Etienne, *Assistant Professor of Urban and Regional Planning, Taubman College of Architecture and Urban Planning, University of Michigan.*

This excellent and comprehensive text serves as an introduction to the field and practice of urban planning. It is exceedingly useful in public administration courses that focus on the complexities of managing urban environments. Like previous editions, authors Cullingworth and Caves address contemporary topics, policy advancements and new challenges of the profession in an understandable and insightful discussion. Topic themes are clearly presented and make this book a great choice for teaching undergraduate students."

Marie R. Wong, Ph.D., *Associate Professor, Institute of Public Service, Seattle University, USA.*

 A range of further resources for this book are available on the Companion Website: www. routledge.com/cw/cullingworth.

PLANNING IN THE USA

Policies, issues, and processes

Fourth edition

Barry Cullingworth and Roger W. Caves

Routledge
Taylor & Francis Group

LONDON AND NEW YORK

First published 1997

Third edition 2009

Fourth edition 2014 by Routledge
2 Park Square, Milton Park, Abingdon, Oxon OX14 4RN

and by Routledge
711 Third Avenue, New York, NY 10017

Routledge is an imprint of the Taylor & Francis Group, an informa business

British Library Cataloguing in Publication Data
A catalogue record for this book is available from the British Library

Library of Congress Cataloging in Publication Data
Cullingworth, Barry.
Planning in the USA / Barry Cullingworth and Roger W. Caves. — Fourth edition.
pages cm
Includes bibliographical references and index.
1. City planning—United States. 2. City planning—Environmental aspects—United States.
3. Land use, Urban—United States. 4. Urban policy—United States. I. Caves, Roger W.
II. Title. III. Title: Planning in the United States of America.
HT167.C85 2013
307.1'2160973—dc23
2013010239

ISBN: 978-0-415-50696-0 (hbk)
ISBN: 978-0-415-50697-7 (pbk)
ISBN: 978-0-203-12656-1 (ebk)

Typeset in Garamond
by Keystroke, Station Road, Codsall, Wolverhampton

To Betty Cullingworth and Carol Caves

Our work would not have been possible without their love and support

Contents

Plates

Figures

Boxes

Preface

This book has two objectives. First, it is intended to give an outline of policies relating to land use, urban planning, and environmental protection. Second, it aims to provide an introduction to the policy-making process in these fields. The central concern is with the way in which policy issues are identified, defined, and approached. The coverage of the book is wide: it includes the nature and limitations of planning and governance, land use regulation, the quality of the environment, growth management, transportation, housing, and community development, as well as an extensive discussion of current environmental issues. The focus is on the problems facing policy-makers in their search for solutions (though the term 'resolutions' is preferred). It also discusses the difficulties of separating facts and values. This is particularly clear with environmental issues where even 'experts' are protesting that their expertise is limited, and that questions of 'risk' have no scientific answers. It is now widely accepted among scientists that determining acceptable degrees of risk is a matter for public policy, not for science. Such professional modesty is increasingly apparent in the professions dealing with environmental hazards, but it is also growing in the professions concerned with urban and land use planning issues. It is against this background, together with an associated mistrust of government that public involvement in the planning process takes on a new meaning.

The first edition of *Planning in the USA* was written while Barry was at the Department of Land Economy at the University of Cambridge, England. This deprived him of the direct help of numerous American colleagues. Some helped him to overcome the problems of distance by sending contemporary materials and even by commenting on draft chapters. Federal and state officials were particularly helpful in responding to the constant barrage of transatlantic letters. All this direct assistance has been supplemented by many who have unknowingly helped him through their writings: the wide coverage of this book implies a huge debt of gratitude.

Barry passed away on February 6, 2005. He leaves a legacy of great works in planning and public policy. As an academic and government consultant, Barry offered everyone a clear understanding of complex issues. The education of planners and public officials continues to be greatly influenced by Barry's writing. He had a unique way of synthesizing a lot of material and discussing it in a clear and understandable fashion. I was honored when he asked me to help him update *Planning in the USA, second edition*. I hope I have continued his legacy in the later editions and that he would like this new, fourth edition.

I would also like to thank Marisa Mangan, Gabriela Fernandez, Aubrey Smith, Michael Viglione, and Afshin Atapour for their research on sustainability, formulating questions, and gathering helpful website information used for all of the chapters.

Roger W. Caves
Professor of City Planning
San Diego State University

Acknowledgments

The authors would like to thank the following for permission to reproduce plates, images, and photographs in this publication:

Plate 8: Wright Runstad & Co.; Plate 14: San Antonio B-cycle; Plate 16: CALTRANS, District 11; Plate 17: James Fawcett; Plates 21 and 22: Seattle Housing Authority; Plate 41: Evan Wasserman; and Plates 44 and 45: Fritz Wagner.

The authors would like to thank Dr Sherry Ryan, San Diego State University, for generating a GIS image; Dr Douglas Stow, San Diego State University, for generating a remote sensing image; the UCLA Urban Simulation Team for providing a Virtual 3D model for Downtown Los Angeles; Public Domain Photos for the Los Angeles and San Francisco photos (http://pdphoto.org).

The authors and publishers would also like to thank Curtis Brown Ltd for permission to reproduce 'Song of the Open Road' by Ogden Nash.

Acronyms and abbreviations

Though acronyms abound in the planning and (even more so) environmental fields, they have been largely avoided in this book: they are confusing and frequently difficult to remember. Nevertheless, they are useful on occasion and, in any case, readers who follow up the references will soon find themselves immersed in them. Hence the following list may be helpful.

ACHP	Advisory Council on Historic Preservation
ACIR	Advisory Commission on Intergovernmental Relations
ACSC	area of critical state concern (Florida)
ADU	accessory dwelling unit
AICP	American Institute of Certified Planners
APA	American Planning Association
ARM	adjustable rate mortgage
ARRA	American Recovery and Reinvestment Act of 2009
ATIS	Advanced Traveler Information System
BART	Bay Area Rapid Transit
BRAC	Base Realignment and Closure Commission
BRIDGE	Bay Area Residential Investment and Development Group (San Francisco)
CAA	Clean Air Act
CAC	citizen advisory committee (Oregon)
CAD	computer-aided design
CAP	citizens' alternative plan (Seattle)
CARB	California Air Resources Board
CCC	California Coastal Commission
CDBG	Community Development Block Grant
CDC	community development corporation
CEQ	Council on Environmental Quality
CEQA	California Environmental Quality Act
CERCLA	Comprehensive Environmental Response, Compensation and Liability Act 1980 (Superfund)
CFCs	chlorofluorocarbons
ChemSTEER	Chemical Screening Tool for Exposures and Environmental Releases
CLG	Certified Local Government
CLUG	Community Land Use Game
COAH	Council on Affordable Housing
COG	Council of Government
CZMA	Coastal Zone Management Act
DenverGIS	Denver Geographic Information Systems
DDT	dichloro-diphenyl-trichloroethane
DOT	Department of Transportation
DRI	developments of regional impact (Florida)
DSS	Decision Support System
DTI	debt-to-income ratio
EA	environmental assessment
EC	enterprise community
EDF	Environmental Defense Fund
EHLP	Emergency Homeowner's Loan Program

EIR	Environmental Impact Report (California)	MEPAS	Multimedia Environmental Pollutant Assessment System
EIS	environmental impact statement (NEPA)	MPDU	moderately priced dwelling unit
EPA	Environmental Protection Agency	MPO	Metropolitan Planning Organization
ESA	Endangered Species Act	MULTIMED	Multimedia Exposure Assessment Model
ESG	Emergency Shelter Grant		
EZ	enterprise zone/empowerment zone	NCCED	National Congress for Community Economic Development
FBC	form-based codes		
FHA	Federal Housing Administration	NCCP	Natural Community Conservation Planning (California)
FHWA	Federal Highway Administration		
FONSI	finding of no significant impact (NEPA)	NEJAC	National Environmental Justice Advisory Council
FTA	Federal Transit Administration	NEPA	National Environmental Policy Act
GAO	General Accounting Office	NH GRANIT	New Hampshire Geographically Referenced Analysis and Information Transfer System
GIS	geographic information system		
GHG	greenhouse gas		
GPO	Government Printing Office	NHPA	National Historic Preservation Act
GSE	Government-Sponsored Enterprise	NIMBY	'not in my backyard'
HAMP	Home Affordable Modification Program	NIMTOO	'not in my term of office'
		NPDES	National Pollution Discharge Elimination System
HARP	Home Affordable Refinance Program		
		NPL	National Priority List (hazardous waste sites)
HCFCs	hydrochlorofluorocarbons		
HCP	Habitat Conservation Plan	NRC	National Research Council
HOPE	Homeownership and Opportunity for People Everywhere	NRHP	National Register of Historic Places
		NTHP	National Trust for Historic Preservation
HOPWA	Housing Opportunities for Persons with AIDS		
		OTA	Office of Technology Assessment
HOV	high occupancy vehicles	PAB	Planning Accreditation Board
HTF	housing trust fund	PBF	public benefit features (Washington)
HUD	Department of Housing and Urban Development	PCBs	polychlorinated biphenyls
		pH	[measure of acidity]
ISTEA	Intermodal Surface Transportation Efficiency Act	PRP	potentially responsible party (hazardous waste)
ITP	incidental take permit	PUD	planned unit development
ITS	Intelligent Transportation System	RC	Renewal Community
LESA	land evaluation and site assessment	RCA	regional contribution agreement
LIHTC	Low Income Housing Tax Credit	RCRA	Resources Conservation and Recovery Act
LOS	level of service standards (Florida)		
LUBA	Land Use Board of Appeals (Oregon)	REAP	Rural Economic Area Partnership
		SANDAG	San Diego Association of Governments
LULUs	locally unwanted land uses		
MEGIS	Maine Office of Geographic Information System	SanGIS	San Diego Geographic Information Source

SAUS	*Statistical Abstract of the United States*	TIF	Tax Increment Financing
SCPEA	Standard City Planning Enabling Act	TIP	transportation improvement program
SCS	Sustainable Communities Strategy	TMA	transportation management area
SHPO	State Historic Preservation Office	TOD	transit-oriented development
SIP	state implementation plan (clean air)	TRI	Toxics Release Inventory
		TSCA	Toxic Substances Control Act
SMSA	standard metropolitan statistical area	TSD	treatment, storage and disposal facility
SOV	single occupancy vehicle	UDAG	Urban Development Action Grant
SSZEA	Standard State Zoning Enabling Act	UGB	Urban Growth Boundary (Oregon)
TDC	transfer of development credits	USGS	United States Geological Survey
TDM	transport demand management	VOC	volatile organic compound
TDR	transfer of development rights	YIMBY	'yes in my backyard'
TEA-21	Transportation Equity Act for the Twenty-First Century	ZBA	Zoning Board of Appeals

Introduction

Standing on its own, the term 'planning' often relates to land use planning but, of course, it can apply to many other areas of public or private activity – economic development, health, housing, social security, defense, energy, and so on. This book focuses on land use, urban, and environmental planning. However, as will be very apparent, problems have a habit of becoming 'interconnected': they refuse to be neatly parceled into separate areas which can be conveniently dealt with by individual government agencies, policies, programs, or budgets. They also refuse to be neatly defined. These conditions pose difficulties for policy-makers and implementers; indeed, it often seems that it is this interrelationship of problems which is the central problem of government. The issue is neatly highlighted in Donna Shalala's lecture on urban policy:

> Every time Treasury changes the Tax Code, every time Congress alters a welfare program, every time the Defense Department awards a military contract, urban policy is being made.

Yet such measures are often not even recognized as constituting 'urban policy'.

Debates on urban policy seem endless. Attempts to tackle 'the problem of the cities' have proved to be of extraordinary complexity because there is not a single problem: there is a host of interconnected problems. These include housing and community development; land use and transportation; employment, training, and economic development; poverty and social security; city finance and local, state, and federal taxation – a complete list would be very long. To exacerbate the situation, although people may agree that a problem exists, they cannot agree on how to define the problem. Nelson (1977) suggests that we examine the nature of the problem. Technical problems are more easily defined and more readily solved. Social problems have many faces or dimensions that cause people to define them in various ways. For example, should we define housing as an economic problem, a sociological problem, a political problem, or what?

To deal with any one of the previous problems is not easy: taken together they are extremely difficult to comprehend, let alone to attempt to solve, if, in fact, they can be solved. Actions taken in one policy area may affect another policy area. Shouldn't we take transportation into consideration when planning for land use? Moreover, political and ethical issues constantly arise: is the problem of housing one of education, employment, neighborhood, culture, poverty, or moral turpitude? When should we intercede in an issue?

As with a policy package, this book attempts to overcome these difficulties and present a clear and reasonably succinct account of the web of issues which constitute planning and public policy. But, again like a policy package, it must fall short of comprehensive goals. In a world where everything is related to everything else, it is impossible to deal with all things at once, and therefore problems have to be broken down into manageable issues. Yet, once this is done, important relationships are separated, and the 'manageable issue' also necessitates a limited approach. The boundaries of this book mirror these wider matters. It focuses on land use planning: land use, the connection between land uses (transport), the interaction between land uses and the natural elements (environment), and the ways in which land uses develop (urbanization) and are controlled ('land use planning' in its narrow legal sense). It also discusses the ways in which controls over land uses interact with systems of government (public participation, support, and prejudice), and with specific aspects of urbanization (economic development, the provision of infrastructure, the demand and supply of housing). This is a formidable list, but it is by no means complete. For instance, it omits natural resources, public finance, architecture and civic design, information systems, demography, and income distribution. The list could easily be lengthened, but the point does not need laboring any further.

This is an important factor to bear in mind constantly when each individual policy issue is under consideration. Contemporary problems are severe in their number, extent, and complexity. Yet, resources are available on an unprecedented scale to deal with them; and public opinion presses for quick solutions. Why do so many solutions prove to be inadequate and, seemingly, give rise to additional problems? So great is the disillusionment that strong arguments are canvassed that it is better to leave at least some of the problems to 'free market forces', and that 'government is the problem, not the solution'.

In this book, it is suggested that the answer does not lie in that direction (though a less regulated approach can sometimes help). There is no such thing as 'free' market forces: their freedom depends on a legal and political framework which protects them from contrary forces, and provides a framework of security for action. Without that framework, there would be anarchy, not a market. Even more tendentiously, it can be argued that there is no point in pursuing a 'minimum government' approach: the electorate will not allow it. Politicians frequently face the imperative to 'do something', even if they are unclear what it is they should do.

Many of these are matters which are not to be settled by diktat, appeals to reason, or argument. They involve differing values and beliefs. It would be helpful if the questions of value could be neatly separated from those of fact, but, unfortunately, they cannot.

The reader will already have detected some of the authors' values. Though this book is intended as an academic text, it is inevitably influenced by personal views. Hopefully, these have been made explicit but, such is the interaction between facts and values, this cannot always be so. Perhaps it can be suggested that there is educational value in being asked to sort out what is fact and what is value? This is not a flippant comment: one of the themes of this book is that policy debate inevitably involves both, and that much difficulty is created by their confusion.

Questions of value are important not only because they get mixed up with those of fact, but also because there are so many issues where the 'facts' are limited. When this is admitted, the decks are clear for a confrontation of values. These may be modified by greater understanding, but, in the final analysis, they are matters of individual beliefs. Democratic systems of government allow these to have full play.

If this argument is accepted, it is easier to understand the essentially political nature of the policy process. It needs to be constantly borne in mind that the words 'policy' and 'politics' have the same root. Interestingly, as scientific knowledge increases, uncertainty grows: scientific matters are increasingly becoming concerned with 'chance' and 'risk'. The implications of this are particularly clear in the case of environmental policy; but (so it is argued in the last chapter) they go much further, and demand some rethinking about the formulation and implementation of public policy.

Planning education and careers

Compared to other fields, educating planners is relatively new. As such, it continues to evolve since the first formal city and regional planning program at a university was established at Harvard University in 1923. Earlier, individuals were educated in such fields as engineering and architecture. This is still the case in a number of countries.

The education of today's planners has evolved over the years. Today, students experience an interdisciplinary/multidisciplinary approach in their education. After all, the problems that planners face involve a number of fields. For example, the issue of a lack of affordable housing involves looking at the problem from different viewpoints. This means that we need to investigate more than simply the construction of a dwelling unit. We need to examine issues such as where to locate the housing, the impact of the housing on the surrounding neighborhood, the individual needs of the inhabitants, the financing of the unit, the costs of public services to support the housing, etc.

There are a number of colleges and universities offering undergraduate degrees in planning. These degrees could be in urban planning, city planning, urban studies and planning, urban and regional planning, or community and regional planning. They offer students a broad education in such areas as an introduction to planning, the history of planning, planning theory, quantitative methods, information and communications technologies, design, planning studios, practical experience in planning, and a capstone course. Although most programs at this level offer a generalist approach to planning, a number of programs allow students to specialize in an area of interest to them.

The actual field of study for the undergraduate degree did not have to be in planning. Many students had their undergraduate degrees in other fields such as economics, urban studies, geography, sociology, political science, business and finance, environmental studies, architecture, landscape architecture, women's studies, African-American studies, Latin American studies, urban design, etc. These and other undergraduate backgrounds provide students the opportunity to blend their specialized training with advanced planning education.

For many years, the undergraduate degree was all that was needed. Today, many undergraduate students choose to continue their education to the master's degree level. In actuality, receiving a master's degree is required for many entry-level positions in planning.

Graduate studies in planning offer students the ability to get advanced training in core areas of planning as well as an opportunity to concentrate or specialize in a given area of planning. Core training involves courses in theory, methods, practice, ethics, implementation, law, and urban design. Internships are also an important component of the training. While all programs offer a generalist perspective on planning, students may opt to have a specialty in

housing, community development, urban revitalization, economic development, environmental planning, international development, historic preservations, sustainable development, transportation planning, geographic information systems, or any other area. The specific concentrations available to students vary by university.

Undergraduate and graduate planning programs have the option of seeking out accreditation from the Planning Accreditation Board (PAB). This organization partners with the American Association of Collegiate Schools of Planning (ACSP), the American Institute of Certified Planners (AICP), and the American Planning Association (APA), which examines the areas of program performance, integrity, and quality. Accreditation is a voluntary process. The process involves examining standards and criteria in the areas of program mission and strategic plan; students; faculty; curriculum and instruction; governance; program assessment to meet program goals; and progress in meeting the goals, changes, and future issues.

Who are the planners?

Planners operate on multiple levels: public, private, and non-profit. They work in the federal government, state government, and local governments. They work for international bodies. They work for private think-tanks, consulting firms, banks, real estate developers. They work for non-profit organizations like community or economic development organizations. Planners are not restricted to one specific sector of employment.

The role of the planner is a multi-faceted one. The role may shift as situations or circumstances change. Planners serve as advisors to policy-makers. They serve as problem evaluators and solvers. To many people, they serve as visionaries. They might be policy planners or physical planners. In most cases, they are part of a team developing visions, and programs to achieve a vision or visions.

Regardless of where the planner works, a planner could perform multiple roles. These roles could start by being a researcher who investigates issues to get an understanding of an issue and why it is an issue.

A planner also serves to educate people on an issue or issues being discussed. Planners also may advocate approaches that might be taken to deal with the issue or issues at hand. This might be considered as a political role. The question then becomes whether or not this role is opposite to a role of being value neutral. This is a role that is always being debated by academics and practitioners.

Planners have the option of becoming what is called a certified planner. Some people consider this a form of licensing. In order to become a certified planner through the AICP, a planner must be a member of the APA, be engaged in planning, and have completed a certain amount of education and experience. The education and experience varies according to whether or not an individual has an undergraduate or graduate degree in planning, whether or not the degree is in a different field, and whether or not the individual possesses a degree.

To become a full member of the AICP, an individual must pass an examination covering knowledge in the areas of planning history, theory, and law; plan making and implementation; functional areas of practice; spatial areas of practice; public participation and social justice; and the AICP Code of Ethics. Once the examination is passed, members must pay annual dues to become a full member of the AICP.

The AICP Code of Ethics serves as a guide or standard for an individual's behavior and conduct in the profession. It represents the principles in which a planner practices. According to AICP Code revised October 3, 2009, planners aspire to the following principles:

1. Our Overall Responsibility to the Public – Our primary obligation is to serve the public interest, and we, therefore, owe our allegiance to a conscientiously attained concept of the public interest that is formulated through continuous and open debate. We shall achieve high standards of professional integrity, proficiency, and knowledge.
2. Our Responsibility to Our Clients and Employers – We owe diligent, creative, and competent performance of the work we do in pursuit of our client or employer's interest. Such performance, however, shall always be consistent with our faithful service to the public interest.
3. Our responsibility to Our Profession and Colleagues – We shall contribute to the development of, and respect for, our profession by improving knowledge and techniques, making work relevant to solutions of community problems, and increasing public understanding of planning activities.

Format of the book

The book is divided into four main parts. Part 1 contains four chapters on planning and government. Chapter 1 expands on the discussion started in this Introduction: what is the character of planning? Is it based (as some theories contend) on rationality and, if so, what kind of rationality? Can it be comprehensive, or must it be essentially incremental? What are the roles of the different levels of government, and of the courts? What is the role of the public in planning? How important are interest groups? What are the underlying cultural attitudes? How do private and public planning processes differ? New material includes discussions on the growing roles of foundations and national and local think-tanks; professional and quasi-professional organizations; foundations involved in planning and public policy; and the use of technology and social media for planning and community engagement purposes. Planning students, practitioners, and policy officials need to recognize that various information and communications technologies can be used to improve data gathering, analysis, and plan generation.

Chapter 2 summarizes current urbanization trends. The United States is a uniquely mobile society, always has been, and always will be. There is a restless search for improvement in the quality of life which is seen in many different fields. Discussions of deindustrialization and shrinking cities can be found in this chapter. In environmental terms it has led to a high degree of urbanization and, later, suburbanization. The chapter describes this, and analyzes the role which government policies have played.

Chapter 3 discusses a number of issues relating to the government and planning of urban areas. The first colonial settlements quickly saw the need for a simple system of local government which had sufficient powers to deal with the problems of a relatively simple society. As economic development and urbanization gathered speed, governmental systems had to develop, though the forces of privatism were (and remain) strong. The reform movement had several dimensions, ranging from the battle with corruption to the promotion of the City Beautiful. Planning had a difficult birth, and its present uncertainties have historical roots.

Chapter 4 covers a topic that has grown in importance over the years – sustainability. It is a topic that is on everyone's mind today. Not a day goes by without a mentioning of sustainability and the role planning has within it. This chapter discusses the history, theory, and practice of sustainability. Examples of how the private sector, public sector, non-profit organizations, and citizens are planning for sustainability are provided. Among the topics covered in this chapter are the growing uses of alternative energy sources. Ultimately, the concept of sustainability will be incorporated throughout various chapters in the book.

Part 2 has five chapters devoted to various aspects of land use regulation. Chapter 5 gives a brief history of the emergence of planning and zoning as a method of controlling land use, which served the dominant interests of the time. The institutional and legal framework of planning and zoning, discussed in Chapter 6, was fashioned by the federal government (with the preparation and dissemination of model 'standard state enabling acts'), and by the support of the courts. A discussion of *Stop the Beach Renourishment, Inc.* v. *Florida Department of Environmental Protection* has been added to the chapter. Zoning has developed a profusion of techniques which go far beyond the simple districting of the early ordinances. Chapter 7 discusses the overarching concept of the Comprehensive Plan, its development, and recent comprehensive planning elements around the country. The multiplicity of techniques is illustrated in Chapter 8. Chapter 9 focuses on the schemes which have been devised for financing and apportioning the costs of development. New discussions on budgets, unfunded mandates, bankruptcy, and tax credit programs have been added to this chapter. Interwoven with issues of land use control are

wider ones such as those of economic growth and the exclusion of unwanted social groups.

Though zoning might be regarded essentially as a form of 'growth management', its use for this purpose is relatively recent. Part 3 deals with this in two chapters: Chapter 10 discusses its use by local governments; and Chapter 11 deals with the role assumed by a number of states. Growth control policies can take zoning nearer to the concept of comprehensive planning, though the attainment of comprehensiveness has seldom been sought and even less frequently attained.

Land use planning has been predominantly concerned with quantity rather than quality, but Part 4 examines a number of planning and development issues: environmental policy and planning (Chapter 12), limits of environmental policy (13), transportation (14), housing (15), community and economic development (16), urban design and aesthetics (17), and heritage and historic preservation (18).

Environmental issues have grown in political significance (and in reality) over the last quarter of a century. Chapter 12 deals with the growth of environmental concerns, the National Environmental Protection Act, and three major areas of environmental policy: clean air, clean water, and waste. Chapter 13 is a more discursive discussion of the limits of environmental policy. This takes up and develops points made earlier about expertise, scientific uncertainty, and risk. Policy involves the calculation of the degree of risk which is unacceptable. Since this is essentially a value issue, not a scientific one, it has to be settled in a democratic manner.

Transportation (Chapter 14) is the circulatory system of the economy and, in the metropolitan areas at least, it is suffering from sclerosis. To continue the medical metaphor, a wide range of remedies are being prescribed, including controls, incentives, charges, and comprehensive planning. Unfortunately, neither the diagnosis nor the treatment is proving a different way: the objectives have always been clear. They were set out in the Housing Act of 1949: to provide 'a decent home and a suitable living environment for every American family'. However, reaching that goal presents formidable problems of finance, politics, and planning. The outlook is bleak for many.

A new discussion of HOPE VI, the home foreclosure crisis, and community land banks, and their role in providing affordable housing has been added to Chapter 15. Community and economic development is put forward as one solution, and it presents some promise if adequate resources are made available. New discussions on the use of 'economic gardening' and initiatives put forward by the Obama administration are included in Chapter 16.

The remainder of Part 4 covers two additional areas. Chapter 17 deals with the long history of urban design and the legitimation of aesthetic controls, while Chapter 18 chronicles the evolution of policies relating to heritage and historic preservation. Discussions on urban design, community character, digital billboards, and landscape planning have been added to Chapter 17. Aesthetic controls started with billboards, at the end of the nineteenth century. These gave rise to a long battle between planners and the powerful billboard lobby, which still rages. The control of good design may seem to some to be an oxymoron: certainly good design is an elusive quality which is difficult to define, let alone to control. Historic preservation has been a legal battleground, in which New York's Grand Central Terminal looms large. The scales have been further tipped in favor of this special aspect of land use control with the realization that historic areas have economic benefits for tourism. New material for Chapter 18 includes discussions on federal and state historic rehabilitation tax credits, the Main Street programs, and Neighborhood Conservation Districts.

The final chapter of the book, Chapter 19, attempts to bring together significant points made in the body of the book. An updated discussion on what has transpired in New Orleans since the third edition of this book was published, as well as an update on Yucca Mountain, and a discussion of global planning problems facing us have been added to this chapter. Since it is not the authors' intention to enter the lists of those who attempt to provide programs of reform, the discussion is focused on the framing of questions and the manner in which they might profitably be debated.

Web addresses found in each chapter's 'Further reading' section were current at the time of writing but may become unavailable if sites have upgraded.

PART 1

PLANNING AND GOVERNMENT

Planning is a purposive process in which goals are set and policies elaborated to implement them. Such is the theory; but the theory is affected in many ways. There are problems with the concept of planning, with the forces of urbanization which it seeks to regulate, and with the very nature of government. This first part of the book introduces this complex of issues which make planning such a difficult, frustrating, and fascinating subject to study.

Chapter 1 discusses the nature of planning. How far is it a rational activity akin to mathematics? The question appears absurd at first sight: surely a planner would not proceed in an irrational way? Do cities always engage in rational comprehensive planning? The discussion shows that the issues are much more complicated than this suggests. Rational goals are elusive, and apparently sensible methodologies are strangely difficult to implement. Moreover, though it may seem intelligent to attempt to plan comprehensively, experience shows that this is an ideal which faces formidable obstacles. Rational planning is a theoretical idea. Actual planning is the practical exercise of political choice that involves beliefs and values. It is a laborious process in which many people and private agencies are concerned. These comprise a wide range of conflicting interests. Planning is a means by which attempts are made to resolve these conflicts. This is particularly difficult in land use planning because of the cultural, legal, and constitutional aspects of property rights. This chapter also discusses the importance of informing citizens on planning matters and engaging citizens in planning through traditional mechanisms and through the use of various technologies that enhance opportunities for participation.

Since this book is focused on urban planning issues, it is appropriate to examine the nature of world urbanization and, specifically, its trends in the United States. Chapter 2 does this in a summary way. The original settlers formed a rural society, and towns were very small. Urbanization accelerated in the second half of the nineteenth century, and by 1920 a half of the population was urban. The proportion increased to almost 70 percent in 1960, to 75 percent in 1990, and to approximately 81 percent in 2005. Urbanization was followed by suburbanization, largely as a result of developments in transport, highways, and innovations in the finance of home ownership. In this the federal government played a major role. Suburbanization eventually led to inner-city decline as people and (later) shops and jobs moved out. By 1990, over a half of the American population lived in the suburbs. Over a sixth of the population moves every year. Though most of these moves are short-distance, huge regional movements have taken place. These have resulted in enormous growth in states such as California and Florida. The nation's center of population in 2000 was in Missouri, some 32 miles west of the center of population in 1990; in 1850, it was in West Virginia. Migration is still taking place on a larger scale, but its character is now complex and volatile. This chapter also examines current issues surrounding deindustrialization and the problems associated with shrinking cities.

Having discussed the nature of planning and the history of urbanization, Chapter 3 extends the historical account and examines the development of urban government and planning. This helps in an understanding of the historical and intellectual heritage of urban planning. Previous generations have battled with questions of how to make planning effective in a democratic society: their experience is of relevance to the contemporary scene. Issues highlighted in this account include a discussion on federalism, the persuasiveness of privatism, the reform movement, the City Beautiful movement, and the growth of planning.

The idea of sustainability is covered in this new edition as Chapter 4. It starts with a discussion of the definition of sustainability and why it is important. The chapter includes discussions on international activities regarding sustainability, federal actions in the US regarding sustainability, state and local programs devoted to sustainability, green building regulations, and the increasing role of community gardens in sustainability.

The nature of planning

If we can land a man on the moon, why can't we solve the problems of the ghetto?

Nelson 1977: 13

We must first exorcize the ghost of rationality, which haunts the house of public policy.

Wildavsky 1987: 25

The character of planning

Planning is a process of formulating goals and agreeing the manner in which these are to be met. It is a continuous process by which agreement is reached on the ways in which problems are to be debated and resolved. It is a process involving multiple participants with multiple perceptions, beliefs, and objectives.

Definitions of planning abound: there is a large literature devoted to exploring the meaning of the term. One generally common element in these definitions is that planning is forward-looking; it seeks to determine future action. At the simplest, one may plan to go to the library tomorrow. Such a 'plan' involves a choice between alternatives – *not* to go to the library tomorrow, to go elsewhere, or to stay at home. The plan may also be based on explicit assumptions; for example, the decision to visit the library may be dependent upon finishing the books that have already been taken out of the library, or on the weather being fine. On the other hand, if the books will be overdue, and subject to a fine if they are not returned tomorrow, the plan may override other considerations.

This is, perhaps, a trite example, but it does contain important elements which are present in more sophisticated forms of planning: forethought, choice between alternatives, consideration of constraints, and the possibility of alternative courses of action dependent upon differing conditions. Of course, when a plan involves other people (which it usually does), the plan must incorporate an acceptable way of reconciling differences among the participants: this is a major feature of any type of planning; and the more numerous and diverse the participants, the greater the difficulties of planning. At the extreme, fundamental clashes in outlook, beliefs, or objectives may make planning impossible. At the worst, there is a resort to violence – of which there are, tragically, all too many examples around the world.

This underlines another important aspect of planning: there has to be a sufficiently sound basis of agreement for planning to be possible. In democratic societies, large numbers of diverse interests not only have to be considered but also have to be involved in the planning process. Much of 'planning' then becomes a process of reaching agreement on objectives. But, as will be shown repeatedly through this book, objectives and ways of reaching the objectives are not easily separable. Many may agree that a comprehensive system of health care is needed, but it may prove impossible to fashion an acceptable method of providing this – a point dramatically illustrated by the collapse of

President Clinton's health proposals. In the debate, differences appear in both means (such as methods of financing) and ends (such as the extent to which health care is to be 'comprehensive' in terms of both the people to be included and the health conditions to be covered). Very speedily, ends and means become confused.

By contrast, where there is full agreement on a planning objective (putting a person on the moon, for example), the debate focuses on methods. When the nation agreed that it was a national priority to devote the necessary resources for this incredible feat, there was no problem with defining the problem, or of obtaining the necessary funding. Though the objective was incredibly difficult, it was simplified by the agreement that supported it. There was, for example, no argument on whether it might not be better to build a transoceanic tunnel, or to build a 10-mile high city, or to attempt any other seemingly impossible enterprise. More realistically, there was little serious debate as to whether the resources could not be put to better use in, for example, eliminating poverty.

The planning of wars contains many lessons on the problems of planning, but consider how much more difficult is the planning of a 'war on poverty'. In his first State of the Union message in 1964, President Lyndon Johnson declared such a war: 'This Administration today, here and now, declares unconditional war on poverty in America.' Sad though it is, no such 'war' was possible: the single aim of destroying the enemy of poverty inevitably broke up into a myriad of problems concerning a proliferation of programs aimed at constituent parts of 'poverty', their financing, their administration, their adequacy, and their effects. Poverty proved to be a hydra-headed monster, encompassing an incredible number of issues – from food stamps to regional development, from model cities to education, from health to income maintenance: 370 new programs of assistance to states and local governments were introduced between 1962 and 1970.

The issues are discussed further in Chapter 16. Here it is important to note the types of questions that the plans raised. Is poverty an economic issue (in which case the answer would lie somewhere in the policy area of maintaining incomes)? Or is it a matter of personal inadequacy (in which case, what scope is there for remedying this)? Or is it a market failure (which might be dealt with by market incentives)? Then there were questions as to how far the state could – and should – interfere with the market. Do public programs destroy individual initiative? Where should resources be concentrated: on individuals, communities, urban redevelopment, or job creation? And so the questions multiplied. Distinctions between ends and means proved baffling. Poverty became seen as an umbrella term for a wide range of problems of modern postindustrial society. The problems were difficult to define, let alone to resolve.

Similar problems arise with any form of planning where there is not a single, clear, and accepted objective – which is usually the case. To take a further example, which is a major focus of this book: land use planning. How does one plan urban development? The first question is why it should be planned at all. The answers to this are legion, and they are usually expressed in very general terms: to achieve 'orderly' development, to minimize the loss of agricultural land, to reduce transport needs to the minimum, to encourage economic development, or to facilitate private investment in property.

Typically, there are several objectives. There may be a general desire to provide a spacious environment while, at the same time, maximizing the use of public transport and safeguarding rural land adjacent to the built-up area. These objectives involve conflicts: spacious environments consume a greater amount of land (often previously in agricultural use); and low densities present problems for public transit which operates most effectively in high-density corridors.

Planning necessarily involves restraint on the actions of individual landowners and residents. Such restraints arouse opposition and claims that property rights are being infringed. This is an important limitation on the scope of planning, more so in the United States than in those countries where a high degree of public command over land development is politically acceptable. In fact, there is a considerable amount of control over the use of land in the United States and, though it is a source of continual controversy, the principle of some degree of regulation is generally accepted. (To use the customary example, no one wants a glue factory to be located in their neighborhood.) The issue then is not whether there should or should not be planning, but how much of it there should be, and how it should operate. There is an extensive literature on this, replete with a wide variety of concepts. Immediately apparent is the divorce between planning theory and planning practice.

Planning theory

Ideology plays an important role in planning. It defines planning and forms the body and basis of planning, our values, how we frame issues, and how we set and achieve goals. Our ideology poses a number of questions on how we plan. For example, how do our views influence the identification of issues? How do our views influence the development and implementation of plans? Are our goals to improve the public interest? How do we measure the goals and progress towards achieving the goals? Are there competing ideas and goals that must be taken into consideration? How do we choose among the competing ideas? Should we advocate public intervention in an issue or should the issues be addressed by the private sector? The question of public sector involvement in various issues continues to be debated by those involved in planning and public policy.

Rationality

A central theme to planning theory is the concept of rationality. Since rationality requires all relevant matters to be taken into account, the use of the concept readily leads to a comprehensive conception of planning. This stems from the simple (and valid) idea that, in the real world, everything is related to everything else, and the planning of one sector cannot properly proceed without coordinated planning of others. Rationality also requires the determination of objectives (and therefore – though not always explicitly – of values), the definition of the problems to be solved, the formulation of alternative solutions to these problems, the evaluation of these alternatives, and the choice of the optimum policy. It is a time-consuming

process. Moreover, much of the difficulty of this approach (quite apart from matters of implementation, discussed below) is that it can mask the essentially political nature of the process. The overriding consideration easily becomes procedural efficiency, which places planning on a 'scientific' level 'above politics'.

The persuasiveness of the concept of comprehensive rational planning is persistent: it can be seen in a succession of federal governmental initiatives based on concepts of coordination and systematic targeting of resources, as with the Model Cities Program for instance (see Box 1.1). Of course, the planning process produces 'objectives', definitions of problems, and proposed 'solutions'. But all this is done in the context of the politics of the place and the time, and against the background of public opinion and the acceptability or otherwise of governmental action. Some important issues may be regarded not as problems capable of solution but as powerful economic trends which cannot be reversed. Others may be of a nature for which possible solutions are conceivable but untried, too costly, too administratively difficult, too uncertain, or even dangerous to the long-term future of the area. And, as will be apparent from later chapters, these acutely difficult problems (of urbanization, congestion, inner-city decay, for example) have continually proved beyond the powers of governments to solve, at least in the short run; the long run is unpredictable. Major differences of opinion exist among experts, politicians, and electors on these matters. As a result, there are severe constraints operating on the planning process, and there is little resembling a logical, calm set of procedures informed by intellectual debate.

Perhaps the most misleading concept of planning is the theatrical analogy. Planning is often likened to the production of a play – involving the coordination of many roles: the actors, the backstage hands, the management, the marketing, and so on. But the analogy is false in that with the theater there is a common objective to which all are committed: the production of a play – 'the show must go on'. Of course, there may be cross-currents and disagreements, but these are subservient to the overall objective (if not, there are resignations, replacements, or – at the very worst – abandonment of the play). The same holds true with

BOX 1.1 RATIONAL COORDINATION – THE MODEL CITIES ATTEMPT

In the 'war on poverty', a Model Cities Program was proposed which would involve (in the words of the legislation) 'concentration and coordination of federal, state and local public and private efforts'. As originally conceived, the program was to be of a 'demonstration' character restricted to the poverty areas of a very small number of central cities. This implied more than the coordination of programs: it explicitly envisaged the redistribution of resources.

The 'demonstration' concept did not survive the political process: the need to obtain political support for the legislation led to an increased number of cities – eventually to 150. More than this, Congress was not willing to see funds diverted from other programs into model cities; and so the congressional commitment became largely to another categorical program rather than to a coordinative mechanism for reforming other grants-in-aid.

While there was support for the idea of better coordination among urban programs, there was also a fear about 'a concentration of power within any single executive agency'. Bureaucracies can be viewed not only as machines for the efficient implementation of policy but also as a dangerous concentration of power in 'monolithic organizations where a few powerful men at the top concentrate control over a vast range of activities. The implication of this view is that efforts to strengthen coordination among agencies are potentially dangerous, because they may upset the existing balance of power that permits considerable freedom of action for many interest groups.'

In case this is thought to be an extreme view, the reader is cautioned that 'after Watergate and after Vietnam, the dangers of excessive White House power are all too obvious'. It is concluded that 'if the designers of future urban policies take away any single lesson from model cities, it should be to avoid grand schemes for massive, concerted federal action'.

Sources: Frieden and Kaplan 1977; Downs 1967

the more complex productions of opera and movies. But it is not complexity that is the crucial factor. It is difficult to think of anything more complex than putting a person on the moon, but agreement on that single objective, coupled with the provision of the necessary resources, enabled problems of great complexity to be resolved.

Incrementalism

The obvious failure of comprehensive planning to attain the goals that are theoretically possible has led to a number of alternative theories. Many of these revolve around the problem of making planning effective in a world where market and political forces predominate. Meyerson (1956) proposed a 'middle-range bridge' (between ad hoc decision-making on minor issues and long-range comprehensive planning) which would monitor and interpret market and community trends. Lindblom (1959) went further, and dismissed rational–comprehensive planning as an impractical ideal for dealing with complex problems. In his view, it was important to simplify problems and necessary to accept the realities of the processes by which planning decisions are taken: for this he outlined a 'science of muddling through'. Essentially, this incrementalist approach restricting the number of alternatives replaces grand plans by a modest, step-by-step approach which aims at realizable improvements to an existing situation. This is a method of 'successive limited comparisons' of circumscribed problems and actions to deal with them. Lindblom argues that this is what happens in the real world: rather than attempt major change to achieve lofty ends, planners are compelled by reality to limit themselves to acceptable modifications of the status quo (see Box 1.2). On this argument, it is impossible to take all relevant factors into account or to separate means from ends. Rather than attempting to reform the world, the planner should be concerned with incremental practicable improvements.

Mixed-scanning

An alternative is provided by Etzioni's 'mixed-scanning' model: this incorporates elements from both comprehensive and incremental planning theories trying to incorporate the strengths of both models and eliminating their weaknesses. Etzioni (1967: 8) claimed that 'this approach is less demanding than the full search of all options that rationalism requires, and more "strategic" and innovative than incrementalism'. He likened this approach to scanning by satellites with two lenses: wide and zoom lenses. According to Etzioni (1986: 8), 'instead of taking a close look at all formations, a prohibitive task, or only at the spots of previous trouble, the wide lenses provide clues as to places to zoom in, looking for details'. It holds that decisions on 'fundamental' issues – such as primary goals – are followed by detailed examination of alternative programs of implementation (Etzioni 1967). This is an attractive theory, though skeptics are not convinced! Many argue that political forces are stronger than those of rationality. Altshuler, for example, categorically states that 'the city planner like almost everyone in American politics controls so little of his environment

---

BOX 1.2 MODELS OF DECISION-MAKING

Rational–comprehensive

1a) Clarification of values or objectives distinct from and usually prerequisite to empirical analysis of alternative policies.

2a) Policy formulation is therefore approached through means–end analysis: first the ends are isolated, then the means to achieve them are sought.

3a) The test of a 'good' policy is that it can be shown to be the most appropriate means to the desired ends.

4a) Analysis is comprehensive; every important factor is taken into account.

5a) Theory is often heavily relied upon.

Successive limited comparisons

1b) Selection of value goals and empirical analyses of the needed action are not distinct from one another but are closely intertwined.

2b) Since means and ends are not distinct, means–end analysis is often inappropriate or limited.

3b) The test of a 'good' policy is typically that various analysts find themselves directly agreeing on a policy (without their agreeing that it is the most appropriate means to an agreed objective).

4b) Analysis is drastically relevant: (i) important possible outcomes are neglected; (ii) important alternative potential policies are neglected; (iii) important affected values are neglected.

5b) A succession of comparisons greatly reduces or eliminates reliance on theory.

Source: Lindblom 1959: 154–5

that unquestioning acceptance of its major features is a condition of its own success' (Altshuler 1965).

The few practicing planners who have written about planning theory point to the validity of this statement.

Even the exceptional Norman Krumholz, who explicitly placed 'equity planning' at the top of his personal agenda, makes it clear that his high ideals had to be mediated through the rapids of blatantly political forces. After describing the Cleveland political scene in the 1970s, he asks what the implications were for planning:

> First, there was probably little interest in city planning above the level of project planning. Any deals that had been cut between developers and politicians would be difficult or impossible for planners, speaking the language of 'consistency with the general plan', 'long-range significance' or 'the public interest', to modify. Second, Council took a great interest in zoning because it might be market-able . . . Third, there was little interest in general medium- to long-range planning, since its impli-cations and marketing opportunities were unclear . . . Fourth, appeals to 'rationality' had little capac-ity to stir action or support. Who cared how rational a policy was if it didn't produce patronage? Finally, new physical developments of all kinds were welcomed, and if the city had some subsidies to offer, they would be made available on generous terms.
>
> (Krumholz and Forester 1990: 14–15)

Though this may read like an indictment, Krumholz's intention is to show the framework within which he and his planning staff had to operate.

The practice of planning

Planning is not under the purview of any one depart-ment or agency. Planning is undertaken by a vast array of actors. Of course, the most common thing is to think that planning is done only in a local planning agency. This is not the case. Planners can be found in other public agencies dealing with economic development,

schools, public services, recreation, and transportation. Planners can also be found in consulting firms, real estate agencies, non-profit organizations, foundations, etc.

Practitioners are quick to point out that planning involves deciding between opposing interests and objectives: personal gain versus sectional advantage or public benefit, short-term profit versus long-term gain, efficiency versus cheapness, to name but a few. Planning touches on a number of issues, directly or indirectly, and should not be conducted in isolation from any group. It entails mediation among different groups and compromise among the conflicting desires of individual interests. Above all, it necessitates the balancing of a range of individual and community concerns, costs, and rights. It is essentially a political as distinct from a technical or legal process, though it embraces important elements of both.

To illustrate, a small town may wish to preserve its character and to 'protect' it from further development, but individual local businesses may look to the advantages of increased trade, and landowners may see the profits to be made from additional development. Complicating matters further, the school board may welcome growth because additional students will provide the rationale for improving the range of educational provision, while utilities may oppose growth since they are already stretched to the limit and incapable of expanding services at reasonable cost. Additionally, there could be local issues relating to road capacity, wetlands, scenic beauty, waste disposal, or parking. A full list would be very long and, though not all issues will arise with every development proposal, it is not at all uncommon for many conflicts to arise.

Sectoral and comprehensive planning

The concept of comprehensive planning in theory may be contrasted with the narrowly focused planning which takes place in practice. Each administrative agency takes its decisions within its particular sphere of interest, understanding, resources, and competence. How can it be otherwise? The task of any agency is to undertake the task for which it is established, not to take on the complicating and possibly conflicting responsibilities of others (which in any case would be resistant to a takeover). Thus, a conservation agency will take decisions of a very different character from an economic development agency: they have separate and potentially conflicting goals. The idea that there is some level of planning (presumably to be administered by superhuman planners?) which can rise above the narrow sectionalism of individual agencies is not only inconceivable in terms of implementation but also assumes that an overriding objective can be identified and articulated. This is typically expressed in terms of the public interest; yet there are very many 'publics'. They have conflicting interests which are represented by, or reflected in, different agencies of government. Moreover, the publics will vary from issue to issue.

The example of metropolitan government is a case in point. Such a tier of government could rise above local interests and take decisions for the benefit of the region as a whole. There are some good theoretical arguments in favor of this, but the practical point is that people live locally, not regionally. They view any regional policy in terms of its local impact. Thus, it is in the interests of the metropolitan area that adequate provision of affordable housing is made. It is part of the metropolitan planning process to ensure that sufficient sites are identified for this purpose. However, even if the electorate agrees in principle to the provision of affordable housing, they may well – and typically do – object to it being located in their particular neighborhood. The same issues arise with a wide range of provisions which, while necessary for the metropolitan area as a whole, are unpopular locally. It is for this reason that proposals for metropolitan government are usually defeated, and why there are so few metropolitan governments in the United States. The various constituencies in a metropolitan area have such a wide range of conflicting interests that any agreement is very difficult to achieve.

It might also be noted at this point that a number of factors are leading to an increasing privatization of space in urban areas: this trend, seen with the growth of so-called 'common interest communities' and, more generally, of 'private governments', is explicitly aimed

at safeguarding and promoting very local interests (Barton and Silverman 1994; McKenzie 1994). Indeed, the result – if not the aim – may be isolation from the troublesome problems of the adjacent areas.

Interest groups

The resolution of differences of interest (and the establishment of acceptable means of dealing with them) is a central problem of planning. Obvious differences of interest arise along lines of economic position, age, race, occupation, and a host of others. Their variety is illustrated by the enormous number of organized interest groups. These groups are the organizations of a democratic society. Individuals separately can exert little influence (unless they are of great wealth or extraordinary charisma). Influence is gained by combining with other like-minded people to pursue shared goals. Indeed, interest groups form the core of political activity in general and of planning activity in particular.

Interest groups are of extraordinary variety. At the national level are organizations of the professional bodies (planners, architects, engineers, and experts in water systems, environmental pollution, soil science, and so on). Developers, builders, suppliers of building materials have their own organizations, as do various forms of local government. Then there are 'lobbies' based in Washington and other locations who keep a close eye on the legislative process and on any proposal that may affect their interests. Think-tanks may be allied to these, or they may have varying degrees of independence, though generally they will exhibit some political or philosophical leaning. Similar bodies exist at the state level and, to a much lesser extent, at the local level. Not surprisingly, local organizations tend to be preoccupied with issues affecting their locality, and take a broader geographical interest only when cooperation with bodies in other areas promises more effective action.

This short selection illustrates the range of interest groups which operate in the field of land use planning. An important part of the planning process consists of recognizing and negotiating with such organizations. This is not only because wide participation is a hallmark of a democratic society: it is also efficient to bring into the process those who are to be affected by it and those whose cooperation is needed if it is to be effective. Indeed, without the support of the more powerful groups, planning will not work. (As will be shown later, this is a situation that is frequently encountered.) It does, however, present a difficulty for those who are weakly organized and who may include the poorest members of society; without effective organization and power, the interests of these groups are often ignored or overridden.

A selection of interests represented organizationally at the national level is given in Box 1.3. The list is only illustrative, and the divisions are not as clear-cut as the headings may suggest since many organizations fulfill a range of functions, from research to lobbying, and from professional concerns to political action. Nevertheless, the list gives some idea of the huge range of organizations whose influence is brought to bear, with varying degrees of effectiveness, on the planning process at the national level.

BOX 1.3 SOME NATIONAL INTEREST GROUPS IN LAND USE PLANNING

Governmental

Advisory Council on Intergovernmental Relations; Council of State Governments; National Association of Counties; National Association of Regional Councils; National League of Cities; National Association of Towns and Townships; National Governors Association; United States Conference of Mayors.

Professional

American Planning Association; American Society for Public Administration; American Institute of Architects; American Society of Landscape Architects; American Society of Civil Engineers; Council of American Building Officials; National Association of Housing and Redevelopment Officials; International City Management Association; American Public Health Association; International Association of Chiefs of Police; International Association of Fire Chiefs; Institution of Transportation Engineers; American Park and Recreation Society; American Bar Association; Association of Environmental Professionals.

Developmental

National Association of Home Builders; Urban Land Institute; National Association of Real Estate Brokers; National Association of Realtors; Manufactured Housing Institute; National Council for Urban Economic Development; Partners for Livable Places; American Road and Transportation Builders Association; Waterfront Center.

Public works

American Public Works Association; American Water Resources Association; National Solid Wastes Management Association; Airport Association Council International; American Association of Port Authorities; Association of Metropolitan Sewerage Agencies.

Research

Urban Institute; Brookings Institution; Regional Science Association International; Transport Research Board; Environmental Design Research Association; National Center for Preservation Law; Environmental Law Institute; Housing and Development Law Institute; Resources for the Future; Rand Corporation.

Disadvantaged groups

National Association for the Advancement of Colored People; National Urban League; National Urban Coalition; National Association for State Community Service Programs; Center for Community Change; National American Indian Housing Council; Community Transportation Association of America.

Property rights groups

Defenders of Property Rights; Property Rights Congress of America; Property Rights Foundation of America; Coalition for Property Rights; American Land Rights Association.

Environmental groups

American Farmland Trust; Izaak Walton League; National Audubon Society; National Resources Defense Council; Sierra Club; League of Conservation Voters.

Immigrant and refugee groups

Citizens and Immigrants for Equal Justice; New York Association for New Americans; American Civil Liberties Union Immigrant Rights Project; Immigrant Legal Resource Center; National Immigration Forum; National Immigration Law Center.

Local interest groups

At the local level matters are much more complicated. Each area will have its complement of permanent organizations. Many of them are related to national bodies. In addition, there will be a myriad of local groups of varying degrees of permanence established to influence particular neighborhood issues, or to campaign for the provision of some local amenity, or to organize an opposition to some unwanted development. Some groups are single-issue organizations while others deal with multiple issues. To complicate matters a bit more, interest groups occasionally change names and have a way of appearing on the scene with the new name.

By definition, interest groups share a concern about a specific issue or range of issues, but they may have very differing ideas in other directions. Thus, a local group organized to protect the character of a suburb may contain a diversity of attitudes concerning taxes, car

parking, aesthetics, street lighting, schools, and recreation. Moreover, since any one individual has many
interests, some of these may present internal personal
conflicts. An individual member of our suburban protection organization is not only a suburbanite, but can
be also a motorist, a parent, a shopper, a business person,
a golfer, a commuter, a gardener, a member of a political
party, and a property-tax and income-tax payer.

Individuals are quite capable of living with internal
conflicts – often rationalizing them in such a way that
they do not appear to be in conflict. Thus a suburban
resident may be a strong supporter of a local growth
management policy which severely restricts new
development in the area; but, if he is prevented from
developing a piece of land which he owns, he may argue
that this is a case where the policy is being imposed
far too stringently and insensitively. Human beings
have a remarkable ability to reconcile conflicting views
when their own interests are at stake.

Since land use planning is essentially a local matter,
local interest groups naturally exert a considerable
influence on planning policy. Above all, given the high
proportion of home owners, their interests tend to
predominate, particularly in suburban areas. Indeed,
pressures to introduce or to extend planning controls
often come from home owners concerned about the
effect of change on their property values. The widespread use of exclusionary zoning policies is the result.
Homogeneous communities, with no low-income
housing, are seen as the guarantee of stable property
values. The morality of this is, of course, questionable,
but morality is often not a primary consideration in a
process as political as land use planning. This political
system responds to the views of the powerful constituents. Home owners and development interests are
frequently the most powerful. Where developers go
against the interests of existing home owners (for
example in proposing to build low-income housing),
the home owners are likely to win. On the other hand,
where development and environmental issues clash, the
development interests tend to win, particularly when
jobs are at stake. (This is still generally true, even
though there is now a heightened concern for the
environment: real though this is, it typically takes
second place to the need for local employment.)

Advocacy planning

Interest groups which have influence are typically well
organized, well funded, and highly articulate. So who
speaks for the unorganized, inadequately funded, and
powerless? Who represents minorities, the poor, the
disadvantaged? The answer, of course, varies from place
to place. Nationally, there are bodies such as the
National Association for the Advancement of Colored
People and the National Urban League (see Box 1.3).
Many national organizations have a local organization.
But there are also countless organizations which have
been formed locally to represent local interests.
Sometimes these are short-life bodies set up to deal
with some specific local issue (opposition to a road
scheme, for instance). Others keep a watching brief on
some aspect of local conditions or politics. Some are
active in promoting the development or improvement
of local opportunity or living conditions through
neighborhood organizations such as community development corporations. (A few examples are given in
Chapter 16.)

The role of the planning profession in promoting the
interests of the disadvantaged is a problematic one on
which there is continual debate. The professional code
of ethics, however, seems quite clear:

> We shall seek social justice by working to expand
> choice and opportunity for all persons, recognizing
> a special responsibility to plan for the needs of the
> disadvantaged and to promote racial and economic
> integration. We shall urge the alteration of policies,
> institutions, and decisions which oppose such needs.
> (American Institute of Certified Planners,
> *AICP Code of Ethics and Professional Conduct*,
> adopted March 19, 2005, effective June 1,
> 2005, revised October 3, 2009)

This clause was added to the ethical code of the
American Institute of Certified Planners (now the
American Planning Association) as a result of strong
and eloquent pressure from Paul Davidoff. He was a
planner and lawyer who founded the Suburban Action
Institute as a pressure group to increase access to
suburban jobs. He set a pattern, which many followed,

of combining research and action. Research establishes the needs of the disadvantaged and the ways in which they are being denied, while action is promoted by making these issues known, by organizing communities, and by legal and political initiatives.

Davidoff stressed the need to openly invite debate on the political and social values that underlie plans. The planner should not, so he argued, be a mere technician: the planner must act as an advocate. Davidoff certainly aroused the conscience of the planning profession, though there have been more words than actions. Planners have employers who pay their salaries and define what their jobs are. Employers are often unsympathetic to 'alternatives' for the disadvantaged. Developers are concerned with profitability, politicians with majority votes. Though they may have social concerns, these are unlikely to be predominant and, in any case, may well be interpreted somewhat differently than they are by the people affected by their actions. Planners may be squeezed between conflicting groups. The few planners who have nevertheless been able to follow an active 'equity agenda', such as Norman Krumholz, are the exception.

Equity planning is based on the belief that planners and public officials need to consider the social implications of their actions. Metzger (1996: 113) defines equity planning as

> A framework in which advocacy planners in government use their research, analytical, and organizing skills to influence opinion, mobilize underrepresented constituencies, and advance and perhaps implement policies and programs that redistribute public and private resources to the poor and working class in cities.

It is about expanding the opportunities to those that have been left out or ignored in planning. Instead of focusing primarily on land use and physical planning, we need to address social and economic issues. In other words, in earlier years, we have had a bias towards the former two areas while ignoring the latter two areas.

Experience has also shown that open advocacy can be self-defeating. It may raise overwhelming opposition that might possibly have been avoided by more subtle methods. A classic case is integration, where explicit initiatives have so often failed. By contrast, schemes such as the Gautreaux experiment in Chicago and the Moving to Opportunity program (noted in Chapter 19), which have been termed 'stealth programs', have had positive results. The essential feature of these programs is that they operate on a small scale (with only a small number of black families moving into a white area): they thus avoid raising resistance.

Such an argument may raise moral doubts, but the intense opposition to open programs has created increasing pessimism about the viability of traditional approaches. Another alternative has been proposed (and implemented) by Chester Hartman. This involves the stimulation and utilization of research which is focused on issues that enable activists to be effective in carrying out an advocacy agenda. His own organization, the Poverty and Race Research Action Council, has had some success with such an approach (Hartman 1994: 159).

Foundations and giving to communities

Community groups no longer rely on the availability of state and local funds to combat their urban ills. Many community groups are turning to foundations and local non-profit organizations for funding projects and programs facing their jurisdictions.

There are a plethora of private foundations that partner with community groups in efforts to solve community problems. Each private foundation has its own guidelines, subject areas that its funds, geographic areas where it funds, etc. Community groups have become more knowledgeable about seeking out funding from national private foundations.

Eligibility for funding from private foundations varies by foundation. Private foundations require that organizations be recognized as a 501 (c) (3) organization. This means they must be a tax-exempt non-profit charitable organization and cannot participate in any campaign for or against any political candidates.

The Ford Foundation is a leading private foundation that partners with groups in the US and around the world in such areas as accountable government, economic fairness, educational opportunity and scholarship, metropolitan opportunity, and sustainable development. Founded in 1936, the Ford Foundation recognizes the importance of working with others to promote social change. As noted in the 1949 Report of the Study for the Ford Foundation on Policy and Program (p. 9):

> The people of this country and mankind in general are confronted with problems which are vast in number and exceedingly disturbing in significance. While important efforts to solve these problems are being made by government, industry, foundations, and other institutions, it is evident that new resources, such as those of this Foundation, if properly employed, can result in significant contributions.

Over the years, the Ford Foundation has funded important programs that have improved the quality of life for African Americans and other disenfranchised communities in Mississippi; helped to cultivate, develop, and retain a pool of young leaders to build the capacity of organizations actively working to shape the Detroit of tomorrow; strengthened community organizing in low-income communities of color in North Carolina; identified issues affecting efforts to revitalize distressed cities, neighborhoods, and rural areas; and developed a smart growth training program for African-American public officials and administrators and other community-based partners in the Gulf Coast.

Another leading foundation that provides grants to and partners with groups committed to help meet the needs poor individuals and communities is the Kresge Foundation. Founded in 1924, the Kresge Foundation, based in Troy, Michigan, has a special emphasis on helping non-profits and community initiatives in Detroit and provides grants to other areas around the US in the areas of arts and culture, community development, education, environment, health, and human services. Among recent projects and programs that it

has funded are a program making financing available to small businesses and non-profit organizations in low- and moderate-income communities; a training program to help under-served communities in Idaho and Tennessee develop local forest and water climate adaptation plans; and a program to provide workshops and technical guidance to communities in an effort to strengthen their resilience to the impacts of climate disruption.

The John D. and Catherine T. MacArthur Foundation is a leading private foundation based in Chicago, Illinois. Established in 1970, the MacArthur Foundation provides funding for domestic and international programs and projects that seek to support creative people and organizations that make cities better places. Among the subject areas funded are conservation and sustainable development, community and economic development, housing, and the preservation of affordable housing.

There are over 700 community foundations in the United States that provide funding to non-profits dealing with problems facing our communities. These foundations can be found in such locations as Boston, Massachusetts; Atlanta, Georgia; Dayton, Ohio; San Diego, California; and Sioux Falls, South Dakota. As is the case with national private foundations, community foundations differ in what they fund.

One of the biggest community foundations in the United States is the Seattle Foundation. Founded in 1946, the Seattle Foundation funds organizations dealing with issues facing Seattle and King County. According to its website (http://www.seattle foundation.org), its mission is 'to create a healthy community through engaged philanthropy, community knowledge and leadership'. Examples of recent funding include providing operating support for community organizations, funding for community engagement and the development of transit-oriented neighborhoods in Seattle and King County, facility renovation and equipment purchases for non-profit organizations serving low-income persons and seniors, and funding for writing and after-school tutoring programs.

One of the earliest community foundations is the New York Foundation. Established in 1909, it focuses its resources on the city's neighborhoods and the many community organizations working to improve the neighborhoods. Funding can be devoted to such activities as leadership training and development, community organizations, and managerial assistance. Recent projects that have been funded include the following:

- Improving access to higher education and creating equal opportunity for immigrant youth;
- Providing legal and advocacy services to elderly New Yorkers;
- Providing funding to an organization mobilizing residents and volunteer groups to improve their neighborhoods;
- Providing training for public-interest professionals and leaders of non-profit organizations; and
- Providing capacity-building assistance to emerging and well-established neighborhood organizations and institutions.

Planning vs. implementation

It might be reasonably assumed that plans are prepared in order to be implemented. Though this may often be the intention (even if a vain one), it is not always the case (see Box 1.4). In fact, some plans may never have been intended to be implemented. One level of government may provide funds for developing a plan while assuming that another level of government would continue the work by providing funds for implementing the plan. Some plans are basically pieces

of propaganda intended to boost the attraction of an area (usually for development), or to promote one type of future over another (such as one with greater leisure provision, or one which is more ecologically sustainable), or to press for some particular character of development (as with the classical architecture of the City Beautiful movement). Plans can serve many functions: inspiration may be more important than implementability. Or the preparation of a plan may be the short-term answer to a particular political pressure 'to do something' about the future of an area: the plan is thus seen as the first step; but by the time the plan is completed, the enthusiasm for change may have dissipated, or the plan may be seen as impracticable or too costly. Again, a plan may be required as part of a submission to a higher level of government for grant-aid: once the grant is obtained, the plan has served its purpose. Not infrequently, plan-makers indulge in a dream: they know that they cannot forecast what influences will exert themselves in the future, but they feel compelled to try; they thus 'resolve the conflict by making plans and storing them away where they will be forgotten' (Banfield 1959).

The rational model of planning embraces the simplistic view that there is a logical progression through successive stages of 'planning', culminating in implementation. The beguiling logic does not translate into reality. On the contrary, it is highly misleading – and dangerous – to separate policy and implementation matters. In fact, sometimes policy emanates from ideas about implementation rather than the other way round. Thus, a policy of 'slum clearance' or 'redevelopment' focuses on the clearly indicated

BOX 1.4 PROBLEMS OF IMPLEMENTATION

1 Many policies represent compromises between conflicting values.
2 Many policies involve compromises with key interests within the implementation structure.
3 Many policies involve compromises with key interests upon whom implementation will have an impact.
4 Many policies are framed without attention being given to the way in which underlying forces (particularly economic ones) will undermine them.

Source: Barrett and Hill 1993: 105

types of action. The implementation becomes the policy, and the underlying purpose is left in doubt. If the objective is to improve the living conditions of those living in slum areas, there might be better ways of doing this, such as rehabilitation, or area improvement through local citizen action. With such an approach, demolition might be merely an incidental element in the local program. With clearance as a policy, however, there is a danger that different objectives might be served (such as central city commercial interests). Demolition might even be detrimental if it reduces the quantity of affordable housing. With hindsight, it is not surprising that this is what happened with the urban renewal policy. Clearance and redevelopment (later expanded to urban renewal – 'the renewal of cities') became a policy of economic development of central city areas.

Even when policy is not framed with a particular form of implementation in mind, it is frequently modified by implementation; it may even be transformed. The crucial difficulty is the void between the purpose and hopes of a paper plan and the realities on the ground. In Banfield's words, policy 'is an outcome which no one has planned as a "solution" to a "problem": it is a resultant rather than a solution' (Banfield 1959). The interrelationship between policy-making and implementation arises from the necessity of collaboration among a multiplicity of public, private, and voluntary agencies. But since each of the agencies has its own agenda, and even its own way of looking at problems, such collaboration necessitates compromise; and compromise means that the policy is changed (Pressman and Wildavsky 1984).

Local vs. central control

Land use planning in the US is largely a local matter. Though there are important exceptions (the state and federal governments have specific planning functions in relation, for example, to environmental and coastal concerns), the scope and character of land use planning is mainly determined locally. This means that there is a great variety in the ways in which planning is carried out. At one extreme are areas where there is virtually

no planning at all: landowners are free to build where and what they wish. There are few regulatory controls present. At the other extreme are areas where there is a highly sophisticated planning machine which controls the location, character, quality, and design of all development. These are areas with a high degree of regulatory control. In some other countries there is much more uniformity. For instance, each of the Canadian provinces has planning laws which operate in a generally consistent way, and are subject to certain controls operated by the provincial governments. In Britain, a planning code operates over the whole country, subject to mandatory central government requirements, and coordinated by an elaborate governmental apparatus.

No such central control exists in the United States. On the contrary, the governmental system was explicitly designed to prevent centralization. It is characterized by 'checks and balances' which other countries find baffling since they clearly reduce the efficiency of government – which is precisely what was intended. In the words of Richard Hofstadter, what has emerged is 'a harmonious system of mutual frustration' (Hofstadter 1948: 9). This deep concern about the dangers of government, together with the high esteem accorded to the Supreme Court (and the legal system generally), is distinctively American, and it profoundly affects the character of US planning.

There are other features of US planning which are distinctive. One is the limited amount of discretion which the constitutional framework allows to local governments. Discretion implies differential treatment of similar cases, and therefore runs foul of the equal protection clause of the Constitution. The Bill of Rights guarantees that individuals are to be free from arbitrary government decisions. This is a major constraint on planning in the United States. By contrast, the British planning system provides for a great deal of discretion. This is further enlarged by the fact that the preparation of a local plan is carried out by the same local government that implements it. In the United States, the 'plan' is typically prepared by the legislative body – the local government – but administered by a separate board. The British system has the advantage of relating policy and administration, but to American eyes 'this institutional framework blurs the distinction between policy making and policy applying, and so enlarges the role of the administrator who has to decide a specific case' (Mandelker 1962: 4).

Another striking characteristic of US land use planning is its domination by lawyers and the law. In this, it is different only in degree from other areas of American public policy. All government is assumed to be 'an intrinsically dangerous and even an evil thing, to be tolerated only so long as its disadvantages are not outweighed by its defects' (Nicholas 1986: 11). By contrast, the law is a thing of great reverence. The particularly strong presence of law in land use planning derives, of course, from the strong attachment to property – an attachment that is enshrined in, and protected by, the Constitution. Land use planning is thus inherently a matter of law.

Underlying attitudes to land and property

Perhaps the most tangible illustration of the American attitude to property is the constitutional safeguard. The Fifth Amendment provides that:

> No person shall be . . . deprived of life, liberty or property without due process of law; nor shall private property be taken without just compensation.

Many countries have nothing equivalent to this; and they have very varying attitudes to property. The Netherlands and Britain, for example, have a positive attitude to government controls over land. There is a popular support for the preservation of the countryside and the containment of urban sprawl. Without these attitudes, the systems of land use planning that operate in these countries would be impossible. In the United States, land has historically been viewed as a replaceable commodity that could and should be parceled out for individual control and development; and if one person saw fit to destroy the environment of his valley in pursuit of profit, well, why not? There was always another valley over the next hill. Thus the seller's concept of property rights in land came to include the

right of the owner to earn a profit from his land, and, indeed, to change the very essence of the land, if necessary, to obtain that profit. 'Cheap land has as one of its consequences that of stimulating and universalizing acquisitive instincts and respect for property rights' (Philbrick 1938: 723). However, the time came when the ever-receding frontier ceased to be so; it was overtaken, and land became more valuable.

One might have expected the growth of a conservationist ethic, as is prevalent in western Europe. However, though this happened to a limited extent, particularly with environmentally valuable resources, the main effect was in the opposite direction: to increase the attractiveness of land as a source of profit. Speculation has never been frowned upon in the United States: on the contrary, it has been a notable feature of the economic landscape. In many countries, land is regarded as something special, to be preserved and husbanded. In the United States, the dominant ethic regards land as a commodity, no different from any other. This, of course, is related to the sheer abundance of land. Indeed, until recently, it seemed limitless; and even now there are many parts where this still seems true. In some areas, however, the rate of urbanization has given rise to the emergence of 'growth management' policies which seek to channel growth into areas judged to be acceptable. That these have failed to prevent urban growth is unsurprising: the forces at work are strong, and the governmental powers of control are weak (as is discussed at length in Chapters 10 and 11).

Private and public planning processes

On a simple view of the development process, the private sector is responsible for development proposals, while the public sector is responsible for regulating them. This ignores the important role which the public sector plays in the provision of infrastructure, and in rendering essential services; but it approximates the reality, even if in a somewhat distorted way. Certainly, the two sectors have different functions and methods of operation; it is worth examining these briefly at this stage.

The private development process is a series of stages by which a proposed development is brought to fruition. At its simplest, the process is conceived and followed through by the person or company which undertakes the construction: that person decides that there is a market demand for a particular development; a site is selected; finance is arranged; permits are obtained from the appropriate regulatory agencies; the project is constructed; and the finished product is marketed either for sale or for rent.

Of course, developments vary greatly. The initiator may be the client (a home buyer for example), who has the necessary finance, but requires a site, an architect, and a builder; or the client may already own a site and require a developer to coordinate all the planning and building operations. The prime mover could be an insurance company seeking an investment, a farmer wanting to convert the farm into a subdivision (the 'last cash crop'), or a municipality attempting to expand its tax base.

Another variation arises when the body providing the finance wishes to be a joint partner in the venture and share in the profits (and the risks). Here the finance company would be looking at a balance between risk and high profits – a very different situation from that of a long-term equity investor who seeks security and plays a passive role in the development process.

The private planning process typically involves a degree of risk. There therefore needs to be good judgment about the state of the market, the likely demand for development, the prospects for specific locations, the trend in interest rates, the political outlook, and a host of similar uncertainties. Good judgment can pay off handsomely; so can good luck. Bad judgment or bad luck can be disastrous. The success of the private developer is measured simply: does the development return a profit?

The public planning process is very different. It is essentially reactive to private initiatives. (Even the provision of infrastructure often follows, rather than leads, private investment.) It is concerned to ensure that development accords with the standards set out in legislative instruments. Its regulatory character means that it is on the lookout for deviations, misinterpretations, and errors. Its chronic shortage of resources, and its perpetual concern about future costs of maintaining the public estate, leads it to try to secure the maximum amount of public benefit from developers. The public process also involves a multiplicity of agencies each with its own objectives, plans, finances, and concerns. These numerous agencies operate in their specific areas of competence and responsibility (water, roads, schools, parks, libraries, waste disposal, clean air, fire services, and so on). A major problem for a developer is finding a way through the maze of agencies that have to be satisfied, or at least consulted, about development proposals. (The developer may not know that the agencies may be equally confused about the overall process.)

The public sector has no equivalent to the developer's profit (though dissatisfaction with local and state services can be bluntly registered through the ballot box). Indeed, all too often, it is by no means clear where blame – or praise – lies. Even if the responsible agency can be identified, it may be difficult to identify the responsible person.

There are thus two different worlds, with different objectives. One is characterized by willingness (and necessity) to take risk; the other, being publicly accountable, is averse to taking risks. One is opportunistic; the other is bureaucratic. One seeks financial reward; the other good husbandry and probity.

Given these very different frameworks, there is a serious communication gap between the two sectors (Peiser 1990). Bridging this gap is a major part of both the private development and the public planning processes. How it works in practice is discussed at length in the following chapters. Box 1.5 provides an overview of the formal steps in the planning process. (The adjective 'formal' is used to indicate that there may be informal ways of dealing with problems as they arise.)

The elements of the planning process

The planning process encompasses the preparation of a plan and its implementation. However, the process is much more complicated than simply developing a plan and then implementing it. Issues need to be identified and defined. A variety of players, ranging from citizens, public officials, and private sector, to non-profit groups, should have the ability to participate in this phase. Disputes might surface in this or any other phase of the planning process. Nevertheless, agreeing on the issue is critically important in the process. Failure to gain consensus on the nature of the issue could result in a great deal of wasted time, energy, and resources. Agreeing on what should be done and how to prioritize goals and objectives is equally as critical. It is important at this stage of the process to look at an agency's mandate. Information will be needed from a host of resources to help determine the history and current status of an issue and to determine the existence of any trends or patterns. This information should assist planners and decision-makers in developing alternative policies and programs. A plan and alternatives would then be based on the available knowledge of the participants. Once the decisions have been reached as to the plan that will ultimately be forwarded for adoption by the appropriate legislative body, it will be reviewed by the public and interested groups. After the plan has been publicly debated and adopted, it goes through a 'web' of approvals.

The implementation process which follows is of crucial importance. It will involve not only the planning department but a wide range of agencies including such departments as economic development,

transportation, information technology, parks, and housing.

In theory, a local government has a comprehensive plan (discussed in Chapter 7) which forms a framework for its zoning and subdivision ordinances. Approval of the plan and the ordinances are legislative acts which are the responsibility of the elected legislative body (the council), though they may be prepared by an advisory planning commission (see Box 1.6). The implementation of the ordinance rests with the planning department, but applications from owners for changes in zoning are dealt with by a separate zoning board of appeals (sometimes termed a 'board of adjustment'). This rather complicated system is a product of history, which is discussed later. (State enabling acts differ, and there are variations on this model.)

As previously indicated, in addition to these planning and zoning controls, there are many other areas of control which may fall to the responsibility of other departments of the local authority or to other agencies. These include public utility connections and building codes. A large development might also involve negotiations with the departments or agencies responsible for transportation, education, and environmental protection. A successful developer will have considerable skills in maneuvering a route through this network.

Not surprisingly, 'regulatory barriers' are a common target of criticism, and measures are intermittently taken to reduce them. However, they perform an

---

BOX 1.5 STEPS IN THE PLANNING PROCESS

Step 1: Identify issues and options
Step 2: State goals, objectives, priorities
Step 3: Collect and interpret data
Step 4: Prepare plans
Step 5: Draft programs for implementing the plan
Step 6: Evaluate potential impacts of plans and implementing programs
Step 7: Review and adopt plans
Step 8: Review and adopt plan-implementing programs
Step 9: Administer implementing programs; monitor their impacts

Source: Anderson 1995

BOX 1.6 DIVISION OF PLANNING RESPONSIBILITIES

Policy: Legislative body (council) approves plans and ordinances
Advice: Planning Commission holds hearings and makes recommendations to the legislative body concerning
    plans and policy matters
Implementation: Planning Department
Appeals: Zoning Board of Adjustment

---

important role in modern society, and there are strong public pressures which may lead to an increase in regulation rather than a decrease.

Citizen participation

The ability of citizens to participate in deciding public matters has had a long tradition in the United States. The degree and nature of citizen participation varies widely. In some areas, the normal electoral process is considered to be sufficient. In others there can be extensive meetings. Cunningham (1972: 595) defines citizen participation as 'a process wherein the common amateurs of a community exercise power over decisions related to the general affairs of a community'. The range of participation is neatly illustrated in Arnstein's (1969) classic 'ladder of participation' in which the various 'rungs' illustrate the range from degrees of active participation to mere tokenism.

In order to participate in the public policy process, citizens need information. They need information that is understandable. If they don't understand what they have been given, how are they expected to be able to participate in a meaningful way? While they must be able to understand the information they have been provided by the public agency, they must have the information in a reasonable time period. Giving an individual a 100-page document a few days before a meeting merely creates frustration on the part of the reader. It may serve as an indicator of just how seriously an agency views or wants citizen participation.

Creating a citizen-participation program is no easy task. A number of factors must be considered. First,

who are the citizens? Is it a neighborhood, a city, a certain group of the population? Second, when does a citizen-participation program start? It needs to start early in the planning process, when an issue is first being discussed. Third, how and when will the citizens be notified? Citizens should be able to participate in how the issue is being defined, how alternatives are being addressed, deciding what is going to be done, and in determining how the chosen course of action is faring.

Citizen-participation requirements are a common feature in federal policy and programs. In 1964, President Johnson's plea for a 'War on Poverty' called for the maximum feasible participation of minority and low-income residents in areas. Two years later, the Model Cities Program called for widespread participation of people affected by the proposed actions. According to Section 1506.6 of the National Environmental Policy Act (NEPA) of 1969, agencies had to do the following:

(a) Make diligent efforts to involve the public in preparing and implementing their NEPA procedures.
(b) Provide public notice of NEPA-related hearings, public meetings, and the availability of environmental documents so as to inform those persons and agencies who may be interested or affected.
(c) Hold or sponsor public hearings or public meetings whenever appropriate or in accordance with statutory requirements applicable to the agency.
(d) Solicit appropriate information from the public.
(e) Explain in its procedures where interested persons can get information or status reports on

environmental impact statements and other elements of the NEPA process.

(f) Make environmental impact statements, the comments received, and any underlying documents available to the pursuant.

Other federal environmental and transportation programs, such as the 1980 Resource Conservation and Recovery Act (RCRA) and the 1991 Intermodal Surface Transportation Efficiency Act (ISTEA), have requirements for timely notice, public hearings, information, the ability to comment on the information, and the ability to access officials throughout the process.

States have followed the call for opportunities for citizen participation. In its 1998 Growing Smarter legislation, the state of Arizona required jurisdictions to adopt written procedures to provide effective, early, and continuous public participation when developing and adopting general plans and major amendments. According to Arizona Revised Statutes, there must be consultation with individuals, organizations, and agencies. Written procedures must provide for: 1) broad dissemination of proposals and alternatives; 2) opportunities for written comments; 3) public hearings after effective notice; 4) open discussions, communications programs, and information services; and 5) consideration of public comments.

In that same year, Mesa, Arizona, enacted, via a resolution, citizen participation guidelines for most major land development proposals including rezonings, site plan modifications, use permits, and variances. The guidelines were later adopted by ordinance and placed into the Mesa Zoning Ordinance on November 4, 2002.

Mesa's Citizen Participation Ordinance requires a Citizen Participation Plan and Citizen Participation Report. The process begins with a pre-application meeting to familiarize the land use applicant with the required procedures and the requirements that must be met by the applicant. The Plan, which must be implemented prior to the first public meeting, is designed to:

1 Ensure that applicants pursue early and effective participation in conjunction with their application,

providing the applicant with an opportunity to understand and address any real or perceived impacts their development may have on the community.

2 Ensure that citizens, property owners, and neighbors have an adequate opportunity to learn about applications that may affect them and to work with applicants to resolve potential concerns at an early stage of the process.

3 Facilitate ongoing communication between the applicant, interested citizens and property owners, and city staff, throughout the application review process.

The Plan must also include information on who will be affected, how these individuals will be notified and informed about the proposed action, and how they will be given an opportunity to participate in the process.

The Report documents the results of implementing the Plan. The Report details specific techniques that were used and a summary of the concerns and issues that were identified during the process. The Report must be provided at least ten days prior to the first scheduled public hearing.

Planners today cannot simply rely on traditional methods of citizen participation. Many have realized that relying on such traditional techniques as public meetings and public hearings fails to obtain the views of a number of interested publics. As Klein (2000: 425) has indicated:

The old-style processes of involving the public – particularly public hearings – often result in perfunctory, stilted, 'go through the motions' styles of engagement. By any stretch of the imagination, these practices rarely provide meaningful public participation or engagement. Typically, they are organized and run from the top down and are scheduled at the end of the process, immediately before adoption of the measure being considered. If citizens come at all, they often leave the hearing feeling ineffectual, co-opted, or manipulated; they often leave believing that 'the fix' was in.

Consensus building offers one way of hearing and responding to groups that have different points of view

BOX 1.7 TEN CONSENSUS-BUILDING PRINCIPLES

1   Involve interests as early as possible.
2   Tailor the process.
3   Be inclusive.
4   Identify and nurture shared interests.
5   Share credible information.
6   Provide impartial and collaborative leadership.
7   Consider using professional help.
8   Maintain momentum.
9   Validate results.
10 Involve the media.

Source: Klein 2000

(see Box 1.7). It provides a means to help planners collaborate with officials, citizens, development interests, and other groups where the possibility of conflict exists. For example, planners could employ a neutral party to act as a mediator or facilitator to run a consensus-building exercise to develop recommendations to resolve issues involving such controversial topics as developing goals for future growth or determining the site of a waste disposal facility.

By involving multiple parties, consensus building has the potential to be a long and time-consuming process or, as Salsich (2000: 739) notes, it is 'a messy process because of the large number of meetings and other activities that must be held in order to fully engage the community'. An effective facilitator or mediator is critical for the success of a consensus-building exercise.

Charrettes are often used in consensus-building efforts. Lindsey, Todd, and Hayter (2003: 1) define a charrette as 'an intensely focused activity intended to build consensus among participants, develop specific design goals and solutions for a project, and motivate participants and stakeholders to be committed to reaching those goals'. A charrette could be convened in one session or multiple sessions depending on the issue being addressed. They can be used for formulating alternatives to a proposed highway site, a city growth plan, or a controversial public facility.

A good example of how multiple groups were involved in a planning process can be found in 'Forward Dallas! Vision'. In this effort, over 2,000 Dallas residents (including high school students and other youth) and business leaders participated in multiple workshops and meetings. The process involved polling residents and business leaders for their ideas; establishing advisory committees; holding community-based workshops to generate input through consensus and group discussion; creating community maps on desired future development. Alternative scenarios were created to provide views of what Dallas might be in the future. These scenarios were then evaluated to determine the best course of action for Dallas' future. Ultimately, a shared vision for Dallas was created.

Technologies and planning

Planners are using more technologies in their work than ever before. Tools such as GIS, remote sensing, visualizations, social media, websites, interactive media, etc., provide planners with various technologies to improve data analysis, data gathering, and plan generation. The various technologies enable them to become more efficient in the use of their time and funds, enhance a planner's communications with the public, and offer them opportunities to increase citizen participation.

Cities have certainly determined that using various forms of information and communications technology is advantageous in their operations. Some communities quickly embraced its use while other communities did not. More than twenty years ago, Moss (1987: 535) noted:

> The telecommunications infrastructure – which includes the wires, ducts, and channels that carry voice, data, and video signals – remains a mystery to most cities. In part, this is due to the fact that key components of the telecommunications infrastructure, such as underground cables and rooftop microwave transmitters, are not visible to the public. Unlike airports and garbage disposal plants, telecommunications facilities are not known for their negative side effects, and until recently, have not been the source of public disputes or controversy.

Over ten years ago, Graham and Marvin (1996: 51) furthered the discussion by indicating that 'many city planners and managers do not even know what the telecommunications infrastructure is in their cities; very few have the power, influence or conceptual tools to reshape it to their desired impacts'.

We have certainly come a long way in regard to the use of technology in cities. The use of technology has taken on an important role in recent years as citizen demands for more and better information have increased. Cities are using municipal websites and other forms of information technology to make a wide range of material available to the public. For example, comprehensive plans and municipal codes are available online (the internet). Minutes of various meetings and applications to serve on various governmental boards, committees, and commissions are available online. The City of South St Paul, Minnesota, like other cities around the US, provides citizens the opportunity to view City Council and Planning Commissions 'on demand'. Development applications or permits are routinely available online so that citizens and other organizations don't have to venture to the various locations to pick them up. Prospective contractors for city projects and service can access Requests for Proposals (RFP) and submit them online. Some areas provide developers the opportunity to track their development applications online and to monitor and make changes and updates on their proposals. Cities around the US have employed citizen call systems or 311 Citizen Relationship Management (CRM) systems that allow citizens to connect with government. These systems can be traced to the late 1990s when the Federal Communications Commission reserved the use of the three-digit – 311 – for national non-emergencies such as loss of water, park closures, library closures, road closures, rescheduling trash and recycling collection, etc.

Many communities throughout the US have turned to various social media techniques such as Facebook, Twitter, YouTube, MySpace, Flickr, podcasts, blogs, etc., to engage the public as a public service. As a definition, the City of Iowa City, Iowa, defines social media as: 'any facility for online publication and commentary, including without limitation blogs, wiki's, content hosting sites such as Flickr and YouTube, and social networking sites such as Facebook, LinkedIn, and Twitter' (http://www.icgov.org/?id=2128).

The use of social media devices is not designed to replace traditional public meetings or forums, calling a complaint line, or writing the proverbial letter to a public official. These forms of citizen engagement are time consuming and are often felt to be very ineffective. Social media devices are to offer different opportunities for citizens to have their voices heard and to offer constructive ideas or criticisms on issues. Citizens have access to the ideas of other people and can comment on the various views. Responses by government representatives are able to be viewed by everyone participating in the discussion. Social media techniques are also often touted as good means to improve transparency in government. Box 1.8 provides a summary of the reasons why Arlington County, Virginia, uses social media.

The uses of the various social media techniques vary by community. The City of New York has asked residents to make suggestions online for recommendations to make the city greener and more sustainable. In California, a community is using an iPhone app to

BOX 1.8 ARLINGTON COUNTY, VIRGINIA, REASONS FOR USING SOCIAL MEDIA

Social media platforms offer many advantages – and help us open up government to encourage citizen participation, strengthen our democracy and support a civic culture.

- Increase transparency of government.
- Enable rapid response.
- Identify trends before they take off.
- Listen to residents, customers – enabling us to improve County services, programs, practices.
- Reach out – go to the conservation, don't wait for it to come to us.
- Respond directly to the community without a filter.
- Expand communication tools with vast distribution system for content.
- Answer what folks want to know.
- Fill the vacuum being created by the death of newspapers.

Source: Arlington County, Virginia, 2009

---

alert people trained in cardiac pulmonary resuscitation (CPR) when a nearby citizen is suffering a heart attack. Communities commonly use social media to promote economic development, emergency preparedness, and job opportunities. A community can use a social media device to ask residents to consult on proposed rule changes and to prioritize issues facing their neighborhoods. The City of Evanston, Illinois, uses a Twitter page for emergency alerts. Areas around the US also employ YouTube videos on emergency preparedness and use social media devices to inform residents of severe weather alerts. The City of Phoenix uses a daily podcast video to report on crimes committed in the area and progress on existing criminal cases. Other communities offer so-called 'Feedback' sections on their community websites so that residents can help to identify issues regarding economic development opportunities and constraints, land use planning, infrastructure planning, and the environment. While many of the various social media devices are used to connect with local residents, they can be used at the regional, state, national, and international levels.

The increasing use of social media tools has led to the development of government policies governing their use by employees and residents. An individual

or committee may be placed in charge of developing local policies for their use and for placing a post on a city social media site. Policies have been developed for employees because they serve as representatives of the city. The policies may consider any posting made during works hours and away from the office during non-business hours. In some cases, violating city internet and social media postings could result in disciplinary action or loss of job. Recognizing that their use is to reach a broader audience, the City of Seattle, Washington, outlines standards for public postings:

Users and visitors to social media sites shall be notified that the intended purpose of the site is to serve as a mechanism for communication between City departments and members of the public. City of Seattle social media site articles and comments containing any of the following forms of content shall not be allowed:

a. Comments not topically related to the particular social medium article being commented upon;
b. Comments in support of or opposition to political campaigns or ballot measures;
c. Profane language or content;

d. Content that promotes, fosters, or perpetuates discrimination on the basis of race, creed, color, age, religion, gender, marital status, status with regard to public assistance, national origin, physical or mental disability or sexual orientation;

e. Sexual content or links to sexual content;

f. Solicitations of commerce;

g. Conduct or encouragement of illegal activity;

h. Information that may tend to compromise the safety or security of the public or public systems; or

i. Content that violates a legal ownership interest of any other party.

(http://www.seattle.gov/pan/SocialMediaPolicy.htm)

The idea of some individuals not being able to take advantage of the increased use of information technologies represents a constant challenge to planners. This so-called 'digital divide' indicates that some people do not have access to computers or other forms of information technology. These individuals are unable to take advantage of the increasing amounts of information made available to the public. The reasons for this lack of access may vary from community to community. For example, lack of income may prevent individuals from access to computers if technology centers are not available to residents. Another factor to consider is that community websites may be available only to those individuals who read English. This dilemma has led many governments to develop their websites in several languages. The provision of sites in several languages allows people who previously lacked access to this information to now take advantage of the opportunity to use the community websites. Various foundations and government agencies have provided funding to help lessen this 'divide' in access to technology through the development of community technology centers. Although the access gap is still there, it has lessened over the years.

Land use simulation games

Individuals can learn a lot about various aspects of cities and city-building through the use of computer simulation games. They create an opportunity to build a city, learn about the interconnections of a city, its different land uses, how services should be provided, and how to place road systems.

An early simulation game that many students used was the Community Land Use Game (CLUG). Developed by Allan G. Feldt, University of Michigan, CLUG helped players understand the various factors affecting the growth of a region. Within this game, players have a certain amount of capital that can be used to purchase land, construct buildings, provide municipal services, etc. Playing the game offered individuals an opportunity to learn about how decisions are made.

A later city-building simulation game called 'SimCity' was released in 1989. This game brought the idea of city-building to the masses. Individuals playing the game could build and design their own cities. They could experiment with different land uses (residential, commercial, industrial, open space) and see what would happen if different lands were too close to each other. Roads, power plants, airports, etc. could be placed at various locations in the city. As the game progressed, players could get interesting calculations on the area's population, crime rates, etc. Different versions of the game, including one that uses a 3D engine for its graphics, have been developed over the years. In SimCity 4, to develop a successful city, a player must be able to manage a city's finances, and protect its environment and quality of life for the citizens.

Geographic information systems and remote sensing

As illustrated throughout this book, planners use and need a great deal of information in their daily tasks. Much of this information is spatial information – information linked together by geography. In past years, planners had to go through a laborious process to gather information.

They visited numerous offices and collected numerous pieces of information and maps. In the end, they returned to their offices and had to plot the information they obtained on maps. They would also be able to combine various disparate pieces of information into

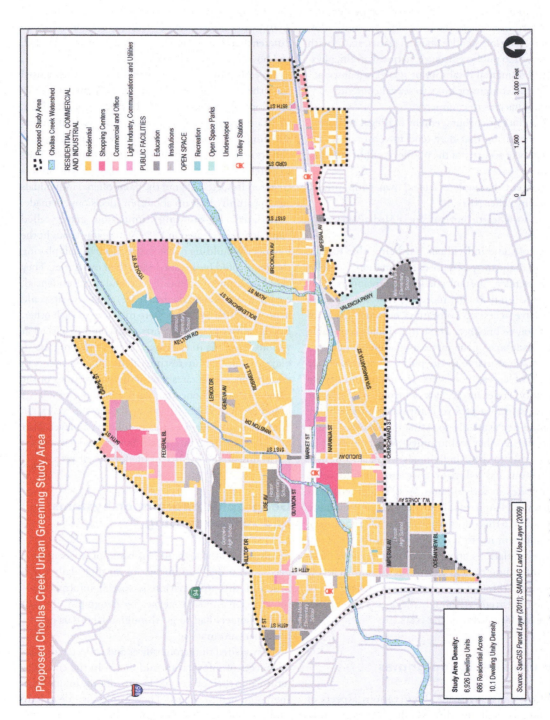

Plate 1 Example of a geographic information system map

SANDAG, Census Bureau 2000

one map. Information was accessible but it had to be manually input and manipulated.

The advent of geographic information systems (GIS) has made the ability to create maps a great deal easier than in previous years. A GIS is an organized collection of computer hardware and software that is designed to display, store, and manipulate information about spatially distributed geographically referenced data/information. Various software packages provide the GIS user with a variety of functions.

GIS is much more than simply a map-making system. It operates as a tool to provide planners and decision-makers with information needed to make a decision. It can be used to provide information on median household income, crime rates, voter turnout rates, and code violation rates by census tract (see Plate 1). It does not make decisions. It simply enables planners and other individuals to make better and more informed decisions.

A major strength of a GIS is its ability to incorporate layers of data (see Box 1.9). It could be likened to a deck of cards with each card representing different types of information. For example, one layer of data may contain a parcel's property value. Other layers of data may contain information on such items as housing tenure, zoning classification, condition of the structure, and age of the structure. A GIS provides the user with the capability to map an individual characteristic or to have the ability to analyze various categories of information to develop a type of 'what if' alternative scenario map. For example, planners could examine housing code violations in relation to age of housing and zip code to see whether the variables are linked. They would also be able to present the information at different scales. For instance, a planner might be required to provide a map at a 1:5,000 scale. This means that one inch on the map would equal 5,000 inches on the ground.

The use of GIS has drastically increased over recent years. A number of states have developed offices that seek to guide the development of GIS within their state. For example, the Office of GIS (MEGIS) coordinates a statewide GIS in Maine. This office provides services to encourage the development of GIS within the state and promotes data sharing among the various users of GIS. The New Hampshire Geographically Referenced Analysis and Information Transfer System (NH GRANIT) is a cooperative project designed to create, maintain, and make available a statewide geographic database serving the needs of the various jurisdictions within New Hampshire.

Local and regional uses of GIS vary by jurisdiction (see Box 1.10). The DenverGIS manages the area's comprehensive spatial GIS information and related databases. It manages data layers from various city departments, agencies, etc. It provides maps on property, parks and recreation, cultural, public safety, government, sewers and public utility projects, and land surveys. An interactive mapping service is now available on the public website for residents', businesses', and tourists' usage.

Denver's Animal Control Division and Environmental Services Division have worked together on a GIS to track rat and mosquito breeding sites. The Police Department uses GIS to show the location of police stations, police districts, and to develop crime reports and map criminal activity 'hot spots'. Residents can search for crime data and generate maps of crime around their home, neighborhood, schools, and other

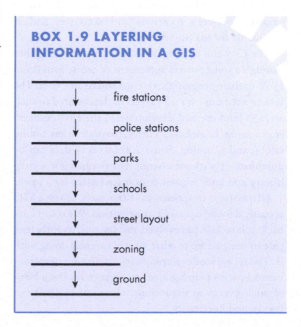

BOX 1.9 LAYERING INFORMATION IN A GIS

- fire stations
- police stations
- parks
- schools
- street layout
- zoning
- ground

BOX 1.10 POSSIBLE INFORMATION INCLUDED IN A GIS

land use	police stations	fire stations
property values	bus routes	recreation centers
medical facilities	educational facilities	churches
historic properties	road network	libraries
socio-economic data	criminal incidents	childcare facilities
building permits	telecommunications	natural resources
accidents	pet care facilities	census tracts
voting precincts	population density	vacant properties
polling locations	housing quality	group homes
public health incidents	public service complaints	adult entertainment sites

areas. Parks and Recreation uses GIS to show the location of city parks, golf courses, recreation centers, bike paths, etc. Other uses of the system include showing the location of various public offices, historic landmarks, capital improvement projects, transit-oriented developments, and health providers. Individuals using the system can also create custom maps to meet their data needs. As noted on the DenverGIS website (http://www.denvergov.org/tabid/426400/default.aspx), 'this information can be used to understand the interrelationship between locations and activities and is invaluable for making decisions'.

The City and County of San Diego joined together through a joint powers agreement to create SanGIS in 1997 to share geographic systems and related data. The system contains over 450 layers of data related to such areas as land use and development, the environment, engineering, homeland security, transportation, public safety, and planning. Some 200 layers of data can be downloaded without charge. Users can view a map library and later request customized maps for a price.

Ultimately, it appears as if the ways to use a GIS actually depend upon our imagination. We could look back at how GIS has evolved and say that its early uses pale in comparison to what we are currently doing with it. Today, we are developing various disaster preparedness scenarios helping areas to prepare for the advent of such events as explosions, chemical spills, fires, floods, and hurricanes.

GIS has also been used in combination with remote sensing data. Remote sensing provides a unique way of observing the Earth (see Plate 2). Kemp (1998: 348) defines remote sensing as:

The observation of the surface of the earth from a distance by means of sensors. Aerial photography was the earliest form of remote sensing, but satellite observation is now most common, involving the creation of direct photographic images or the collection of data in digital form. Remote sensing can provide information for large areas very rapidly, but accurate analysis requires ground control.

In 1972, a series of Earth-observing satellites, called Landsat, were launched. The satellites were jointly operated by NASA and the US Geological Survey. Reflectance information provided by the satellites gave planners and decision-makers digital photographs of the Earth and allowed them to examine man-made or natural changes occurring on Earth. Additional government and privately owned satellites continue to provide information today.

Mapping information images from satellite data over several periods of time and incorporating it with related geographic information can provide powerful and valuable information to planners and decision-makers. It allows direct observation of the land and

Plate 2 Remote sensing image of San Diego

Panchromatic Landsat ETM+ image for San Diego, collected October 7, 2002, USGS/EROS, Sioux Falls, South Dakota

water surface at various intervals of time. It is a decision-making tool, not a decision-maker.

Determining land cover, vegetative conditions, water conditions, etc. via remote sensing and the information's incorporation into a GIS are commonly used today by planners. We know that they change but, without temporal data, we do not know the extent of the changes. We can determine stream clarity and map aquatic vegetation by various scales. More recently, governments in the west and south have used remote sensing and GIS to map incidents of wildfires and the land destruction caused by the wildfires.

Risk assessment models

The Webster (1989: 1236) *New Universal Unabridged Dictionary* defines 'risk' as 'exposure to the chance of injury or loss; a hazard or dangerous chance'. We can separate risks into two categories: voluntary or involuntary. A voluntary risk occurs when you make a decision to work in an industrial plant that uses chemicals in its processes and you have been warned about exposure, or you purchase a home in an area where it was disclosed that it was once a dumpsite. An involuntary risk would occur if you were to work in an industrial plant and management never told you about the use of certain chemicals in its processes, or if you moved into an area and it was never disclosed to you that it was a former dumpsite. Concerns surrounding the exposure to hazards (from pesticides) have been documented by Rachel Carson in her 1962 book *Silent Spring*. This book brought the idea of 'risk' to the forefront of discussions on environmental policy.

Findley and Farber (2000: 145) define 'risk assessment' as 'the use of scientific data to define the probability of some harm coming to an individual or a population because of exposure to a substance or situation'. National environmental laws such as the Clean Water Act, the Clean Air Act, and the Comprehensive Environmental Response, Compensation and Liability Act all require risk assessment. A number of tools and techniques are available to evaluate the risks and potential health hazards associated with exposure to various chemicals or events. It comes down to a question of exposure and the outcomes of the exposure.

A variety of computer models are available for risk assessment. The American Petroleum Institute has developed a Risk and Exposure Assessment Decision Support System (DSS) to measure human exposure, carcinogenic risk, and effects of petroleum products releases above or below surface. The DSS measures the carcinogenic risks associated with such events as drinking the contaminated water, ingesting contaminated soil, and absorbing the contaminant while bathing. EcoFATE is another environmental risk assessment model that helps determine the cumulative impacts of chemical concentrations in the water, sediment, and the entire ecosystem. The US Environmental Protection Agency (EPA) also uses an assessment system called RISKPRO. This system incorporates models used to evaluate the effects of a toxic chemical when released on the land, in the air, or in water. The Multimedia Environmental Pollutant Assessment System (MEPAS) examines the exposure to and risks associated with environmental emissions in the air, water, and soil. The US EPA also employs a Multimedia Exposure Assessment Model (MULTIMED) to examine how contaminants leach from a waste disposal facility. The model assesses how the contaminants move and transform once they have been released. The US EPA also employs an exposure assessment tool called the Chemical Screening Tool for Exposures and Environmental Releases (ChemSTEER) that estimates the effects of inhalation of and skin exposure to a chemical released in the air, land, or water on workers in industrial and commercial operations. The California Department of Toxic Substances Control uses a model called CALTOX to assess the risks of an adverse health effect on people living near or working near an area with contaminated soils. It determines the concentration of a chemical in the air, drinking water, and food supply. Countless other risk assessment models currently exist and are constantly being refined to produce more accurate figures on the effects of being exposed to various contaminants.

Using other technologies

Computer-aided drafting/design (CAD) is a software commonly used by civil engineers, facility planners, and physical planners. Work tasks that previously took days to complete can now be done in a fraction of the time. The software can help planners modify site plans, create subdivision designs, design highways, develop contour plans, develop sewer projects, create minimum lot sizes, and create maps. A 3D version can help planners with such activities as site development and road design. Golf courses can be designed with the use of CAD software. In addition, CAD software can be used in conjunction with GIS and aerial photography to create footprints of communities.

Advances in technology have also enabled the development of virtual models of projects, central business districts, neighborhoods, and cities. Virtual 3D models of urban spaces can be used to illustrate how individual buildings or a combination of buildings would fit with existing buildings in a redevelopment project.

An Urban Simulation Team at the University of California, Los Angeles, is engaged in a massive project designed to build a real-time virtual reality model of the entire Los Angeles Basin. The model incorporates satellite images with 3D models and street-level video. It integrates CAD and GIS with visual simulation.

Virtual reality models can be used for a variety of purposes. They can examine areas as they currently

Plate 3 Virtual Los Angeles model of downtown

Image courtesy of Urban Simulation Team @ UCLA

exist. If any changes are proposed, the model can show how different project designs or a transit station fit into the existing fabric of the area. A model was developed to show how a sports and entertainment district would fit into downtown Los Angeles (see Plate 3). Other models have also been developed by the Urban Simulation Team to illustrate how they can be used in transit projects and to illustrate beautification opportunities at the Los Angeles International Airport. Only our imagination can constrain the type of models and their uses.

Further reading

Because land use planning is a matter of law, the Internet Law Library (http://www.lawguru.com/ilawlib) is a good resource for city planners. Recent Supreme Court cases and regulations appear regularly on the site, which is part of the Office of the Law Revision Counsel of the US House of Representatives.

Many land use and law sites refer to the Cornell University Law School Wex (http://www.law.cornell.edu/wex). Wex is a 'collaboratively-edited legal dictionary and encyclopedia' for 'law novices'.

The major professional organization for planning and planners in the United States is the American Planning Association (APA) (http://planning.org). Its website provides planning information on innovative land use controls, community participation techniques, planning research, and planning conferences. This information is available to members and non-members alike. The APA also offers information about different APA state chapters or chapters that are created by local planners for a particular topic.

For an open dialogue with urban planners or the interested public, the Urban Planning Research website (http://planningresearch.blogspot.com/feeds/posts) provides an online forum for comments on urban planning issues. Contributors mainly consist of professors, authors, and other city planners and the posts are sorted by topic, but all members are able to post comments. Cyburbia

(http://www.cyburbia.org) is an international version of the same type of forum. Its threads include comments on urban documentaries and recent controversial land use planning.

For a directory of everything urban planning related, see About Planning (http://www.aboutplanning.org). Along with a listing of over 300 planning websites, it also has news, announcements, and jobs directly on the site.

To learn more about some private national foundations and community foundations, see: Ford Foundation (http://www.fordfound.org), Kresge Foundation (http://www.kresge.org), MacArthur Foundation (http://www.macfound.org), Seattle Foundation (http://www.seattlefoundation.org), and the New York Foundation (http://www.nyf.org).

Citizen participation plans actively in use today are commonly found within local government plans. For example, the plan for Oregon's City of Ashland (http://www.ashland.or.us/Page.asp?NavID=116) clearly lists the City's goals for citizen participation; City of Spokane, Washington, Citizen Participation Plan (http://www.spokane citycd.org/citizen/index.htm); Berkeley, California, General Plan – Citizen Participation Element (http://www.ci.berkeley.ca.us/contentdisplay.aspx?id=474)

The following websites offer interesting insights into consensus building and other means of involving the public in resolving controversial issues: Association for Conflict Resolution (http://www.acrnet.org) – an organization dedicated to enhancing the practice and public understanding of conflict resolution; US Institute for Environmental Conflict Resolution (http://www.ecr.gov) – a site designed to provide resources to help people develop workable solutions on tough environmental conflicts; and the US EPA Conflict and Mediation and Resolution Center (http://www.epa.gov/adr/cprc_faq.html) – offers a frequently asked questions section on alternative dispute resolution, conflict prevention, facilitation, and consensus building.

For additional information on charrettes and how they can be used, see NCI Charrette System™ (http://www.charretteinstitute.org/charrette.html).

For a good overview to GIS, see the GIS Lounge website (http://gislounge.com). This website offers information on current studies using the software as well as downloadable data. It also includes GIS training links and career outlook information. A more in-depth study of GIS would include the Federal Geographic Data Committee (http://www.fgdc.gov), whose purpose is to aid in the distribution of geospatial data throughout the United States. The website maintains the National Spatial Data Infrastructure network which is hosted by the National Geospatial Programs Office of the US Geological Survey. The leading retailer and supporter of GIS is ESRI (http://www.esri.com) and its website includes updates, tutorials, and conferences for GIS.

Land Care Research, an environmental research organization in New Zealand, offers an excellent discussion on a variety of risk assessment models – http://contamsites. landcareresearch.co.nz/risk_assessment_models_reviews. htm.

The University of Tennessee, Knoxville, Research Corporation has developed a site on spatial analysis and decision assistance – http://www.tiem.utk.edu/~sada/ index.shtml. It offers free software that incorporates various environmental assessment tools (e.g., geospatial analysis, human health risk assessment, cost-benefit analysis, and decision analysis) to solve problems.

The UCLA Urban Simulation Team website (http:// www.ust.ucla.edu/ustweb/projects.html) offers a fascinating glimpse into the history of the virtual LA model, images, and descriptions of projects.

For examples of social media policies, see: http://www.ca-ilg.org/post/sample-socialmedia-policies; http://social. media.biz/social-media-policies/#governments.

A good discussion of concepts of planning, together with an excellent bibliography, is Alexander (1992) *Approaches to Planning: Introducing Current Planning Theories, Concepts, and Issues*, particularly Chapter 4. (Another version of this chapter, published in 1981, is 'If planning isn't everything, maybe it's something'.) Useful collections of papers on planning theory are Faludi (1973) *A Reader in Planning*

Theory; Burchell and Sternlieb (1978) *Planning Theory in the 1980s*; and Mandelbaum *et al.* (1996) *Explorations in Planning Theory*. Other important books are Friedmann (1987) *Planning in the Public Domain: From Knowledge to Action*; Wildavsky (1987) *Speaking Truth to Power: The Art and Craft of Policy Analysis*.

For further information on planning theory, see: Taylor (1998) *Urban Planning Theory since 1945*; Reynarsson (1999) 'The planning of Reykjavik, Iceland: three ideological waves – a historical overview'; Brooks (2002) *Planning Theory for Practitioners*; and Fainstein and Campbell (2011) *Readings in Planning Theory*.

The classic studies on implementation include: Meyerson and Banfield (1955) *Politics, Planning and the Public Interest: The Case of Public Housing in Chicago*; Altshuler (1965) *The City Planning Process*; Pressman and Wildavsky (1984) *Implementation*; and Levy *et al.* (1973) *Urban Outcomes*. A most interesting case study is Derthick (1972) *New Towns In-Town: Why a Federal Program Failed*. On the problems of defining problems see Rittel and Webber (1973) 'Dilemmas in a general theory of planning'; and Nelson (1977) *The Moon and the Ghetto*.

The classic paper on advocacy planning is Davidoff's 'Advocacy and pluralism in planning' (1965). See also his 'Working toward redistributive justice' (1975). An interesting and useful set of essays on 'Paul Davidoff and advocacy planning in retrospect' is edited by Checkoway (1994). The major text on 'equity planning' is Krumholz and Forester (1990) *Making Equity Planning Work: Leadership in the Public Sector*. (There is an extract from this in Stein (2004) *Classic Readings in Urban Planning*.) Questions of equity also lead into questions of ethics, on which see Hendler (1995) *Planning Ethics: A Reader in Planning Theory, Practice and Education*; Howe (1994) *Acting on Ethics in City Planning*; and Wachs (1985) *Ethics in Planning*. Varady and Raffel (1995) *Selling Cities: Attracting Homebuyers through Schools and Housing Programs* argue that cities need to achieve a balance between greater equity and maintaining their social and economic ability through educational and housing programs to attract and hold middle-income families. See also Brenman and Sanchez (2012) *Planning as if People Matter: Governing for Social Equity*.

For an excellent bibliography on equity planning, see Metzger (1996) 'The theory and practice of equity planning'.

On 'private governments', see Barton and Silverman (1994) *Common Interest Communities: Private Governments and the Public Interest*; and McKenzie (1994) *Privatopia: Homeowner Associations and the Rise of Residential Private Government*.

For a discussion of the different outlooks and perspectives of planners and developers, see Peiser (1990) 'Who plans America? Planners or developers?' A standard text on real estate is Miles *et al.* (1996) *Real Estate Development: Principles and Process*. On the methodology of plan preparation, see Anderson (1995) *Guidelines for Preparing Urban Plans* (Box 1.5 is taken from this most useful book). The classic text on this subject is Kent (1964 and 1990) *The Urban General Plan*.

For discussions on citizen participation, see Arnstein (1969) 'A ladder of citizen participation'; Moynihan (1969) *Maximum Feasible Misunderstanding: Community Action in the War on Poverty*; Cunningham (1972) 'Citizen participation in public affairs'; Cole (1974) *Citizen Participation and the Urban Policy Process*; Forester (1999) *The Deliberative Practitioner: Encouraging Participatory Planning Processes*; Fischer (2000) *Citizens, Experts, and the Environment: The Politics of Local Knowledge*; Beierle and Cayford (2002) *Democracy in Practice: Public Participation in Environmental Decisions*; and Duerksen, Dale, and Elliott (2009) *The Citizen's Guide to Planning*.

For discussions on consensus building and alternative dispute resolution see Susskind and Cruikshank (1987) *Breaking the Impasse: Consensual Approaches to Resolving Public Disputes*; Fisher *et al.* (1991) *Getting to Yes*; Innes *et al.* (1994) *Coordinating Growth and Environmental Management through Consensus Building*; Innes (1996) 'Planning through consensus building'; Salsich (2000) 'Neighborhoods: grassroots consensus building and collaborative planning'; Margerum (2002) 'Collaborative planning: building consensus and building a distinct model for practice'; Booher and Innes (2002) 'Network power in collaborative planning'; Innes and Booher (2003)

'The impact of collaborative planning on governance capacity'; Innes (2004) 'Consensus building: clarification for the critics'; and Margerum (2011) *Beyond Consensus: Improving Collaborative Planning and Management*.

Condon (2007) offers a step-by-step guide to building consensus and cooperation in *Design Charrettes for Sustainable Communities*. A discussion on the use of the 311 CRM system can be found in ICMA (2008) *Call 311: Connecting Citizens to Local Government Final Report*.

For a wide discussion on the role and impacts of information technologies in planning, see Caves and Walshok (1999) 'Adopting innovations in information technology'; Graham and Marvin (1996) *Telecommunications and the City*; W. J. Mitchell (1999) *e-topia*; Wheeler *et al.* (eds) (1999) *Cities in the Telecommunications Age: The Fracturing of Geographies*; Graham and Marvin (2001) *Splintering Urbanism: Networked Infrastructures, Technological Mobilities and the Urban Condition*; Mitchell (2004) *Me+ +: The Cyborg Self and the Networked City*; Graham (2004) *The Cybercities Reader*; Laguerre (2006) *Digital City: The American Metropolis and Information Technology*; and Sheller *et al.* (2006) *Mobile Technologies of the City*.

For an informative compilation of articles on technology, the digital divide, and low-income communities, see Schon *et al.* (1999) *High Technology and Low-Income Communities*; and Servon (2002) *Bridging the Digital Divide: Technology, Community, and Public Policy*.

The literature on geographic information systems (GIS) has grown substantially over the years. There are a number of informative sources to investigate, including: Budic (1994) 'Effectiveness of geographic information systems in local planning'; Onsrud and Rushton (1995) *Sharing Geographic Information*; Monmonier (1996) *How to Lie with Maps*; Mitchell (1997/1998) *Zeroing In: Geographic Information Systems at Work in the Community*; Foresman (1998) *The History of Geographic Information Systems*; Heyward *et al.* (1999) *An Introduction to Geographic Information Systems*; McGuigan and Downey (1999) *Technocities: The Culture and Political Economy of the Digital Revolution*; Brail and Klosterman (2001) *Planning Support Systems: Integrating Geographic Information Systems, Models, and Visualization Tools*; Clarke (2000) *Getting Started with Geographic Information Systems*; DeMers (2000) *Fundamentals of Geographic Information Systems*; O'Looney (2000) *Beyond Maps: GIS and Decision Making in Local Government*; Berhardsen (2002) *Geographic Information Systems*; Huxhold *et al.* (2004) *ArcGIS and the Digital City: A Hands-on Approach for Local Government*; Solove (2006) *Digital Person: Technology and Privacy in the Information Age*; and Tomlinson (2007) *Thinking about GIS: Geographic Information System Planning for Managers*.

Questions to discuss

1 Much has been written about the importance of factors other than rationality in planning. Discuss whether this amounts to an argument that planning is *irrational*.

2 Do you think that US land use planning would be different if the country contained the same number of people on 5 percent of the land area?

3 Discuss the merits of an incrementalist approach to problems of public policy. Is such an approach compatible with long-term planning?

4 What role do interest groups play in the planning process?

5 Discuss the extent to which planners are able to advocate planning approaches which favor the disadvantaged.

6 Discuss how 'Issues of implementation are crucial in the policy-making process.'

7 How do national private foundations and community foundations help non-profit organizations that work in communities?

8 Discuss reasons for and against citizen participation in deciding public issues.

9 How is consensus building different from the traditional techniques of citizen involvement?

10 What is 'equity planning' and why is it important?

11 What is a GIS and how could planners use it?

12 How do cities use various forms of information technology in their day-to-day activities?

13 What is remote sensing?

14 Why would a city want to develop virtual models of communities?

15 How can social media devices aid cities in engaging the public?

2

Urbanization

In the heart of the continent arose a new homo Americanus more easily identified by his mobility than by his habitat. He began to dominate the scene in the years between the American Revolution and the Civil War, and he was now shaping the new nation into a New World.

Boorstin 1965: 49

World urbanization

Urbanization is a long-term process of people migrating from rural areas to urban areas. Although the rate of urbanization may vary by country, every country has gone through the process of urbanization. Early populations settled in areas close to water supplies and sources of food. They would move from location to location as water supplies and the availability of food supplies lessened. Settlements were created on elevated sites for defensive reasons. Eventually, individuals started to group together.

The reasons for urbanization are numerous. Many people move from rural areas and farms for job opportunities – the industrialization reason. Others move for family reasons. Some may move to be closer to better health care, better schools, and better transportation opportunities. Other might migrate to urban areas because of government plans and policies.

There is no guarantee that moving to an urbanized area will result in a better quality of life. Many may have to move to segregated areas, thus perpetuating the problems associated with inequality. Overcrowding of dwelling units may increase due to a lack of employment. Overall, an individual's standard of living could actually decrease on arrival at the urbanized area.

The world has become increasingly urbanized over the years. This can be clearly seen in Africa, China, and India. The United Nations has consistently noted that the amount of the world's urban population has increased and will continue to increase. Rural flight is still occurring throughout the world. However, as of 2010, over half of the world's population lives in urbanized areas. According to the UN's Population Division of the Economic and Social Affairs Department 2010 estimates, the percentage of those living in urbanized areas is expected to increase to almost 70 percent by 2050.

A culture of mobility in the United States

If one word were to be chosen to describe the character of the United States, it might well be 'mobility'. The land was settled by migrants from other continents – first Europe, later Asia and South America. From 1901 to 1910, approximately 8.8 million immigrants were admitted into the United States. The number admitted decreased to 5.8 million from 1911 to 1920; to 4.1 million from 1921 to 1930; and to 528,000 from 1931 to 1940. From 1991 to 2000, approximately 9.1 million immigrants were admitted into the United States.

The number dropped drastically from 2001 to 2005, when only 4.9 million immigrants were admitted. The proportion of the US population that was foreign born reached a high of 15 percent in 1910, to 10 percent in 1997. This meant an estimated one out of ten people in the United States was foreign born in 1997. In 2006, 12 percent of the US population was foreign born.

The growth of the United States has long been characterized by movement: movement from the coastal settlements to the interior, from the North to the South, from the East to the mid-West and then to the far West. And the 'non-migrants' did not stay still: they moved constantly in search of new opportunities or better employment or improved housing conditions (and later for improved environments for leisure or retirement). The same restlessness or, more accurately, the same keenness to discover, to initiate, and to experiment, the same desire for advancement, the love of the new, is deeply ingrained in the American psyche. The United States was born of the search for better ways of living, and the search continues.

In this chapter some of the main features of this continual mobility will be summarized. It is important for the subject of this book, since it is the underlying motive power of land development and urbanization.

Three centuries of urban growth

In the colonial era the United States was a rural country: apart from native habitations, settlements were small and scattered. This was an agrarian subsistence economy. Though the first settlements were in the nature of towns, they were very small and, as farming developed, the growth of the rural population soon outstripped that of the towns. In 1690, the urban population made up about a tenth of the total; a century later the urban proportion had halved, with only twenty-four places being 'urban' (defined generously as having populations in excess of 2,500). It was not until 1830 that the urban population attained the 1690 level – whereafter the growth was phenomenal. By 1860, there were a hundred cities with populations in excess of 10,000, of which eight exceeded 100,000. The second half of the nineteenth century saw an acceleration of urban growth, and by 1910 the number of 100,000+ cities had increased to fifty. Boston increased in population from 43,000 in 1820 to 251,000 in 1870, to 748,000 in 1920 and to 770,000 in 1940. The comparable figures for Philadelphia were 64,000, 674,000, 1.8 million, and 1.9 million in 1940, while New York topped the league with 137,000 in 1820, 1.5 million in 1870, 5.6 million in 1920, and 7.4 million in 1940.

The maritime cities of the colonial period looked more to their colonial masters across the Atlantic than to the hinterland. They were outposts of empire and had more association with England than with each other. This was to change, of course, particularly after Independence. The rate of change was phenomenal, and was caused by a number of interacting forces. The rapid development of commerce and international trade, the development of manufacturing and transport networks, and the massive immigration from Europe worked together to transform a scattering of maritime centers into a complex, industrialized nation.

Urbanization represented a major feature of this metamorphosis. Its driving force was a dynamism of enterprise, mobility, experimentation, and exploration that played itself out in many ways. On the geographical dimension it was seen in the growth of existing towns and the settlement of new places that, it was always hoped, would expand and prosper. Town promotion became highly competitive. The winners were those who, by chance of geography, luck, or success in securing new transport links, founded a base that could attract further investment. The competition for transport investments was intense. Particularly successful were those places that, by one means or another, secured the terminus of a canal or railroad, though these new routes also became 'life-promoting arteries all along their way' (Boorstin 1965: 168).

Some of the initiatives were huge in scale, and many had strong federal or state governmental backing. Plans for the National Road were started in 1806, and it was later constructed by the federal government roughly along what became US Highway 40. 'In 1808, Albert Gallatin, Jefferson's imaginative Secretary of the Treasury, gave the Senate his remarkable *Report on Roads and Canals*, promoting a comprehensive scheme he had worked out with Jefferson for a federally aided transportation system to cover the nation, connecting the eastern rivers with the Mississippi basin' (Boorstin 1965: 252). Though delayed by the war with Britain, most of Gallatin's projects were built over the next sixty years. The 363-mile-long Erie Canal – 'the wonder of the age' – was constructed with state and private support. Its construction started in 1817 and it was opened in 1825. Philadelphia tried to repeat the success with the creation of the Pennsylvania canal system, while Baltimore opted for the Baltimore and Ohio Railroad. These and many other schemes opened up the continent and provided avenues for a population growing yearly by immigration. They signaled the beginning of a population and trade movement westward.

Town development

Urbanization not only spread westward, but also changed in character. While the early cities were loosely structured, with little distinct separation of land uses (and social classes), great change accompanied the development of the industrial city; and as the cities grew, a pattern of land uses emerged. The central area became the location of religious, administrative, political functions, together with housing for the wealthy. The suburbs became less desirable environments. (The word 'sub-urb' originally meant a settlement on the urban fringe: 'a place of inferior, debased, and especially licentious habits of life', according to the *Oxford English Dictionary* (Fishman 1987: 6).)

As the economy grew, the inner parts of the cities became less attractive places for living. Commercial activities increased in the center as central business districts began to develop. Workers were crowded into tenements in the surrounding zone, close to the factories (an efficient location since it minimized the walk to work). The factories themselves created noise and dirt, which was considerably increased with the coming of the railroads (which were not allowed to penetrate into the hearts of the towns). As economic activity mushroomed, land values and land speculation

rose, and those who could afford to escape to quieter surroundings did so. The opportunities for this were increased as 'transport' developed: horse-drawn buses in the 1830s, the first commuter train services in the 1840s, horse-drawn streetcars in the 1850s, a rapid growth of commuter trains in the 1860s, and, above all, the electric trolley routes of the late 1880s.

Transportation for commuters

It was the electric trolley services that really unleashed the forces of suburbanization. Trolley services created the possibility of living longer distances from the central city. The growing middle class could now move well beyond the limits imposed by the earlier and poorer forms of transit, and they had the incomes to enable them to do so (leaving their former residences for multiple occupation renting by working-class families). Suburbs grew at a remarkable rate along a narrow band parallel to the streetcar lines. The same happened with the commuter railroads that spread further out from the cities and served both middle- and upper-class housing markets. New towns sprung up along the railroads. At the end of the century, the United States 'had more miles of railway track than the rest of the world combined' (Jackson 1985: 91).

One interesting feature of the later suburbanization was the promotion of 'streetcar suburbs' by entrepreneurs who developed both transport and land. These 'streetcar suburbs' could be found throughout the United States. In Fort Collins, Colorado, a streetcar began operation in 1907. It had a dramatic effect on the residential and commercial development of the city. The two initial lines started in the downtown commercial core and branched into the western and southern parts of the city. In Milton, Massachusetts, the development of streetcar lines contributed to the rapid expansion of the community. Moreover, in Shrewsbury, Massachusetts, streetcars played a major role in its development as a suburb of Worcester, Massachusetts, and today of Boston. The development of the 'streetcar suburb' also contributed to the changing nature of many communities throughout the United States.

Subsidization of the streetcars to attract families to the suburbs enabled large profits to be made from the land. These land/transport developers often had little or no long-term interest in their transport operation, and services deteriorated after the housing development was completed. In some areas, the services suffered from overextension and the collapse of subdivision plans; in others, short-term policies (and worse) caused services to flounder.

Many continued only by municipal annexations and extensive subsidies. Later, of course, the private car released the constraints imposed by mass transit, thus still further worsening the difficulties of mass transit.

The development of transport (automobiles, street-cars, railways, etc.) thus played a key role in urbanization (as it has continued to do). Each new technological advance made possible further extension of the suburbs, until in recent years the outward movement of a whole range of urban activities has transformed outer suburbs into totally new urban forms. Advancements in communications technology have also enabled individuals to move further out into the periphery while still working in the cities. These urban forms are continuing to evolve.

Immigration and urbanization

In the decades from 1851–60 to 1871–80 migration into the United States averaged two and a half million. In the following decades it rose to five and a quarter million (1881–90), to three and two-thirds million (1891–1900), and to nearly nine million in 1901–10. In the single year 1907 it reached the staggering height of one and a quarter million. Between 1890 and 1920, the population of the United States rose by over 42 million. Urban areas grew at an incredible rate, quite overpowering the ability of city governments to provide basic public services. During the same three decades, the urban population of the United States increased from 22 million to 54 million; the proportion of the population living in cities rose from 35 percent to 51 percent (Miller and Melvin 1987: 79); and the number of cities with a population of 50,000 or more rose from 50 to 144. The growth of individual

cities was even more dramatic. Between 1880 and 1920, New York grew from 1,478,000 to 5,620,000; Philadelphia from 847,000 to 1,823,000; Baltimore from 362,000 to 748,000; and Boston from 332,000 to 748,000. The difficulties created by these huge increases in population were exacerbated by the fact that the newcomers were different from previous immigrants:

> In the thirty years from 1890 to 1920, more than eighteen million immigrants poured into America's cities. These new immigrants were more 'foreign' than those who arrived before, coming mainly from Italy, Poland, Russia, Greece, and Eastern Europe. Overwhelmingly Catholic or Jewish, they came to cities that were already industrialized and class conscious. They made up the preponderance of the working force in the iron and steel, meatpacking, mining, and textile industries. They shared no collective memories of the frontier or the Civil War, much less of the American Revolution. Few spoke English, and many were illiterate even in their native language.
>
> (Judd 1988: 118)

They were therefore perceived as a threat to public health as well as to the sensibilities of middle- and upper-class residents of the outer city who had to pass the ghettoes on their way to work. Even more ominously, they threatened the entrenched urban political systems.

Public policies and suburbanization

Housing

It is commonly held that suburbanization results from an innate desire on the part of Americans to own spacious single-family homes built at low density in pleasant, peaceful green surroundings, separated from the bustle and problems of the city. Though there is no doubt about the compelling attractiveness of this idyllic image, there are other issues that are highly significant in the development of the suburbs.

Foremost among these are public policies. A number of public policies have contributed to the development of the suburbs and beyond. Kenneth Jackson, in his classic book on suburbanization, *Crabgrass Frontier* (1985: 11), has gone so far as to state that 'suburbanization has been as much a governmental as a natural process'. One could go further and argue that the promotion of suburbanization has been among the most successful public policies ever pursued in the United States. This, however, would be considered a distortion of the truth since suburbanization has typically been more a consequence than an objective of policy.

The most important contribution of public policy to the suburbanization process has been federal mortgage insurance. Before this was introduced in 1933, mortgages were commonly negotiated (for a third to a half of the value of the house) over a period of five to ten years, at the end of which the balance was renegotiated at current rates (which, of course, were unpredictable and could be so high that the buyer might be unable to obtain a new mortgage and thus face foreclosure). Buyers who could not afford the required deposit would take a second mortgage at a higher rate of interest. The system worked fairly well until the Great Depression, when its risky nature became all too apparent. Foreclosures became common, rising to a quarter of a million in 1932. In 1933, the rate was even higher – over a thousand a day (Jackson 1985: 193). The Depression seemed set to completely destroy the home-financing system, and federal action became imperative.

The initial solution to this dilemma was the introduction of federal funds to home loan institutions with insurance against risk for depositors. The structure of home-financing was changed: loans were made for a period of twenty years, with amortization over the life of the loan. This coped with part of the immediate problem, but it failed to bring about the hoped-for recovery of the housing market, and extended provision was made in 1934 when the Federal Housing Administration (FHA) was established and mortgage insurance extended significantly. It worked; and it worked so well that it is still in operation today after being consolidated into the US Department of Housing and Urban Development in 1965. Its impact on house

production was substantial: housing starts rose from 93,000 in 1933 to 216,000 in 1935, and 619,000 in 1941, when the war sharply curtailed house-building. The new financing system considerably reduced the cost of house purchase and, in fact, it was often cheaper to buy than to rent. The FHA program assisted in the financing of military housing for returning veterans and their families after the war. It should be noted, however, that though this system is seen as a major instrument of national housing policy, it was introduced for economic reasons, not for the production of housing: above all to reduce unemployment, which reached 12 percent in 1934 (and even higher in the construction industry).

The changes brought about by the FHA were accompanied by – and accelerated – others, such as real estate finance and the organization and scale of the development process. Together, these changes provided a new base for large-scale housing development that, though held back by the war, burgeoned as soon as the war was over. The expansion of the federal role in the mortgage market was especially important. FHA policies were liberalized and, for example, allowed thirty-year mortgages with a mere 5 percent deposit. Financing under the Veterans Administration (the GI Bill) was even more generous in providing for mortgages without any deposit!

Most of these mortgages went to suburban houses, partly because it was in the suburbs that the majority of new houses were built, and partly because the federal agencies followed conventional business practices in relation to mortgages. These included favoring 'economically sound' locations over more doubtful inner-city areas, owner-occupied instead of rental dwellings, and racially homogeneous (i.e. white) districts. Attitudes such as these, which predominated in the private market, were shared by the public agencies: their interest was essentially in supporting the real estate and banking interests (from whose ranks their staff often came). All these factors favored the suburbs; and there were no official policies that directed otherwise. As such, public action thus followed market forces.

Another benefit bestowed on home buyers by public policy came about almost by accident – favorable tax treatment. The home mortgage deduction became a significant and tangible assistance to home buyers, and was buttressed by other advantages discussed in Chapter 15.

It should be noted that mortgages were truly not available to everyone. A wealth gap was clearly evident between Black Americans and White Americans. Many people did not have access to mortgage lending. As Novek (2001) has indicated, 'between 1930 and 1960, fewer than 1 per cent of all mortgages in the nation were issued to African Americans'. Sugrue (2011) echoes this idea by indicating, 'Whites, meanwhile benefited from enormous homeownership subsidies through the Federal Housing Administration and the Veterans Administration; blacks did not, at least until the late 1960s, when local, state and federal laws that forbade housing discrimination were passed.'

Discrimination didn't only exist in lending. There was discrimination in where individuals chose to live. Many areas wanted to reserve suburban areas for particular racial groups. As LaCour-Little (1999: 17) noted, 'by some estimates, over 80 per cent of the suburbs developed in the late 1930s through late 1940s contained racial covenants preventing Black Americans (and other non-white groups) from purchasing homes in those neighborhoods'. Thus, racial inequality was clearly evident.

Other forms of discrimination also existed. Mortgage lending institutions used a personal characteristic such as gender as a reason to reject a mortgage application. Ability to repay the loan was not a factor in considering the loan. Many were denied a loan, strictly based on their gender.

Though changes in the organization of housing production and its finance were extremely important in the suburbanization process, they were not sufficient in themselves to achieve the scale of suburbanization that took place. They provided the motive power; it was the highways program which (literally) provided the track on which the suburbanizing engine could run.

Highways

The suburbs built prior to World War II were still part of the cities that they surrounded: they were only a short distance from the city center, and they could take advantage of its facilities. They extended the city; and though they might keep it at bay (for example by rejecting annexation), they did not threaten it. But a break had developed between the central city and the suburb and, to use Gelfand's analogy, this 'was one of the many fault lines' which would be apparent later (Gelfand 1975). The enormous suburban expansion of the postwar period was different: the suburbs grew in scale and character and, eventually, became a new type of urban settlement. This was more than peripheral to the city: it was increasingly independent of it, and threatened its viability.

This change would not have been possible without the huge programs of highway building which have characterized the last half-century. Paradoxically, these began as an attempt to rescue the cities from decline: they were seen as providing easy access to the city. Unfortunately, it was not perceived that they might equally well provide easy exit. The initial programs began with civic concern to adapt the cities to the changed transport era. If urban congestion was to be reduced, thus allowing cities to fulfill their traditional functions, radial freeways had to be constructed to a beltway around the urban area; these would distribute traffic to desired entrances to the central area, which would be generously provided with the necessary car parks. In this way, so it was commonly thought, city streets would be freed of congestion, and urban blight would be banished.

With hindsight, it is not easy to understand how such misconceived policies could ever have been adopted, let alone implemented on a scale that devastated so many acres of city land. Yet, for many years, there was widespread backing for the program: 'across America, superhighway proposals won broad popular support, for they seemed to promise greater mobility and a new boost for the central cities burdened by a horse-and-buggy street system' (Teaford 1990: 97). Local resources, state aid, and federal aid on a grand scale provided the funding for this that was later greatly increased by the Interstate Highway Act of 1956.

The interstate highway program (pressed with vigor by a well-organized lobby) was one of the greatest public works ever. It committed the federal government to an expenditure of $33,500 million for 41,000 miles of highway that would be completed by 1975. The highways would link cities to other cities. The program also established a Highway Trust Fund into which revenues from fuel taxes were automatically siphoned, thus providing a continual replenishment of the resources for new road building. (Diversion of some of these enormous funds into public transit was resisted until the Nixon era.)

The importance of this highway program is difficult to exaggerate: it confirmed the ascendancy of the car over all types of personal transport; it enabled the development of truck and (until its later replacement by cheap air travel) long-distance bus travel – and thereby accelerated the decline of rail; it provided an unprecedented accessibility for employment, trade, leisure, and shopping. It represented the ultimate in the extension of the possibilities of mobility, for which Americans had long shown an addiction. Though it later suffered the excesses of its own achievements (which are discussed in Chapter 14) it transformed the urban scene. The suburbs were no longer peripheral: they had become the new center.

Decentralization

In this remarkable transformation of urban America, population movement was both followed and led by the movement of employment and urban services. Industrial decentralization has a long history which pre-dates the rise of the truck – spurred first by rising central land prices, then by the greater locational freedom presented by the advent of both electrical power and the motor truck, and finally by the construction of the highway system. The truck and the highway system had the most dramatic effect on decentralization, although there had been considerable outward movement of industry very much earlier. Indeed, concern was being expressed about the effect of the 'decentralization of industry' on the cities as early as the first decade of the twentieth century – while others saw the resultant residential movement as a means by which the housing problem of the cities might be relieved (Scott 1969: 130). Industrial

decentralization (and warehousing) accelerated in the 1920s with the growth of both cheap electricity and transportation. Though most industrial traffic was by rail, truck traffic increased in importance as roads were built under the Federal Highway Act of 1916. The number of trucks increased from 150,000 to 3.5 million between 1915 and 1930. A major push for industrial suburbanization came with World War II and its requirements for huge plants that for reasons of safety (from the expected enemy attacks) and for logistics could not be located elsewhere. By the early 1960s, a half of industrial employment occurred in the suburbs; by the end of the 1970s, the proportion had increased to around two-thirds. The trend of decentralization of employment continues to occur. Jobs are following people to the suburbs.

Retailing followed a similar pattern. Decentralization started with the shopping 'strips', aimed at the motorist-shopper. Suburban shopping centers catering for the new suburbanites came later. Sears and other major retailers built additional stores in suburban locations in the early 1920s, using the highly successful formula of generous parking space. The first modern-type suburban shopping center was Country Club Plaza in Kansas City, built 1922–25 by J. C. Nichols. Its followers might well have been more numerous had it not been for the Depression (only eight had been built by 1946). The postwar boom, however, led to many more, bigger and bigger, shopping centers.

Eventually, the effect on many city stores was fatal. Perhaps the most symbolic demise was Hudsons of Detroit, the third-largest store in the United States (after Macy's in New York, and Marshall Field in Chicago). As its customers moved out to the suburbs, Hudsons provided them with local stores until, by their success, the city store became redundant. It was no longer 'a simple fact that all roads in the Motor City led to Hudsons': its suburban branches succeeded all too well, and it closed in 1981 (Jackson 1985: 261).

Shops have for long followed people, but a new feature of the huge enclosed malls that began to be developed in the late 1950s is that they became development catalysts themselves. A good example is Cherry Hill Mall in Delaware Township, New Jersey, developed by the Rouse Company in 1961. This provided a center for a center-less suburban spread, and an identity for a diffuse area. It acted as such a catalyst to further building that the township changed its name to that of the shopping mall – a highly symbolic act! There are, however, many other malls that have become centers of activity in the suburbs, whether they followed the population or vice versa; it can sometimes be difficult to be sure which came first.

By the mid-1980s, there were 20,000 large shopping centers, accounting for almost two-thirds of the national retail trade (Jackson 1985: 259). That number had more than doubled by 2005. Shopping centers of all sizes have become so much a part of American life, and so tied in with suburbanization and personal car transport, that they are now the norm: it is the thriving city center which is remarkable.

The suburbs have for long ceased to be an appendage to the city. They have assumed a character of their own, as part of a regional mosaic of development which contains within its area most, if not all, of the functions formerly performed by cities (Palen 1995). So great has been the change that the term 'suburban' is no longer appropriate to many of these areas. They are not suburbs in the traditional sense of the term: they are a new type of decentralized city. Some might call them perimeter cities. Fishman (1987) has termed them 'technoburbs' but the term 'edge city', coined by journalist Joel Garreau (1991), has proved to be more endearing (see Box 2.1).

An 'edge city' is an area created by and shaped by transportation. They are areas adjacent to existing areas containing shopping and employment opportunities.

As cities have grown outward, new forms of development have evolved. We have witnessed population movements from the central city, to the suburbs, and to the exurbs. According to Garreau (1991: 6–7), an 'edge city' is any place that:

- has five million square feet or more of leasable office space – the workplace of the Information Age;
- has 600,000 square feet or more of leasable retail space;
- has more jobs than bedrooms;
- is perceived by the population as one place; and
- was nothing like 'city' as recently as thirty years ago.

Classic examples of an 'edge city' can be found at Tyson's Corner, Virginia; Scottsdale, Arizona; Buckhead, Georgia; Sandy, Utah; Silicon Valley, California; and along Route 128 in Boston, Massachusetts.

The scale and speed of suburbanization is breathtaking. During the 1950s and 1960s, the suburban population increased from 35 to 84 million. Further increases in the following two decades resulted in the suburbs being home to virtually a half of the population in 1990.

Dramatic events have served to either accelerate or decelerate population movement. Had it not been for the Depression and World War II, the *speed* of suburbanization would have been slower (though it is hazardous to speculate whether its character would have been any different). The Depression caused a general and far-reaching slow-down in activity (including house-building and household formation) that meant

BOX 2.1 EDGE CITIES

Edge Cities represent the third wave of our lives pushing into new frontiers in this half century. First, we moved our houses out past the traditional idea of what constituted a city: this was the suburbanization of America, especially after World War II. Then we wearied of returning downtown for the necessities of life, so we moved our marketplaces out to where we lived. This was the malling of America, especially in the 1960s and 1970s. Today, we have moved our means of creating wealth, the essence of urbanism – our jobs – out to where most of us have lived and shopped for two generations. That has led to the rise of Edge City.

Source: Garreau 1991: 4

that, when better conditions arrived, there would be a pent-up demand to be met. This backlog was further increased by the war years, which involved the diversion of enormous resources to war-related purposes. The end of the war released the pent-up demand (together with a baby boom) and led to a long period of growth and prosperity of which suburbanization was both the outcome and the hallmark.

Part of the speed of suburbanization was facilitated by changes in house-building techniques, of which the mass-production Levittown developments were prototypical. William J. Levitt was the Henry Ford of the building industry: by means of prefabrication, preassembly, site planning, and scheduled delivery of 'components', he transformed house-building into a manufacturing process and achieved an incredible rate of production. When it was in full operation, this reached as many as 150 completed houses a week, at a rate of one every sixteen minutes. When the development was completed, over 17,000 homes had been produced at below market value. The mass production resulted in lower-cost housing. The development had helped thousands of families realize the American dream of home ownership. Levitt, who did not start the mass production trend in housing, is the best known of the mass builders, but he was not unique. Many others experimented with various forms of inexpensive prefabrication. Concomitantly, if the houses were basic, small, and cheap, they also fulfilled an urgent need and proved capable of adaptation to later changes in the incomes and aspirations of their owners – as Barbara Kelly's 1993 study has demonstrated.

Current trends

The people of the United States are constantly on the move – to seek better homes, better jobs, better climate, or even just somewhere new. It is a fact of life. The amount of mobility continues to be incredible. Over a sixth of the population moves every year. Many of the moves are for housing reasons: households move to another house to improve their housing conditions (or, less frequently, to obtain cheaper accommodation).

This type of 'residential mobility' is conceptually different from movements caused by a change of job that necessitates a change of house, possibly in another county or state. In practice, the distinction is often not clear-cut. A change of job may follow a change of dwelling, as well as vice versa; and sometimes a mixture of motives may be involved.

Statistically, most moves are short-distance (well over a half within the same county), but there are also huge long-distance movements, particularly over a period of time. The biggest of these has been a move to the west that, at various rates, has persisted since the frontier was breached. There were, however, other notable migrations. For example, the movement of blacks from the rural South to the urban North during and after World War II for employment and other reasons continued on a large scale a long-established pattern of migration – the theme song of the fast-paced jazz laments for the 'Double Diaspora' suffered by African Americans (Boorstin 1973: 299). Approximately 90 percent of the African-American population lived in the South in 1900. Many individuals moved to the Midwest in search of economic freedom. The population figures would continue. About 200,000 blacks moved north between 1890 and 1910. The rate stepped up after the outbreak of World War I, when immigration of unskilled laborers ceased, and Henry Ford and other northern manufacturers actively recruited southern blacks (even hiring special freight cars for their passage): half a million moved between 1914 and 1919.

In the 1960s and beyond, much of the African-American population was concentrated in metropolitan areas and central cities. Population movement from the rural areas to urban areas occurred for employment opportunities. A movement to suburban communities was taking place in areas throughout the United States. A loss of employment opportunities due to globalization and the restructuring of the economy has helped fuel the growth of many suburban areas in the United States.

Large population movements also characterized the farming scene. There has been a steady decline in farm population for over a century. The movement of the farming population in the 1920s was around 1.5

million. Since then there have been huge changes: between 1920 and 1970, the farm population fell from over 30 million to less than 10 million, and by 1990 it had fallen still further to 4.6 million – a mere 2 percent of the population. By 2002, the farm population remained at about 2 percent of the total US population. Advances in agricultural technology and global competition have been factors in the decreasing of the US farm population.

The movement of workers in World War II was immense: over a period of four years, some 20 million moved house in response to changes in the wartime economy, and another 12 million left home to join the armed forces (Brogan 1986: 584). The years following the end of World War II saw equally dramatic movements – from the farms; from older industrial areas to the new areas developing in response to government defense contracts (later to be dubbed the gunbelt); and to the rapidly growing suburbs (see Box 2.2).

These movements have added up to striking patterns of population redistribution. Demographers have identified three dominant patterns: the movement to the west, which persists in its regional primacy; the redistribution from rural to urban places, and from non-metropolitan to metropolitan areas; and, within the metropolitan shifts, the movement 'up the size hierarchy', with the largest metropolitan areas gaining the most (Frey 1989: 34).

There was a consistency in these patterns until the 1970s, when changes occurred on such a scale as to give rise to the question as to whether there had been a 'turnaround' in migration trends. A major feature of this change was the growth of the non-metropolitan areas at a faster rate than that of the metropolitan areas. This became known as the 'rural renaissance'. Some industries needed more land. The available land happened to be in these areas. Jobs soon followed. Others moved to these areas to escape the hectic pace of city life. Others moved for health reasons, while others simply wanted a better quality of life for their families and children.

Some of the largest metropolitan areas, instead of continuing to exhibit the 'up-the-size-hierarchy' redistribution, actually experienced population loss. This 'counterurbanization' was a reversal of the secular

trends of increased urban growth and western movement and, interestingly, it had its counterparts in many other countries. For a while, there was a spirited controversy on the reasons for this change and its likely continuation. Now, with the benefit of the passage of time, it is possible to evaluate the changes and to assess the likelihood of their continuation in the future. Figures from the 1990 census show that there was indeed a change in the 1980s, but it was not a simple return to the pattern of earlier years. Economic conditions caused metropolitan areas to be hard hit by recessionary forces. People picked up and moved to where employment opportunities existed. Advances in information technology have afforded some people the opportunity to live and work in rural areas. Nevertheless, the percentage of people residing in urban areas in the United States has increased from approximately 65 percent in 1950 to an estimated 81 percent in 2005.

Americans continue to be a mobile population. Population continues to go to metropolitan areas in the South and in the West. By the year 2000, Perry and Machun (2001) indicated, one-half of all Americans lived in suburban areas. Their research also concluded that the population of our metropolitan areas outside central cities had increased by nearly 13 percent, while growth inside central cities had grown by only 4 percent (Perry and Machun 2001). Today the older suburbs are losing population to developments even further out than the suburbs. This trend exists in many areas.

These developments occur on the periphery of cities and rural areas for a variety of reasons. Many people continue to be pushed away from cities due to increasing congestion levels, increasing crime rates, poor infrastructure, and so on. They are being attracted to a so-called 'country style' living made possible by improvements in technology and the transportation systems that enable individuals to commute further and further distances – moves to the so-called 'exurbs'.

The term 'exurbia' has generally been attributed to Spectorsky (1955). In his book, *The Exurbanites*, Spectorsky concluded development was occurring between the suburbs and the rural areas, but still within commuting distances to jobs. These areas are in

BOX 2.2 FEDERAL POLICY AND THE GUNBELT

Because of the size and singularity of the gunbelt, its rise ranks among the most powerful of changes in American settlement patterns in the postwar period, rivaling other momentous changes like the continued movement from central city to suburb. A whole new set of industries, arrayed around aerospace production and including electronics, communication equipment, and computing, and populated by a set of insurgent firms, has led to an extraordinary shift in the nation's industrial center of gravity away from the heartland.

Labor pools have been built with ease around new emerging gunbelt cities – able to attract substantial new contingents of professional and technical labor. In other words, people follow jobs: new firms, industries, and military facilities fashioned, often very deliberately, the labor market institutions that would generate an ongoing supply of labor. The Pentagon facilitates this lopsided recruitment, out of the heartland and into the gunbelt, by paying for the relocations of scientific and technical personnel as a part of the 'cost of doing business'. Unintentionally, this mechanism has financed one of the greatest selective and for-profit population resettlements in the nation's history.

Source: Markusen *et al.* 1991

BOX 2.3 GROWTH AND POPULATION OF EXURBS

As of 2000, approximately 10.8 million people live in the exurbs of large metropolitan areas. This represents roughly 6 percent of the population of these large areas. These exurban areas grew more than twice as fast as their respective metropolitan areas overall, by 31 percent in the 1990s alone. The typical exurban census tract has 14 acres of land per home, compared to 0.8 acres per home in the typical tract nationwide.

Source: Berube *et al.* 2006: 1

the midst of transitioning to a more urban area (see Box 2.3). Exurbs have also been referred to as periurban areas, the rural–urban fringe, the metropolitan orbit, the extended urban areas, and the urban shadow (Clark *et al.* 2005).

The definitions of an exurb have been debated over the years. Daniels (1999) suggests that an exurban area has the following characteristics:

- Located 10–50 miles from urban centers of 500,000 people or 5–30 miles from cities of at least 50,000;
- Commute time is at least 25 minutes each way to work;

- Communities have a mix of long-term and newer residents; and
- Agriculture and forestry are active, but declining industries in the community.

Other researchers, such as Nelson (1992a), have developed definitions of an exurb incorporating distances, population, and percent of growth.

Three issues stand out. In the first place, the situation has become much more complicated (Frey 1994a; Frey and Speare 1992). A major characteristic of the new situation is the speed and volatility of change. Employment in urban areas is being affected by unprecedented national and international forces. There is now a real sense in which one can speak of the global economy. The geographical impact has varied with the strength or diversity of economic structure and, as a result, differences in growth and decline have accelerated and widened. Unfavored areas included those with outdated manufacturing base economies, and those that suffered from cutbacks in mining or military expenditure. Favored areas were those that had growing financial, service, educational, health, and leisure centers. Some localities attracted very high rates of growth; for example, the resort and retirement areas such as Las Vegas and Phoenix and much of southern Florida.

Second, a dimension of increased importance is the growth and distribution of minority populations, especially Blacks, Hispanics, and Asians. All have natural increases above the rate of Whites, and immigration from Latin America and Asia has been significant. Of the 7.3 million immigrants in the decade 1981–1990, 2.8 million came from Asia, and 1.7 million from Mexico (US Census Bureau 2007: Table 8). Of the 3.8 million immigrants during the 2001–2004 period, 1.3 million came from Asia, and 715,000 came from Mexico (US Bureau of Census 2007: Table 8). Though there has been greater 'dispersal' than in earlier decades, there is a marked concentration in certain areas. A striking example is the Los Angeles metropolitan area, which had over a fifth of the total growth in minorities during the 1980s (numbering 2.8 million). As a result, the area housed 12 percent of the entire country's minority population in 1990.

Third, employment continues to decentralize. The suburbs continue to be the favored location for new urban employment and residential growth. They have become growth areas in their own right, rather than simply recipients of people and jobs from the cities. Most suburbanites now work within the suburbs (though not necessarily the one in which they live), and the pattern of commuting has therefore changed and become more complex. It is not only car-based, but also car-dependent: public transport could not cope with the new patterns of commuting.

Minority movement to the suburbs has grown significantly, particularly in the West, which experienced the greatest increase in minority populations.

The suburbs have become more differentiated by race and also by economic structure, and they 'represent the arena of future growth in most metropolitan areas' (Frey 1994a: 132). During the 1980s, suburbs grew on average at over twice the rate of the central cities.

These patterns are considerably complicated by regional differences which themselves are associated with the differential impacts of economic change. Regional restructuring plays itself out clearly in metropolitan areas. Thus, while advanced service-based economies such as those of New York and Boston were able to build on these (offsetting some of the decline due to deindustrialization), areas heavily dependent on manufacturing, such as Detroit, Cleveland, and Pittsburgh, suffered further decline. (These cities are now exhibiting varying degrees of revitalization, due in no small part to the proactive efforts of city and state governments.)

The pattern of urbanization has become more complex. It now defies simple characterization, and has become both more volatile and unpredictable. This presents difficult problems of comprehension, let alone policy-making. When it is unclear at the local level what is happening and what forces are at work, it is not easy to forge relevant and workable policies. Yet there is nothing new in this, as history abundantly shows. The various phases of postwar urban policy have more frequently failed than succeeded, often because the underlying causes were not appreciated. But that there is scope for effective governmental action is apparent from the influence that this has had in shaping urbanization.

Deindustrialization

Economies are not static. They have a way of changing over time. In earlier times, the United States was an agrarian (agriculturally based) society. Over the years, as technology advanced, the country embarked on an industrial revolution. Workers flocked to the northeast and upper midwestern parts of the country, the so-called 'Frostbelt', for manufacturing jobs in such industries as the automotive industry, textiles, steel, electronics, and machines tools. This event led to the growth and transformation of many areas, including Detroit, Michigan; Flint, Michigan; Milwaukee, Wisconsin; Youngstown, Ohio; Akron, Ohio; and Pittsburgh, Pennsylvania. The economies of these cities were built on the jobs created by manufacturing.

The industrial base of these and other cities has eroded over time. We entered a period in the 1970s where cities began losing jobs in the manufacturing sector. This loss of employment as a proportion of the total employment of an area is generally referred to as 'deindustrialization'. We have witnessed a changing of the economy from a manufacturing-based economy to a service-based economy, with jobs being created in such fields as retail, banking, tourism, entertainment, financial services, legal services, etc.

The loss of manufacturing jobs represents a compelling issue for cities. It has ripple effects through-out the community. Individuals and families may lose their residences. The city's tax base suffers. An increase in the amount and cost of providing social services to these individuals and their families may occur.

Deindustrialization is a global issue. Today, jobs are moving from country to country for reasons of cheaper labor costs, worker training costs, production costs, and regulatory costs. Countries are competing with each other for jobs. American companies have started operations abroad, benefiting such countries as India, China, Malaysia, South Korea, Indonesia, and the Philippines.

Shrinking cities

A great deal of attention has been devoted to the problems associated with areas experiencing population growth. What about those cities throughout the world that have been dramatically affected by deindustrialization and the accompanying decrease in population? Only recently has there been attention directed to areas experiencing continued economic decline and population losses. These are the so-called 'shrinking cities'. The problems facing these areas are just as compelling as the problems facing growing areas.

Natural disasters like hurricanes, tornadoes, or wildfires could also contribute to a shrinking city.

Many people are hesitant to return to their areas, out of fear that a disaster could strike again. If they go back, they might go back to no home or no job. Ultimately, one event acts as a trigger for subsequent events. For example, unemployment might lead to an inability to make monthly rent or mortgage payments. The latter could lead to possible foreclosure. All of these events could lead to a need for increased assistance.

Many cities could be labeled a 'shrinking city'. For instance, when examining city population loss from 1950 to 2010, the following cities come to mind: Detroit, Michigan (1,849,568 to 713,777); Buffalo, New York (580,000 to 270,000; St Louis, Missouri (856,796 to 319, 294); Cleveland, Ohio (914,808 to 396,815); Pittsburgh, Pennsylvania (676,808 to 305,704). These cities are located in the 'Frostbelt' or the 'Rustbelt'. However, a 'Sunbelt' city such as New Orleans, Louisiana, could also be considered a 'shrink-ing city'. Hurricane Katrina has contributed a great deal of the area's economic and population decline.

Population loss remains a challenge to many cities. Some groups are more affected than others. The poor are especially hard hit. As Hollander and Nemeth (2011: 363) note:

> During times of growth and vitality, the poor and politically marginalized need to fight for their seat at the table and attempt to pick up the crumbs of capital that fall to them. During times of decline and disinvestment, there is often not a table at all and certainly few crumbs. When jobs are scarce and city services meagre, the poorest segment of a community often need to struggle to meet their basic needs and are less likely to be able to focus on urban planning processes.

Urban infill development

We are constantly looking for alternatives to the haphazard and wasteful use of resources associated with sprawl. Cities around the world have realized there isn't an endless supply of land. Past development patterns are not sustainable. Over the years, development has continued to move to the suburbs and exurbs. Cheap land was a prime factor in this outward growth. Leap-frog development was commonplace.

Many areas are conducting inventories of their vacant and underutilized parcels. Developing these properties within city limits offers an alternative to sprawling development patterns. Using existing infrastructure in these areas is a more efficient use of tax dollars. Nevertheless, developers have seemed to gravitate toward outlying areas. As noted by the Local Government Commission (2001: 6):

> There are many good reasons why developers prefer to build on raw land, and some of these reasons relate to local government policy. Even though there are greater economic, social and environmental costs to sprawl development than infill, our public policies have stacked the incentives in the wrong direction.

While urban infill development represents an attractive alternative to sprawling development, there are reasons why a city might not want to use it. Some properties are contaminated. These are the so-called 'brownfields'. It might be more costly to remediate these sites instead of building on a 'greenfield' site. Some properties may have size and site constraints that prevent them from being developed. Community opposition might prevent the use of infill development. Some residents might oppose infill development out of the belief that the proposed development is out of character with the surrounding area. While it might be a common belief that infrastructure exists in all of the areas, there is no guarantee that the existing infrastructure can handle the added use.

The role of government in urbanization

It is abundantly clear that the federal government has directly or indirectly exerted a major influence on the scale and character of urbanization. Financial aid for housing, the creation of military bases and other government installations, the building of a huge highway network, and the indirect funding of development in the gunbelt are three particularly important ways in which the federal government has, at a minimum, at least facilitated, if not created, the modern suburbs. Many local governments have tried, with varying levels of success, to redevelop downtowns in the hopes of attracting people to both live and work in the downtown. At the same time, a number of cities are also trying to influence the direction of growth and development by extending mass transit lines in the suburbs and by encouraging development along transit corridors.

What is clear is that government's influence has been a consequence of a multiplicity of policies directed at other goals, and that there has been no implicit policy related to urbanization. Though the federal government has been a major force in urbanization, it has not attempted to guide this – or even acknowledge it. As we shall see later, a few states and localities have attempted to influence the rate or nature of urbanization, though not with a great deal of success. The majority, however, have not even tried. Whether they might have done, should have done, could have done, or might do in the future are questions to which we will return later.

Further reading

There are a number of good internet websites available to readers that can add to their understanding of the material contained in this chapter. The following are recommended.

The Urban Land Institute (http://www.uli.org) is primarily a non-profit real estate forum, but also creates an arena to discuss land use research and new development techniques. The website for this development interest group provides information on industry topics such as finance, housing, civic engagement, public policy, and smart growth, to name a few.

The Urban Development Timeline (http://www.urban timeline.org) provides a national and international look at urbanization for the twentieth to the twenty-first centuries, highlighting the connections between policies, methods, roles, and lessons learned. The Timeline was initiated by Planning and Development Collaborative International (PADCO) and its partners, which include, among others, the US Housing and Urban Development (HUD) department.

Founded in response to President Johnson's call for an 'independent, nonpartisan analysis of the problems facing America's cities', the Urban Institute (http://www.urban.org) is a scholarly research organization for problems facing urban Americans. Visitors can browse by author or by topics such as sectors, issues, places, and demographic groups. This website also has a Metropolitan Housing and Communities Policy Center that focuses on communities in urban areas.

For an international version of the Urban Institute, see Urbanicity (http://www.urbanicity.org). Urbanicity provides links to city and government websites as well as research and conferences. To its credit, this website maintains partnerships with the UN Habitat Best Practices and the Local Leadership Programme.

The Peopling North America: Population Movements and Migrations (http://www.ucalgary.ca/applied_history/tutor/migrations) website offers interesting insights into early migration to the United States and European migrations to North America.

The Edward J. Blakely Center for Sustainable Suburban Development (http://cssd.ucr.edu) focuses on suburban development and maintenance issues. Established in 2003, the Center is part of the Riverside campus of the University of California. Its focus is to make suburbs long-lasting communities while also accommodating the acceleration of growth. The center biannually publishes an electronic journal of suburban and metropolitan studies called *Opolis*.

A visual and archival documentation of one of the first American suburbs may be found on the website 'Levittown: Documents of an Ideal American Suburb' (http://tigger.uic.edu/~pbhales/Levittown.html). Compiled by Professor Peter Bacon Hales, visiting artist at the Art History Department at the University of Chicago, this website provides a detailed history and his commentary about the post-World War II suburb.

The US National Park Service supplies a history of suburban development on its website (http://www.nps.gov/history/nr/publications/bulletins/suburbs/intro.htm). Many original photos, including examples of gridiron plans, are found in its study of suburban land development practices from as early as 1830.

With photos taken from satellites, a NASA website illustrates the history of American sprawl (http://science.nasa.gov/headlines/y2002/11oct_sprawl.htm). An accompanying article focuses on satellite-based tools for urban planners along with hypothetical projections of city growth.

Sprawl Watch Clearinghouse (http://www.sprawlwatch.org) was founded in 1998 to develop strategies for those who manage or are concerned about growth. Based in Washington, DC, this organization collects and compiles data about urban development patterns.

The study of highway growth today is concerned with modern environmental problems. The Green Highways Partnership (http://www.greenhighways.org) is a public-private partnership designed to assist in the creation of more environmentally friendly highways in terms of recycling, reusing, and watershed planning. Current public members include the US Environmental Protection Agency and the US Federal Highway Administration.

Many states publish their own highway information online. Examples include Pennsylvania Highways (http://www.pahighways.com), which provides updates to interstate construction and traffic resolutions, and California Highways (http://www.cahighways.org), which provides a history of the state's numbered highways.

The vigor of the interstate highway program encountered a backlash of opinion demanding representation for other modes of transportation. One such group, called Transportation Alternatives (http://www.transalt.org), is a non-profit, New York-based organization that promotes bicycling, walking, and public transit, and acts as a government watchdog. Groups such as these are countered by highway interest groups such as the non-profit American Highway Users Alliance (http://www.highways.org), which promotes the interests of highway users in Washington, DC, as well as providing traffic information and links to research on roadway innovations.

There are many good general books on American history that deal well with urban issues. The three-volume *The Americans* by Boorstin is very readable (and each volume has an extensive bibliography). The three volumes are: *The Colonial Experience* (1958); *The National Experience* (1965); and *The Democratic Experience* (1973). Equally enthralling reading is provided by Hugh Brogan's one-volume *The Pelican History of the United States of America* (1986).

The major histories of planning are Reps (1965) *The Making of Urban America: A History of City Planning in the United States*; and Scott (1969) *American City Planning since 1890*. Reps is concerned mainly with physical planning, and his volume contains a unique set of reproductions of town plans. Scott's book sets the development of planning within a wider social framework. Peter Hall's *Urban and Regional Planning* (1992) has a very useful chapter on planning in the United States since 1945 as well as an analysis of the intellectual background in both the United States and Europe (which is more thoroughly developed in his 1988 *Cities of Tomorrow: An Intellectual History of Urban Planning and Design in the Twentieth Century*). Sies and Silver (1996), in *Planning the Twentieth-Century American City*, offer a number of readings on the foundations of twentieth-century planning organization and process of planning, the federal presence in planning, and broadening the planning agenda.

Urban history is thoroughly dealt with by Glaab and Brown (1983) *A History of Urban America*. Sam Bass Warner's *The Urban Wilderness: A History of the American City* is a marvelous study of the historical roots of the major urban problems of today; first published in 1972, it was reprinted in 1995. See also his *Streetcar Suburbs* (1978): this is a detailed study of the growth of Boston that gives a fascinating insight to the general process of suburbanization. A history covering the period from 1900 to 1990 is Teaford (1993) *The Twentieth-Century American City* (this has an excellent bibliographic essay). Miller and

Melvin (1987) *The Urbanization of Modern America* is an overview, with telling illustrations and guides to the relevant literature. Gelfand's (1975) *A Nation of Cities: The Federal Government and Urban America 1933–1965* provides a particularly good account of the role of the federal government during the period covered. An excellent brief account is Gerckens (1988) 'Historical development of American city planning'. Monkkonen (1988) offers an interesting historical view of urban America in *America Becomes Urban: The Development of U.S. Cities and Towns, 1780–1980*. Peterson's (2003) *The Birth of City Planning in the United States, 1840–1917* offers another perspective on the evolution of city planning in the United States. Corey and Boehm (2010) offer an excellent reader on urban history in their *The American Urban Reader: History and Theory*.

For a discussion on discrimination in mortgage lending, see: Jackson (1985), *Crabgrass Frontier: The Suburbanization of the United States*; LaCour-Little (1999), 'Discrimination in mortgage lending: a critical review of the literature'; Turner and Skidmore (1999), 'Mortgage lending discrimination: a review of existing evidence'; Novek (2001) 'You wouldn't fit here: the subtle and blatant forms of communication that keep segregation going'; Katznelson (2005) *When Affirmative Act Was White: An Untold History of Racial Inequality in Twentieth Century America*; and Kaplan and Valls (2007) 'Housing discrimination as a basis for Black reparations'.

The classic account of suburbanization is Jackson (1985) *Crabgrass Frontier: The Suburbanization of the United States*. Palen (1995) *The Suburbs* is a useful textbook that synthesizes material from a wide range of sources, with a mainly sociological orientation.

Fishman (1987) *Bourgeois Utopias: The Rise and Fall of Suburbia* is a study of the origins of suburbia in the United States and Britain and a discussion of 'the rise of the technoburb'. A journalistic discussion of the technoburb (with a more elegant term) is Garreau (1991) *Edge City*.

Markusen *et al.* (1991) give an interesting glimpse into the 'gunbelt' in *The Rise of the Gunbelt: The Military Remapping of Industrial America*.

Some recent manuscripts on 'suburbs' include: Baxandall and Ewen (2000) *Picture Windows: How the Suburbs Happened*; Lucy and Phillips (2000) *Confronting Suburban Decline: Strategic Planning for Metropolitan Renewal*; Marshall (2000) *How Cities Work: Suburbs, Sprawl, and the Roads Not Taken*; and Hayden (2003) *Building Suburbia: Green Fields and Urban Growth, 1820–2000*. Hayden offers interesting sights into the role of Levitt and Sons in building housing and contributing to early suburbia. Dunham-Jones and Williamson (2011) offer an interesting glimpse into the issue in *Retrofitting Suburbia, Updated Edition: Urban Design Solutions for Redesigning Suburbs*.

For a discussion of exurbs and exurbia, see Clark *et al.* (2005) 'Spatial characteristics of exurban settlement pattern in the U.S.'; Daniels (1999) *When City and County Collide*; Mikelbank (2004) 'A typology of U.S. suburban places'; Nelson (1992a) 'Characterizing exurbia'; Spectorsky (1955) *The Exurbanites*; and Sutton *et al.* (2006) 'Mapping exurbia in the conterminous United States using nighttime satellite imagery'.

For a discussion of employment decentralization, see Ihlanfeldt (1995) 'The importance of the central city to the regional and national economy: a review of the arguments and empirical evidence'; and Glaeser and Kahn (2001) 'Decentralized employment and the transformation of the American city'.

The literature on shrinking cities has grown in recent years, see: Gallagher (2010) *Reimagining Detroit: Opportunities for Redefining an American City*; Hollander (2011) *Sunburnt Cities: The Great Recession, Depopulation and Urban Planning in the American Sunbelt*; Hollander and Nemeth (2011) 'The bounds of smart decline: a foundational theory for planning shrinking cities'; Ryan (2012a) *Design After Decline: How America Rebuilds Shrinking Cities*. A good website discussing shrinking cities from an international perspective is the Shrinking Cities International Research Network, http://www.shrinkingcities.org.

Urban infill development has garnered a great deal of attention in recent years. For a further discussion of urban infill see: Local Government Commission (2001) 'Building livable communities: a policymaker's guide to infill development'; Suchman (2002) *Developing Successful Infill Housing*; Municipal Research Services Center (1997) 'Infill development: strategies for shaping livable neighborhoods'; and Wheeler (2008) 'Smart infill: creating more livable communities in the Bay Area: a guide for Bay Area leaders'.

Further reading on transport issues is given in Chapter 14.

Questions to discuss

1   **What are the main reasons for urbanization in the United States? Have these changed over time?**

2   **To what extent can it be said that the suburbs have been created by federal policies?**

3   **Do you think that it is appropriate to term the modern suburb 'a new urban form'?**

4   **Discuss the argument that in the postwar years 'cities had no realistic alternative but to embark on road building programs that were suicidal'.**

5   **Do you consider that current trends in decentralization, population, and employment, are fundamentally different from those of earlier years?**

6   **Is it possible that edge cities will replace central cities as the 'heart' of American metropolitan areas? What would be the effects of such a shift?**

7   **What is 'exurbia' and why is growth occurring in exurbs?**

8   **What is urban infill development and why are cities turning to it?**

9   **Discuss some of the problems commonly found in 'shrinking cities'.**

Governing and planning urban areas

The 'city problem' in the United States was, as President A. Lawrence Lowell of Harvard University said, like a jellyfish. You could not pick up a part here and a part there and succeed. You had to lift it altogether.

Scott 1969: 110

Basic needs for government

The original colonial cities were small both in terms of population and in area. They were intimate walking cities, usually less than a mile across. Their government was likewise modest, with the leading families taking control of matters in a natural way. Such services as were needed – for protection against hazards – were provided cooperatively. Life was based on family and community relationships, with little distinction between public and private enterprise. There was, however, a remarkable degree of regulation both by the English government (to keep the colonists under control) and by the new communities themselves (for their very existence). The former, of course, ultimately proved self-defeating when the onerous regulations (and 'taxation without representation') led to the Revolution and the birth of the United States. The latter, on the other hand, were self-imposed and had an acceptable rationale. Despite the abundance of land, the colonists quickly found that they had to plan and control the growing of certain crops. Without this, there was a danger that individual colonists, intent on maximizing their profits, would overproduce crops that were valuable for export and grow insufficient of the crops that were essential for local needs. Virginia restricted the growing of tobacco and required 'each white adult male over 16 to grow two acres of corn, or

suffer the penalty by forfeiting an entire tobacco crop. An Act of 1642 required the growing of at least one pound of flax and hemp, and an Act of 1656 required landowners to cultivate at least ten mulberry bushes per 100 acres in order to stimulate the production of silk' (Bosselman *et al.* 1973: 82).

In urban areas, the regulations were designed to promote health and safety. Following the great fire of Boston, laws were passed requiring the use of brick or stone in buildings. 'No dwelling house could be built otherwise, and the roof had to be of slate or tile upon penalty of a fine equal to double the value of the building' (Bosselman *et al.* 1973: 82).

There are many other examples of the extensive amount of regulation in colonial times. Though the settlers had the freedom to use their land unfettered by the restrictive feudal-type controls which operated in the country which they had left, the sheer necessities of surviving in a strange, undeveloped country, with its new hazards of climate, disease, and relationships with native Indians, forced a remarkable degree of self-regulation. There was also a religious strand which reinforced the dictates of the physical environment. Freedom to follow the religion of their choice was one of the reasons for the move to the new land, and each religious group had its own convictions and dogma. Settling in a new country meant not only building a new physical environment but also new communities in which religious beliefs could be practiced freely and in accordance with the observances which they demanded. The Puritan philosophy in particular stressed the divine joy of building a new society: the 'city upon the hill', to use the phrase immortalized by John Winthrop in the famous sermon delivered on board ship bound for the new world; but it was an orderly city which would serve God according to His rules.

As in any Utopia, the rules were not always followed, but neither were they ignored: they provided a framework for individual behavior and community action – and, unlike most Utopias, the city upon the hill was built. Similarly with William Penn's city of brotherly love: Philadelphia was a 'holy experiment' in the Quaker tradition that grew rapidly from the date of its foundation in 1683 as Friends emigrated from persecution in England (and later from Germany). The city was planned in detail (its legacy can still be clearly seen today), though Penn was unable to stem the pressures of land speculation which rapid growth engendered. Such speculation had no place in the earliest settlements: the problems of establishing and developing the community were of prime concern. In addition to these matters of high principle, there were more practical issues which the settlers had to consider. Since everything was to be provided from scratch, it was simple common sense to give some thought to the general layout of the settlement. At the least, it was necessary to decide upon the broad pattern of the streets and the location of public buildings. In fact, the early town charters listed requirements for such matters as

the basic road layout, and the reservation of land for the church, the town house, and the market place. Such charters continued to be drawn up in later years to guide the development of the western territories. Planning was therefore no stranger to early America, but, like other features of these times, the pace and character of growth brought about great change.

Privatism

In particular, problems of survival gave way to problems of development, first of trading, and later of manufacture. The towns became, above all, places for economic growth and money-making. This was an individualistic activity, in which government was seen as a facilitator of private enterprise rather than as a mechanism for order and control. The dominant philosophy became one of 'privatism': the free operation of individual initiatives in the search for private profit. Change took place as a result of private action and competition. Public controls were viewed as restraints on progress and, though they have developed significantly over time, privatism remains a powerful force in contemporary society. Indeed, much of the debate on current urban policy can be seen as a battle between philosophies of privatism and public planning. The philosophy of privatism has been well articulated in the writings of Sam Bass Warner (see Box 3.1). It is a distinctive, though not unique, feature of American urban development. The cities of the older world grew (as the very word 'city' indicates) as centers

---

BOX 3.1 PRIVATISM

Psychologically, privatism meant that the individual should seek happiness in personal independence and in the search for wealth; socially, privatism meant that the individual should see his first loyalty as his immediate family, and that a community should be a union of such money-making, accumulating families; politically, privatism meant that the community should keep the peace among individual money-makers and, if possible, help to create an open and thriving setting where each citizen would have some substantial opportunity to prosper.

Source: Sam Bass Warner 1968

of civilization (Mumford 1961). They were places of religious, cultural, and political power where governments determined public policy. The American experience, born of its different history, was essentially entrepreneurial and disdainful of government.

The growth of public powers

In spite of this, issues arose which demanded the use of public powers. These were broadly of two sorts: economic and social. Economic development itself required governmental support. The building of roads, canals, and railroads involved large amounts of capital, and was important in the competition between towns for key developments. The Erie Canal is the classic case of a huge public investment, which reaped enormous benefits for New York. Even if private capital might have sufficed for many more modest ventures, there was a mutual interest of private and public bodies in the promotion of economic development. Railroads were quick to see how important they were to the future success of individual cities, and they took good advantage of this in obtaining concessions and benefits. Entrepreneurship was a hallmark of city as well as private behavior.

The second type of problem was an outcome of the very success in the growth of the new towns. These gave rise to novel problems of public health and sanitation, overcrowding and congestion, public order, fire protection, education, and poverty. The problems were of an extreme nature (particularly as immigration expanded to incredible proportions) for which no ready solutions were apparent. One writer has suggested that these problems were on such an unprecedented scale as to create 'nearly irresolvable political and social tensions' (Judd 1988: 13) – an assertion which has a disturbingly contemporary relevance.

Machine politics

The development of public policies has been an erratic one, beset with continual controversies about the relative roles of public and private action and of the organizational structures which are required. The early

towns were 'omnibus' authorities, and new functions (if not undertaken by private or voluntary bodies) were simply added to the existing local government or, in many cases, given to ad hoc boards. (Philadelphia had thirty separate boards at one time.) The lack of a clear line between individual and public enterprise led to rampant corruption which, in a curious way, oiled the city governmental machine. It did this by explicitly serving the political interests of the time through 'machine politics'. This grew as a response to the problems of the cities: the growth in population and economic activity required a huge provision of new streets, water and sewage systems, transport, and other public utilities; and when complete they needed many thousands of workers to run them. These were often under the control of local political machines which were as attuned to profit-making as any industrial entrepreneur: they needed a regular flow of income from graft to keep their machines running. The system worked and was accepted (if not acceptable); in the words of the infamous Boss Tweed of New York, 'This population is too hopelessly split up into races and factions to govern it under universal suffrage, except by the bribery of patronage or corruption.' The system also served the interests of business whose financial contributions bought influence in the awarding of contracts and franchises.

Machine politics is now looked back on as a corrupt form of government (and it led to a distrust of city government which has persisted), but it acted as a mediator between the multiplicity of conflicting interests in the cities. It provided direct support for ethnic groups, and particularly for newcomers who had to familiarize themselves with a culture which was foreign in so many ways to them. Today, with the large numbers of minorities and recent migrants in some cities, the question is being raised as to whether a contemporary version of machine politics would benefit these groups (Erie 1988).

The reform movement

The reform of municipal government and a later movement for municipal land use planning emanated from broader concerns for improving the character of urban life and the instruments of control. This was the 'Progressive Era' in US history, marked by a desire to bring about radical change both in the arrangements for planning and administering local affairs and in the worsening conditions under which so many of the urban population were now living. As with all such movements for reform, there were many strands. These included the revelations of successive inquiries and reports on urban living conditions (including the best-selling book by Jacob Riis, *How the Other Half Lives* (1890), and the massive six-volume Pittsburgh Survey, which was the most extensive of many such surveys). From the 1870s through to the outbreak of World War I, reports on social conditions appeared in profusion from religious organizations, welfare societies, researchers, settlement house workers, journalists, and many others. Even the federal government was moved to study some of the problems: the Commissioner of Labor produced reports in 1894–95 on slum housing (though funding restricted the investigation to only four cities: Baltimore, Chicago, New York, and Philadelphia), and on the experience in European cities of providing housing for 'the working people'. Better known was Upton Sinclair's *The Jungle*, which depicted the horrors of Chicago's meatpacking industry; this was one piece of writing which led to clearly connected and tangible results – the establishment of the federal Food and Drug Administration in 1905. Sinclair's moving account of the wretched existence of Chicago's immigrants had no parallel effect, but it added to a growing concern about living conditions.

The overcrowded slums brought several responses, though public action was restricted by firmly held political beliefs. The direct provision of housing was seen as essentially a matter for market economics and, since incomes were low, housing standards were low. Though New York introduced its first tenement house law in 1867, following some disastrous fires, building codes were not generally introduced until the last two decades of the century. Even then, they were primarily concerned with fire hazards, and not with other controls such as height (the absence of which later made possible the development of the office skyscraper). New York was again in the lead (in 1895) with legislation relating to slum clearance, but few other cities followed, and little was achieved, though by the turn of the century it could be said that a national housing movement was under way (Lubove 1962).

Parks

One of the powers granted by the 1895 New York legislation was for the provision of playgrounds and parks in crowded districts. This reflected another strand in the reform movement that was concerned with landscape and parks. There were two overlapping concerns: one related essentially to natural and landscape beauty, the other to the desperate need of urban neighborhoods for some relief from their congested environment by way of the provision of recreational areas. The protagonists of both schools saw beautiful recreation areas as sources not only of rest from labor but also, more romantically, of moral rejuvenation.

New York's Central Park was (literally) the major landmark in the campaign for urban parks. Designed by Frederick Law Olmsted and Calvert Vaux, Central Park has been seen by some people as a miracle – large amounts of open green space surrounded by massive buildings of all shapes and sizes. Olmsted and Vaux were chosen as the ultimate winners of a public competition in 1857 to design the new, large public park.

Getting the park developed was certainly no easy task. Political battles were common over various aspects of the plan. In fact, Olmsted and Vaux had resigned on numerous occasions. The evolution of Central Park is definitely a fascinating study of planning and public policy.

The emphasis on the development of urban parks was on natural beauty (considerably helped by human hands). The concern stemmed in part from the agrarian origins of the country and the high regard for the countryside and rural life, epitomized in Jeffersonianism, and later further idealized by Ralph Waldo Emerson, Henry Thoreau, and others. This rural dream was by no means uniquely American but, as in Europe, it drew additional strength from the unlovely urbanization of the nineteenth century. Unlike European parks, however, the objective was not the landscape garden beloved by the upper classes of Europe but the preservation of wild land: Central Park dramatically demonstrated that this could be provided close to the developing urban areas. Similar parks, and even systems of parks, followed. The planning of these led to an elementary form of planning: judging how a park would relate to the development of the urban area and its transport provision. This was the route along which a number of landscape architects traveled to become planners.

Though aesthetic and romantic in origin, the parks movement was also concerned with meeting the needs for recreation and for relief from the overcrowding of city neighborhoods. The need for local parks assumed a high profile in the mid- and late nineteenth century. Perhaps this was because the need was tangible, the provision relatively cheap, and the opposition weak? Certainly, the provision of neighborhood parks was less daunting than housing reform, where the hegemony of market principles made progress extremely slow. Be that as it may, the parks movement was a force of significance and, coupled with other concerns about the preservation of natural and landscape resources, it widened to embrace an attack on all forms of ugliness and the inefficiencies of contemporary cities.

The City Beautiful

The widespread concern for parks constituted one of the elements which made up a concern for urban grace and beauty which became known as the City Beautiful movement. This reached its zenith with the World's Columbian Exposition: a 'plaster fantasy' which celebrated, in classical architectural terms, the four hundredth anniversary of the landing of Columbus. It demonstrated the confidence of 'a nation grown rich by the development of its natural resources and its industries, a nation at last critical of its municipal institutions, and determined to remold them to serve broader public purposes' (Scott 1969: 45). Located on the shore of Lake Michigan, the 'White City' was the occasion of a national celebration (though, because of unavoidable delays, one year later than the anniversary). It was, in some ways, an irrelevance (an 'anachronistic

symbol of accomplishment' in Mel Scott's phrase), but it represented in a tangible way the merging of a number of concerns and ideologies which had developed over the last decades of the nineteenth century. It marked a desire to make American cities places of beauty, set in an artificial naturalistic landscape. Such dreams could not survive the realities of a growing industrial society: they were a reaction to it, not a solution for its problems. However, the Exposition has the important historical significance of being clearly placed in time and space as the marker of a coming together of numerous attempts to create a more humane and livable environment.

It was also responsible for a legacy of beautiful buildings which are to be seen throughout the country, in countless civic centers, boulevards, college campuses, railroad stations, banks, and other public buildings: all reflecting the Beaux Arts tradition which was embraced by the Exposition. (Their current fame often stems from their success in resisting redevelopment in the name of historic preservation – as with the Penn Central Railroad's Grand Central terminal in New York, which was the subject of a landmark zoning case. This case is discussed in Chapter 18.)

The strands which joined together to create the City Beautiful movement were many – some of which have already been touched upon. Following Wilson's analysis, seven of them can be singled out as being of particular importance. First, as the name suggests, there was a desire to make cities beautiful. The origins here were the landscape, park, and municipal art movements, which, at least initially, were essentially aesthetic in conception, though they sometimes exhibited a degree of social awareness and even ideas of municipal efficiency. This was particularly apparent in a second strand: a perception of beauty which incorporated some concept of public or private profit. The idea that 'good design pays' is only a short step from the contention that 'good design is not more expensive than bad design' – a contention that is frequently heard in architectural circles. The argument went further, however, since it incorporated the idea that beautiful designs were more efficient. Black smoke pouring from a factory chimney was both ugly and inefficient (an idea which has achieved a new formulation in contemporary environmental policy). Similarly, a graceful design has palpable utilitarian features; an imposing boulevard has an effectiveness in accommodating traffic; an elegant road scheme is an efficient distributor of traffic. One enthusiast went so far with the conceptual marriage of beauty and utility as to coin the term 'beautility'. This mercifully failed to gain currency, but it was a neat epitome.

A third strand was the importance attached to expertise. Efficiency required experts, and there was a rapidly growing number of them at the end of the nineteenth century. It is not too much of an exaggeration to describe the time as 'the age of the expert'. The achievements of nineteenth-century capitalism had led to a belief in the great potential of business-like methods of production and control. This extended to the rapidly growing middle-class cadre of professionals: doctors, dentists, teachers, social workers, architects, and planners. The early beginnings of a technocratic society needed, and could afford, these new skills. Leading later to the conceptual transformation of the City Beautiful into the City Efficient, this belief in the expert, wedded to ideas of progress, had important implications for municipal government and planning.

There was a class element in this (which constitutes the fourth of our selected strands): the expanding middle class attracted a respect and achieved a position which gave them a power to influence the course of events unsurpassed in later times. They were the high priests of the cult of expertism; they might not have an answer to every technical problem, but they knew that there was one to be found. They could also advise on what provisions should be made for the lower working classes: whether these be in the form of parks for recreational relief from the toil of everyday labor, or of beautiful landscapes to raise their spirits. The professional classes knew what was best, and they made some attempt to bridge the chasm between their ideals and the realities of the nineteenth-century city. One element in this paternalism was fear of open class conflict. Industrial strife was well known, and there were fears that this might turn into something more sinister. But the prevailing philosophy was essentially confident and optimistic. This fifth strand in the current ideology prevailed over fears of revolt, though not always easily. At least for New Yorkers, there was the vivid memory of the riots of 1863, 'when the poor streamed out from their gloomy haunts to burn, murder and pillage' (Lubove 1962: 12).

More broadly, there were widespread concerns about the waywardness, the unruliness, the depravity of the working class. The belief in individual responsibility, the antagonism to socialistic ideas, the fears of immigrants were all too clear to see. These views played themselves out mainly in other arenas, but they impinged upon the City Beautiful movement by way of a belief in righteousness and reform. There was a fervor in this which might have belied deeper fears. Certainly, some of the language used was exaggerated, to say the least. Charles Mulford Robinson was perhaps the most florid in his *Modern City Art, or the City Made Beautiful*, where he foresaw:

> the adjustment of the city to its needs so fittingly that life will be made easier for a vast and growing proportion of mankind, and the bringing into it of that beauty which is the continual need and rightful heritage of men and which has been their persistent dream.
>
> (Robinson 1903: 375)

Every movement needs its poet. Robinson, however, was an effective poet, not only in inspiring an awareness of the quality of what we would now call the environment, but also in inspiring large numbers of people to do something about it. His 1907 book, *The Improvement of Towns and Cities*, was a best-seller, and stimulated the formation of large numbers of 'local improvement' societies.

Part of Robinson's beguiling effervescence stemmed from the sixth strand: the 'American rediscovery of Europe'. Though huge numbers of immigrants had forsaken the beauties of Europe for the more prosaic benefits of the New World, its architectural treasures were models to copy. So were some of its city governments: European cities were seen to work in a way in which American cities did not. Frederick C. Howe's (1913) *European Cities at Work* extolled the superiorities of German expertise, though later others, more realistically, were critical of German, enterprise-crushing bureaucracy.

Finally, there was the new acceptance of the American city. With a heavy dose of wishful thinking, American cities were regarded as being capable of major improvement: all that was needed was the same dynamism in civic improvement that had proved so successful in the industrialization of the economy.

These and other influences that created the City Beautiful movement are of more than historical interest. The movement itself was only a name, not a concerted campaign; but its elements remain important not only for an understanding of the historical background to planning but also for an appreciation of the forces which still affect the conception and the operation of US planning.

Municipal reform

The City Beautiful movement also focused on city government and its inadequacies. This was a major plank in the wider scene of critical analysis which characterized the times. On this the 'muckrakers' had a field day. Lincoln Steffens' exposé of corruption, *The Shame of the Cities* (1904), was one of the most influential; but there were many others in the period from the 1880s to World War I who remorselessly attacked the blatant corruption of city governments. The issue figured continuously in the newspapers and magazines of these years. At the same time as the excesses of municipal corruption bred increasing discontent among businesses, the corruption of big business was itself a target of criticism. Both fed into the wider reform movement. Reform of city government thus attracted the support of those seeking to safeguard business profits as well as those who saw them as excessive, particularly in the context of widespread poverty. The forces of change were more complicated than this might suggest since, in supporting municipal reform, there was a common purpose among those concerned with beauty, with efficiency, with physical conditions, and with other concerns of what one historian (Brogan) has termed the 'Progressive Adventure'.

It was from this wide ferment of ideas that action emerged to rectify the inadequacies of municipal government. Here, visions of the City Beautiful merged with, and became dominated by, those of the City Efficient. A particularly important connecting thread was the promise offered by 'scientific management'. This had been perceived to be successful in fields as diverse as engineering and factory organization; and it was now seen as being equally relevant to the management and engineering of cities. The validity of 'the scientific method' became part of the religion of the age. The High Priest of this was Frederick Winslow Taylor, whose *Principles of Scientific Management* (1911) was, like all such texts, more widely quoted than actually read. So famous were his ideas that they gave rise to the eponymous creed of Taylorism. Taylor is remembered as the inventor of time-and-motion study, but the important aspect of his theories was the separation of planning from implementation: identifying the problems involved in a process, establishing a scientific way of resolving them, and then implementing the new system. Applied to cities, it spelled the separation of politics from administration, and the rule of the expert.

A streamlined commission type of municipal government seemed to be the answer to the 'weak mayor' form of government which was a notable feature of cities dominated by machine politics. The structural weakness lay in the division of responsibilities and financial power among many sectors of municipal government, and the lack of coordinated control. Accountability was confused, and the control of patronage was of greater importance than the good government of the municipality. Administrative fragmentation both was engendered by, and supported, machine politics. The system could not have provided a greater contrast to the business-like management of industry (or at least the common image of this).

One popular remedy was the 'strong mayor' type of administration where, though elected, the mayor operated as a chief executive, coordinating all municipal government functions and maintaining an authority over departmental heads (see Figure 3.1a). Power was concentrated in the hands of the mayor. The mayor was not a member of the city council and had

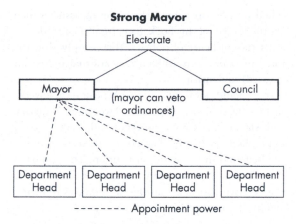

Figure 3.1a **Council with strong mayor**

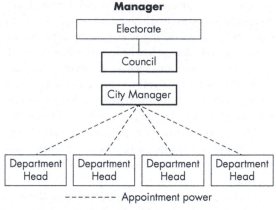

Figure 3.1c **Council with manager**

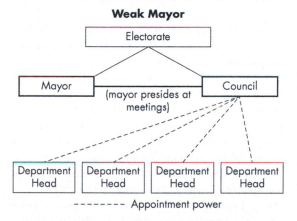

Figure 3.1b **Council with weak mayor**

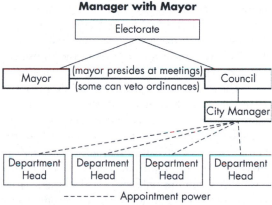

Figure 3.1d **Council manager with mayor**

veto power or authority over the city council. The mayor possessed extensive administrative authority and appointed a Chief Administrative Official to assist in policy and administrative functions. This individual served at the pleasure of the mayor. Moreover, appointment and removal of departmental heads was at the mayor's discretion.

Some cities operated under a 'weak mayor' type of administration (see Figure 3.1b). In this form of local government, the mayor did not possess any executive power. The position was more of a ceremonial position with limited powers. The mayor shared the policy-making role with the city council but had no vote.

While the mayor could veto council decisions, the council could override the veto by a two-thirds majority vote.

An alternative form of local government was the commission form of government. This form of local government was propelled into prominence by an accident of history: the devastating flood of 1900 which destroyed much of the town of Galveston, Texas, killing 6,000 of the town's 37,000 inhabitants. The commission consisted of five members, who took responsibility for both making local policy and its administration. Each member of the commission was responsible for various administrative departments of the local government. It was deemed remarkably successful in its restoration and improvement of the town and its services, and it was widely admired and copied. Nevertheless, it had its shortcomings, of which an important one was seen to be that it was not sufficiently business-like: the commissioners were both policy-makers and administrators; and there was no guarantee that they would coordinate their respective fields of responsibility. As experience was gained with newer forms of municipal government, the commission style fell out of favor except in a few smaller municipalities.

A solution to this perception of the problem was the transfer of all administrative functions and power to a professional/expert city manager who would serve at the discretion of the council and would be able to coordinate the work of the separate departments (see Figure 3.1c and d). In other words, the city manager supervises, coordinates, and administrates the functions of a local government by carrying out the implementation of policies made by the locally elected governing body. The city manager system also had the perceived advantage of further separating the government of municipalities from political 'interference'. City managers function most effectively in areas where the politics are not too divisive and where there is a broad agreement on the significant policy issues. This is unusual (though not unknown) in the largest cities with more volatile politics, where the strong mayor system often works better. However, as with all generalizations about the United States, such statements fail to cover all circumstances.

Today, most municipalities work under the 'strong mayor' or the city manager system. Nevertheless, cities can change their form of local government. Colorado Springs, Colorado, and San Diego, California, have in recent years changed from a council manager form to a strong mayor form of government, while Cedar Rapids, Iowa, and El Paso, Texas, moved from a mayor council form to a council manager form of government.

What is notable about much of the debate on the form of municipal government is that a preoccupation with controlling the excesses of corruption has marginalized the importance of the political process. It refuses, however, to be neutralized; and it often attains

salience in debates about the planning issues which are the subject of this book. Structural changes do not avert the influence of politics: they simply alter the form in which this influence is allowed (or is not allowed) to flourish.

We should not forget another type of local government – the county government. Two states, Louisiana and Alaska, have equivalent types, called parishes and boroughs respectively. The power to create a county can be found in the state constitution. The powers of these bodies of government can also be found in the state constitutions, as can the form of government a county chooses to use.

Reform and the planning function

Urban planning emerged as a promising field of professional activity supported by public opinion and governmental capabilities in several countries in the early years of the twentieth century. The year 1909 stands out as a particular high point in this emergence of planning: it was in this year that the first national conference on city planning was held. (It was also the year in which Wisconsin passed a state law providing powers for the creation of city planning commissions; Benjamin Marsh published what is arguably the first textbook on planning, *An Introduction to City Planning*; Harvard introduced the first university course on city planning; Burnham completed his Plan of Chicago and a Chicago Plan Commission was established to implement it; and, illustrating the international character of the planning movement, Britain passed its first planning act.)

These were promising times, and in 1917 it seemed that the promise might be fulfilled: in that year the planning interests felt sufficiently bold to establish a new professional organization, the American City Planning Institute. The membership of this was, in all likelihood, the most diverse any professional body has ever witnessed: the fifty-two charter members included fourteen landscape architects, thirteen engineers, six attorneys, five architects, four realtors, two publishers, two 'housers', and an assorted group of writers, tax specialists, land economists, educators, and public officials (Scott 1969: 163). A common denominator was a recognition of the fact that cities had increasing problems which existing institutional structures were unable to deal with, or even to comprehend. The spirit of reform was still in the air, but it wore a many-colored cloak.

It soon became apparent that more was needed than scientific management: the city planning commissions which were set up to give substance to the drive for efficiency were severely constrained in what they could actually do – despite the signs of purposive activity by such bodies as municipal information clearing-houses, and a burgeoning of planning courses, conferences, publications, and the like. The forces of privatism were too strong to be contained by public officials. Indeed, urban planners seldom did more than follow residential and commercial developers with transportation and sewer systems. Despite a desire to imitate some of the trends which were emerging in Europe, the US planning movement was, in fact, unique. This stemmed from the uniqueness of the United States itself: the dynamism of its urban system, the pace of its growth, the strength of its private enterprise, and the general reluctance to place fetters on the forces of development. In this, the gridiron plan played a significant role.

The gridiron plan

The previous discussion has outlined the character of the social framework within which a movement favorably disposed to planning was surfacing. Foremost, of course, was the need to facilitate urbanization: here the favored 'plan' had for long been established as the gridiron. The gridiron plan was particularly useful in land transactions, and it was simple in the extreme. (Its disregard for topography sometimes made it literally 'extreme' – as was later so dramatically illustrated in San Francisco.) It provided for uniformity of lots, thus easing description both for legal deeds and for land sales (particularly when sight-unseen), and it enabled urban growth to proceed in an orderly fashion. New York, in 1811, explicitly accepted the 'decisive' advantages of the gridiron for the undeveloped part of the

Manhattan Island. Commissioners appointed to propose a plan concluded:

> In considering the subject, they could not but bear in mind that a city is to be comprised principally of the habitations of men, and that straight sided and right angled houses are most cheap to build, and the most convenient to live in.
>
> (Glaab and Brown 1983: 252)

Even in towns where the gridiron was not wholly adopted (for example, in Buffalo, Indianapolis, and Detroit), its influence was generally all too apparent, and even when natural areas were safeguarded as open spaces, market pressures often led municipalities to free them for development (Reps 1965). The gridiron also worked easily in the western urban promotions, since it was 'the natural tool of the land speculator' and fitted in neatly with the lines of the land ordinances (each township was divided into 36-square sections of a square mile: the checkerboard is still clearly visible to the air traveler).

However, the gridiron had its limitations (in addition to its blatant disregard for contours), especially when urban development spread over very large areas. Access became increasingly problematic, and land values in peripheral areas were affected. Difficulties were increased where one town ran into another. The problem clearly pointed to the need for some type of planning – but this raised the central problem with which planning always has to contend: how to balance the public interest against the rights of private property. The gridiron system did this in a way which was judged acceptable; an alternative was not to be readily found.

One approach which had only limited success was the establishment of boards of survey charged, as in Boston in 1891, with making 'plans showing the location of highways which present and future interests of the public require' (Scott 1969: 3). Sensible though this might seem, it adversely affected some property owners who were prepared (then as now) to seek a judicial remedy. In the Boston case, the Massachusetts Supreme Court invalidated the procedure on the ground that there had been a taking of property without due compensation. The attitude of the courts in most states was that property owners could not be required to conform to paper plans for new streets (as distinct from streets that were actually in existence). Though some states (such as Pennsylvania) took a more liberal line, it was some time before it was generally accepted that rights of way could be established by the use of the police power. In the meantime, the gridiron plan was almost universally accepted both in the peripheral extension of existing towns and in the establishment of new towns.

If such elementary planning as that concerned with the placing of new streets was problematic, it was hardly to be expected that more ambitious land use planning would be welcomed. At every turn, public action was hampered by the importance attached to property rights. Paradoxically, as more thought was given by planners to the subject of their emerging profession, it became increasingly apparent that planning was indeed a most difficult matter. Not to be daunted, however, an attempt was made in line with the spirit of the age to make it more scientific.

City planning as an exact science

City Beautiful plans were concerned above all with appearances. In this, they were precisely the opposite of the burgeoning industrial cities, where the over-riding object was production and profit. But to pit the beautiful against the ugly was not enough in this practical and increasingly 'scientific' age: what was needed was a plan which would increase the efficiency of the city. The new planners realized this, and increasingly turned their attention to the physical workings of the city. Efforts to understand aspects of city life had become increasingly scientific. Studies such as those of the New York Council of Hygiene in the 1860s had carefully and methodically studied housing and sanitary conditions: they provided a model which other investigators copied and which set a standard to which planners aspired. But they faced difficulties which continue to beset those who attempt to make planning a scientific process: how to isolate and analyze the objects of study, how to coordinate the findings of separate studies, how to make planning 'comprehensive', and how to prevent political factors spoiling well-laid plans.

Few would be bold enough today to claim that the difficulties involved in such a task are surmountable: more likely it would be seen as a search for the Holy Grail. Some had an inkling of this even in the heady days before World War I. Speaking at the second national conference on city planning in 1910, Olmsted gave voice to a general apprehension about 'the complex unity, the appalling breadth and ramification, of real city planning'. The prospect of understanding, let alone controlling, the forces at work was daunting. Yet Olmsted was not overwhelmed by this prospect and, like many after him, considered that an attempt had to be made to understand 'the complex web of the city'.

Ironically, the search for a rational basis for planning led quickly away from the concerns of those who had shown how to study the social life of the city. Even a long-standing regard for housing waned as it became apparent that it was impossible to do anything significant within the existing social and political framework. What emerged was a preoccupation with physical controls by way of the separation of land uses: zoning became the focus of planning action. Plans were still commissioned, but they were typically superficial, glossy productions which, while of some use for promotional purposes, were largely irrelevant to the problems which planners had initially glimpsed. Further, with its focus on legally enforceable uses of land, zoning lost the essential planning concern for future patterns of development. What was essentially a legal and administrative device for regulation (akin to a building or sanitary code) took the place of the vision which, even if remote from reality, inspired the plans of the City Beautiful era. In place of dreams of the future city came detailed regulation to prevent unwanted uses invading desirable residential areas. Thus the distinctive character of US land use planning was established.

Regional planning

Though zoning moved to center stage in the 1920s, wider concepts of planning were prominent in planning debates. Some of these were aimed at providing the machinery for dealing with the problems of servicing large areas undergoing subdivision, or the protection of natural and recreational resources, and of coordinating the inherently limited capabilities of existing agencies. Such practical considerations led to the establishment of regional planning organizations such as the Los Angeles County Regional Planning Commission (the first of its kind) in 1922 and the Chicago Regional Planning Association in 1923. Such bodies had some success in road planning, particularly as growing car ownership dictated expanded road-building programs.

Alongside these efforts of practical persons were the visions of Henry Wright, Lewis Mumford, Clarence Stein, Catherine Bauer, Benton MacKaye, and others, who founded the Regional Planning Association of America in 1923. This espoused the cause of self-contained communities set in natural environments (what would today be called 'sustainable environments'). They achieved little success: their main practical experiment, the garden city of Radburn, New Jersey, fell victim to the Depression; and the governmental realization of their dream – the New Deal program of greenbelt towns – was axed by an antipathetic Congress in 1938 just as it was getting under way. Though utopian, the ideas of these thinkers have persisted and form part of a tradition of planning thought which emerges from time to time as a vision of a comprehensive regional planning system.

Attempts to foster regional planning have a long history. Sadly, the story is not a thrilling one: with some notable exceptions (such as Portland, Oregon, and the Twin Cities of Minneapolis–St Paul) it is typically a succession of false starts and disappointed hopes. Nevertheless, some progress was made. The Tennessee Valley Authority was established as early as 1933, and a number of other economic planning agencies followed the Public Works and Economic Development Act of 1965 (such as the multi-state Appalachian Regional Commission).

State governments set up regional physical planning bodies such as the New York Adirondack Park Agency, New Jersey's Pinelands Commission, California's Coastal Commission. The Adirondack Park Agency was created by New York State law in 1971 to develop long-range Park policy by balancing local government interests with statewide concerns. The Pinelands Commission of New Jersey came into being in response to the growing concern with urban sprawl. Recognizing the national and international importance of the Pinelands, the US Congress created the Pinelands National Reserve. New Jersey took the lead in balancing its protection with the needs of new development and established the Pinelands Commission in 1979. A Pinelands Protection Act was passed by the New Jersey Legislature that same year, requiring county and municipal plans be consistent with the Pinelands Comprehensive Management Plan. The 1954 Housing Act introduced federal financial assistance for metropolitan planning, and a large number of metropolitan organizations were established. Additionally, special acts were passed creating such agencies as the Twin Cities Metropolitan Council (Minneapolis–St Paul). These bodies differ enormously in function, power, and performance, but all demonstrate a degree of willingness to look at problems on a regional scale.

A thrust for creating a means of cooperation between the constituent parts of metropolitan areas came in President Kennedy's 1961 *Housing Message to Congress* in which he argued that the old jurisdictional boundaries were no longer adequate (see Box 3.2). Without entering into a historical account, it is of relevance to this chapter to give some indication of the course of events.

Particular progress was made through the Bureau of Roads, which required local governments to cooperate in a regional planning exercise as a condition for highway construction grants. The system was gradually extended and, in 1965, urban areas with a population of more than 50,000 became ineligible for federal grants for highway construction unless they had a 'comprehensive transportation process for the urban area as a whole, actively being carried on through cooperative efforts between the states and the local communities' (Advisory Commission on Intergovernmental Relations 1964: 106).

The Urban Renewal Administration followed suit with its requirement that states and local governments produce comprehensive plans. Then, in 1968, the Bureau of the Budget issued Circular A-95, which sought to establish a 'network of state, regional, and metropolitan planning and development clearing houses' to receive and disseminate information about proposed projects; to coordinate applicants for federal assistance; to act as a liaison between federal agencies contemplating federal development projects; and to perform the 'evaluation of the state, regional or metropolitan significance of federal or federally-assisted projects' (Mogulof 1971: 418; Elazar 1984: 186).

The continued growth and formation of municipalities and other governmental entities in California increased the debate over localism versus regionalism. Local governments have been vested with the power to develop policies and plans within their jurisdictional boundaries. As is commonly known and reported throughout the literature, many local policies and plans ignore jurisdictional boundaries and impact upon neighboring municipalities. As such, municipalities have been unable to solve a number of problems.

For years, areas have practiced functional planning – planning on an issue-by-issue basis. Fragmentation represented a major dilemma. There were calls for creating regional planning agencies to deal with the larger-than-local issues affecting metropolitan areas. Localities were concerned about losing some of their land use powers.

In 1963, the California Legislature responded to these concerns by passing legislation that created Regional Planning Districts. The reasons for the legislation can be found in California Government Code, Planning and Zoning Law, Chapter 2, Section 65060.2:

(a) That the State has a positive interest in the preparation and maintenance of a long-term, general plan for the physical development of each of the State's urban areas that can serve as a guide to the affected local governmental units within such areas and to the state departments and divisions that are charged with constructing state-financed public works within such urban areas.

(b) That continuing growth of the State, and particularly urban areas within the State, present problems which are not confined to the boundaries of any one single county or city.

(c) That the planning activities of counties and cities can be strengthened and more effectively performed when conducted in relation to studies and planning of an urban regional character.

(d) That in order to assure, insofar as possible, the orderly and harmonious development of the urban areas of the State, and to provide for the needs of future generations, it is necessary to develop a means of studying, forecasting, and planning for the physical growth and development of these areas.

If a voluntary regional or metropolitan association already existed, there was no need to create a new agency.

In the 1960s and 1970s, twenty-six regional organizations were established. These organizations were called 'Councils of Government' (COGs). These organizations were not regional governments. They were voluntary organizations of local governments. They dealt with the myriad of issues that cross boundaries of local governments – issues requiring regional attention. They are multi-service and multi-

BOX 3.2 PRESIDENT KENNEDY ON THE CITY AND ITS SUBURBS

The city and its suburbs are both interdependent parts of a single community bound together by the web of transportation and other public facilities and by common economic interests . . . This requires the establishment of an effective and comprehensive planning process in each metropolitan area embracing all activities, both public and private, which shape the community.

Source: President Kennedy's *Housing Message to Congress* 1961

jurisdictional regional/areas-wide agencies functioning as a regional planning organization. In later years, additional COGs were created in such areas as Calaveras County, Orange County, San Mateo County, and Ventura County. The Southern California Association of Governments (SCAG), established in 1965, has become the largest of some 700 COGs in the United States.

Between 1968 and 1970, the number of COGs increased from 100 to 220. Almost all the 233 Standard Metropolitan Statistical Areas had regional councils of some type: COGs, economic development districts, regional planning commissions. These varied greatly in the extent to which they became involved in regional planning: many did as little as was possible to meet the federal conditions. Others became actively committed, particularly after HUD introduced yet another regional planning scheme: the comprehensive planning assistance program, popularly known (by reference to the relevant section in the Housing Act) as the 701 program. This program allowed jurisdictions to engage in comprehensive planning for their areas through the availability of federal funding assistance.

Regional planning was not as effective as its protagonists had hoped. It was mainly a creature of federal initiatives; and frequently it did not receive more than nominal support from the member governments who were apprehensive about the growth of an independent source of regional influence. Instead, they typically saw it as 'a service giver, a coordinator, a communications forum, and an insurance device for the continued flow of federal funds to local governments' (Mogulof 1971: 418). Moreover, though one of the major objectives was to ensure that individual federally funded projects were in harmony with metropolitan or regional plans, such plans often did not exist. Above all, there was no effective machinery for promoting and implementing them.

Weak though the COGs were, they constituted a point from which regional thinking could develop; hopefully, action would have followed. But, in 1982, the Reagan administration rescinded Circular A-95, and halted the system of federally funded regional clearinghouses. In their place, states were encouraged to establish their own machinery. Some particular

examples of this are illustrated at a number of points in this book. The indications are that, in response to increasing problems of environmental pollution and urban growth, and, above all in transportation, there is a reawakening of interest in forms of regional planning.

Studies such as those of David Rusk (1993) and Henry Cisneros (1995) have made persuasive statements of the need for regional planning. Even more eloquent is the establishment of a directly elected metropolitan government for Portland, Oregon (see Box 3.3). Metro, as it is called, serves over 1.3 million residents in three counties and twenty-four cities in the Portland region. According to Metro, it is the 'only regional government in the United States with a home-rule charter and directly elected officials'. Its primary responsibility is regional land use planning, with other responsibilities in transportation planning, natural resources planning, recycling, parks, trails, and green spaces, garbage and hazardous waste, etc. The region's voters approved its Charter in 1992 and amended the Charter in 2000. The amendments consolidated the Executive and Council offices. Among the new changes were: the Council President would be elected by a region-wide vote, the number of Council districts would be reduced from seven to six, and the Council would appoint a Chief Operating Officer to handle management duties that were previously performed by the Executive Officer. Finally, in December 2002, the Metro Council approved an expansion of its urban growth boundary (UGB). Additional discussion of the UGB can be found in Chapter 10. The land inside the UGB represents a twenty-year supply of land for future residential development and is designed to protect farms and forests from urban sprawl. The twenty-year supply figure is required to be updated every five years.

Minnesota's Twin Cities area, a seven-county metropolitan area, has a regional planning agency called the 'Metropolitan Council'. The Metropolitan Council has sixteen members with each member representing a geographic district of the metropolitan area. A Chair serves as an at-large member of the Council. Council members are appointed by the Governor of Minnesota and confirmed by the State Senate.

According to a history of the Metropolitan Council, it was created to:

- Plan for the orderly and economical development of the seven-county metro area, and
- Coordinate the delivery of certain services that could not be effectively provided by one city or county.

Among the activities of the Metropolitan Council are the following: operate the region's largest bus system, collect and treat wastewater, plan for the future growth of the region, plan for a regional system of parks and trails, etc. The body's coordination of planning and development function has been strengthened over the years with the merger of various agencies into the Metropolitan Council.

Another form of regional or subregional planning is watershed planning. The US Environmental Protection Agency (2005a: 1–2) defines a watershed as 'the land area that drains to a common waterway, such as a stream, lake, estuary, wetland, or ultimately the ocean'. A watershed could potentially cover thousands of square miles. It could encompass a single state or group of states.

Watershed planning is important for a number of reasons. The numerous benefits and challenges associated with water resources are illustrated, using California's resources as an example, in Box 3.4.

Watershed issues transcend local boundaries. Multiple stakeholders must be involved – federal, state, and local governments, the private sector, non-profit organizations, universities, etc. Collaborative planning among these stakeholders is crucial to the success of any watershed planning effort. Activities must be designed to address any real or potential threats to water quality. These efforts must also be integrated into other local or regional plans that address issues of water quality.

The continued growth of cities and metropolitan areas has led many people to ponder the future spatial development of these areas. Cities have appeared to simply merge together. Jean Gottman (1961) recognized this issue when he noted that groups of densely populated metropolitan areas were blending together to combine a new urban form – a 'megalopolis'. To the common eye, cities and metropolitan areas have appeared to merge. It is impossible to tell where one city begins and ends and another city starts. An example of Gottman's new urban form would be the Boston to Washington, DC, 'megalopolis'.

Variations of Gottman's idea of a 'megalopolis' have occurred over the years. Peirce, Johnson, and Hall (1993) wrote about areas undefined by political boundaries, so-called 'citistates' – 'a region consisting of one or more historic central cities surrounded by cities and towns which have a shared identification, function as a single zone for trade, commerce and communication, and are characterized by social, economic and environmental independence'.

Kenichi Ohmae (1995: 12) has claimed that in a global economy, nation-states have become little more than bit-part actors. Instead, 'region states' have

BOX 3.3 PORTLAND METRO: A DIRECTLY ELECTED REGIONAL GOVERNMENT

We, the people of the Portland area metropolitan services district, in order to establish an elected, visible and accountable regional government that is responsive to the citizens of the region and works cooperatively with our local governments; that undertakes, as its most important services, planning and policy making to preserve and enhance the quality of life and the environment for ourselves and future generations; and that provides regional services needed and desired by the citizens in an efficient and effective manner, do ordain this charter for the Portland areas metropolitan services district, to be known as 'Metro'.

Source: Portland Metropolitan Services District Charter 1992

BOX 3.4 BENEFITS AND CHALLENGES FACING WATER RESOURCES

California's rivers, lakes, and estuaries provide a host of public benefits, including commercial and sport fishing, drinking water supplies, recreation and scenic values. Increasingly, our public agencies face challenges in managing these public trust resources in ways that protect them and allow for other important uses. Challenges include: managing polluted storm water in urban centers, managing floods, restoring native salmon stocks and other threatened species, reducing toxicity from pesticides, reducing sediment impacts from working forestlands, managing hydropower plants, and protecting the quality and supply of drinking water.

Source: California Resources Agency 2002: 7

BOX 3.5 DEFINITION OF 'MEGAPOLITAN AREA'

- Combines at least two, and sometimes many dozens of existing metropolitan areas.
- Totals more than 10,000,000 projected residents by 2040.
- Derives from contiguous metropolitan and micropolitan areas.
- Constitutes an 'organic' cultural region with a distinct history and identity.
- Occupies a roughly similar physical environment.
- Links large centers through major transportation infrastructure.
- Forms a functional urban network via goods and services flows.
- Creates a usable geography that is suitable for large-scale regional planning.
- Lies within the United States.
- Consists of counties as the basic units.

Source: Lang and Dhavale 2005: 4–5

become more important. Examples of 'region states' would be the San Diego–Tijuana, Mexico, region, or the Research Triangle Park area in North Carolina. These natural economic zones do not respect boundaries in the global information society and economy.

Lang and Dhavale (2005) have also noticed the merging of cities and metropolitan areas. People are able to commute further distances. Improvements in transportation systems are making areas, once inaccessible to workers, now accessible. These areas are called 'megapolitan areas' by Lang and Dhavale (see Box 3.5). Examples of such areas would include the Arizona Sun Corridor, which would be anchored by the Phoenix and Tucson metropolitan areas; the Florida

Peninsula, which would be anchored by the Miami and Orlando metropolitan areas; and the Piedmont, which would be anchored by the Atlanta and Charlotte metropolitan areas and would cover the states of Alabama, Georgia, North Carolina, Tennessee, South Carolina, and Virginia.

Regional planning initiatives will undoubtedly continue to be subjected to legal challenges. For example, in April 2002, in *Tahoe-Sierra Preservation Council, Inc.* v. *Tahoe Regional Planning Agency* (535 US 302), the US Supreme Court ruled that two moratoria on residential development imposed by the Tahoe Regional Planning Agency for a total of thirty-two months while the agency was devising a comprehensive land use plan were held not to constitute per se taking, requiring the payment of just compensation under the Fifth Amendment. Nevertheless, calls for regional planning continue to be heard throughout the United States in individual states and through metropolitan regions in two or more states. In California, there are even calls for binational regional planning for the San Diego–Tijuana, Mexico, metropolitan area. It is too early to pass judgment on these new endeavors, but the issues are discussed further in other chapters.

Federalism

The United States operates under a federal system of government with a federal government, state government, and local governments. The executive branch of federal government is headed by a President, Vice President, and the now fifteen Cabinet-level departments of Agriculture, Commerce, Defense, Education, Energy, Health and Human Services, Homeland Security, Housing and Urban Development, Interior, Justice, Labor, State, Transportation, Treasury, and Veterans Affairs. The President implements and enforces the laws enacted by the United States Congress. The legislative branch of the federal government, comprised of the Senate and House of Representatives, has the power to make laws and regulations. The judicial branch, headed by the United States Supreme Court, resolves disputes and rules or decides on the constitutionality of the laws. The federal government is granted both explicit powers by the United States Constitution and implied powers to carry out the explicit powers.

The Tenth Amendment of the United States Constitution stipulates that the federal government cannot exercise any power not delegated to it by the states. Those powers are reserved for the states and the people. Each state has its own constitution, government, and code of laws. Each state has an executive, legislative, and judicial branch of government. However, it cannot perform any function that would undermine the federal government. States also create the means to create various categories of local governments and the authority these local governments possess.

The state government also allows for the creation of local governments. In essence, cities are 'creatures of the state'. They possess only the powers granted to them by the state. These powers include those granted in express words, those necessarily or fairly implied in or incident to the expressed powers, and those powers essential to the creation of the cities. This viewpoint is commonly known as Dillon's Rule, named after Chief Justice John Forest Dillon of the Iowa Supreme Court in the 1860s.

Cities can be characterized as being a 'general law' city or a 'home rule or charter' city. A 'general law' city is regulated by state law. It is limited by the powers and duties authorized or given to it by the state. They are bound by the aforementioned Dillon's Rule. A 'home rule or charter' city establishes a charter with its own ordinances and resolution under the procedures established by the state government. These cities possess all of the legislative powers not prohibited by state law or charter. Voters must approve the charter and any changes made to it. The charter is, in essence, a constitution for the city. These cities have less interference from the state than a 'general law' city.

We must be cognizant of other areas such as the unincorporated United States territories of Guam, American Samoa, Puerto Rico, the Northern Mariana Islands, and the United States Virgin Islands. Each has a governor. Moreover, Washington, DC, was established by the United States Constitution to serve as the nation's capital. It is not part of any state and has an elected mayor and city council.

A common feature found in the governmental structure of the United States is the ever-changing roles of the various levels of government. Tensions have surfaced over the years. Today, the federal government gives less money to the states. It is also true that the states give less money to the local governments. Unfortunately, the federal government requires states to comply with federal laws without federal funding, while the states require local governments to do things without any state funding – the ever-present 'unfunded mandate' dilemma.

Neoliberalism

Questions as to the role of government in the area of economic growth are constantly being debated. Some individuals feel public sector involvement is needed, while others feel the public sector is too inefficient. They feel more comfortable with less reliance on the public sector and more reliance on the private sector.

Relying solely on the public sector could be problematic. Too many rules and regulations are simply stifling the economy. As Ronald Reagan noted, in his January 20, 1981 presidential inaugural address, 'government is not the solution to our problem; government is the problem'. Moreover, to many individuals, the public sector is clearly inefficient and wasteful of public dollars.

One school of thought is that the government should not be involved in regulating the economy because its intervention interferes with market forces. To proponents of this way of thinking, free markets are a good thing that should not be hindered. Another line of thought suggests that the government needs less regulation and strongly encourages regulatory reform.

Neoliberalism, according to David Harvey (2005) and others, is not an easy topic to define. Capital is the dominant component theme of this view since it dominates all facets of our lives – cultural, economic, political, and social. Government should work with people and companies to promote the economy. We should advocate free markets. We should maximize profits at all costs by eliminating the restraints that governments put on private property owners and businesses. This means removing barriers to free trade, such as tariffs and regulations. It also may mean privatizing some functions of local government such as education, prisons, and infrastructure.

The federal government has tended to overspend in periods of bad economic times. This leads to potential budget deficits. These problems are felt not only in the US but throughout the world, in such countries as the United Kingdom, Canada, Chile, and New Zealand.

There are potential problems associated with adhering to the belief in neoliberalism. It redistributes wealth. Privatizing lands has led to many people being forced off their lands. While we have witnessed many individuals and companies making massive sums of money in various countries, some countries have a negative income trend among the overall population.

Further reading

The website entitled The Next American City (http://www.americancity.org) is dedicated to alleviating city problems through social progress. The Next American City offers a magazine written by urban planners and the progressive public, although many of its articles are also available online. Partnering with two other organizations, The Next American City helped establish an Urban Innovation Symposium Series in 2007 which is also available online (http://americancity.org/groundup).

The New Colonist (http://www.newcolonist.com) is a web publication for people who live in and love cities. They call themselves colonists because they live in a 'new territory', namely, the 'once-neglected city'. The articles are contributed by city dwellers and deal with urban issues.

The Project for Public Spaces (http://www.pps.org) is a non-profit organization dedicated to the creation of public spaces, including parks for recreation. Since 1975, the Project has worked with communities to create public spaces that are active community centers. The website includes design tips as well as advice for guiding the political process of public space making.

Central Park is a big part of New York's identity. Founded in 1980, the Central Park Conservancy (http://www.centralparknyc.org) provides information about what is happening daily in Central Park, as well as the history of the park from the 1800s.

The interactive Digital History Collection (http://columbus.gl.iit.edu) showcases the World's Columbian Exposition, the so-called 'White City' of the Chicago World's Fair in 1893. Created by the Paul V. Galvin Library, the site features original guidebooks, photos, and commentary. Original maps of the Exposition contain points of interest that, when clicked upon, take the visitor to the original guidebook pages or photos of the famous, temporary city.

The rise of the importance of experts necessitates a support system for those experts. The International City/County Management Association (http://www.icma.org) is one such support system. This non-profit organization provides publications, training, and updates to public managers.

Many universities conduct urban research and work closely with city planners and local government. One such center at the University of Washington is the Municipal Research and Services Center (MRSC) (http://www.mrsc.org), a non-partisan, non-profit organization that shares information with cities and counties in Washington State. The center offers a 'help desk' to answer questions for local government.

The department of American Studies as the University of Virginia calls William Penn's Philadelphia an early attempt at utopian city planning (http://xroads.virginia.edu/~cap/PENN/pnplan.html). A commentary on the original gridiron plan of Philadelphia can be found on the website and is one section of a four-part website about the early American city.

Many of the titles recommended for the previous chapter are equally relevant here, particularly those concerned with nineteenth-century history.

The major historical study of the land use regulation issue is Bosselman *et al.* (1973) *The Taking Issue*.

To understand the development of Central Park and the backgrounds and ideas of its developers, the following sources are recommended: Kinkead (1990) *Central Park, 1857–1995: The Birth, Decline, and Renewal of a National Treasure*; Kowsky (1998) *Country, Park and City: The Architecture and Life of Calvert Vaux*; Rosenzweig and Blackmar (1998) *The Park and the People: A History of Central Park*; Rybczynski (1999) *A Clearing in the Distance: Frederick Law Olmsted and America in the Nineteenth Century*; and Cedar Miller (2003) *Central Park, An American Masterpiece: A Comprehensive History of the Nation's First Urban Park*.

On the Regional Planning Association of America, see Spann (1996) *Designing Modern America: The Regional Planning Association of America and Its Members*.

On privatism see Warner (1968) *The Private City: Philadelphia in Three Periods of its Growth*; and the more extensive discussion in Barnekov *et al.* (1989) *Privatism and Urban Policy in Britain and the United States*.

Good political science texts include Judd (1988) *The Politics of American Cities: Private Power and Public Policy*; Judd and Swanstrom (1994) *City Politics*; Ross *et al.* (1991) *Urban Politics: Power in Metropolitan America*; and Harrigan (1993) *Political Change in the Metropolis*.

Good discussions on watershed planning and management include Davenport (2002) *Watershed Project Management Guide*; Randolph (2003) *Environmental Land Use Planning and Management*; France (2005a) *Facilitating Watershed Management: Fostering Awareness and Stewardship*; France (2005b) *Introduction to Watershed Development: Understanding and Managing Impacts of Sprawl*; Sabatier *et al.* (2005) *Swimming Upstream: Collaborative Approaches to Watershed Management*; and Heathcote (2009) *Integrated Watershed Management: Principles and Practices*.

For recent writings on metropolitan and regional planning, see Rusk (1993) *Cities Without Suburbs*; Cisneros (1995) *Regionalism: The New Geography of Opportunity* (further references on state and regional growth management are given in Chapter 10); Orfield and Luce, Jr. (2010) *Region: Planning the Future of the Twin Cities*; and Seltzer and Carbonell (2011) *Regional Planning in America: Practice and Prospect*.

Questions to discuss

1 **What were the strands in the nineteenth-century reform movement?**

2 **Describe the City Beautiful movement. How did it originate?**

3 **Discuss the benefits and problems of gridiron plans.**

4 In what ways were the 'principles of scientific management' thought to be relevant to city planning?

5 Does the history of American city planning offer any lessons for the debate about the relative merits of comprehensive and incremental planning?

6 What arguments could be made for and against regional planning?

7 What forms of local government are there in the US?

4

Planning and sustainability

We have met the enemy and he is us.

Walt Kelly, Earth Day Poster, 1970

Concerns over the effects of population growth and decline are not new. The idea of sustainability has permeated planning and policy discussions for many years. Rachel Carson (1962), in her book, *Silent Spring*, detailed the problems for animals, birds, and humans associated with the use of the deadly pesticide DDT. Ehrlich (1968), in his book, *The Population Bomb*, sounded an alarm heard around the world that warned us about the dangers and effects on the environment of overpopulation. Meadows, Meadows, Randers, and Behrens III (1972), in their book, *The Limits to Growth*, examined different scenarios of global development and produced computer simulations, for alternative scenarios, of the effects of economic and population growth on the Earth's resources. The interactions of these scenarios were indeed disturbing to the world.

We have continued to use and abuse our resources. The burning of fossil fuels has contributed to changes in global climate patterns. We continue to abuse the soil by destroying farmlands and forestlands. We have paved over land to build buildings and developments of all types. Wetlands have been destroyed. Our growing population continues to consume more resources and to generate more waste. These and other reasons have contributed to the increasing need to plan for sustainability in the US.

What is sustainability?

Finding an agreed-upon definition of anything is difficult. Terms mean different things to different people. Sustainability is a classic example of a fluid concept. Perhaps the most commonly used definition of sustainable development originates from a report, *Our Common Future*, published by the World Commission on Environment and Development. This body was an independent commission created by the United Nations (UN) in 1983 and charged with investigating various environment and development problems facing the planet and to suggest ways to solve identified problems. It was chaired by the Prime Minister of Norway, Gro Harlem Brundtland. The final report, better known as the Brundtland Report, defined sustainable development as 'development which meets the needs of the present without compromising the ability of future generations to meet their own needs' (World Commission on Environment and Development 1987: 27).

There are numerous definitions of the terms 'sustainability' and 'sustainable development'. Environment Canada (2013) enlarges the Brundtland Commission's definition and defines sustainable development as being 'about meeting the needs of today without compromising the needs of future generations. It is about improving standards of living by protecting human health, conserving the environment, using resources

Three-legged Sustainability Stool

Figure 4.1 Three-legged stool depiction of sustainability

efficiently and advancing long-term economic competitiveness. It requires the integration of environmental, economic and social priorities into policies and programs and requires action at all levels – citizens, industry, and governments.' President George W. Bush, in Executive Order 13423, January 24, 2007, indicated that the word sustainable meant 'to create and maintain conditions, under which humans and nature can exist in productive harmony, that permit fulfilling the social, economic, and other requirements of present and future generations of Americans . . .' The US Environmental Protection Agency (EPA) acknowledges that:

> Sustainability is based on a simple principle: Everything that we need for our survival and well-being depends, either directly or indirectly, on our natural environment. Sustainability creates and maintains the conditions under which humans and nature can exist in productive harmony, that permit fulfilling the social, economic and other requirements of present and future generations.

Figures 4.1 and 4.2 depict the two most commonly held views of sustainability. The three-legged stool figure suggests that sustainability is supported by three legs: environment, economy, and equity (occasionally referred to as social equity or just social) – the Three Es. Within the environment leg, we are focusing on such topics as climate, energy, air quality, loss of open

space, resource conservation, air and water quality. The economy leg examines such topics as jobs and employment benefits, living wages, fuel prices, workforce training, market development, financial systems, and infrastructure. The equity leg covers such topics as well-being, social justice, equal opportunity, neighborhoods, diversity, discrimination, and gender equity.

It might be suggested that each leg is interrelated with the other legs. The overlapping circles depictions may better illustrate the notion that the environment, economy, and equity are interrelated and must be discussed together. In an October 18, 2001 speech at Oregon State University, Oregon Governor John Kitzhaber described the overlapping circles idea:

> Imagine, if you will, three overlapping circles – one representing our economic needs, one representing our environmental needs and one representing our social or community needs. The area where the three circles overlap is the area of sustainability – the area through which run all the elements of a good quality of life: a healthy, functioning natural environment; a strong economy with jobs and job security; and safe, secure communities where people have a sense of belonging and purpose and a commitment to each other. These elements – these

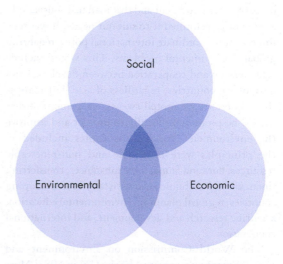

Figure 4.2 Overlapping circles depiction of sustainability

threads, which together weave the fabric of sustainability – are things we hold in common. They represent a common set of desires and aspirations that add value and quality to our lives.

For example, discussions on energy subsidies involve the economy and environment. The subsidies would also affect individuals. In another example, issues such as climate change are felt more by the low-income population, thus bringing in the idea of environmental justice.

Additional depictions of sustainability are certainly available. A third image of sustainability could show it as a group of three concentric circles. The first circle would represent the economy. It needs, and operates within, the society or equity circle. Ultimately, the economy and social circles require the environment in order to exist and operate.

International actions

A number of international meetings have contributed to the global awareness of sustainable development. Concerns over the environment took a front seat at the 1972 UN Conference on the Human Environment in Stockholm, Sweden. Delegates recognized the importance of the environment and that man had consistently altered or transformed it to suit our needs. It was now important to coordinate international policy regarding global environmental problems. This would include coordination and cooperation between developed and developing countries, regardless of size. Ultimately, the Conference promoted twenty-six principles that should be promoted globally to protect and improve the environment. Among the topics included in the principles were renewable and non-renewable resources, flora and fauna, toxic substances, transferring financial and technological assistance to developing countries, regional planning, environmental education, scientific research and development, and international cooperation.

The World Commission on Environment and Development was convened by the UN in 1983. More commonly known as the Brundtland Commission, this Commission, as noted above, provided us with the most widely accepted definition of sustainable development. The Commission convened multiple public hearings and meetings throughout the world to garner global perspectives on environment and development issues. It issued the report, *Our Common Future*, in 1987.

The meetings confirmed the idea that environment and development issues were interrelated and should be discussed together, not separately. The Commission advocated global, long-term environmental strategies and cooperation between developed and developing countries. It also acknowledged the increasingly important role that non-governmental organizations at all levels must play in promoting public awareness of and political change on sustainable development.

Five years after the publication of *Our Common Future*, the UN Conference on Environment and Development was convened in Rio de Janeiro, Brazil in 1992. This Conference is commonly referred to as the Earth Summit. It was attended by heads of state or government leaders and representatives from non-governmental organizations from around the world, and delegates reaffirmed the interrelatedness of our global problems and concluded that there were no easy solutions to the problems.

The world's population continued to grow. We continued to degrade the environment. We continued to produce enormous amounts of waste. The realization of global climate change soon became a hot-button topic around the world. It was also clearly evident that non-governmental organizations were becoming increasingly vocal at the Conference. Whether this was the result of continued government inaction on environment and development problems or because of continuing tensions between developed and developing countries was certainly debatable.

Two notable documents were created at the 1992 Conference. A global action agenda or blueprint called Agenda 21 on how to achieve global sustainable development at international, national, regional, and local levels was created. It recognized that cooperation between nations was essential to the realization of global sustainable development. Although it was adopted by delegates at the Conference, it was merely voluntary in nature, not a binding agreement.

Agenda 21 was made up of forty chapters that covered a host of topics that would affect the environment. The forty chapters were divided into four sections: 1) social and economic dimensions; 2) conservation and management of resources for development; 3) strengthening the role of major groups; and 4) means of implementation. It was a policy document with recommendations for action on the areas addressed in the various chapters.

Agenda 21 recognized the vital part that local governments will play in promoting and achieving sustainable development. As such, a process called Local Agenda 21 calls for the development of a long-term strategic plan involving a number of local groups to achieve sustainable development. Local governments around the world have developed such plans.

The second notable non-binding document was the Rio Declaration on Environment and Development, which laid out a set of twenty-seven principles that would help guide responsible development throughout the globe. It also stressed the need for increased cooperation and partnerships between countries. Among the areas covered by the principles were the right to live in a healthy environment; the recognition that resource development can cause repercussions beyond borders; eradicating poverty throughout the world is essential; the development process must include environmental protection as one of its components; and that, in order to achieve our goal of sustainability, nations should share knowledge and technologies.

The 2002 World Summit on Sustainable Development was held in Johannesburg, South Africa. Better known as the Rio +10 Summit, the Summit sought to reinvigorate the need for public-private partnerships and a global partnership between nations for achieving sustainable development. It also sought to renew prior commitments made to the aforementioned Agenda 21 and Rio Declaration on Environment and Development, at a time when tensions continued to flare up over the disagreements between developed and developing nations. However, acknowledging the limited progress that had been made in achieving sustainable development, Secretary-General of the UN, Kofi A. Annan, indicated:

Ten years ago at the 'Earth Summit' in Rio de Janeiro, Governments committed themselves to just such a transformation, and to Agenda 21, as the comprehensive plan of action for getting there. But commitments alone have proven insufficient to the task. We have not fully integrated the economic, social and environmental pillars of development, nor have we made enough of a break with the unsustainable practices that have led to the current predicament.

(UN, 2001: 1)

The 2012 Conference on Sustainable Development (Rio +20) in Rio de Janeiro continued the long-running discussion on global sustainable development. Once again, countries renewed their commitments on the topic and examined high-priority issues like jobs, energy, sustainable cities, food security and sustainable agriculture, water, oceans, disaster readiness, and the continuing need to close the gap between developed and developing countries. Eradicating poverty was a main topic of discussion, with delegates acknowledging the fact that it represented the greatest global challenge facing the realization of sustainable development.

US national actions

Concerns over the interactions and interrelationships between the environment, the economy, and social equity are not new to the US. As noted in the opening paragraph of this chapter, the 1962 book *Silent Spring* represented a wake-up call to the country about the deadly dangers of pesticides to animals, birds, and humans. The creation and celebration of Earth Day on April 22, 1970 informed and educated residents on the many values of the natural environment. As the years passed, we have witnessed continued debates on the virtues of economic growth versus environmental protection and debates on who actually benefits from economic growth – the equity argument.

President Richard Nixon was inaugurated as the thirty-seventh president of the US on January 20, 1969. In November of 1969, Nixon created a President's

Advisory Council on Executive Organization to advise him on whether or not all federal environmental activities should be housed in one agency. On April 29, 1970, the Advisory Council's Chair, Roy Ash, sent a memorandum to President Nixon recommending that anti-pollution programs be housed in a separate agency of the Executive Branch called the Environmental Protection Administration. Ash recognized the myriad of environmental problems associated with population growth and economic growth. Their interaction can clearly be seen in the following quote:

> Pesticides have increased the yield of our crops and made it possible for less land to produce more food. They have also polluted the streams and lakes. Automobiles have broadened our economic and social opportunities, even as they dirtied the air and jammed our highways. Some means must be found by which our economic and social aspirations are balanced against the finite capacity of the environment to absorb society's wastes.

Instead of attacking pollutants on a piecemeal basis through different federal agencies and departments, a new approach was needed. On July 9, 1970, President Nixon sent a special message to Congress with his recommendation to establish an Environmental Protection Agency (EPA), earlier called Administration, which would be the federal focal point for the attack on pollutants. The EPA, established in December 1970, would undertake environmental research, monitor pollution levels, set pollution levels, and enforce federal environmental laws.

Seven months before Nixon's recommendation to establish the EPA, a national policy for the environment was enacted – the National Environmental Policy Act of 1969 (NEPA). NEPA sought to create harmony between man and the environment. In Section 101 (42 USC Section 4331) it further declared the following of the federal government:

> in cooperation with State and local governments, and other concerned public and private organizations, to use all practicable means and measures, including financial and technical assistance, in a manner calculated to foster and to promote the general welfare, to create and maintain conditions under which man and nature can exist in productive harmony, and fulfill the social economic, and other requirements of present and future generations of Americans.

On June 29, 1993, President Bill Clinton signed Executive Order 12852 creating the President's Council on Sustainable Development. This Council would advise the President on matters related to sustainable development. The Council's goals, as seen in Box 4.1, echoed the prior economic, environmental, and social equity dimensions of sustainable development that were identified in the Brundtland Report of 1987. In its 1996 report, the President's Council stressed the importance of creating dialogues among all parties in an attempt to foster the integration of the economy, equity, and the environment into national policy.

The President's Council issued two reports in 1997. In its first report, *Building on Consensus*, the Council acknowledged the importance of working together through partnerships to achieve sustainable development. It recognized the crucial leadership role that the federal government needed to play in promoting sustainable development. It also encouraged the President to extend the life of the Council. The life of the Council would be later extended to the end of June 1999.

The second report, *The Road to Sustainable Development*, offered examples of some of the progress made on sustainable development activities in the US. These activities were undertaken by the business community, non-governmental organizations and academic institutions, the federal government, and regional, state, and community efforts. It was hoped that the lessons learned from the various activities would contribute to the growing global dialogue on sustainable development.

In 1999, the President's Council issued *Towards a Sustainable America*. This report offered ideas and examples of how a sustainable America could be achieved. It also showed how the discussion on sustainability at the federal level could not be confined

BOX 4.1 GOALS OF CLINTON COUNCIL ON SUSTAINABLE DEVELOPMENT

Goal 1: Health and the Environment – Ensure that every person enjoys the benefits of clean air, clean water, and a healthy environment at home, at work, and at play.

Goal 2: Economic Prosperity – Sustain a healthy U.S. economy that grows sufficiently to create meaningful jobs, reduce poverty, and provide the opportunity for a high quality of life for all men in an increasingly competitive world.

Goal 3: Equity – Ensure that all Americans are afforded justice and have the opportunity to achieve economic, environmental, and social well-being.

Goal 4: Conservation of Nature – Use, conserve, protect, and restore natural resources – land, air, water, and biodiversity – in ways that help ensure long-term social, economic, and environmental benefits for ourselves and future generations.

Goal 5: Stewardship – Create a widely held ethic of stewardship that strongly encourages individuals, institutions, and corporations to take full responsibility for the economic, environmental, and social consequences of their actions.

Goal 6: Sustainable Communities – Encourage people to work together to create healthy communities where natural and historic resources are preserved, jobs are available, sprawl is contained, neighborhoods are secure, education is lifelong, transportation and healthcare are accessible, and all citizens have opportunities to improve the quality of their lives.

Goal 7: Civic Engagement – Create full opportunity for citizens, businesses, and communities to participate in and influence the natural resource, environmental, and economic decisions that affect them.

Goal 8: Population – Move toward stabilization of U.S. population.

Goal 9: International Responsibility – Take a leadership role in the development and implementation of global sustainable development policies, standards of conduct, and trade and foreign policies that further the achievement of sustainability.

Goal 10: Education – Ensure that all Americans have equal access to education and lifelong learning opportunities that will prepare them for meaningful work, a high quality of life, and an understanding of the concepts involved in sustainable development.

Source: President's Council on Sustainable Development 1993

or restricted to any one agency or department. Any discussion must involve multiple agencies and departments at all levels of government as well as the private sector, non-profit sector, community organizations, and citizens.

President George W. Bush also acknowledged the importance of sustainability and the importance of federal leadership. In his Executive Order 13423 (January 24, 2007), Bush instructed federal agencies working on environmental, transportation, and energy-related activities to conduct them in an environmentally, economically, and fiscally sound manner. Among the goals to be achieved were: increasing alternative fuel, hybrid, etc., vehicles; reducing petroleum consumption; increasing alternative fuel consumption; reducing energy intensity; reducing greenhouse gas emissions; reducing water consumption; and purchasing environmentally sound goods and services.

The Obama administration continues to promote sustainability. A partnership between the US Department of Housing and Urban Development (HUD), the US Department of Transportation (DOT), and the US Environmental Protection Agency (EPA) was announced in June 2009 as a means of assisting areas to protect the environment, to improve access to affordable housing, and to increase transportation options in communities. Instead of having a variety of uncoordinated programs dealing with sustainability, coordinated efforts between departments and agencies would achieve more results and be a more efficient use of federal money. Grants would be available for such program areas as redeveloping brownfields, creating livable communities, providing technical assistance on topics being addressed by the Partnership for Sustainable Communities, and providing capital investment in surface transportation projects that focus on livability and sustainability improvements. The partnership incorporated the following livability principles into their programs, policies, and legislative proposals: provide more transportation choices; promote equitable, affordable housing; enhance economic competitiveness; support existing communities; coordinate and leverage federal policies and investment; and value communities and neighborhoods.

President Obama has stressed the importance of the federal government serving as a leader in the US in promoting sustainability. In Executive Order 13514, October 5, 2009, Section 1, he noted:

In order to create a clean energy economy that will increase our Nation's prosperity, promote energy security, protect the interests of the taxpayers, and safeguard the health of the environment, the Federal Government must lead by example. It is therefore the policy of the United States that Federal agencies shall increase energy efficiency; measure, report, and reduce their greenhouse gas emissions from direct and indirect activities; conserve and protect water resources through efficiency, reuse, and stormwater management; eliminate waste, recycle, and prevent pollution; leverage agency acquisitions to foster markets for sustainable technologies and environmentally preferable materials, products, and services; design, construct, maintain, and operate high performance sustainable buildings in sustainable locations; strengthen the vitality and livability of the communities in which Federal facilities are located; and inform Federal employees about and involve them in the achievement of these goals.

Among the energy, water, and waste reduction targets for federal agencies are reductions in vehicle fleet petroleum use; improvement in water efficiency; use of environmentally responsible products and technologies; and recycling and waste diversion. Agencies are required to make plans to achieve the targets for review and approval.

State and local actions

States and local governments are playing an increasingly important role in planning for sustainable development. Over the years, many of them have developed sustainable development plans or component climate change plans and programs.

The State of Oregon has always been recognized as a leader in land use and environmental planning. In 2000, Governor John Kitzhaber issued an Executive Order that promoted sustainability in state government operations. Recognizing the importance of sustainability, one year later, Oregon passed the Oregon Sustainability Act – setting objectives for state agencies in their operations and mission. According to Oregon Revised Statutes 184.421, sustainability 'means using, developing and protecting resources in a manner that enables people to meet current needs and provides that future generations can also meet future needs from the joint perspective of environmental, economy, and community objectives.'

More recently, the State of Hawaii embarked on a state sustainability effort. The State Legislature formed a twenty-five-member Sustainability Task Force to develop a statewide sustainability plan. In order to get public input, in 2005, state officials asked stakeholders, through a number of different community engagement tools, what type of Hawaii they wanted and how they envisioned the future.

Hawaii recognized the importance of developing a balance between the State's economic, community, and environmental goals and aspirations. Sustainability was defined as achieving the following:

• Respects the culture, character, beauty and history of our state's island communities.

- Strikes a balance among the economic, social and community, and environmental priorities.
- Meets the needs of the present without compromising the ability of future generations to meet their own needs.

(State of Hawaii 2008: 1)

The 2050 Sustainability Plan was published in 2008; a year after the Task Force issued a report to the Governor and Legislature. As an outgrowth of the Plan, an annual report card would be produced along with indicators to show how the State was doing in regards to the five identified goals: 1) Living sustainably is part of our daily practice in Hawai'i; 2) Our diversified and globally competitive economy enables us to meaningful [sic] live, work and play in Hawai'i; 3) Our natural resources are responsibly and respectfully used, replenished and preserved for future generations; 4) Our community is strong, healthy, vibrant and nurturing, providing safety nets for those in need; and 5) Our Kanaka Maoli and island cultures and values are thriving and perpetuated (State of Hawaii 2008: 2).

Each of the goals contained a series of indicators that would measure the progress in meeting the goal. For example, a number of indicators were developed to measure the State's progress in achieving a diversified and globally competitive economy. One indicator used to measure the diversity of the economy is the percentage of science- and technology-based workers. Another measure used to examine economic self-sufficiency is the amount of dollars spent in locally owned businesses. A proportion of food produced and consumed locally is an indicator of progress toward food self-sufficiency – a key indicator of sustainability.

The City of New York issued a sustainability plan titled 'PlaNYC' in 2007 and updated it in 2011. It made it clear that it should not be considered an isolated or stand-alone plan. Instead, it should complement other City efforts to strengthen the economy, combat climate change, and enhance the quality of life for all New Yorkers. It would serve as a leader for others to follow, not just city government agencies and departments. The plan identified goals for

ten areas: 1) housing and neighborhoods, 2) parks and open space, 3) brownfields, 4) waterways, 5) water supply, 6) transportation, 7) energy, 8) air quality, 9) solid waste, and 10) climate change. Each goal had associated metric(s) or indicator(s) to track the progress being made in achieving the goal. In total, a set of twenty-nine indicators were created.

The City of El Paso, Texas clearly recognized that sustainability was more than simply an environmental issue. It is a balancing act involving the environment, social equity, and economic issues of the area. Individual strategic plans must be integrated into the citywide sustainability effort. It required communication and collaboration among the public sector, private sector, non-profit sector, and citizens. Each party plays a role in communicating and cooperating with the other parties. Ultimately, it is hoped that El Paso's effort can serve as a model to promote the need for a more sustainable region.

The vision contained in the City's *Livable City Sustainability Plan* (City of El Paso 2008) is 'we balance what we have, what we use, and what we want for today and tomorrow'. This vision is consistent with the Brundtland Commission's definition of sustainability. It focuses on issues surrounding air, community, development, energy, transportation, and waste resources. Each focus area has identified targets to achieve. The City is also cognizant of the fact that the Plan should be fluid and respond to changes in technology and the community. The City must be able to adapt to changes and make adjustments accordingly.

The City of Annapolis, Maryland released a *Sustainable Annapolis Community Action Plan* in 2009. The Plan noted that the City had engaged in many efforts promoting sustainability prior to 2009, like creating a green purchasing ordinance, a recycling program, an energy efficiency task force, a yard composting program, and offering mass transit options to residents and visitors.

It is important to note that Annapolis integrates climate action targets into its sustainability goals. An annual report card is issued describing strategies undertaken to achieve the various targets. Targets are provided for both government reduction targets and community reduction targets. This clearly indicates that sustainability requires actions by all parties, not simply the government. For example, under the climate action goal, strategies to achieve the greenhouse gas emission targets are given for both government and community. Among the strategies offered by government are purchasing 25 percent green energy, replacing current street lights with LEDs, utilizing fuel-efficient low carbon scooters, and creating carpool incentives for workers. Among the strategies for the community to reduce greenhouse gas emissions are purchasing electricity from renewable energy sources, promoting recycling by reducing trash pickup to once a week, and providing low-cost energy audits.

A number of cities around the US have also created separate offices to lead local efforts on sustainability, or advisory bodies to advise city councils on recommended courses of actions. The City of Seattle, long a leader in local sustainability efforts, has promoted climate protection through the development of a climate action plan, promoted the use of clean energy, promoted alternative modes of transportation, promoted recycling and composting, and developed a sustainable buildings and sites policy. It also created an Office of Sustainability and Environment (OSE) under the direction of the mayor to provide leadership and ideas on how to achieve sustainability.

The City of Long Beach, California, passed an ordinance establishing an advisory body called the Sustainable City Commission. Chapter 2.38.020 of the Long Beach Municipal Code details the purpose of the Commission:

> The purpose of the commission shall be to make advisory policy recommendations to the City Council on issues relating to the environment, a sustainable City plan, efforts on programs to address environmental issues such as air quality, water quality, resource conservation relating to the protection and integrity of the natural environment, programs to increase education and awareness of the environment, to serve as a forum for community discussion of the environmental issues, and to encourage input and participation from all sectors of the community on issues of sustainability and the environment.

Mayors have recognized the pivotal role that cities play in promoting sustainability. They recognize the pivotal role cities play in reducing global warming pollution levels in their cities. In 2005, at the seventy-third Annual US Conference of Mayors meeting in Chicago, the mayors endorsed the US Mayors Climate Protection Agreement.

The Agreement called for federal and state governments to develop and implement measures to meet or exceed reducing global warming pollution levels to 7 percent below 1990 levels by 2012. They did not advocate simply reaching minimum targets, they wanted to exceed them. It also endorsed the development and passage of national greenhouse gas (GHG) emissions reduction. Finally, the mayors recommended a variety of actions at the local level to help meet or exceed the Kyoto Protocol targets (additional information on the Kyoto Protocol can be found in Chapter 12). The actions recommended were:

1　Inventory global warming emissions in City operations and in the community, set reduction targets and create an action.
2　Adopt and enforce land use policies that reduce sprawl, preserve open space, and create compact, walkable urban communities.
3　Promote transportation options such as bicycle trails, commute trip reduction programs, incentives for carpooling and public transit.
4　Increase the use of clean, alternative energy by, for example, investing in 'green tags', advocating for the development of renewable energy resources, recovering landfill methane for energy production, and supporting the use of waste to energy technology.
5　Make energy efficiency a priority through building code improvements, retrofitting City facilities with energy-efficient lighting and urging employees to conserve energy and save money.
6　Purchase only Energy Star equipment and appliances for City use.
7　Practice and promote sustainable building practices using the US Green Building Council's LEED® program or a similar system.
8　Increase the average fuel efficiency of municipal

fleet vehicles; reduce the number of vehicles; launch an employee education program including anti-idling messages; convert diesel vehicles to bio-diesel.

9 Evaluate opportunities to increase pump efficiency in water and wastewater systems; recover wastewater treatment methane for energy production.

10 Increase recycling rates in City operations and in the community.

11 Maintain healthy urban forests; promote tree planting to increase shading and to absorb CO_2.

12 Help educate the public, schools, other jurisdictions, professional associations, business and industry about reducing global warming pollution.

As of 2009, over 1,050 mayors had joined the US Mayors Climate Protection Agreement.

Green building regulations

The need to encourage environmentally friendly or environmentally responsible building in the design of homes and buildings is clearly evident in cities throughout the US. Communities have turned to developing and implementing regulations that incorporate healthy, resource- and energy-efficient materials and building methods.

The standard for a rating system that measures building sustainability in the US and around the world is the Leadership in Energy and Environmental Design (LEED®) standard developed by the US Green Building Council. LEED® is a voluntary program that allows communities to determine how 'green' a structure is built. The structures could be individual homes, individual government and commercial buildings, neighborhoods, or communities. They could be newly constructed or in the process of being renovated.

LEED® offers four certification levels: Certified, Silver, Gold, and Platinum. Each level is based on the number of credits or points that are obtained in the following categories: sustainable sites, water efficiency, energy and atmosphere, materials and resources, and indoor environmental quality. Among the attributes

seen in certified buildings are: gardens with drought resistant or native plants and shrubs, cooling towers using recycled wastewater, low-flow toilets, minimization of paved areas, water- and energy-efficient fixtures, use of solar shades, indoor air monitoring systems, use of recycled materials in the construction or renovation process, filtered air quality, and on-site sewage treatment. The more points received by a project, the more 'green' the project is considered. Plate 4 shows the David L. Lawrence Convention Center in Pittsburgh, the only convention center to receive Gold (new construction) and Platinum (existing building: operations and maintenance) ratings. Waste from the original convention center was recycled and used as fill material. The new building uses an environmentally friendly natural ventilation system, practices recycling, and has an on-site water treatment facility.

Another voluntary rating system for the construction and operation of a variety of building types is the method developed by the Building Research Establishment (BRE) and called the BRE Environmental Assessment Method (BREEAM). It pre-dates the establishment of the LEED® standards. It examines the performance of a building in such areas as management, energy use, health and well-being, pollution, transport, land use, ecology, materials, and water consumption and water efficiency.

While BREEAM and LEED® are voluntary green rating systems, California has recently enacted a state-mandated code for green building called CALGreen. It sets minimum green standards for new construction projects. Counties and municipalities can develop more stringent standards.

States and communities throughout the US have realized how the built environment affects areas such as climate change, air quality, and rising energy costs. The phrase 'going green' can be seen in cities of all sizes. Cities are developing sustainable design and construction programs to make buildings more environmentally friendly and more energy efficient. They recognize that such programs can result in environmental benefits, economic benefits, and health benefits. In its 2005 Green Building Regulations, Section 24.15.010, the City of Santa Cruz, California, acknowledges the benefits:

Plate 4 David L. Lawrence Convention Center, Pittsburgh, Pennsylvania
Photo by author

The city finds that green building design, construction and operation can have a significant positive effect on energy and resource efficiency, waste and pollution generation, and the health and productivity of a building's occupants over the life of the building. The second purpose is to create healthy work and living environments increasing the productivity of workers and residents and visitors to the city by improving indoor air quality and lighting.

. . . The city also finds that green design and construction decisions made by the city in the construction and remodeling of city buildings can result in significant cost savings to the city over the life of the buildings. The city also recognizes that it must lead by example in order to have the general populace follow suit and therefore commits itself to the practice of green building for all new and remodeling construction on city owned buildings and structures.

The State of Minnesota developed sustainable building design guidelines in 2004 and required new buildings and buildings undergoing major renovations receiving funding from bond proceeds to meet the guidelines. The legislation was expanded in 2009. In 2005, the State of Washington required new buildings and renovation projects that receive state funding to be constructed to various green building standards. The standards to be followed depended on the type of structure being constructed or renovated.

Recognizing that jurisdictions in California have adopted a variety of green building regulations, the State of California decided to develop a uniform standard that offered consistency throughout the state. The California Green Building Standards Code, also known as the CALGreen Code, seeks to minimize the impact of construction activities through sustainable construction practices in planning and design, energy efficiency, water efficiency and conservation, material conservation and resource efficiency, and environmental quality. In 2010, the provisions found in CALGreen became mandatory for all new public and private structures built in California.

Green building strategies will vary by city. One city may mandate that certain green technologies be used, such as reducing energy consumption. Another city may offer incentives to go green, such as expediting the project review process and getting the necessary building permits. Some cities have adopted LEED® standards for city-owned properties; all new private residential and commercial construction; and for all new single-family attached and unattached residential units.

Washington, DC, established a Green Building Act in 2006 as a part of the city's sustainable development strategy. It decided to phase-in the regulations over a period of time. All public buildings would be required to meet the regulations first, and then all publicly financed buildings. Private construction would then be subject to the regulations. The city was the first city in the US to require privately constructed buildings to meet LEED® standards.

Other cities have adopted various types of green building regulations. For example, Scottsdale,

Plate 5 City of Chicago City Hall green roof
Photo courtesy of the City of Chicago

Arizona's green building program incorporates residential buildings as well as newly constructed and renovated public facilities. The City of Santa Cruz's green building regulations are mandatory for any construction, remodeling, or adding to a building that requires the issuance of a building permit.

Green roofs, also called garden roofs or ecoroofs, are becoming an increasingly popular component of sustainability. They offer opportunities for energy efficiency, serve as a noise buffer, protect a roof from extreme temperatures, serve as a place to grow food, and can be used as an outdoor park space. Plate 5 shows the green roof of the Chicago City Hall Building. Many cities offer incentives to home owners, businesses, and developers to install green roofs. Portland, Oregon, offers an expedited permitting process as well as loans or grants for capital costs associated with developing green roofs. New York, the home of many green roofs, offers a one-year property tax abatement for developing a green roof.

Community gardens

Exactly what is a community garden? Cities may vary in defining the term. Cincinnati's definition of a community garden, as found in its Zoning Ordinance (Section 1401-01-C14), is 'a site operated and maintained by an individual or groups to cultivate trees, herbs, fruits, vegetables, flowers, or other ornamental foliage for the following uses: personal use, consumption, donation or off-site sale of items grown on the site'.

Community gardens have been around for many years. They can be found throughout the US and all continents. Hou, Johnson, and Lawson (2009) indicate that they could be called by many names: allotment gardens, vacant-lot gardens, and garden patches. The creation and use of community gardens has skyrocketed over the years.

Many cities around the world are promoting community gardens as a part of their sustainability efforts. Once thought of as temporary gardens, they are now permanent features in the inner city, the suburbs, and in rural areas (see Plate 6).

Community gardens serve a multiple of purposes. Historian Thomas Bassett (1981) observed that they serve the cultural system in times of social or economic crisis. They can contribute to urban sustainability by creating environmental, social, and economic benefits. They can help eliminate so-called 'food deserts' – areas where fresh fruits and vegetables are not readily available in low-income neighborhoods. Community gardens can also serve in a therapeutic capacity. They are also a part of an area's green infrastructure system and represent an important component of an area's open space system.

The values of community gardens have been widely acknowledged. They have become an important component of a city's fabric. In Seattle, the noted P-Patch program, a program overseen by the Seattle Department of Neighborhoods and a non-profit organization – the P-Patch Trust – notes the widespread benefits of community gardens: growing community, nurturing civic engagement, practicing organic gardening, fostering an environmental ethic and connecting nature to peoples' lives, improving access to local, organic, and culturally appropriate food, transforming the appearance and revitalizing the spirit of their neighborhoods, developing self-reliance and improving nutrition through education and hands-on experience, feeding the hungry, preserving heirloom flowers, herbs, and vegetables, and budding understanding between generations and cultures through gardening and cooking. Recognizing their importance, the APA adopted a Policy Guide on Community and Regional Food Planning that elaborates on the importance of supporting a comprehensive food planning process at the community and regional levels. The Policy Guide indicates that they strengthen local and regional economies by promoting community and regional food, support food systems that improve the health of the region's residents, support food systems that are ecologically sustainable, food systems that are socially equitable and just, systems that preserve and sustain diverse traditional food cultures, and the development of state and federal legislation that facilitates community and regional food planning, including barriers.

The location of community gardens varies by city. Cleveland established an Urban Garden District to

Plate 6 Community garden in Salem, Massachusetts
Photo by author

make sure urban garden areas are appropriately located and protected so that they are of the highest and best use to the community. They could be located on publicly owned sites like schools and parks, on rooftops (possibly as hydroponic gardens), or on vacant lots. The City of Detroit is promoting the creation of community gardens on vacant lots. It is putting the vacant lots into constructive use. In 2011, the City Council of City of Austin, Texas, created a Sustainable Urban Agriculture and Garden Program to streamline the process of establishing community gardens on city land. In 2012, the City of New York adopted amendments to the city's zoning code allowing, subject to certification, the creation of rooftop farms and greenhouses. The Zone

Green, as the rules have been labeled, exempts any greenhouse on residential buildings from height and floor-area limits.

It should come as no surprise that regulations for community gardens also vary by jurisdiction. Commonly found in the regulations are the following:

- obtaining a license to create a community garden;
- hours of operation;
- permitted structures on the site;
- parking for the gardeners;
- water use on the site;
- composting;
- need for organic gardening;

- trash/recycling receptacles;
- sale of produce and plants; and
- prohibited plants.

Some areas, like the states of New York, Illinois, and New Jersey, have recognized community gardens as a permissible use of state and local land. Cities might allow them in all zones where row and field crops are permitted. Another, city like Minneapolis, might find that they represent a permitted use in all zones.

Maintenance issues represent an important concern for gardeners and communities. Failure to properly maintain it could result in the garden becoming neglected and becoming an eyesore. This is troublesome in that creating the garden was meant to replace a vacant lot. To help alleviate this possibility, some areas are requiring periodic assessments/evaluations of the sites. In Chicago, to participate in the Greencorps Program, a program providing assistance to those groups wanting to start and maintain community gardens, groups are required to undergo periodic assessments. In other areas, the responsibilities for maintaining a community garden may be covered by the city and by the individual gardener. For example, in some areas, the city might be responsible for such items as providing water, collecting refuse from the site, and building pathways and exterior fences. Individual gardeners might have to provide their own gardening materials, water the site, and maintain the site.

Further thoughts

There have been a number of factors that have hindered progress on achieving sustainability. One common issue has been that of politics. People might agree on the existence of a problem but fail to come to any consensus on what to do about it. Political differences between developed versus developing countries had hindered progress on achieving sustainability. We have witnessed the role of politics in virtually every area of inquiry.

Economics ranks as another major issue in lack of progress. Some areas seek economic development benefits for some individuals at the expense of the environment and other people. In other words, some groups will benefit from economic development activities while other groups are essentially excluded from reaping any of the benefits. This is the so-called 'rich get richer' dilemma.

There are certainly additional reasons for areas failing to promote or achieve sustainability. Some areas may simply not want to change their current ways of doing things. Other areas pursue their self-interest and ignore the ramifications of their actions on people and neighboring jurisdictions. Ultimately, some areas might even fail to grasp the simple idea that what we do today can greatly affect future generations. Until this idea is acknowledged, any hopes of achieving sustainability will remain a dream.

The quote at the beginning of this chapter says it all – we are the enemy. Our actions today have serious implications for the future. Until this fact is recognized by all countries, the following passage by Anthony D. Cortese (1999: 1) is still applicable:

Human beings and the natural environment are on a collision course. Human activities inflict harsh and often irreversible damage on the environment and on critical resources. If not checked, many of our current practices put at serious risk the future that we wish for human society and the plant and animal kingdoms, and may so alter the living world that it will be unable to sustain life in the manner that we know. Fundamental changes are urgent if we are to avoid the collision our current course will bring about.

Further reading

For various definitions of sustainability, see: http://www.jal.cc.il.us/green/pdfs/definitions_of_sustainability.pdf; http://www.ssfindex.com/sustainability/ notes-and-definitions/; http://www.eco-officiency.com/what_is_sustainability.html; and http://www.ec.gc.ca.

A discussion on the US HUD, US DOT, and the EPA Partnership for Sustainable Communities, can be found at http://www.sustainablecommunities.gov.

The American Planning Association has adopted and updated a Policy Guide on Planning for Sustainability and a Policy Guide on Climate Change. They can be found at http://www.planning.org/policy/guides/pdf/climatic change.pdf and http://www.planning.org/policy/guides/adopted/sustainability.htm.

The International Institute for Sustainable Development is an international public policy research institute for sustainable development. Its mission and activities can be found at http://www.iisd.org.

Discussions on the Leadership in Energy and Environmental Design (LEED®) and the BRE Environmental Assessment Method (BREEAM) can be found at www.new.usgbc.org and http://www.breeam.org.

Various reports from the United Nations have discussed global sustainable development. These reports include: UN (1972) *Report of the United Nations Conference on the Human Environment*; UN (1992) *Report of the United Nations Conference on Environment and Development*; UN (2001) *Johannesburg Summit 2002: World Summit on Sustainable Development*; UN (2002) *UN Report of the World Summit on Sustainable Development*; UN (2012) *Report of the United Nations Conference on Sustainable Development*; and the World Commission on Environment and Development (1987) *Our Common Future*.

Information on Executive Order 13423 can be found in Office of the Federal Environmental Executive (2007) 'FACT Sheet: Executive Order 13423'.

The four reports produced by the President's Council on Sustainable Development are: *Sustainable America: A New Consensus for Prosperity, Opportunity, and A Healthy Environment for the Future* (1996); *Building on Consensus: A Progress Report on Sustainable America* (1997a); *The Road to Sustainable Development: A Snapshot of Activities in the United States of America* (1997b); and *Towards a Sustainable America: Advancing Prosperity, Opportunity, and A Healthy Environment for the 21st Century* (1999).

There is a large literature on sustainability and sustainable development: Campbell (1996) 'Green cities, growing cities, just cities?: urban planning and the contradictions of sustainable development'; Dernbach (2002) *Stumbling Toward Sustainability*; Portney (2003) *Taking Sustainable Cities Seriously*; Wheeler (2004) *Planning for Sustainability: Creating Livable, Equitable and Ecological Communities*; Blackburn (2007) *The Sustainability Handbook: The Complete Management Guide to Achieving Social, Economic, and Environmental Responsibility*; Chifos (2007) 'The sustainable communities experiment in the United States: insights from three federal-level initiatives'; Saha and Paterson (2008) 'Local government efforts to promote the "Three Es" of sustainable development: survey in medium to large cities in the United States'; Dernbach (2009) *Agenda for Sustainable America*; Wheeler and Beatley (2009) *Sustainable Urban Development Reader*; Fitzgerald (2010) *Emerald Cities: Urban Sustainability and Economic Development*; Grodach (2011) 'Barriers to sustainable economic development'; APA Planning Advisory Service (2012) *Incorporating Sustainability into the Comprehensive Plan*; and Elliott (2012) *An Introduction to Sustainable Development*.

A number of publications have been written on LEED®. Among the recent articles are: Jackson (2009) 'How risky are sustainable real estate projects? An evaluation of LEED and ENERGY STAR development options'; Retzlaff (2009) 'The use of LEED in planning and development regulation: an exploratory analysis'; and Meisal (2010) *LEED Materials: A Resource Guide to Green Building*.

For additional information on various aspects of community gardens, see the following: Bassett (1981) 'Reaping the margins: a century of community gardening in America'; Kaufman and Bailkey (2000) 'Farming cities: entrepreneurial urban agriculture in the United States'; Lawson (2004) 'The planner in the garden: a historical view into the relationship between planning and community gardens'; Lawson (2005) *City Bountiful: A Century of Community Gardening in America*; American Planning Association (2007) *Policy Guide on Community and Regional Food Planning*; Hou, Johnson, and Lawson (2009) *Greening Cities Growing Cities: Learning from Seattle's Urban Community Gardens*.

Questions to discuss

1 What is sustainability and why is it important?

2 Describe the importance of the Brundtland Commission.

3 Discuss the conclusions of several of the international conferences on sustainability.

4 What was the purpose of the Clinton Presidential Council on Sustainable Development?

5 How have states and local governments dealt with sustainability issues?

6 What are green building regulations?

7 Discuss the significance of community gardens in planning for sustainability.

8 What roles do community gardens play in community development?

PART 2 LAND USE REGULATION

Planning arose from the need to protect property. Following the tradition of nuisance law, the favored technique since the 1920s has been zoning: the division of a local government area into districts which are subject to differing regulations regarding the use of land and the height and bulk of buildings that are permitted. A major reason for the introduction of zoning was the huge influx of immigrants into the cities (18 million in the thirty years from 1890 to 1920) and their impact on middle- and upper-class neighborhoods.

Chapter 5 examines the evolution of planning and zoning in the US. It summarizes why property protection is needed and our early land use controls. Discussions on the Standard State Zoning Enabling Act and the Standard State Planning Enabling Act are also included in the chapter. The famous 1926 case of *Euclid*, in which the Supreme Court declared zoning to be constitutional, thus paving the way for its rapid spread across urban America, is highlighted in the chapter.

The institutional and legal framework of planning and zoning which developed is discussed in Chapter 6. Much if not most of the land use planning in the United States is not planning but zoning and subdivision control. The former implies comprehensive policies for the use, development, and conservation of land. Zoning represents the division of an area into districts with differing regulations; subdivision is the legal division of land for sale and development. In operating these regulatory controls, local governments have the ability to impose a range of conditions. Most of the discussion of this chapter relates to zoning, but distinctive aspects of subdivision control are dealt with

separately at the end of the chapter. A discussion of the 2009 US Supreme Court decision in *Stop the Beach Renourishment, Inc.* v. *Florida Department of Environmental Protection*, the latest takings decision reached by the US Supreme Court, has been added to the chapter.

Chapter 7 introduces the reader to the Comprehensive Plan. This document contains the vision of a community; the blueprint for its physical development. Among the topics covered in this chapter are the purposes of the plan, its elements, adoption procedures, and other issues associated with the development and implementation of the Comprehensive Plan, and a discussion on recent comprehensive planning efforts.

Moving on from how zoning has developed, and the institutional and legal framework within which it operates, Chapter 8 examines in more detail some of the many 'tools' of zoning which are available to land use planners. The account is confined to the more important ones. Local governments have been known to impose conditions which are considered by developers to be onerous or unreasonable, and thus the courts are involved in settling disputes. Since most issues are dealt with at the state level, there can be a wide variation in judicial opinion on some issues. The case law is immense and sometimes incoherent. Only seldom does the Supreme Court establish a clear lead.

Among the conditions imposed on developers by municipalities are a range of charges and imposts. (Terminology varies, often confusingly: in Chapter 9 the inclusive term 'development charges' is used.) These charges have come about partly in response to increases in the costs of providing infrastructure; they emerged at the same time as municipal budgets began to come under severe pressure. Today, many municipalities are facing declining budgets, unfunded mandates, and possible bankruptcy. Developers can shoulder these development charges only in certain circumstances: when, for instance, market conditions allow them to pass the costs on to buyers, or they can accept a reduced rate of profit, or they can negotiate a lower price for land. Alternatively, a municipality may offer them an incentive or 'bonus' which improves their profitability. There is considerable scope in this area for ingenuity on the part of both municipalities and developers. Discussions on tax increment financing, Capital Improvement Programs, and the use of bonds to finance infrastructure can be found in this chapter. This chapter also raises the complex question of who, in the final analysis, bears the cost of charges.

The evolution of planning and zoning

Urban America was in something of a zoning crisis in the early 1920s. Like a patient who could endure his fever until he suddenly learned that there was a new remedy for it and who was then impatient to be cured, urban America was now sure that it would perish if it did not have zoning.

Scott 1969

The need for property protection

The history of land use controls is as old as history itself. In this chapter, after a short reference to colonial times, a rapid review is given of the foundations and early development of American controls, from the nineteenth-century 'nuisance' cases up to the time of the classic *Euclid* case, which laid upon zoning the imprimatur of the US Supreme Court. The full story of this case is a fascinating one, particularly with the virtually cliff-edge climax of the Supreme Court's deliberations. By way of introduction it is useful to list some of the more important factors which gave rise to zoning.

Public health was a major problem, which grew rapidly as unbelievable numbers of immigrants crowded into cities which were totally unprepared to cater for their basic needs. Technological factors also played a major role: electricity increased the spread of the streetcar suburbs – the escape route of the middle class from the horrors of the insanitary and congested city. But they themselves contributed to this congestion. Even more so did the two technological innovations of the steel frame and the elevator, which made towering skyscrapers both possible and practical (Goldberger 1981: 5). Central city uses intensified as the middle class sought semi-rural respite by new means of transport. Later the wizardry of Henry Ford escalated problems of traffic congestion to huge dimensions.

Other changes were in progress or in the wind: widespread regulation of election procedures, and the reform movement, which was aimed at securing sound, engineering-type solutions to problems of municipal administration. Even 'planning' was debated as a rational solution to the problems of the city. This started as a City Beautiful movement, but soon changed its character into a concern for the City Efficient. Neither got very far: they were, then as now, too long term ('visionary' was the word) for practical men.

But one problem above all demanded attention: the safeguarding of the new suburbs from the blight which had stimulated their development. The solution was found in the extension of the law of nuisance to land uses, by way of zoning. Zoning provided long-term security against change: industry, garages, apartments, corner shops – indeed, anything which might threaten the sanctity of the single-family-dwelling suburb – could now be excluded. In Mel Scott's words: 'zoning was the heaven-sent nostrum for sick cities, the wonder drug of the planners, the balm sought by lending institutions and householders alike. City after city worked itself into a state of acute apprehension until it could adopt a zoning ordinance' (Scott 1969: 192). While few might understand what 'planning' involved, the protection provided by zoning was immediately apparent; and it spread at an incredible speed.

Early land use controls

As already noted, land use controls have a long history in the United States. Moreover, there were few problems with the taking of land for public purposes. Land was in abundance: so much so that questions of compensation hardly arose. Undeveloped land was perceived to be in such plentiful supply as to have no significant value. However, where developed, improved, or enclosed land was physically acquired, compensation was normally payable. The power of eminent domain was accepted as an inherent power of government for which specific legislation was not required. The taking issue, which became of such importance later, received scant attention. Indeed, there is a paucity of evidence on the reasons why the taking clause became a part of the Constitution.

Matters changed dramatically with the adoption of the Constitution and the Bill of Rights, particularly when (under John Marshall) the Supreme Court claimed the singular power to determine the consti-tutionality of legislative acts. So far as the taking issue was concerned, it was accepted by both the federal and the state courts that a regulatory action could not involve a taking. The term 'taking' was applied only to the physical acquisition of land by government – an approach encapsulated in the phrase: 'no taking without a touching'. Where the use of property was restricted by regulatory controls, no compensation was payable. This was so even if landowners were deprived of all use of their land, as is illustrated by an 1826 case involving a cemetery in New York City. Land which had originally been in the country well outside the urban area had been conveyed to the City for a church and cemetery. Over time, the City grew and surrounded the cemetery. A bylaw, passed by the City, which prohibited cemetery use was appealed by the cemetery. Since it was generally believed at the time that burying the dead produced unhealthy vapors, the court held that it would be extremely unreasonable to endanger the public by the cemetery's use, despite the terms of the lease. In such cases, since the physical property (as distinct from the property rights) had not been invaded, no compensation was appropriate.

In a much later case, that of *Mugler* v. *Kansas*, decided by the US Supreme Court in 1887, the court ruled that a Kansas Act rendered Mugler's brewery virtually worthless (Box 5.1). (There must have been numerous

---

BOX 5.1 REGULATION IS NOT A TAKING

In *Mugler* v. *Kansas*, decided by the US Supreme Court in 1887, Mugler's brewery was made virtually worthless by a Kansas Act which prohibited the manufacture and sale of intoxicating liquor. Mugler still retained his premises and could use them for any legal purpose – that is, excluding the formerly legal brewery use! In its judgment, the court argued:

> there is no justification for holding that the State, under the guise merely of police regulations, is here aiming to deprive the citizen of his constitutional rights; for we cannot shut out of view the fact, within the knowledge of all, that the public health, the public morals, and the public safety, may be endangered by the general use of intoxicating drinks . . . A prohibition simply upon the use of property for purposes that are declared, by valid legislation, to be injurious to the health, morals, or safety of the community, cannot, in any sense be deemed a taking or an appropriation of property for the public benefit.

Source: *Mugler* v. *Kansas* 1887

BOX 5.2 USE OF THE POLICE POWER

In the exercise of the police power, the uses in a municipality to which property may be put have been limited and also prohibited. Thus, the manufacture of bricks; the maintenance of a livery stable; a dairy; a public laundry; regulating billboards; a garage; the installation of sinks and water closets in tenement houses; the exclusion of certain business; a hay barn, wood yard or laundry; a stone crusher, machine shop or carpet beating establishment; the slaughter of animals; the disposition of garbage; registration of plumbers; prohibiting the erection of a billboard exceeding a certain height; regulating the height of buildings; compelling a street-surface railroad corporation to change the location of its tracks; prohibiting the discharge of smoke; the storing of oil; and generally, any business, as well as the height and kind of building, may be regulated by a municipality under power conferred upon it by the legislature.

Source: Lincoln Trust Co v. Williams Building Corporation 1920

Muglers in the United States during the prohibition years.) The court recognized the limits of the police power by advancing a 'harm/benefit' test, where it held that if government acts to prevent a harm to the public health, safety, and welfare, then it is an exercise of the police power (see Box 5.2). Accordingly, no compensation will be provided to the property owner. On the other hand, if government acts to obtain various benefits by appropriating or taking private property for a public use, then it is an exercise of the power of eminent domain. Government is then constitutionally required to pay the property owner just compensation.

There were many such cases of the use of the police power. One further important example can be given here: the 1915 case of *Hadacheck* v. *Sebastian*. Hadacheck had owned and operated a brickworks in the open countryside since 1902; but in the following years residential development spread, and the area was annexed by the City of Los Angeles. There had never been any previous attempt to regulate it. The brickworks now became a nuisance to the local inhabitants, and the city passed an ordinance which effectively prohibited Hadacheck from continuing to operate his brickworks (which gave the land a value of $800,000), though he could use it for other purposes (value $60,000). Hadacheck claimed that the ordinance deprived him of the use of the property and that a taking of private property had occurred. The court held that 'vested interests' could not be asserted against the ordinance because of conditions which previously existed: 'To so hold would preclude development and fix a city forever in its primitive condition. There must be progress, and if in its march private interests are in the way they must yield to the good of the community.' In the court's view, absent a clear showing that the government had acted in bad faith, the ordinance was a proper exercise of the police power.

Underlying these regulations was the English common law concept of nuisance, which held that no property should be used in such a manner as to injure that of another owner. These were largely 'negative' instruments, but gradually land use controls developed into more positive tools of planning. For instance, in 1867 San Francisco passed an ordinance which prohibited the building of slaughterhouses, hog storage facilities, and hide-curing plants in certain districts of the city. Though clearly in the tradition of nuisance law, the ordinance was notable because it was 'preventive rather than after the fact and restricted land uses by physical areas of the city'; it thus 'set the stage for further evolution of land use zoning in the United States' (Gerckens 1988: 26). Such cases increased as the problem of urbanization escalated at a phenomenal rate.

The movement for planning

Planners in the first two decades of the twentieth century had few tools with which they could retune the urban system. Zoning, however, was one tool which offered great promise: it had a particular appeal which extended beyond those whose essential concern was with planning (and who saw zoning merely as an instrument of planning) (see Box 5.3). The crucial feature of zoning, then as now, was its utility in excluding unwanted neighbors.

There was one major difficulty: it was unclear whether zoning would be accepted by the courts as constitutional. That the fears were justified was clearly illustrated several years later, after the passing of the New York Ordinance, by the rejection of the *Euclid* ordinance by the lower court (noted later in this chapter). Considerable effort and skill were employed by planners and lawyers in drafting ordinances which would stand judicial scrutiny. The battle — the word is appropriate — was between those who saw zoning as 'a protection of the suburban American home against the encroachment of urban blight and danger', and those who saw it as 'the unrestrained caprice of village councils claiming unlimited control over private property in derogation of the Constitution' (Brooks 1989: 7). However, the first major zoning ordinance emerged in New York, where the forces in favor of zoning were exceptionally strong.

The New York zoning ordinance of 1916

The 1916 New York City zoning ordinance is usually regarded as the first comprehensive zoning ordinance in the United States. It was the successful outcome of an open campaign to stop changes that were taking place on Fifth Avenue. It was a war on two fronts: one between carriage trade merchants and the invading garment industry, the other between wealthy residents and the invading retail trade. In Toll's words: 'If this was war of sorts, it was in truth a double war: garment manufacturers fighting retail merchants fighting wealthy residents. The entire conflict was much closer in spirit to social Darwinism than to the Geneva Convention. There were no rules and only one objective, survival by any means' (Toll 1969: 110). But it was the encroachment of the Jewish garment makers and their immigrant workers which formed the central issue. Property values fell by a half in the five years up to 1916 (Feagin 1989: 81).

A Commission on Heights of Buildings reported in 1913, and recommended that height, area, and use should be regulated in the interests of public health and safety, and that the regulations should be adapted to the varying needs of the different districts – a radical innovation. This reflected the increasing criticism that tall buildings cast shadows and deprived surrounding properties of sunlight and air. The city and the state legislature accepted these proposals and the city

BOX 5.3 THE ATTRACTION OF ZONING

Nothing appeared so destructive of urban order as garages and machine shops in residential areas, or loft buildings in exclusive shopping districts, or breweries amid small stores and light manufacturing establishments. Nothing caused an investor so much anguish as the sight of a grocery store being erected next door to a single family residence on which he had lent money. Nothing made whole neighborhoods feel so outraged and helpless as the construction of apartment houses when the private deed restrictions expired and there was no zoning to prevent vacant lots from being used for multifamily structures. Zoning was the heaven-sent nostrum for sick cities, the wonder drug of the planners, the balm sought by lending institutions and householders alike.

Source: Scott 1969: 192

charter was amended to include 'districting' provisions. In 1916, a comprehensive zoning code was adopted for the whole city. The code also separated incompatible land use such as factories from residential areas and the encroachment of industrial uses on the office and department store district.

There was another difference between the new zoning controls and the well-established police power regulation of buildings and factories (see Box 5.4). Whereas the latter was intended to solve existing problems and to promote health and safety, zoning applied only to new development. So far as Fifth Avenue was concerned, further incursion by the garment industry was preventable, but nothing could be done about the changes already brought about, at least not through zoning itself: political action was another matter. Existing uses were hallowed as 'nonconforming': 'In any building or premises any lawful use existing therein at the time of the passage of this resolution may be continued therein, although not conforming to the regulations of the use district in which it is maintained.' Thus, ironically, the very problem which gave rise to the zoning ordinance remained untouched. The reason is not far to seek: there was so much concern that the newfangled zoning system would reduce property values that the com-

mission was most anxious to allay the fears. Indeed, huge areas were zoned for business and industrial use. Protection of 'investments' was, and remains, a major objective of zoning.

Successful though the campaign for the New York ordinance appeared, it was, in one crucial respect, a dismal failure: in contrast to the hopes of the proponents of the planning movement, it lacked any 'planning' component. It was a substitute for a plan: it was concerned with protecting existing property interests rather than with providing for future needs. Nevertheless, it was rapidly copied by numerous cities throughout the United States.

But the most significant indication of progress was the appointment, by Secretary of State Herbert Hoover, of the Advisory Committee on Building Codes and Zoning. This committee drafted a Standard State Zoning Enabling Act (SSZEA), which rapidly became the model for a large number of zoning ordinances (see Box 5.5).

The Standard State Zoning Enabling Act

Zoning was a part of the scientific management movement which was sweeping America in the first quarter

---

BOX 5.4 THE NOVELTY OF ZONING (1916)

The novel feature of zoning as distinguished from building code regulations, tenement house laws, and factory laws was that suitable regulations for different districts were established. We have become so accustomed to zoning regulations that it is difficult to understand how fixed the popular notion was that all land should be regulated in the same way throughout a municipality. On this account imposing different regulations on different areas appeared to many to be a discriminatory, arbitrary, and therefore an unlawful invasion of private rights. To counteract this impression it was considered important that the regulations within each district should be uniform for the same kind or class of buildings. A provision to this effect was placed in the original zoning clauses of the charter of New York City, and there can be no doubt that courts which early passed upon these regulations were to a considerable extent persuaded to favor them on account of this requirement of uniformity. If it had been possible to make different regulations for the same sort of buildings in different parts of the same district, it is unlikely that zoning would have received the court approval that it now has.

Source: Bassett 1940: 26

BOX 5.5 PURPOSES OF THE STANDARD STATE ZONING ENABLING ACT

1 For the purpose of promoting health, safety, morals, or the general welfare of the community, the legislative body of cities and incorporated villages is hereby empowered to regulate and restrict the height, number of stories, and size of buildings and other structures, the percentage of lot that may be occupied, the size of yards, courts, and other open spaces, the density of population, and the location and use of buildings, structures, and land for trade, industry, residence or other purposes.

2 For any or all of said purposes the local legislative body may divide the municipality into districts of such number, shape and area as may be deemed best suited to carry out the purposes of this act; and within such districts it may regulate and restrict the erection, construction, reconstruction, alteration, repair, or use of buildings, structures, or land. All such regulations shall be uniform for each class or kind of building throughout each district, but the regulations in one district may differ from those in other districts.

BOX 5.6 ZONING: A NEW SYSTEM OF ORDER

In the search for a new order to the American city, the division of land uses and regulations restricting building heights and bulk became tactical rearrangements . . . Zoning, it was claimed, embodied and exemplified the idea of orderliness in city development; it encouraged the erection of the right building, in the right form, in the right place. 'What would we think of a housewife who insisted on keeping her gas range in the parlor and her piano in the kitchen?' Yet these were commonplace anomalies in the American city of the 1920s: gas tanks next to parks, garages next to schools, boiler shops next to hospitals, stables next to churches, and funeral parlors next to dwelling houses.

Source: Boyer 1983: 155

of the twentieth century. Herbert Hoover was an important figure on this stage. He was instrumental in creating a new area of federal responsibility: one which Christine Boyer has termed the 'cooperative state'. Central to this was scientific study of the facts (and the collection of scientific data), and the establishment of 'a central clearinghouse for social and economic reforms'. Boyer documents some of the areas in which Hoover applied this philosophy; these included the standardization of industrial parts and of the plans, designs, materials, and structural elements of houses; and the coordination of information concerning the housing market. Zoning clearly fell into this kind of thinking (see Box 5.6).

Hoover's philosophy was that the role of the state was not to interfere with market forces, but to make them more efficient by, for example, facilitating the production of better market information, advancing the acceptance of standardization, and (in the area of housing and urban development) assisting with the introduction of a system for orderly development which would be safe as an investment for both lenders and borrowers. Zoning was seen as the instrument for providing the necessary security against both unwanted development and legal challenge. In particular, it provided protection to home owners from uncongenial neighboring uses which would affect both amenity and market value.

The SSZEA gave state legislatures 'a procedure, based upon an accepted concept of property rights and careful legal precedent, for each community to follow' (Boyer 1983: 164). A crucial element in the rationale here was the belief that a single legal code would pass legal muster in a way which a multiplicity of individual local ordinances would not. A carefully crafted ordinance, based on this universal model and embodying the fruits of planning expertise, supported by local citizens, would provide a defensible framework for an extension of the hard-to-define limits of the police power.

The SSZEA contained eight sections. The first section contained the grant of power that enabled local governments to regulate land use (see Box 5.7). Meck (1996: 1) describes enabling legislation in the following manner:

Enabling legislation is a mechanism by which a state delegates its inherent police power authority, which includes the power to plan and zone, to local government. It permits the local governments to do something, but in a certain way and through certain mechanisms. Sometimes, the local government will already have this power delegated directly to it through the state constitution (as in a municipal home rule provision) or through a statutory grant of power by the state legislature (this also known as statutory or legislative home rule).

Section 2 allowed localities to divide the land area into districts. Section 3 required that any regulations be developed in conformance with a comprehensive plan. Section 4 mandated that any proposed regulation or restriction should go before a public hearing prior to becoming effective. Section 5 acknowledged the fact that regulations and restrictions could be amended, modified, or repealed and discussed what to do if the public protested against any change. Section 6 called for the creation of an appointed zoning commission that would recommend changes and conduct public hearings over zoning matters.

Section 7 discussed the creation of a board of adjustment that would hear any appeals that might arise. Section 8 authorized localities to enforce the SSZEA through the creation of zoning ordinances.

The model Act was hugely popular. The first edition, published in 1924, became a best-seller, with sales of more than 55,000 copies. Within a year, nearly a quarter of the states had passed enabling acts which were modeled substantially on the Standard Act.

And so, though planning languished, zoning boomed: by 1926, forty-three of the (then forty-eight) states had adopted zoning enabling legislation; some 420 local governments, containing nearly a quarter of the population, had adopted zoning ordinances, and hundreds more were in process of preparing them (Mandelker and Cunningham 1990: 166). By 1929, 754 local governments had adopted zoning ordinances: these contained about three-fifths of the urban population of the country (Hubbard and Hubbard 1929: 166).

The list of purposes of zoning set out in the Standard Act (see Box 5.7), which is constantly paraded before the courts, omits the one which is by far the most important: the exclusion of unwanted people or uses, and thus the preservation of the status quo. These exclusionary objectives are seldom much below the surface, even when they are not explicit. Of course, all zoning is exclusionary; by definition zoning excludes *some* uses. The one exception (now rarely used) is where a zone is 'unrestricted'. Thus, in the 1916 New York ordinance, areas not zoned as residential or business were unrestricted. The village of Euclid also had an unrestricted zone, as does any similar 'cumulative' zoning system. Euclid provided, first, for an exclusively single-family house zone. The second use zone provided additionally for two-family houses; the third further included apartment houses; the fourth offices and shops; and so on until the final zone could accommodate all uses (and was therefore, in effect, an unrestricted zone). In each zone, development could take place that accorded not only with its specific categorization but also with all 'higher' uses. Under this system, a single-family house could be built in any zone, but elsewhere only the uses specified for the particular zone *and* all higher zones were permitted.

There is occasional confusion over the relationship between zoning and land use. Land use sets the stage for development by classifying property – open space,

BOX 5.7 THE PURPOSES OF ZONING

The Standard State Zoning Enabling Act listed the following purposes of zoning:

- to lessen congestion in the streets;
- to secure safety from fire, panic, and other dangers;
- to promote health and the general welfare;
- to provide adequate light and air;
- to prevent the overcrowding of land;
- to avoid undue concentration of population;
- to facilitate the adequate provision of transportation, water, sewerage, schools, parks, and other public requirements.

residential, commercial, or industrial. Zoning lays out the specific standards for developing the property. It regulates private property and makes sure the uses are compatible with each other.

The *Euclid* case

Despite the growing popularity of comprehensive zoning, it was not until 1926 that the Supreme Court dealt with its constitutionality. In that year, in the *Euclid* case, the Supreme Court upheld the constitutionality of the Euclid zoning ordinance. Some important background information will illustrate how the court reached its decision.

The City of Cleveland was undergoing a period of industrial expansion. Its industrial development was quickly approaching a portion of the village of Euclid. In 1922, Euclid adopted a comprehensive zoning plan. The village's 12 square miles of land area was divided into six use districts, three classes of height districts, and four classes of area districts. The U-1 district was 'restricted to single family dwellings, public parks, water towers and reservoirs, suburban and interurban electric railway passenger stations and rights of way, and farming, non-commercial greenhouse nurseries and truck gardening'. The U-2 district allowed the same uses as did the U-1 district but now allowed two-family dwellings. In other words, if district A allowed six uses,

district B allowed the same uses, but with an additional seventh use.

An analogy to a pyramid might illustrate what has become known as 'Euclidian Zoning'. The top or tip of the pyramid allows only a few exclusive uses. Those are the only uses allowed at the top. As we descend from the top to the bottom, the uses allowed become cumulative to where all of the uses allowed at the upper parts of the pyramid are allowed at the base of the pyramid.

Ambler Realty owned sixty-eight acres of land that it wanted to develop. The village rejected Ambler's proposal. Ambler objected to the decision by claiming the village was blocking the natural course of industrial development by directing it to other, 'lesser suited sites'. Ambler felt the decision destroyed the value of its property and artificially increased the value of other property.

Ambler challenged the passage of Euclid's zoning ordinance, arguing that it was not done for a public purpose and therefore was not a valid use of the police power:

The ordinance does not, in fact, pursue any rational plan, dictated by considerations of public safety, health and welfare, upon which the police power rests. On the contrary, it is an arbitrary attempt to prevent the natural and proper development of the land in the village prejudicial to the public welfare.

This property in the interest of the public welfare, should be devoted to those industrial uses for which it is needed and most appropriate.

The question facing the court was indeed difficult. It needed to distinguish between a legitimate use of the police power and an illegitimate use of the police power. The court noted that Euclid was a separate municipality with its own powers to govern itself. This included the ability to decide, through the public policy-making process, where industrial development would occur within its boundaries. Cleveland had no power in this case to decide for Euclid. The court did not, however, rule out the possibility of cases 'where the general public interest would so far outweigh the interest of the municipality that the municipality would not be allowed to stand in the way'.

The question then became whether or not Euclid could create residential use districts where commercial uses, including hotels and apartments, were excluded. The court acknowledged the need for zoning to adapt to the changing conditions and changing needs of society and the fact that the zoning ordinance must be related to the public health, safety, and welfare. To support its beliefs, the court acknowledged the findings of numerous earlier reports on zoning:

These reports, which bear every evidence of painstaking consideration, concur in the view that the segregation of residential, business, and industrial buildings will make it easier to provide fire apparatus suitable for the character and intensity of the development in each section; that it will increase the safety and security of home life; greatly tend to prevent street accidents, especially to children, by reducing the traffic and resulting confusion in residential sections; decrease noise and other conditions which produce or intensify nervous disorders; preserve a more favorable environment in which to rear children, etc.

The US Supreme Court failed to be persuaded by Ambler's arguments. It denied Ambler's attempt to have the court stop Euclid from enforcing the zoning ordinance because it was arbitrary and capricious. The court had put its seal of approval on comprehensive zoning. This represented a significant extension of the police power in that it enabled a municipality to prohibit uses which were not 'nuisances' in the strict sense of the term. In particular, shops, industry, and apartments were excluded from single-family zones. Apartments in particular were greatly feared by home owners:

Once a block of homes is invaded by flats and apartments, few new single family dwellings ever go up afterwards. It is marked for change, and the land adjoining is forever after held on a speculative basis in the hope that it may all become commercially remunerative, generally without thought for the great majority of adjoining owners who have invested for a home and a home neighborhood only.

(Cheney 1920)

The *Euclid* decision had the important social implication that apartment living could be a 'use' category for the purposes of land use planning. Thus another dimension was added to the exclusionary nature of land use regulation. Box 5.8 provides a sampling of major zoning-related US Supreme Court cases.

The narrowness of zoning

Initially, zoning was concerned essentially with 'districts'. Though lip-service was paid to comprehensive plans (which were supposed to form the rational basis for the operation of zoning) this did not amount to much in practice. The term 'comprehensive' came to mean little more than a zoning provision which covered all or most of the districts in a local government area. However, as we shall see later, the overriding concern with exclusion led to zoning policies which were concerned with the whole of an area. In this way, the term 'comprehensive' took on a new meaning: safeguarding the status quo of a neighborhood was simply writ large.

It is important to stress that zoning is an inherently rigid instrument. This remains so in spite of the extraordinary ingenuity which has been displayed in

BOX 5.8 SAMPLING OF MAJOR ZONING-RELATED US SUPREME COURT CASES

Village of Euclid v. *Ambler Realty Co.*, 272 US 365 (1926)
Nectow v. *City of Cambridge*, 277 US 183 (1928)
Goldblatt v. *Town of Hempstead*, 369 US 590 (1962)
Village of Belle Terre v. *Boraas*, 416 US 1 (1974)
City of Eastlake v. *Forest City Enterprises, Inc.*, 426 US 668 (1976)
Moore v. *City of East Cleveland*, 431 US 494 (1977)
Agins v. *Tiburon*, 447 US 255 (1980)
City of Cleburne v. *Cleburne Living Center, Inc.*, 473 US 432 (1985)
City of Renton v. *Playtime Theatres, Inc.*, 475 US 41 (1986)

adapting it to the real, moving world; and in this rigidity lies its enormous popular appeal. The planning ideal of flexibility is anathema to protectionist home owners. Rigidity provides a degree of certainty and security. But zoning is not planning: it is a restricted instrument for districting.

Complaints over zoning have been voiced for years. Immigrants coming into the United States were attracted to ethnic enclaves in cities across the country. Many moved into apartments or homes with extended families. Other, unrelated individuals from the same countries lived together because individually they couldn't afford the rent. Overcrowding of residential units in which immigrants resided was soon perceived

to be a problem. Cities that favored or embraced zoning appeared to have large numbers of immigrants. In fact, many researchers today have questioned and continue to question whether or not racial segregation was and continues to be an unstated goal of zoning. Individuals feel the zoning process favors some people (the affluent) over other people (the poor and minorities being two groups). In addition, many people feel zoning is too complex and contains too much 'legalese'. They don't understand definitions and zoning procedures. Many feel it is too restrictive and fails to take into consideration the free market system. Clawson (1971: 253) provides a unique take on the problems of the zoning process (see Box 5.9).

BOX 5.9 PROBLEMS WITH ZONING

The zoning process, and the role of the zoning authorities, is like the conquest of a lady of easy virtue. She must be approached by the right man, properly dressed, bringing suitable gifts, and the language must follow established patterns; but in the end, she yields, as everyone, including the lady herself, knew from the beginning that she would. If one retains the right lawyer, dresses up his rezoning proposal in attractive language, perhaps makes a gift of land for schools or parks or otherwise appeases some local opposition, properly emphasizes the employment, the result is really not in doubt. The costs of such concessions and gifts, and the delays of getting favorable action, are less predictable and may prove onerous. But it seems clear that local zoning in an expanding urban or suburban area is not really an effective barrier to most kinds of development.

Source: Clawson 1971: 253

There has been a lot of discussion on the relationship between zoning and sprawl. Zoning has separated incompatible land uses, causing people to have to drive to get to shopping centers, grocery stores, etc. Low density development has led to the spreading out of suburbs. In the past, everything needed its own zone.

The Standard City Planning Enabling Act

Once the Standard State Zoning Enabling Act was completed, Herbert Hoover's Advisory Committee turned its attention to developing a city planning enabling act. This would lay the basic foundation for planning. The Act (p. 4) would 'provide for city and regional planning; the creation, organization, and powers of planning commissions; the regulation of subdivision of land and the acquisition of right to keep planned streets free from buildings; and providing penalties for violation of this act'. State legislatures would grant the authority to plan to the cities. The Standard City Planning Enabling Act (SCPEA) was published in 1928. In the Foreword of the Act, Hoover (p. iii) spoke of the importance of planning:

In several hundred American cities and regions planning commissions are working with public officials and private groups in order to obtain more orderly and efficient physical development of their land area. They are concerned partly with rectifying past mistakes, but more with securing such location and development of streets, parks, public utilities, and public and private buildings as will best serve the needs of the people for their homes, their industry and trade, their travel about the city, and their recreation. The extent to which they succeed affects in no small degree the return, in terms of practical usefulness now and for years to come, of several hundred million dollars to taxpayers' money spent each year for public improvements, as well as the value and serviceability of new private construction costing several billion dollars each year.

Knack *et al.* (1996: 6) noted that the SCPEA covered six important subjects: the organization and power of the planning commission, which was directed to prepare and adopt a 'master plan'; the content of the master plan for the physical development of the territory; provision for adoption of a master street plan by the governing body; provision for approval of all public improvements by the planning commission; control of private subdivision of land; and provision for the establishment of a regional planning commission and a regional plan.

Under Section 6 of the Act, the planning commission has the power and duty to make and adopt a master plan which is 'a general design of the city's development, so that development may take place in a systematic, coordinated, and intelligently controlled manner'. It goes on to state that the general design 'may be broadly classified as dealing with (a) streets, (b) other types of public grounds, (c) public buildings, (d) public utilities, and (e) development of private property (zoning)'. The design would represent a broad and comprehensive view of the area. It is interesting to note that in a footnote in Section 6, the meaning of a master plan is 'a comprehensive scheme of development of the general fundamentals of a municipal plan and that an express definition has not been thought desirable or necessary'.

Section 8 of the Act discusses the procedures under which the planning commission will operate. The process for adopting the plan or amendments to the plan is discussed. In order to provide for public input, the planning commission must hold at least one public hearing. The hearing must be advertised with the notice of time and place in a newspaper of general circulation in the municipality.

Title 2 of the Act covers the area of subdivision control. This title covers the jurisdiction the planning commission has over the subdivision of land within and outside the municipality. It also notes that no subdivision of property can be filed until it has been approved by the planning commission. Title 2 goes on to discuss the content of subdivision regulations and their adoption.

Title 3 discusses the areas of regional planning and planning commissions. It noted that the planning commission of a municipality or a county commission can request the governor of the state to appoint a regional planning commission. The organization of a regional planning commission, its powers and duties (including the development and adoption of a master plan for the region), and the legal status of the master plan are contained in Title 4.

Further reading

Because New York was one of the first cities to create a comprehensive zoning ordinance, the New York planning department (http://www.nyc.gov/html/dcp) is proud of its history. A short description of the original 1916 ordinance is on its website (http://www.nyc.gov/html/dcp/html/zone/zonehis.shtml) along with definitions of zoning and its use in the city today.

The full text and a brief description of the impacts of the Standard State Zoning Enabling Act and the Standard City Planning Enabling Act can be found on the American Planning Association's website (http://www.planning.org/growingsmart/enablingacts.htm).

Examples of current zoning practices can be found on many government websites. The Washington District of Columbia has an Office of Zoning (http:// dcoz.dcgov.org) that offers updates and mapping applications that allow landowners to find their properties in the specified zone. Community groups also become involved in zoning practices, such as the Florida Planning and Zoning Association (http://www.fpza.org). This is an example of a statewide non-profit organization which facilitates the cooperation of planning and zoning boards with local citizens.

There is some fascinating reading on the *Euclid* case, which involved far more than appears at first sight: a major battle on the desirability and legality of zoning raged behind the scenes. The fullest and most accessible account is given in Haar and Kayden (1989a) *Zoning and the American Dream*, but see also Toll (1969) *Zoned American*, and McCormack (1946) 'A law clerk's recollections'; and Flack (1986) '*Euclid* v. *Ambler*: a retrospective'. A brief overview is given in Scott

(1969) *American City Planning since 1890*. See also, Wolf (2008) *The Zoning of America: Euclid v. Ambler*.

For an interesting discussion of the *Euclid* decision, its historical and legal context, and the current state of zoning and land use law, see Bogart's 'Symposium on the seventy-fifth anniversary of *Village of Euclid* v. *Ambler Realty Co.*', in *Case Western University Law Review* 51 (Summer, 2001). The following contributions comprise the Symposium: Lee (2001) 'Introduction'; Chused (2001) 'Euclid's historical imagery'; Korngold (2001) 'The emergence of private land use controls in large-scale subdivisions: the companion story to *Village of Euclid* v. *Ambler Realty Co.*'; Durchslag (2001) '*Village of Euclid* v. *Ambler Realty Co.*, seventy-five years later: this is not your father's zoning ordinance'; Callies and Tappendorf (2001) 'Unconstitutional land development conditions and the development agreement solution: bargaining for public facilities after Nollan and Dolan'; and Bogart (2001) '"Trading places": the role of zoning in promoting and discouraging intrametropolitan trade'.

More generally, in addition to Haar and Kayden, a succinct account of the 'Historical development of American planning' is Gerckens (1988). For an interesting group of articles on planning the American city, see Sies and Silver (1996) *Planning the Twentieth-Century American City*. Also recommended is Boyer (1983) *Dreaming the Rational City: The Myth of American City Planning*. A view of California's planning history can be found in Pincetl (1999) *Transforming California: A Political History of Land Use and Development*.

For discussion on race in planning and zoning, see Silver (1991) 'The racial origins of zoning: southern cities from 1910–1940'; Thomas and Ritzdorf (1997) *Urban Planning and the African-American Community*; and Barry (2001) 'Land use regulation and residential segregation: does zoning matter?'

Questions to discuss

1   Describe the police power; what does it have to do with zoning?

2   Discuss the constitutionality and fairness of the Mugler and Hadacheck cases.

3   In what ways was zoning unique?

4   What are 'standard state enabling acts'? Why were they important?

5   Why did New York pass a zoning ordinance in 1916? Why did it have implications nationwide?

6   Consider the fairness of the Euclid decision.

7   'Zoning is not planning.' Discuss.

6

The institutional and legal framework of planning and zoning

American cities seldom make and never carry out comprehensive plans. Plan making is with us as idle exercise, for we neither agree upon the content of a 'public interest' that ought to override private ones nor permit the centralization of authority to carry out a plan into effect if one were made.

Banfield 1961

Planning and zoning

All fifty states have passed legislation enabling municipalities (and often counties) to plan and operate zoning controls. The enabling legislation provides the basic legislative foundation for planning and zoning. Most is based on the Standard State Zoning Enabling Act (SSZEA) issued by the Department of Commerce in the mid-1920s. Planning and zoning are exercises of the police power: the inherent power of a sovereign government to legislate for the health, welfare, and safety of the community. The Constitution confers the police power upon the states, which in turn delegate it to the local governments.

States grant the power to the local legislative body to accomplish certain functions like creating a planning agency, engaging in comprehensive planning, developing regulations and procedures, and creating a planning commission. The State of New York provided for the creation of city planning agencies through the State General Municipal Law of 1913. In the State of Texas, the State Local Government Code provides enabling legislation to create councils of governments. In Louisiana, the enabling legislation provides parishes and communities with the power for planning, zoning, and subdivision regulation.

Paradoxically (in view of what was said earlier about the distinctions between planning and zoning), another section provides that zoning regulations 'shall be made in accordance with a comprehensive plan'. In fact, zoning was conceived (at least by planners, if not by lawyers) as a tool of planning. But generally the part became the whole, and (with notable exceptions considered later) practice does not follow the text of the Act.

The phrase 'consistent with a comprehensive plan' has been the focus of many policy and legal discussions over the years. Some states require zoning to be consistent with a comprehensive plan. For example, California, in California Government Code Section 65350, requires cities and counties to adopt general plans and requires the zoning ordinances of those local governments to be consistent with the general plan. Furthermore, under California Government Code Section 65860(a):

A zoning ordinance would be considered consistent with a city or county general plan only if both of the following conditions are met: (1) the city or county has officially adopted such a plan; and (2) the various land uses authorized by the ordinance are compatible with the objectives, policies, general land uses, and program specified in the plan.

Arizona has a similar requirement regarding zonings and rezonings. It also provides guidance when there is any uncertainty regarding a proposed rezoning:

In the case of uncertainty in construing or applying the conformity of any part of a proposed rezoning ordinance to the adopted general plan of the municipality, the ordinance shall be construed in a manner that will further the implementation of, and not be contrary to the goals, policies and applicable elements of the general plan. A rezoning ordinance conforms with the land use element of the general plan if it proposes land uses, densities or intensities of the land use element of the general plan.

(Arizona Revised Statutes, 9-462.01(F))

Consistency is, however, rarely defined. And, even when it is defined, the machinery for enforcing it is generally weak. There are some exceptions, particularly in states that have developed a strong growth management policy. These are discussed at length in Chapter 10. For the most part, however, 'in accordance with a comprehensive plan' does not mean what the words suggest; instead, it means that zoning should be carried out comprehensively rather than in a piecemeal manner. In some states, the requirement has come to mean little more than that the zoning laws shall be reasonable! Moreover, where there is a separate comprehensive plan, it is the zoning ordinance that usually carries the force of law, not the plan. The judgment in a 1987 Maryland case captures the essence of the matter (see Box 6.1). Nevertheless, an increasing use is being made of comprehensive plans (or master plans, or general plans: the terms are used interchangeably) by both local governments and the courts. Zoning decisions are much easier to defend before the courts if a strong planning framework can be demonstrated.

Zoning as a local matter

It is important to appreciate that zoning in the United States is essentially a *local* matter. Even the decision on whether to operate a zoning system is usually a local one. Some localities have highly sophisticated zoning systems; some have none at all. But however complex a zoning system may be, it typically remains what it always has been: 'a process by which the residents of a *local* community examine what people propose to do with their land, and decide whether or not they will permit it' (Garner and Callies 1972: 305).

The distinction between the ideal of planning and the reality of zoning is an important one. Planning is concerned with the long-term development (or preservation) of an area and the relationship between local objectives and overall community and regional goals. Zoning represents a major instrument of this; but it is more. Indeed, it has taken the place of the function to which it is supposedly subservient. One of the reasons for this is that responsibility for land use controls has been delegated to the lowest level of local government. These local authorities have traditionally been concerned with attracting development to their areas but, since the 1970s, there has been increasing pressure from electors for their communities to be preserved as they are (or at least safeguarded from unwelcome uses such as industry, apartments, and low-income housing). The

---

BOX 6.1 COMPREHENSIVE PLANS AND ZONING

Comprehensive plans . . . represent only a basic scheme generally outlining planning and zoning objectives in an extensive area, and are in no sense a final plan; they are continually subject to modification in the light of actual land use development and serve as a guide rather than a strait jacket . . . The zoning as recommended or proposed in the master plan may well become incorporated in a comprehensive zoning map . . . but this will not be so until it is officially adopted and designated as such by the District Council.

Source: *West Montgomery County Citizens Association v. Maryland National Capitol Park and Planning Commission* 1987

powers of zoning provide a very effective tool for this – a tool that can be wielded with a skill that thwarts judicial action. The contrast with planning is a sharp one: a comprehensive plan would deal not only with the needs of the existing inhabitants of an area, but also with its role in meeting the needs for housing newcomers, whatever their income or color. Additionally, it would make provision for such undesirable land uses as power stations, landfills, and a host of other uses which have given rise to the acronym NIMBY ('not in my back yard') and its more recent progeny, NIMTOO ('not in my term of office'). This can be, and is, done by some local governments; but there are many more that employ zoning as a means of precluding comprehensive planning. Some flavor of the action at the local level is given in the 1991 report of the Advisory Commission on Regulatory Barriers to Affordable Housing, more popularly known as the NIMBY report (see Box 6.2). This often-cited report suggested that 'excessive regulation' has reduced the supply of affordable housing in the United States. Downs (1991) has noted that the various regulatory barriers raise housing costs in three ways: direct restrictions on housing supply, direct cost increases, and delay-causing requirements. These barriers continue to be the subjects of many heated debates throughout the United States.

The local managers of zoning

Local governments carry out their zoning and planning powers within the framework of powers conferred on them by the individual states, either by constitutional home rule authority or by a specific enabling Act. There are thus fifty different systems of local government – which fortunately it is not necessary to analyze here. What has to be said, however, is that though some states exercise varying degrees of control over local governments, most do not.

The fifty states contain 89,476 local governmental units as of 2012. Over half of these are school districts and other 'special districts' for particular functions such as natural resources, fire protection, and housing and community development. In 2012, there were 3,033 counties, 19,492 municipalities, and 16,519 townships (see Box 6.3). The growth in the number of local government units has led to problems of intergovernmental (between different levels of government) communications and coordination, and intragovernmental (between agencies within the same level of government) communications and coordination.

The variation among states is exemplified by a few statistics from the *2012 Statistical Abstract*. Of the municipal governments, 273 have populations of 100,000 or more, while 1,542 have populations of fewer than 10,000. Illinois, with 1,298 such governments has more municipalities than any other state; Texas has 1,214, and Pennsylvania 1,015. At the other extreme are states such as Connecticut and Massachusetts that have fewer than 50 municipalities. The Massachusetts Community Preservation Act, signed in 2002, is a recent piece of statewide enabling legislation that allowed cities and towns to exercise control over local planning decisions. Voters in over 145 communities had voted to adopt it. Other communities continue to vote on it.

BOX 6.2 NIMBY

In addition to lobbying elected officials, NIMBY groups regularly participate in the regulatory process through vocal input at public forums and hearings dealing with land use and development issues. Unlike the strict rules governing judicial proceedings, many localities have no specific rules regarding who can testify at public hearings or what rules of evidence apply. Participants often represent ad hoc groups that coalesce around a particular development issue. They can be very effective at packing hearing rooms and leaving the impression that public opinion is strongly against whatever project they oppose.

BOX 6.3 UNITS OF GOVERNMENT

Type of government	Number of units
TOTAL	89,527
US Government	1
State governments	50
Local governments	89,476
County	3,033
Municipal	19,492
Township	16,519
School district	13,051
Special district	37,381

Source: US Census Bureau, *Statistical Abstract of the United States*, 2012

Finding the appropriate local department, agency, or division that is responsible for planning can be problematic. The planning function could be housed within a Department of City Planning, a Department of Planning and Community Development, a Redevelopment Department, a Planning and Building Department, a Community Development Department, a Development Services Department, an Economic and Community Development Department, or a Community Planning and Building Department. It is also possible that the planning function could be housed as a division within another department.

Organized county governments are common, though their powers and functions vary. Twenty states have 'townships' that have powers similar to those of municipalities, except that their boundaries are defined without regard to the concentration or distribution of population. Counties play a role in zoning and planning in parts of the country, though the nature of this differs widely.

The doctrine of the separation of powers is an important feature of the US system of government. In brief (and therefore ignoring deviations from the normal rule) a zoning ordinance is passed by the legislative body (e.g., a municipality); applications for rezoning or variances are reviewed by an independent commission (the planning or zoning commission/

board); and appeals are to a board of adjustment (or appeals), and sometimes to the legislative body, and finally – on legal or constitutional grounds – to the normal courts. Furthermore, the role for discretion is severely limited (in theory at least). Indeed, zoning was originally conceived as being virtually 'self-executing': the zoning ordinance (the written regulations) and the zoning map would spell out the permitted land uses in such clarity and detail that there would be little room for doubt or discretion. Thus 'policy' is seen firmly as the responsibility of the legislative body, while the commission deals with its execution through the issuance of permits and the occasional variance or exception.

The names of bodies having powers over planning and zoning issues vary by state. These bodies could be named board of adjustment, planning commission, planning and zoning commission, or any other name. They are generally constituted by the local legislative bodies such as the city council, town council, board of aldermen, county commission, or board of supervisors. They can be of different sizes, members can be either appointed or elected, and the number of terms an individual can serve also varies. State law and municipal charters establish their roles and responsibilities. Functions could range from advising city councils on plans, goals, and policies affecting an area's physical development, subdivision review, design review, preparing Capital Improvement Programs, preparing comprehensive plans, to ensuring consistency with an area's adopted comprehensive plan. Moreover, new issues or circumstances may arise that call for their functions to be expanded. For example, the Philadelphia City Planning Commission has broadened its functions to include such areas as human service delivery, housing policy, economic development, and urban design. Non-physical development issues such as the above are being considered by virtually all planning commissions throughout the United States. The decision-making bodies and officials responsible for zoning administration and enforcement in the City of Norfolk, Virginia, can be found in Box 6.4.

Conflict of interest

Appointed or elected public officials are required to make decisions using their best judgment. Their judgment cannot be subject to outside influences. As such, conflict of interest statements have become increasingly important in the wake of various political scandals. The State of Florida Statutes, Title 10, Section 112.3143 contains language on voting conflicts for county, municipal, other local public officials, or appointed public officers. Section 112.3143(4) states:

> No appointed public officer shall participate in any matter which would inure to the officer's special gain or loss; which the officer knows would inure to the special private gain or loss of any principal

---

BOX 6.4 ZONING ADMINISTRATION AND ENFORCEMENT DECISION-MAKING BODIES AND OFFICIALS IN NORFOLK, VIRGINIA

18–2 City Council

a. To initiate amendments to the text of this ordinance and to the zoning map.
b. To consider and adopt, reject, or modify amendments to the text of this ordinance and to the zoning map, including conditional zoning amendments as defined herein.
c. To approve or disapprove special exceptions.
d. To take such other actions not delegated to other bodies which may be desirable and necessary to implement the provisions of this ordinance.

18–3 Planning Commission

a. To initiate amendments to the text of this ordinance and to the zoning map.
b. To prepare and recommend a general plan for the physical development of the city.
c. To make comprehensive surveys and studies of the existing conditions and trends of growth and the probable future requirements of the city and its residents as part of the preparation of the general plan.
d. To make, as part of the general plan, long-range recommendations regarding the general development of the city.
e. To review the general plan once every five years to determine whether it is advisable to amend the plan.
f. To review, evaluate, and make comments and recommendations to the city council on proposed amendments to this ordinance.
g. To review, evaluate, and make comments and recommendations to the city council on special exceptions.

18–4 Board of Zoning Appeals

a. To hear and decide appeals from any order, requirement, decision or determination made by the zoning administrator in the administration or the enforcement of this ordinance.
b. To authorize variances from the terms of this ordinance pursuant to the procedures and standards for variances set forth in Article IV, Chapter 22.

18–5.2 Norfolk Design Review Committee

The Norfolk design review committee reviews applications for certificates or appropriateness in Historic and Cultural Conservation Districts . . . the committee has an advisory role with regard to downtown development certificates in certain circumstances.

18–6.4 Zoning Administrator

Primary responsibility for administering and enforcing this ordinance is delegated to the director of the department of city planning and codes administration. Except as otherwise specifically provided in this ordinance, the director may designate a staff person to carry out these responsibilities. The staff person to whom such administrative and enforcement functions are assigned is referred to in this ordinance as the 'zoning administrator.' The zoning administrator shall maintain appropriate records of administrative and enforcement actions.

Sources: City of Norfolk, Virginia (2007), Code of Ordinances, Article IV, Chapter 18, Sections 2, 3, 4, 5.2, and 6.4

by whom he or she is retained or to the parent organization or subsidiary of a corporate principal by which he or she is retained; or which he or she would inure to the special private gain or loss of a relative or business associate to the public officer, without first disclosing the nature of his or her interest in the matter.

In Snohomish County, Washington, the Planning Commission Bylaws Article XI: Conflict of Interest indicates: 'If it shall appear to any member at any time that a conflict of interest may arise which could embarrass the integrity of the Commission, it shall be the member's duty to openly state the nature of such conflict, and shall then refrain from any subsequent Commission participation, deliberation or voting on the subject matter for which conflict arises.'

The principle of due process mandates that impartiality be present. Bias cannot come into play. Favors cannot be granted to anyone. As the previous paragraph indicates, financial considerations cannot be taken into consideration. California's Political Reform Act of 1974 – Government Code Section 81001(b) – sums up a great deal of the issue. It requires that 'public officials, whether elected or appointed, should perform their duties in an impartial manner, free from bias caused by their own financial interests and the interests of persons who supported them'. California Government Code Section 87100 provides further statutory rules for elected or appointed officials – 'no public official at any level of state or local government shall make, participate in making or in any way attempt to use his official position to influence a government decision in which he knows or has reason to know he has a financial interest'.

As previously indicated, 'policy' is usually a matter only for the local government: no higher level of government is generally involved. The courts hear appeals against local decisions and therefore, in one sense, act as a type of policy-imposing body. However, policy enters into the courts' deliberations only to the extent that they do or do not defer to the legislative judgment of municipalities: what is termed the 'presumption of validity' (or 'judicial deference'). The courts are not concerned with planning policy issues but with legal and constitutional matters. As will become apparent, this neat division between policy and law does not work in practice.

The constitutional framework

Both the federal and the state constitutions include provisions which are binding on municipalities. One of the most important of these is the protection of property rights. The Fifth Amendment to the Constitution provides: 'nor shall private property be taken for public use without just compensation' (see Box 6.5). The 'taking issue' (alternatively known as the 'just compensation issue') is at the heart of the major problem facing zoning: when does the exercise of the police power over land use constitute such an infringement of the property right as to become a 'taking'? The crucial matter, of course, is the definition of a 'taking'.

An enormous amount of thought, effort and scholarship has been applied to this question, yet the position is not clear. Postponing fuller discussion until later, here it suffices to note that (to quote the famous words of Justice Holmes in the 1922 case of *Pennsylvania Coal Company* v. *Mahon*) 'the general rule . . . is, that while property may be regulated to a certain extent, if regulation goes too far it will be recognized as a taking'. However, this is not very helpful since we are still left with the puzzle as to where the dividing line is between zoning decisions which are acceptable and those which go 'too far'. In truth, there is none: the Supreme Court has taken the view that (as with the question of obscenity) no generally applicable

---

BOX 6.5 CONSTITUTIONAL PROTECTIONS

Amendment V

No person shall be held to answer for a capital, or otherwise infamous crime, unless on a presentment or indictment of a Grand Jury, except in cases arising in the land or naval forces, or in the militia, when in actual service in time of war or public danger; nor shall any person be subject for the same offense to be twice put in jeopardy of life or limb; nor shall be compelled in any criminal case to be a witness against himself, nor be deprived of life, liberty, or property, without due process of law; nor shall private property be taken for public use without just compensation.

Amendment XIV

All persons born or naturalized in the United States, and subject to the jurisdiction thereof, are citizens of the United States and of the State wherein they reside. No State shall make or enforce any law which shall abridge the privileges or immunities of the United States; nor shall any State deprive any person of life, liberty, or property, without due process of law; nor deny to any person within its jurisdiction the equal protection of the laws.

BOX 6.6 THE TAKING ISSUE – THE *PENN CENTRAL* CASE

The question of what constitutes a 'taking' for the purposes of the Fifth Amendment has proved to be a problem of considerable difficulty. While this Court has recognized that the 'Fifth Amendment's guarantee . . . [is] designed to bar Government from forcing some people alone to bear public burdens which, in all fairness and justice, should be borne by the public as a whole', this Court, quite simply, has been unable to develop any 'set formula' for determining when 'justice and fairness' require that economic injuries caused by public action be compensated by the Government, rather than remain disproportionately concentrated on a few persons. Indeed, we have frequently observed that whether a particular restriction will be rendered invalid by the Government's failure to pay for any losses proximately caused by it depends largely 'upon the particular circumstances [in that] case'.

definition is possible, and each case must be decided upon its merits. The classic statement on this was made in the 1978 *Penn Central* case (see Box 6.6).

The Fifth Amendment also includes what is termed 'the public use doctrine': that property can be 'taken' only for a public use. The interpretation of this doctrine has changed significantly in recent decades (illustrating the changes that can take place in the constitutional framework). Until the early 1950s, it was conservatively interpreted as meaning that property which was taken had to be literally used by a public body. It could not be taken for a joint public-private venture, and still less for a private use. Short shrift was made of this in the 1954 *Berman* v. *Parker* case, where it was held that the public purchase of a slum area and its leasing for redevelopment by private enterprise constituted a public use. The court went further: in magisterial terms it declared that the public use requirement of the Constitution was 'coterminous with the scope of a sovereign's police powers'. It also declared that the concept of public welfare was so broad that it could encompass aesthetic matters: 'it is within the power of the legislature to determine that the community should be beautiful as well as healthy, spacious as well as clean, well-balanced as well as carefully patrolled'.

Later cases have further extended 'the public purpose'. One particularly well-known and controversial case was the clearance of the Poletown neighborhood of Detroit, a community of mostly elderly, retired Polish American immigrants, for the purpose of accommodating a new General Motors plant. The city, faced with the prospect of General Motors moving out of the area, condemned some 465 acres of land and conveyed it on favorable terms to GM. In 1981, the Supreme Court of Michigan, in *Poletown Neighborhood Council* v. *Detroit*, ruled that

> the power of eminent domain is to be used in this instance primarily to accomplish the essential public purpose of alleviating unemployment and revitalizing the economic base of the community. It would benefit a small portion of the public. The benefit to a private interest is merely incidental. The new factory led to the destruction of 1,021 homes and apartment buildings, 155 businesses, churches and a hospital, displaced 3,500 people, and all but obliterated a more or less stably integrated community embodying a century of Polish cultural life.
>
> (Hill 1986: 111)

Other relevant constitutional provisions require that land use controls be operated by 'due process': the Fourteenth Amendment states that 'no person . . . shall be deprived of life, liberty, or property without due process of law'. The due process clause applies both substantively (is the action legitimate?) and procedurally (is it administered fairly?). Substantive due process requires that controls serve a legitimate governmental interest such as public health, safety, and general welfare. (A zoning ordinance which excluded low-income families could be challenged on substantive due process grounds.) Procedural due process requires that fair and proper procedures are followed in relation, for example, to public notice of, and hearings on, zoning ordinances. It further requires that an ordinance be clear and specific: a property owner must be able to ascertain what he may or may not do with his property. If the ordinance is not clear, it can be challenged as being 'void for vagueness'. For example, a provision that authorized a planning commission to permit development on criteria which 'include but are not limited to' those set out in the provision would be void since it would allow the commission to consider unspecified, alternative criteria.

The Fourteenth Amendment also provides that no state 'shall deny to any person within its jurisdiction the equal protection of the laws'. An ordinance which involves racial considerations clearly denies equal protection. However, as so often with zoning matters, cases are often not at all clear. Unequal results can be obtained by devious mechanisms such as the prohibition of multi-family dwellings and the imposition of large minimum lot sizes or large minimum dwelling sizes, or even the regulation of laundries.

The role of the courts

Recourse to the courts is a marked feature of the American system of government. As de Tocqueville noted 150 years ago, 'there is hardly a political question in the United States which does not sooner or later turn into a judicial one'. Constitutional safeguards can transform a small administrative matter into a major judicial issue. It is therefore not surprising that the courts play a major role in the land use planning process. The role is, moreover, an 'active' one: decisions change over time in the light of changing economic and social conditions, and also the political complexion of the court. The Supreme Court of the United States, as its name suggests, is the final arbiter; but it does not stand alone. There are over a hundred federal courts, and each of the fifty states has its own system of courts, including a State Supreme Court. Decisions at the state

level stand unless overturned by the US Supreme Court. There are important implications of this. First, until a matter is settled by the US Supreme Court (and few cases reach this level), the law can differ among the states. At the extreme, it is theoretically possible for there to be fifty different interpretations of a legal issue. This is particularly important in land use planning since the majority of zoning cases are dealt with at the state level. As a result, judgments may vary considerably across the country.

Another issue on the role of the courts needs to be made here. Their function is to ensure that municipalities are acting in a constitutional manner. It is not their role to act as a 'super board of adjustment' or 'planning commission of last resort'. There is a 'presumption of validity' in the actions of a municipality to which the courts give 'judicial deference'. This is nicely illustrated by a Missouri case (*City of Ladue* v. *Horn* 1986). In the zoning ordinance of this city, a family is defined as 'one or more persons related by blood, marriage or adoption, occupying a dwelling unit as an individual housekeeping organization'. The case concerned two unmarried adults who were living, along with their teenage children, in a single-family zone. Clearly, they offended the zoning ordinance, but was the ordinance constitutional? Certainly, concluded the court: the city had seriously considered the matter and had come to a decision that was within their competence (see Box 6.7).

Closely related is the 'fairly debatable' concept: this holds that if a decision is a matter of opinion, i.e. open to fair or reasonable debate, the court cannot and will not substitute its judgment for that of the responsible legislative body. Zoning has long been held a legislative matter in which municipalities have historically been given broad discretion in zoning matters. The role of the courts is not to sit in judgment on the wisdom of a local government's legislative actions: that is the function of the political process. Courts are not zoning boards or planning commissions. The judicial role is circumscribed. Typically, the court can overrule a legislative body only if its actions are shown to be clearly arbitrary, capricious, illegal, discriminatory, and unreasonable. In short, 'a court does not sit as a super zoning board with power to act de novo, but rather has,

BOX 6.7 PRESUMPTION OF VALIDITY

The stated purpose of Ladue's zoning ordinance is the promotion of the health, safety, morals and general welfare in the city. Whether Ladue could have adopted less restrictive means to achieve these goals is not a controlling factor in considering the constitutionality of the zoning ordinance. Rather, our focus is on whether there exists some reasonable basis for the means actually employed. In making such a determination, if any state of facts either known or which could reasonably be assumed is presented in support of the ordinance, we must defer to the legislative judgment. We find that Ladue has not acted arbitrarily in enacting its zoning ordinance which defines family as those related by blood, marriage or adoption. Given the fact that Ladue has so defined family, we defer to its legislative judgment.

Source: *City of Ladue v. Horn* 1986

in the absence of alleged racial or economic discrimination, a limited power of review' (Wright and Gitelman 1982: 527).

While this is the traditional view, there is no doubt that the local zoning process is frequently subject to irresistible pressure, and decisions are often taken which serve narrow interests. Rezonings, for example, are often made in defiance of the policy enshrined in the zoning ordinance. Some state courts have held that zoning decisions can be administrative rather than legislative in character, i.e. they constitute the *application* of policy as distinct from the *making* of policy. This is particularly so where ad hoc decisions are taken on rezoning. Where this is held, there is no presumption of validity (which applies only to legislative acts), and the court requires to be satisfied that the rezoning is needed in the public interest. The state which has been particularly aggressive on this matter is Oregon, and a few states have followed its lead – but most have not. If this seems confusing, that is because it is! There is no generally accepted way of distinguishing between legislative and administrative decisions.

Without in any way denying the importance of constitutional issues in land use regulation, it is important to note that they normally operate as a backcloth to local decision-making, rather than being on the front line. Constitutional issues may arise at any time (often unexpectedly), but it is easy to be misled about their primacy. The voluminous legal textbooks on land use planning contribute to this incorrect impression. The very size of these texts is at least in part due to the fact that the courts frequently differ among themselves or refuse to clarify principles which planning authorities can follow. Readers have to digest the cases (which, thanks to the writers of the textbooks, are reduced to manageable length) and try to establish how the particular issues in which they are interested might be treated.

Legal texts are misleading also in that they give the impression that constitutional and legal matters are all-important in land use planning. In fact, legal issues are normally in the background, and their influence on local governments is typically limited. The proportion of cases that reach the courts is very small – though there is always a danger of this happening where an aggrieved person has the time and money to embark on a legal challenge. But this is (statistically) unusual; and the general experience of those who come into contact with land use planning is of a bureaucratic rather than a constitutional nature.

Kelo v. New London

The issues of whether a city could use eminent domain to assemble property for an economic development plan and whether a city's taking of property for the purpose of economic development constituted the Fifth Amendment's 'public use' requirement were center stage in *Kelo* v. *New London* (545 US 469) in 2005.

In 2000, the City of New London, Connecticut approved an economic development plan that would provide multiple benefits to the city and its residents. These benefits included more jobs and increased tax revenues. Ultimately, the proposed plan would result in the revitalization of the area. Having experienced numerous years of economic decline, the City of New London had the dubious distinction of having been designated a 'distressed municipality' by a state agency a decade earlier.

State and local officials decided on the site of a former US military facility, Fort Trumbull, as the site for the city's economic development activities. The New London Development Corporation (NLDC), a local non-profit development corporation, was named the city's development agent. The economic development efforts were helped when Pfizer Inc., a major pharmaceutical company, announced that it would build a research facility on property adjacent to the proposed Fort Trumbull site.

The NLDC drafted a development plan for 90 acres of Fort Trumbull and submitted it to state agencies for their review. After review by the state agencies, the city authorized the NLDC to purchase the property to carry out the economic development project. It purchased some properties from willing property owners. Unfortunately, not all property owners in the proposed area wanted to sell their properties. At this point, the city chose to invoke its power of eminent domain for the properties it couldn't purchase from the residents.

Not all properties in the project area were considered blighted. Some properties had to be condemned, a decision that angered a number of property owners. One of the property owners, Susette Kelo, had made major improvements in her home and argued that the taking of her property by the NLDC would violate the Fifth Amendment's 'public use' restriction. The New London Superior Court prohibited the taking of some properties in the proposed project area. Both sides of the argument appealed the Superior Court's decision to the Supreme Court of Connecticut. On appeal, the Supreme Court of Connecticut held the city's actions were valid because the proposed taking of private property was authorized by state statute.

The Supreme Court's decision was appealed to the US Supreme Court. One question that had to be decided was whether or not a city's decision to take private property for economic development purposes satisfies the Fifth Amendment's 'public use' requirement. In addition, the Court had to rule on whether or not New London's economic development plan served a public purpose.

Justice Stevens delivered the opinion of the US Supreme Court. He noted that a carefully formulated economic development plan served a public purpose by providing numerous benefits to the community. As such, since the plan serves a public purpose, the 'public use' requirement of the Fifth Amendment is satisfied. In essence, 'public use' was determined to be synonymous with 'public purpose'. He went on to indicate that New London used a State Statute that authorized the use of eminent domain to help promote development.

According to Justice Stevens, the promotion of economic development has long been an accepted function of government and serves a public purpose. The court chose not to 'second guess' the city's economic development plan and its choice of properties needed to carry out the plan. It recognized that some property owners would suffer from the city's decisions. Ultimately, the court chose to decide that eminent domain could be used to take private property from one person and give the property to another under the guise of economic development.

Justice Kennedy concurred with the court's opinion that any taking can be considered valid if it is rationally related to a conceivable purpose. He continued to note, 'while there may be categories of cases in which the transfers are so suspicious, or the procedures employed so prone to abuse, or the purported benefits are so trivial or implausible, that the courts should presume an impermissible private purpose, no such circumstances are present in this case' (545 US 469, 493).

Justice O'Connor disagreed with the court's decision and wrote a dissenting opinion. Justice O'Connor felt the court was going back on a long-held limitation of government power. She claimed that 'under the guise of economic development, all private property is now vulnerable to being taken and transferred to another private owner, so long as it might be upgraded' (545 US 469, 494). She did not agree that one person's property could not be taken because a new owner could make the property more profitable. Private property could not be taken from one person for the benefit of another person.

According to Justice O'Connor, a line must be drawn between 'public' and 'private' use. Were the takings in this case convincing enough to allow a 'public purpose' taking to meet the public use requirement? She felt the economic development takings in this case were unconstitutional. To her, the court broadened the definition of 'public use' by determining that 'the sovereign may take private property currently put to ordinary private use, and give it over for new use, ordinary private use, so long as the new use is predicted to generate some secondary benefit for the public – such as increased tax revenue, more jobs, maybe even esthetic pleasure' (545 US 469, 501).

The court's ruling would undoubtedly confuse the public. Property owners had to question whether their property could be condemned for economic development purposes. Justice O'Connor believed that the court's ruling would cause any private property to be taken for the benefit of another private party, as long as the new use generated some secondary benefit for the public. In this case, the individuals that would benefit from the economic development efforts at Fort Trumbull possessed more influence and power in the process than the individuals whose properties were taken under the guise of economic development. This, according to Justice O'Connor, is not what our Founding Fathers had in mind when the US Constitution was written.

Justice Thomas also wrote a dissenting opinion in which he felt the court went against common sense by indicating an economic development project is a 'public use'. He didn't feel that the court truly examined the history and original meaning of the 'public use' clause and urged it to go back and examine earlier cases and the original meaning of the clause. He agreed with Justice O'Connor that if a taking for an economic development purpose constitutes a 'public use', then any taking would constitute a 'public use'. The consequences of the court's decision bothered Justice

Thomas. He claimed that 'allowing the government to take property solely for public purposes is bad enough, but extending the concept of public purpose to encompass any economically beneficial goal guarantees that these losses will fall disproportionately on poor communities' (545 US 469, 521). This is totally unacceptable. In the end, Justice Thomas felt, 'when faced with a clash of constitutional principle and a line of unreasoned cases wholly divorced from the text, history, and structure of our founding document, we should not hesitate to resolve the tension in favor of the Constitution's original meaning' (545 US 469, 523). He voted to reverse the decision of the Connecticut Supreme Court.

The US Supreme Court's decision in *Kelo* v. *New London* generated a great deal of discussion. It sparked a nationwide debate. Cries of government abuse of the power of eminent domain were heard throughout the country. Action was swift. President George W. Bush issued an Executive Order on June 23, 2006 on protecting the property rights of American people. He proclaimed that:

It is the policy of the United States to protect the rights of Americans to their private property, including the taking of private property by the Federal Government to situations in which the taking is for public use, with just compensation, and for the purpose of advancing the economic interest of private parties to be given ownership or use of the property taken.

Specific exclusions including the taking of private property by the Federal Government for the purpose of acquiring abandoned property, acquiring ownership or use by a public utility, and public ownership or exclusive use of the property by the public, were identified in Section 3 of the Executive Order.

In 2006, state legislatures started debating legislation that restricts the power of eminent domain. Responses were immediate. After all, as the Kelo decision indicated, states are free to develop legislative responses to the decision. In South Dakota, legislation was passed prohibiting government from taking private property by eminent domain and giving it to another person, non-government entity, or a public-private business entity. The State of Florida prohibited taking for a private benefit but still allows government to condemn land for traditional public uses. Colorado enacted legislation which required governments to prove that condemnation of a piece of property was necessary for getting rid of blight. Moreover, it noted that tax revenue and economic enhancement did not constitute blight. Alabama also passed legislation that prohibited cities and counties from using eminent domain for any private development activity or for enhancing tax revenue. A loophole in the legislation that allowed property to be condemned under blight law and afterwards turned over to a private interest was later closed. North Dakota passed legislation in 2007 regarding eminent domain and later voters passed a property rights ballot measure which amended Section 16 of Article 1 of the North Dakota Constitution. The measure, which passed by a vote of 68 percent to 32 percent, indicated that 'the taking of private property for a public use or purpose does not include public economic development benefits'. It also indicated that 'private property could not be taken for private benefit unless necessary for conducting a common carrier or utility business'. States such as Arkansas, New York, New Jersey, Massachusetts, and Mississippi have failed to pass meaningful legislation regarding eminent domain.

Stop the Beach Renourishment, Inc. v. Florida Department of Environmental Protection

Four years after it decided *Kelo*, another major case reached the US Supreme Court. This case examined whether or not beachfront property owners in the City of Destin and Walton County were due just compensation because of a Florida Supreme Court's decision that a renourishment project funded under the Florida's Beach and Shore Preservation Act did not take property without just compensation in violation of the Fifth and Fourteenth Amendments.

Many coastal areas in Florida suffer great devastation from hurricanes. Beaches are eroded and homes and

businesses are destroyed. Recognizing the devastating impacts that hurricanes can cause, Florida enacted a state statute in 2007 that provided financial assistance to areas wanting to repair and replenish/renourish their beaches. The city of Destin and Walton County applied for such assistance.

The State of Florida owns the land between the low-tide line and the mean high-water line or mark, with the high-water line representing the boundary between private littoral property and public land. The State has the responsibility and duty to protect public lands under the public trust doctrine. Shoreline property owners own the shore land that is not submerged under water. Essentially, this means that private property owners possess title to the lands west of the beach renourishment areas and government owns the land east of the erosion control line.

The State Legislature in Title XI, Chapter 161, Section 161.053 of the Beach and Shore Preservation Act clearly states the importance of preserving beaches:

> The Legislature finds and declares that the beaches in this state and the coastal barrier dunes adjacent to such beaches, by their nature, are subject to frequent and severe fluctuations and represent one of the most valuable resources of Florida and that it is in the public interest to preserve and protect them from imprudent construction which can jeopardize the stability of the beach-dune system, accelerate erosion, provide inadequate protection to upland structures, endanger adjacent properties, or interfere with public beach access. In furtherance of these findings, it is the intent of the Legislature to provide that the department establish coastal construction control lines on a county basis along the sand beaches of the state fronting on the Atlantic Ocean, Gulf of Mexico, or the Straits of Florida. Such lines shall be established so as to define that portion of the beach-dune system which is subject to fluctuations on a 100-year storm surge, storm waves, or other predictable weather conditions.

Section 161.088 authorized and funded beach restoration and nourishment projects. The Department of Environmental Protection was vested with the authority to implement the program. Destin and Walton County applied for permits seeking to restore some 6.9 miles of beach that had been damaged and eroded by hurricanes. Financial assistance was available from the state to help restore the damaged beaches. A later section of the Act, 161.141, indicated that before the commencement of any restoration project, the line of mean high water for the area to be restored had to be established. This line represented a jurisdictional line or boundary between State sovereignty and private lands. A beach erosion plan is developed and, once approved, an erosion line is set. Moreover, the additions made as a result of the project were subject to a public easement for public access to the beach. It was not the intent of the Legislature to deprive property owners of any use their properties. In Section 161.191, the Act discussed the vesting of title to the lands:

> Title to all lands seaward of the erosion control land shall be deemed to be vested in the State by right of its sovereignty, and title to all lands landward of such line shall be vested in the riparian upland land owners whose lands either abut the erosion control line or would have abutted the line if it had been located directly on the line of mean high water on the date the trustees' survey was recorded.

A group of beachfront property owners that bordered the proposed project area formed a non-profit corporation called Stop the Beach Renourishment, Inc. to challenge the proposed project. The challenge was denied and the Department of Environmental Protection approved the necessary permits for the proposed project. The group then challenged the permits in the state court. The Florida District Court of Appeal for the First District ruled that several of the property owners' littoral property rights had been eliminated. Furthermore, it felt the property owners suffered an unconstitutional taking as a result of the government's action. It put aside the Department's final order approving the issuance of the permits. It also asked the Florida Supreme Court to determine whether the Act deprived upland property owners of their littoral property rights without just compensation. The Florida Supreme Court reversed the Court of Appeal's

decision, finding that the action didn't violate the Fifth and Fourteenth Amendments' property rights of the property owners and affirmed the decision to restore sand to the eroded beach.

In 2009, the US Supreme Court heard the case and reached an 8–0 vote to affirm the decision of the Florida Supreme Court. It reasoned that if no established right of private property existed, no 'taking' had occurred as a result of the state assuming ownership of the land newly created as a result of the beach restoration project. Property law has held the executive or legislative branches of government can take private property through regulatory takings or through the power of eminent domain. Just compensation to the property owners would be required.

Although agreeing with the Court's decision, several Justices acknowledged the issue of whether or not a court's decision dealing with pre-existing property rights of landowners violated the Fifth and Fourteenth Amendments. Does a judicial decision equate to the same degree as an executive or legislative decision? This decision of whether a court would have to pay just compensation to property owners as a result of its ruling has never been decided by US Supreme Court. It would represent a 'judicial taking' – a decision that takes or transfers rights away from a property owner and gives them to the government. This becomes an issue, since to whom would the property owners turn for recourse from a court's decision? Justices Kennedy, Sotomayor, Breyer, and Ginsburg held that the issue wasn't pertinent to this specific case and that it represented a question to be answered in a later case.

Further reading

A bipartisan organization, the National Conference of State Legislatures (NCSL) (http://www.ncsl.org), is a legislative advocate which conducts research and creates an exchange of ideas between the legislators in all fifty states. Because eminent domain is a national and local issue, NCSL follows the issue closely. Its website (http://www.ncsl.org/programs/natres/EMINDOMAIN.htm) provides an overview of eminent domain and tracks current legislation and court cases related to the topic.

The NCSL is countered by citizen activist groups against eminent domain, such as the Castle Coalition (http://www.castlecoalition.org), a project of the Institute for Justice. The Castle Coalition acts as a government watchdog with news, stories of eminent domain abuse, and a legislative 'report card' which rates local government's use of the eminent domain. Their motto is 'citizens fighting eminent domain abuse'.

The *Planning Commissioners Journal* (http://www.planners web.com) is targeted toward citizen planners such as those on local planning commissions and zoning boards. The website serves as an educational resource explaining how to effectively and efficiently plan an urban development from the inside out. While there is a cost for the magazine, most articles can be previewed online before they are ordered.

An explanation of and introduction to the structure of the US court system can be found on the government website entitled US Courts (http://www.uscourts.gov). Along with circuit or district maps and a Constitutional justification of the courts, the website provides a library and newsroom for recent court cases, including land use and zoning cases.

Local governments have made subdivisions interactive. For example, Clark County in Washington created a mapping website using GIS that allows visitors to view subdivisions by searching for specific properties or zooming into the map. Visit it online (http://gis.clark. wa.gov/applications/gishome/subdivisions) and click on GIS Digital Atlas for a preview.

There are several good texts on American planning law, including Mandelker (1993) *Land Use Law*, which contains much more discussion than is general, and omits the extracts from cases which characterize many law books; the second edition (1994) of *Land Use: Cases and Materials* by Callies *et al*. is more manageable and user-friendly than its predecessor. A volume in the West *Nutshell* series is succinct and makes the subject seem surprisingly comprehensible: Wright and Gitelman (2000) *Land Use in a Nutshell*.

For a discussion of 'public purpose', see Merrill (1986) 'The economics of public use'. A blistering attack on the judicial history of 'public use' is to be found in Paul (1987) *Property Rights and Eminent Domain*.

The tragic story of the destruction of Poletown is set out in Wylie (1989) *Poletown: Community Betrayed*; see also Hill (1986) 'Crisis in the Motor City: the politics of economic development in Detroit'. Two recent legal discussions of the Poletown decision are Kulick (2000) 'Comment: rolling the dice: determining public use in order to effectuate a "public-private taking" a proposal to redefine "public use"' and Werner (2001) 'Note: the public use clause, common sense and takings'.

The role of the courts is discussed at length in Waltman and Holland (1988) *The Political Role of Law Courts in Modern Democracies*. See also Goldman and Jahnige (1985) *The Federal Courts as a Political System*. An interesting insight into the operation of the Supreme Court is given in Tribe (1985) *God Save this Honorable Court: How the Choice of Supreme Court Justices Shapes Our History*. See also Haar and Kayden (1989b) *Landmark Justice: The Influence of William J. Brennan on America's Communities*.

A recent discussion on how the courts have treated single-family home definitions can be found in Brener (1999) 'Note: Belle Terre and single-family home ordinances: judicial perceptions of local government and presumption of validity'.

The NIMBY issue is discussed in O'Looney (1995) *Economic Development and Environmental Control: Balancing Business and Community in an Age of NIMBY's and LULUs*; Inhaber (1998) *Slaying the NIMBY Dragon*; and McAvoy (1999) *Controlling Technocracy: Citizen Rationality and the NIMBY Syndrome*.

For discussions on the *Kelo* and the *Stop the Beach Renourishment, Inc.* cases, see: Cohen (2006) 'Eminent domain after *Kelo* v. *City of New London*'; Meystedt (2011) '*Stop the Beach Renourishment*: why judicial takings may have meant a taking a little too much'; Penalver (2006) 'Property metaphors and *Kelo* v. *New London*'; Penalver and Strahilevitz (2012) 'Judicial takings or due process';

Policicchio (2011) 'Stop the Beach Renourishment, Inc. v. Florida Department of Environmental Protection'; and Rutlow (2006) 'Kelo v. City of New London'.

Questions to discuss

1   How does a local government obtain its powers of zoning?

2   In what way do the courts 'presume' that a municipality's judgment is valid? Why do they do this?

3   What restrictions are there on the zoning power?

4   Why are Kelo v. New London and Stop the Beach Renourishment, Inc. v. Florida Department of Environmental Protection important cases for cities and state governments in the United States?

The Comprehensive Plan

A Comprehensive Plan describes a vision for the future of a community. It offers a vision of what the residents want in a community. It sets a type of roadmap to where we want to go. It seeks to identify how we can ensure a community's quality of life. The reason for having such a plan is evident in the following quote from the 2004 Comprehensive Plan for the City of Moorhead, Minnesota:

> The Comprehensive Plan helps to ensure that development occurs in a manner desired by the community, rather than simply as a result of market trends and patterns. The Comprehensive Plan protects what makes the community unique and what made people initially want to live here.
>
> (P.1-2)

The procedures for developing the document and the contents of the document will vary by state. This chapter provides a discussion of the Comprehensive Plan, also referred to as a Master Plan, a General Plan, or a Municipal Development Plan and the steps taken to develop and implement it. To simplify matters, the term 'Comprehensive Plan' will be used throughout the chapter.

State mandates

State law provides language requiring the development of a Comprehensive Plan. The language will undoubtedly vary from state to state. It might even differ as to which agency or body is responsible for the development of the Comprehensive Plan. For example, is it the responsibility of the planning commission or local legislative body to develop a Comprehensive Plan? Arizona Revised Statutes (ARS 9-461.05 Chapter 204) requires that every city prepare a comprehensive, long-range General Plan for the future development of the community. The planning agency shall develop and maintain the Comprehensive Plan. In Michigan, the Municipal Planning Act of 1931 (125.36, Section 6 (1)) notes that the planning commission shall make and approve a Master Plan for the physical development of the municipality, including any areas outside of its boundaries which, in the commission's judgment, bear relation to the planning of the municipality. Moreover, a planning department, according to 125.38b, Section 8b, established by charter is authorized to submit a proposed plan to a planning commission. The State of Virginia, in section 15.2-2223 of the Virginia Code, requires a community to prepare 'a plan for the physical development of the jurisdiction and the Governing Body shall adopt a Comprehensive Plan'. Minnesota, in Minnesota Statutes, Sections 462.351, Subd.1, enables cities to 'carry out comprehensive municipal planning activities for guiding the future development and improvement of the municipality and may prepare, adopt, and amend a comprehensive municipal plan and implement such plan by ordinance or other official measure'.

Looking into the future is something that scares many individuals. Yet, this is what the Comprehensive Plan seeks to do. It is designed to anticipate the future and guide the growth and physical development of an area. It is proactive and not simply reactive to events. It seeks to create a vision or roadmap of what the jurisdiction can be physically, economically, and socially.

It can be considered a 'blueprint' for the physical development of an area. It is to be used as a policy guide for future decisions. Box 7.1 provides examples of how jurisdictions have defined the term.

Guiding land use and decision-making is a key purpose of a Comprehensive Plan. It is a policy instrument that should be general in nature. It provides a vision (see Box 7.2) or direction to a municipality, or some other jurisdiction for the physical development of an area. This direction could be for the growth, development, and redevelopment of an area. The vision could be as generic as preserving, enhancing, and encouraging a high quality of life. The vision should be long range and comprehensive and focused on the physical development of an area. Comprehensive means covering multiple policy areas. The Comprehensive Plan should be flexible so that it can evolve over time.

It should incorporate analyses, diagrams, and recommendations in such areas as housing, transportation, open space, commercial development, public services, schools, sensitive environmental areas, etc. As stated in the Arizona Revised Statutes 11-806B with regards to the county Comprehensive Plan, 'the purpose of the plan is to bring about coordinated physical development in accordance with the present and future needs of the county'.

The time period covered in a Comprehensive Plan may vary from state to state and is generally covered in state statutes. Some states indicate that the plan should cover a time period of ten to twenty years, while other states may opt for twenty-five-year plan coverage. The requirement for periodic updates of the plan is covered in a later section of this chapter.

BOX 7.1 ALTERNATIVE DEFINITIONS OF COMPREHENSIVE PLAN, GENERAL PLAN, AND MASTER PLAN

Bellevue, Washington, Comprehensive Plan – A comprehensive plan is a broad statement of community goals and policies that direct the orderly and coordinated physical development of a city into the future. A comprehensive plan anticipates changes and provides specific guidance for future legislative and administrative actions. It reflects the results of citizen involvement, technical analysis, and judgment of decision-makers.

La Porte, Texas, General Plan – The La Porte 2020 General Plan is a 20-year Master Plan intended to serve as an official public document, adopted by the City Council, to guide policy decisions relating to the physical and economic development of the community. In general, the plan indicates how a community desires to develop and redevelop over the course of the next twenty years. The comprehensive plan is a physical plan, it is long-range, it is comprehensive, and it is a statement of the goals, objectives and policies of the local government. The comprehensive plan is slightly utopian and also inspirational, enough to challenge the future of the community; but it provides clear direction through specific statements of actions to achieve the desired results envisioned by citizens and the leadership of the community.

Charlemont, Massachusetts – A Master Plan is a long-range plan that guides development in a town through vision of what residents would like their town to be in the future. It is a comprehensive document that looks at all aspects of a community including natural resources, agricultural resources, recreation, historic resources, transportation, public infrastructure and municipal services, economic development, housing, and land use. Additional sections may be added if a community would like to address other issues more specifically. The Master Plan includes mapping, inventory, analysis, and recommended strategies for accomplishing the goals and objectives of the town. It may also include a capital improvement program to coordinate large-scale expenditures with the goals of the Master Plan. This is a plan created by and for the citizens of Charlemont.

Source: City of Bellevue 2006: 7; City of La Porte n.d.: 1–1; Town of Charlemont 2003: 1

BOX 7.2 VISION FOR WILMINGTON, MASSACHUSETTS

Wilmington will be an attractive, family-oriented, and environmentally responsible community as it continues to grow. The Town Common will provide the focus for the town's civic and social life; compact, mixed-use business districts will allow people to conduct business and interact with fellow residents in attractive, village-style environments. A variety of housing types will provide affordable housing choices for families, individuals, and seniors. The town will benefit from a network of protected, easily accessible open spaces, including areas for active recreation. Residents, businesses, and the town will work together to conserve water, protect water quality, and reduce disturbance in wetlands and aquifer recharge areas. Pedestrians and bicyclists will be able to reach key destinations throughout the town via a network of trails and sidewalks that will reduce the burden on the town's roadways. Residents will have meaningful opportunities to participate in town government, and the town as a whole will demonstrate consistency in its approach to planning. Decisions will be based on a comprehensive approach to community development that considers the impacts of town actions on both the human and natural environment.

Source: Planner's Collaborative 2001: vii

Citizen involvement

Citizen involvement has been touched upon in Chapter 1. However, involving the public throughout the development and implementation of a Comprehensive Plan is critical. It provides citizens from all sectors of the community with an awareness of the issues facing a community and the process a community is taking to deal with the issues. Involving the public allows citizens the opportunity to help develop a shared vision and strategy for the city. Developing a community vision and strategy to achieve the vision must be thought of as a collaborative process among multiple parties. It is a challenging process in that there will be multiple interests competing to have their views heard. The importance of citizen involvement can be seen in Box 7.3.

Citizen involvement in the development of a Comprehensive Plan is not a voluntary task. It is required in many states. For example, State law in California (Government Code 65351) indicates that 'During the preparation or amendment of the general plan, the planning agency shall provide for opportunities for the involvement of citizens, public agencies, public utility companies, and civic, education, and other community groups, through public hearings and any other means the city or county deems appropriate.' In addition, mayors and councils in many areas are required to develop and adopt written procedures to provide for effective citizen participation. Box 7.4 illustrates the types of actions that might occur to make sure an open discussion of issues takes place at public meetings.

There are a number of tools and techniques that can be used to involve the public throughout the planning process. These include citizen surveys, public forums, public hearings, mailings, news releases, and public committees. Planning commissions and the local legislative body are required to meet any minimum requirements for public hearings. Using one technique to involve the public will not suffice. Mixing and matching alternative means to involve the public is needed. It should be noted that citizens respond to different involvement techniques. If one technique is not working, another technique should be considered.

A number of items must be considered when developing a citizen involvement program. Citizens must be given information they can understand. On occasion, citizens have been given technical information or a long document that they must read in a short

BOX 7.3 IMPORTANCE OF CITIZEN PARTICIPATION

A wide variety of citizens should be part of the master plan process. With the growing interest in open space preservation, pointing out the connections between the master plan and land conservation can help bring people to the table. The planning board, open space committee, environmental commission, municipal governing body, recreation committee, historic preservation commission, and agricultural advisory board may be involved. In addition, residents not directly involved with municipal government will be concerned with the recreational, environmental and aesthetic impacts on the master plan. All parties interested in saving natural resources should be involved in reviewing and amending the master plan. The more residents buy into the plan, the better the long-term support for local regulation will be.

Source: Association of New Jersey Environmental Commissions 2003: 6

BOX 7.4 PROVISIONS TO ENSURE OPEN DISCUSSION OF ISSUES AT PUBLIC MEETINGS

1   An agenda will be established that clearly defines the purpose of the public meeting, or hearing, the items to be discussed, and any actions that may be taken.
2   The scheduled date, time, and place will be convenient to encourage maximum participation by the town residents and property owners.
3   A facilitator or chair will conduct the meeting or hearing in an orderly fashion to ensure that all attendees have an opportunity to comment, discuss issues, or provide testimony.
4   The facilitator or chair will outline the purpose of the meeting, or hearing, and the procedures attendees should use and how the public input will be used.
5   An overview of documents or proposals to be considered will be discussed.
6   All persons attending the meeting or hearing that desire to participate should be allowed to do so. However, specific factors, such as the meeting or hearing purpose, number in attendance, time considerations, or future opportunities to participate may require that appropriate constraints be applied. These constraints will be clearly outlined by the facilitator or chair if the need arises.
7   All attendees will be encouraged to sign in using a provided sign in sheet.
8   Meetings and hearings will be recorded by appointed committee members.
9   Summaries will be transcribed and made available as soon as possible following a public meeting or hearing.
10 Special arrangements will be made under the provisions of the Americans with Disabilities Act (ADA) with sufficient notice.

Source: Town of Wascott 2003: 2

period of time. Notice requirements prevent this from occurring in many locations. Some areas require a minimum of thirty days' notice before any meeting or hearing can be held. The information must be understandable to the public. There is no need to give a report to the public for their comment if they cannot understand the 'jargon' used in the report. When setting the time and location of a meeting or forum, make sure it is being held when most people can attend. Ultimately, all areas can do is provide the opportunity for citizens to participate in the planning process. We cannot force anyone to attend or participate in the process.

Required and optional elements/parts

All states have language in their laws and regulations that certain subjects/policy areas be contained in a Comprehensive Plan. The exact areas vary from state to state. For example, the State of California, in California Government Code Section 65320, mandates the following elements be contained in a General Plan: land use, circulation, housing, conservation, noise, open space, and safety. In Wisconsin, State Statutes Chapter 66: General Municipal Law requires the following nine elements: issues and opportunities, housing, transportation, utilities and community facilities, agricultural, natural and cultural resources, economic development, intergovernmental cooperation, land-use, and implementation. The State of Indiana, in Indiana Code 36-7-4-503, indicates that a Comprehensive Plan must include at least: 1) a statement of objectives for the future development of the jurisdiction; 2) a statement of policy for the land use development of the jurisdiction; and 3) a statement of policy for the development of public ways, public places, public lands, public structures, and public utilities. Furthermore, South Carolina (South Carolina Local Government Planning Enabling Act of 1994, Section 6-29-510 (D1-7)) mandates the following elements be included in a Comprehensive Plan: population, economics, natural resources, cultural resources, community facilities, housing, and land use.

Box 7.5 illustrates the contents of several required elements.

States do allow for flexibility in developing and presenting Comprehensive Plans. Combining elements is one way of providing flexibility, as long as the requirements are met for each of the required elements. For example, in an effort to allow for flexibility and streamlining, California Government Code Section 65301(a) notes that 'a General Plan may be adopted in any format deemed appropriate by the legislative body as long as all topics are covered'. The exception would be the combining of the housing element with another required element because it is required by State Guidelines to contain certain data and analysis as well as a five-year plan for meeting housing goals and objectives. On the other hand, it is not uncommon to see the open space and conservation elements combined.

States also allow jurisdictions to develop additional elements to a Comprehensive Plan. These are the so-called 'optional' or 'permissive' elements. In fact, municipalities are encouraged to develop other elements that address specific needs of a community. They are adopted in the same fashion as required elements and carry equal weight under the law. In other words, once an optional element is adopted, it is no less important than the required elements.

The State of California, in California Government Code Section 65303, allows jurisdictions to 'include other elements or address any other subjects which in the judgment of the legislative body, relate to the physical development of county or city'. The 1998 General Plan for the City of South Pasadena includes a historic preservation element which includes 'goals, policies and implementation strategies to ensure the continued appreciation and protection of South Pasadena's rich legacy of substantially intact historic buildings, residential neighborhoods, and commercial districts'. The 2003 General Plan for the City of San Clemente contains a nuclear safety element. This element is clearly included because of the nearby San Onofre Nuclear Generating Station. The stated purpose of the element is to guide 'the City's response and reaction to a nuclear emergency including planning for the protection of public health, safety, welfare and the

BOX 7.5 MATERIAL CONTAINED IN A REQUIRED ELEMENT IN A GENERAL PLAN

Vermont – A land use plan, consisting of a map and statement of present and prospective land uses, indicating those areas proposed for forests, recreation, agriculture (using the agricultural lands identification process established in 6 VSA Section 8), residence, commerce, industry, public and semi-public uses and open spaces reserved for flood plain, wetland protection, or other conservation purposes; and setting forth the present and prospective location, amount, intensity and character of such land uses and the appropriate timing or sequence of land development activities in relation to the provision of necessary community facilities and services.

California – A safety element for the protection of the community from any unreasonable risks associated with the effects of seismically induced surface rupture, ground shaking, ground failure, tsunami, seiche, and dam failure; slope instability leading to mudslides and landslides; subsidence, liquefaction and other seismic hazards identified pursuant to Chapter 7.8 (commencing with section 2690) of the Public Resources Code, and other geologic hazards known to the legislative body; flooding; wild land and urban fires. The safety element shall include mapping of known seismic and other geologic hazards. It shall also address evacuation routes, military installations, peakload water supply requirements, and minimum road widths and clearances around structures, as those items relate to identified fire and geologic hazards.

Wisconsin – An Intergovernmental Cooperation Element is a compilation of objectives, policies, goals, maps, and programs for joint planning and decision-making with other jurisdictions, including school districts and adjacent local government units, for siting and building public facilities and sharing public services. The element shall analyze the relationship of the local government unit to school districts and adjacent local governmental units, and to the region, the state and other governmental units. The element shall incorporate any plans or agreements to which the local government unit is a party under Section 66.0301, Section 66.0307, Section 66.0309. The element shall identify existing and potential conflicts between the local governmental unit and other governmental units that are specified in this paragraph and describe processes to resolve such conflicts.

Source: Vermont Statutes, Title 24, Chapter 17, Section 4382; California Government Code, Section 65.302. (g)(1); Wisconsin Statutes, Section 66.1001(2)(g)

environment in case of an emergency'. Other optional elements found in California include agriculture, offshore energy, public buildings, community design, disaster preparedness, and economic development.

Optional elements found in other states are devoted to such topics as solar energy, recreation, economic development, conservation, and agriculture. For example, the 2004 Comprehensive Plan for Bainbridge Island, Washington, contains a cultural element (City of Bainbridge Island 2004: 1) whose purpose is 'to link community cultural planning to larger community issues and to set directions for integrating the arts, humanities and history with urban design, economic development, education and other community development initiatives that shape the quality of life that nurtures Bainbridge Island'. The list of potential topics for optional elements varies from state to state.

Consistency

State laws indicate that conflicts within the Comprehensive Plan and between elements of the General Plan cannot exist. For example, California Government Code Section 65300.5 requires that all elements of a General Plan must be consistent with each other, including text and diagrams (see Box 7.6). Two sections of the same element cannot contradict each other. Terms cannot be defined differently in different parts of the same element. Moreover, an open space element cannot contain a policy that has designated a section of the city to be protected from development while the land use element shows the same area being a future location of affordable housing and commercial development.

The importance of consistency can be viewed in Box 7.7. States have also developed language regarding consistency between the Comprehensive Plan and other policies and land use actions of the jurisdiction. The State of Washington requires that a city's Comprehensive Plan be consistent with the State Growth Management Plan. The State of Virginia (Virginia Code Section 15.2-2232 (A)) indicates that the location, character and extent of public facilities (collectively defined as public areas, public buildings, public structures, public utility facilities, and public service corporation facilities) must be submitted and approved 'as being substantially in accord with the adopted Comprehensive Plan'. Recent changes in Chapter 66, Municipal Law of the Wisconsin Statutes indicate that 'beginning on January 1, 2010, any program or action of a local governmental unit that affects land use shall be consistent with that unit's Comprehensive Plan'. Examples of programs or actions

subject to this consistency requirement in Wisconsin would be municipal incorporation procedures, annexation procedures, cooperative boundary agreements, impact fee ordinances, agricultural preservation plans, local subdivision regulations, zoning ordinances, etc. In the end, these actions cannot prevent the visions, goals, and objectives of the Comprehensive Plan from being realized.

Requirements for examining the consistency of a Comprehensive Plan with the Comprehensive Plans of neighboring communities will also vary by state. New Jersey Municipal Law requires this to be done as well as making sure the plan is consistent with the county in which the municipality is located and that it is consistent with the New Jersey State Development and Redevelopment Plan. Examining consistency between neighboring communities and other jurisdictions is certainly a most challenging task.

Environmental review

An environmental review is needed to determine if there are any potential significant impacts of the adoption and implementation of a Comprehensive Plan. The Minnesota Department of Natural Resources is charged with reviewing documents such as Comprehensive Plans to provide decision-makers with

BOX 7.6 DEFINITION OF CONSISTENCY

Consistency: Consistent with: Free from significant variation or contradiction. The various diagrams, text, goals, policies, and programs in the General Plan must be consistent with each other, not contradictory or preferential. The term 'consistent with' is used interchangeably with 'conformity with.' The courts have held that the phrase 'consistent with' means 'agreement with; harmonious with.' The term 'conformity' means in harmony therewith or agreeable to (*Sec. 58 Ops Cal. Atty. Gen. 21 25[1075]*). California Law also requires that a General Plan be internally consistent and also requires consistency between a General Plan and implementation measures such as the zoning ordinance. As a general rule, an action program or project is consistent with the General Plan if, considering all its aspects, it will further the objectives and policies of the General Plan and not obstruct their attainment.

Source: City of Anderson, California, Planning Department 2007: 151

BOX 7.7 IMPORTANCE OF CONSISTENCY

Consistency is critical, because it . . .

Merges intention and action, translating the community's vision into public policies the local government actually intends to carry out.
Ensures that prepared land development regulations, amendments, or 'land-use actions' are assessed in relationship to the Plan.

Source: Jefferson Parrish, Louisiana 2004: slide 33

information on the impacts of a proposed action on such areas as land, water, flora and fauna, agricultural lands, critical environmental areas, etc. The State of New York, through Part 617 of Title 6 of New York's Codes, Rules, and Regulations, requires the incorporation of environmental factors into the planning and decision-making processes to determine if any significant adverse environmental impacts will result in the adoption of a jurisdiction's land use plan. In California, as identified under the California Environmental Quality Act (Public Resources Code Section 21002.1(a)), the purpose of an Environmental Impact Report is 'to identify the significant effects on the environment of a project, to identify alternatives to the project, and to indicate the manner in which those significant effects can be mitigated or avoided'. This means informing all decision-makers and the public of any potential environmental impacts, any mitigation measures that might reduce the impacts, and any potential alternatives to the project that might lessen any environmental impacts. Public notices of meetings to discuss the environmental impacts of a proposed project are required.

States will have different environmental assessment and environmental review procedures, as evidenced in their state environmental regulations. In California, an Initial Study is required to determine the level of environmental review. If it is determined that no significant environmental impacts will occur, a statement called a 'Negative Declaration' would be required. If some environmental impacts might occur and revisions in the project could mitigate or avoid any significant

environmental impacts, a 'Mitigated Negative Declaration' would be required. If it is determined that the proposed project will have significant environmental impacts, an Environmental Impact Review is required. A Draft Environmental Impact Review would then be written. This document would then be circulated to the public and public agencies for their review. Notice requirements for informing the public and public agencies vary by state. Comments are gathered and the agency prepares responses to the questions raised during the review period. The document is then revised, if needed. If the Lead Agency approves the project and the Final Environmental Impact Review that was prepared, a Notice of Determination will be issued by the agency. Federal environmental reviews may also be needed.

Adopting the Comprehensive Plan

An earlier section of this chapter indicated that states require municipalities and counties to develop a Comprehensive Plan. States also require the jurisdictions to adopt the Comprehensive Plan through a formal process. The general wording found in state regulations or statutes is that 'a community is required to prepare a plan and the governing body shall adopt a Comprehensive Plan'. The processes for adoption will differ according to individual states.

The process for adopting a Comprehensive Plan and even amending the Comprehensive Plan can be the same. This can be found in Utah in Section 17-2703 of the Utah State Code (see Box 7.8). Planning commissions are required to hold at least one public hearing on the proposed Comprehensive Plan. An Official Notice, including time and place of the public hearing, is required before convening any public hearing. Notices can generally be found in a newspaper of general circulation in the jurisdiction. There is nothing to preclude a jurisdiction from advertising the public hearing in multiple outlets. Many states indicate a notice of at least fifteen days before the public hearing. City councils, the local legislative body, are also required to hold at least one public hearing to obtain citizen feedback on the proposed Comprehensive Plan. They are also required to publish the time and place of the public hearing. Official Notice requirements will differ by state.

Comprehensive Plans can be adopted by local resolution or ordinance. Readers will need to consult individual states to determine the specific procedures required to adopt the required Comprehensive, General, or Master Plan. The State of Virginia (Code of Virginia, Title 15.2, Chapter 22) requires a Comprehensive Plan to be prepared and adopted. The procedures employed by Virginia are comparable to the procedures found in other states and will be used here as an example.

The local planning commission is vested with the authority to prepare and recommend the Comprehensive Plan to the locally elected governing body. It serves in an advisory capacity to the elected body. Prior to making any recommendation to the elected governing body, notice must be provided to the public. The plan must be placed on a website either that is operated by the planning commission or where it routinely places information that is available to the public. The commission must also place a notice in a newspaper for two successive weeks that is circulated in the area. The notice includes the time and place where the public has the opportunity to comment on the plan. After the notice requirement has been met, the planning commission can hold a public hearing itself or hold a joint meeting with the elected governing body. The commission then has several options at its disposal. It can decide to approve the plan, amend it and then approve the revised plan, or disapprove it. Its decision is a resolution stating the reasons for its decision.

The next step in the process is for the elected governing body to consider the commission's recommendation. The plan is posted on a website available to the public, generally the official city or town website, so that citizens have the opportunity to read the proposed plan. A public hearing is then required so that citizens can comment on any part of the plan or the entire plan. The elected governing body is required to act within ninety days of the commission's resolution.

Should the elected governing body fail to approve the commission's recommendation, the plan will be sent back to the commission with a statement of findings indicating the reasons for not approving it. The planning commission then has sixty days to reconsider the plan and to resubmit a recommendation to the elected governing body. It then is returned to the elected governing body for its reconsideration.

The decision of the elected governing body is formalized in a resolution. The resolution includes information in the form of 'whereas . . .' statements. For example, the resolution would start with the statement indicating that a Comprehensive Plan, its adoption and periodic review, is required by the state. It would also include a statement indicating that the public was informed of a public hearing, including the time and place of the meeting. The meeting was held where comments were obtained from the public and considered by the planning commission. The planning commission voted and made its recommendation to the locally elected body. The locally elected body then held a public meeting and heard the comments of the public, staff, and recommendations of the planning commission. Assuming the locally elected body is approving the Comprehensive Plan or amendments to the plan, a statement indicating how the plan or amendments will further promote the public health, safety, and welfare is provided. The resolution would conclude with indicating that the locally elected body approves and adopts the plan and/or amendments to the plan. The date of the action and the time when the action would take effect would be included in the resolution. The head of the elected governing body is required to sign the resolution.

The number of votes needed to recommend and adopt the Comprehensive Plan will also differ by state. Planning Commissions generally use a majority vote by the entire body to recommend adoption of the Comprehensive Plan to a city council. City councils use the same two-thirds vote of the body to adopt the Comprehensive Plan. However, some jurisdictions have added stronger requirements for adopting the Comprehensive Plan. For example, the City of Los Angeles (1999: no page) has the following adoption procedures for the city council:

The General Plan and any amendment to it must be adopted by a majority vote of the City Council. A two-thirds vote of the Council is required if its action is contrary to the recommendations of either the City Planning Commission or of the Mayor. A three-fourths vote of the Council is required if the action of the Council is contrary to the recommendations of both the City Planning Commission and the Mayor.

Implementing the Comprehensive Plan

There seems to be a general misconception that once a Comprehensive Plan has been developed and adopted by the local legislative body, the process is complete. That cannot be further from the truth. Those steps in the planning process are just the beginning. The adopted Comprehensive Plan will not accomplish the vision or visions contained within it. Implementation means carrying out, accomplishing, or effectuating the Comprehensive Plan. The policies, programs, and actions identified throughout the Comprehensive Plan

---

BOX 7.8 PROCESS FOR ADOPTING A GENERAL PLAN IN UTAH

a) After completing a proposed general plan for all or part of the area within a county, the planning commission shall schedule and hold a public hearing on the proposed plan.

b) The planning commission shall provide reasonable notice of the public hearing at least 14 days before the date of the hearing.

c) After the public hearing, the planning commission may make changes to the proposed general plan.

d) The planning commission shall then forward the proposed general plan to the legislative body.

e) The legislative body shall hold a public hearing on the proposed general plan recommended to it by the planning commission.

f) The legislative body shall provide reasonable notice of the public hearing at least 14 days before the date of the hearing.

g) After the public hearing, the legislative body may make any modifications to the proposed general plan that it considers appropriate.

h) The legislative body may:

- Adopt the proposed general plan without amendment;
- Amend the proposed general plan and adopt or reject it as amended; or
- Reject the proposed general plan.

i) The general plan is an advisory guide for land use decisions.

j) The legislative body may adopt an ordinance mandating compliance with the general plan.

Source: Utah State Code, Section 17-27-03

BOX 7.9 IMPORTANCE OF IMPLEMENTING A GENERAL PLAN

A good plan goes to waste if it isn't implemented. For its implementation, the general plan primarily relies on regulations, such as specific plans, the zoning ordinance, and subdivision ordinances, and public project consistency requirements. State law requires cities and counties to have subdivision and building regulations and open-space zoning, while most of the other measures . . . are adopted at the discretion of the city or county. If the objectives, policies, and proposals of the general plan are to be served effectively, implementing measures must be carefully chosen, reflective of local needs, and carried out as an integrated program of complementary and mutually reinforcing actions.

Source: Governor's Office of Planning and Research 2003: 149

BOX 7.10 DIFFERENCES BETWEEN GENERAL PLAN AND ZONING

City of La Porte, Texas

The Comprehensive Plan should not be confused with zoning. The plan is a general guide for the long-range growth and development of the entire city. Zoning is a legal mechanism enacted by the City whereby specific parcels of land are classified as suitable for particular land uses. The plan provides the legal justification for development regulations but has no practical effect on its own in the absence of such regulations.

Los Angeles County, California

Is the General Plan the same as zoning?

No, these terms are not interchangeable; however, they are intrinsically linked. Zoning is one of the many planning tools for implementing the General Plan. Regulations, policies, guidelines and agreements are other examples of implementation tools, which are often used in combination.

Sources: City of La Porte n.d.: 1–3; Los Angeles County 2007: 369

must be put into action. Box 7.9 illustrates the importance of implementation.

Tempe, Arizona (2008: 55) has referred to the Comprehensive Plan as 'the umbrella document over many other planning documents'. These other documents would include area plans, specific plans, neighborhood plans, Capital Improvement Programs, and historic preservation plans. All of these plans help implement a Comprehensive Plan.

A vast array of tools, techniques, and actions are also available to jurisdictions to carry out the vision and policies, programs, and actions contained in a Comprehensive Plan. Zoning is commonly viewed as a key tool for implementing a Comprehensive Plan. However, there appears to be a common misconception that a Comprehensive Plan is the same thing as zoning. Box 7.10 may help clear up the difference between the two and any misconceptions. Other city ordinances are also used to help implement a Comprehensive Plan.

Cities can employ historic preservation, hillside protection, and design review ordinances to implement various elements of a Comprehensive Plan. Ultimately, the tools, techniques, and actions chosen to implement the Comprehensive Plan may be regulatory or non-regulatory in nature. Many will be existing tools and techniques, while new programs might be needed to further some implementation efforts.

Monitoring, amending, and updating the Comprehensive Plan

The importance of monitoring, amending, and updating a Comprehensive Plan cannot be denied. Comprehensive Plans are not meant to be static. Plans are not written to be placed on the shelf in a room and forgotten. They are not a finished product. The various elements must be monitored to see if the policies are being implemented. Community conditions (economic, social, environmental) change, as do the problems confronting the community. As noted in the 2025 Lincoln City – Lancaster County Comprehensive Plan (p. F-158):

> The Plan is the community's collective vision. Yet, change is inevitable. Major technologies and new community needs will arise during the planning period which were not foreseen during the Plan's development. Jobs, housing, transportation, goods and services will shift over time. The amendment process to the Plan must accommodate and help manage the inevitable change in a way that best promotes, and does not compromise, the community's core values, health and well being. The Plan amendment process must be an open and fair process, utilizing sound planning, economic, social and ecological principles.

New data may become available that might necessitate amending the document. As such, the Comprehensive Plan has become as many communities acknowledge, a 'living document'. It needs to change or adapt to address current issues.

Individuals must consult the appropriate state and local rules and regulations regarding amending a Comprehensive Plan. Clear policies and procedures for amending a Comprehensive Plan are essential. As noted in the City of Phoenix (2008: 1) General Plan amendment application procedures:

> Clear policies and procedures for amending the Phoenix General Plan are critical to maintaining the integrity and ultimate viability of the Plan. The value of any plan depends on its stability and predictability. Both the business community and the general population of Phoenix benefit from clear guidelines to direct and protect investments.

The rules, regulations, and procedures will vary by state. For example, in Washington, under the State Growth Management Act, a Comprehensive Plan cannot be amended more than once a year. The State of California has ruled that the mandatory elements of a General Plan cannot be amended more than four times in one calendar year. However, each amendment could include more than one change. Any proposed amendment must be consistent with the General Plan.

Amendments can take many forms. One city may categorize them as major or minor, with the appropriate distinctions. More generally, jurisdictions categorize an amendment as a text amendment or as a map amendment. A text amendment might be used to correct information contained in a Comprehensive Plan or to expand on the information already contained in the Comprehensive Plan. For example, if new information indicates that the existing information is flawed for whatever reason, a text amendment would be appropriate. A map amendment would occur if the use of property in an area changes from rural to commercial or rural to residential. A map amendment would also be needed if the intensity or density changes in the area.

Amendments can be initiated by a variety of parties. An individual could initiate the change process or a planning commission, government agency, or city council could do so. Regardless of the individual or body recommending the amendment, an application must be filed and properly processed. Jurisdictions will charge a fee to process the amendment. Additional charges may also be levied by government departments and agencies that review and make recommendations on the proposed amendment. A deposit is also generally required.

Communities may differ in their particular approaches to processing an amendment. However, the following steps could be considered as generic steps in the process. An application for a General Plan amendment must be completed and submitted to the proper agency or department. The information on such an application would include the name and address of the individual or body requesting the change (amendment) to the General Plan. The type of change that is being requested would be listed – map or text amendment, etc. The reason or reasons (e.g., errors in mapping, land use requirement changes, etc.) why the amendment is being sought would need to be delineated. A signature or signatures would also be needed and, in some areas, a notary certificate would need to be attached.

Many communities require a pre-application meeting between the individual or group proposing the amendment and the staff. The individual or group proposing the amendment explains the need for the amendment and the staff provides feedback on the proposed change. The application for amending the Comprehensive Plan is then filed. After the application is filed, a public hearing on the issue would be called by the planning commission. The planning commission would then make a recommendation to the city council. The city council would have a hearing on the issue and make its recommendation. Its recommendation would then be made in an ordinance that is signed by the appropriate city officials – in some areas, the city clerk, the mayor, and city attorney.

As previously stated, the Comprehensive Plan is a 'living document'. Although the Comprehensive Plan covers anywhere from a ten-year to twenty-year time frame, it will need to be updated to remain current with existing issues and concerns and reflect the views of the citizens. Ultimately, there is no one time frame for updating a Comprehensive Plan. The requirements for updating vary by state. For example, municipalities in Virginia are required to update their Comprehensive Plans at least every five years; Wisconsin and Minnesota require updates every ten years; and the Washington Growth Management Act requires municipalities in Washington to fully review and update their plans every seven years.

Updating requirements in California are somewhat different. State law does not mandate how often a General Plan should be updated. It does, however, require municipalities to periodically update their General Plan. Although a General Plan should be comprehensively updated every ten to twenty years, some cities use a recommended every eight to ten years to update their General Plan. Moreover, California Government Code Section 65400(b) (1) requires all jurisdictions to submit to their legislative bodies an annual report on the status of the General Plan and on the progress made toward its implementation. California also requires the housing element to be updated every eight years and approved by the California Department of Housing and Community Development. The continued growth of a jurisdiction may cause an increased demand for housing from various segments of the population. Identifying the new housing needs will cause the jurisdiction to develop new policies and implement actions to achieve these needs.

Recent comprehensive planning efforts

There are a number of municipalities currently engaged in innovative comprehensive planning efforts. Two efforts are currently taking place in Pennsylvania. Interestingly, the two cities currently working on Comprehensive Plans – Pittsburgh and Philadelphia – are not required to produce Comprehensive Plans. They are exempt from the requirements of the 1968 Pennsylvania Municipal Planning Code to develop and implement a Comprehensive Plan. This should be taken to mean that comprehensive planning efforts have not been undertaken prior to this time. They have developed and implemented sub-area or project plans for many years.

The City of Pittsburgh is currently engaged in a comprehensive planning process called PLANPGH, an effort that will take five years to complete. This is unique in that it allows the city more time and consultation with stakeholders than is allowed for other cities in Pennsylvania. For a city over 250 years old, it is the first citywide Comprehensive Plan that has been undertaken. The final product is meant to be a proactive effort. As Mayor Luke Ravenstahl has noted in a message to the public:

> A Comprehensive Plan allows the City to be proactive and seize opportunities instead of reacting to problems after they arise. It is a means for identifying, prioritizing, coordinating and funding projects that may involve many partners and funding from multiple sources.
> (http://planpgh.com/mayor_message.htm)

Once developed, the Comprehensive Plan will be adopted by the city and will act as a roadmap and guide for the city's growth for twenty-five years.

Citizen input is critical to the development and ultimate success of PLANPGH. A variety of opportunities have been available to citizens interested in helping formulate the plan. Community surveys have been conducted to get citizen input. Focus groups, committees, and various public events gather valuable public input for the planning process. Since this planning effort represents a partnership between government and stakeholders, it is essential that a variety of techniques or tools be used to gather citizen input.

While municipalities in Pennsylvania, according to the Municipal Planning Code, are required to include nine components/elements, PLANPGH will examine twelve areas:

1 Open Space, Parks and Recreation
2 Cultural Heritage
3 Transportation
4 Public Art
5 Urban Design
6 Public Facilities and Services (City-owned properties)
7 Energy
8 Infrastructure
9 Economic Development
10 Housing
11 Education
12 Land Use.

(http://planpgh.com/faq.htm)

The first five components have been identified as high priority areas and will be developed first. The remaining elements will be addressed based on priority, available funding opportunities, and the availability of partnering organizations. Each component will be regularly updated and will be required to meet the following six goals:

1 Strengthen Pittsburgh's position as a regional hub and enhance its global significance.
2 Provide equal access and opportunities for all to live, work, play, learn, and thrive.
3 Grow and diversify Pittsburgh's economy and its tax base.
4 Foster a sense of Citywide community while strengthening neighborhood identities.
5 Capitalize on Pittsburgh's diverse natural and cultural resources.
6 Respect and enhance the relationship between nature and the built environment.

(http://planpgh.com/mission.htm)

Philadelphia, like Pittsburgh, is a home rule city. It has the authority to do things that are not denied by the Constitution or the laws of Pennsylvania. The Philadelphia Home Rule Charter mandates the development, adoption, and implementation of a Comprehensive Plan. It is currently engaged in an ambitious comprehensive planning process – Philadelphia2035 (http://phila2035.org). Its last Comprehensive Plan was over fifty years ago. Like Pittsburgh, its population has declined over the years due to deindustrialization and other factors. Through Philadelphia2035, the city is creating a blueprint or vision to guide public and private investments and decisions with goals and objectives over a 25-year period.

The development of Philadelphia2035 is a multi-year endeavor that will be done through the Philadelphia City Planning Commission with the assistance and support of other city agencies, consultants, and expert advisors. It will be accomplished in two phases – 1) a Citywide Vision, and 2) eighteen strategic district plans. Each district is bigger than an individual neighborhood. Four district plans are slated to be prepared each year. The Citywide Vision and district plans will be updated on a 5–10-year cycle.

The Citywide Vision represents the citywide plan. It was adopted on June 7, 2011. In a June 2, 2011 letter, Mayor Michael Nutter extolled the virtues of the effort:

> This Vision is built on Philadelphia's three key strengths: a strong metropolitan center well-positioned for global competition, the preservation and enhancement of the city's diverse and authentic neighborhoods, and the renewal and transformation of its industrial areas . . .
>
> The Citywide Vision lays out a series of plan elements under three forward-looking themes entitled THRIVE, CONNECT, and RENEW. Each plan element contains overarching goals and measurable objectives, with dozens of strategies for achieving them over a 25-year period. The recommendations in this document are geared toward citywide implementation, but will be further developed and locally-tailored in 18 subsequent District Plans to be completed over the next five years.
>
> (http://phila2035.org/pdfs/final2035vision.pdf)

The three themes serve as umbrella labels for the nine elements that comprise the Comprehensive Plan. THRIVE considers the areas of neighborhoods, economic development, and land management. CONNECT refers to transportation and utilities. RENEW focuses on open space, environmental resources, historic preservation, and the public realm. Each element will have goals and objectives, and actions to achieve the goals and objectives. As of early 2012, two District Plans had been completed and adopted and two were in process.

Philadelphia has been proactive in seeking citizen input into community planning efforts. The city recognizes the importance of citizen input into the planning process. Each district will have a steering committee comprised of public officials, property owners, community organizers, and other stakeholders. The ultimate success of the planning efforts will depend on the collaborative effort between all of the parties. Public meetings regarding Philadelphia2035 have been widely publicized through websites and social media sites like Twitter and Facebook.

Conclusions

Comprehensive Plans create a vision for the community. While the content of the plan may vary from state to state, its goal is to create a blueprint for the physical development of the community. It will need to be monitored, updated, adopted, amended, and eventually updated, revised, etc. Involving the public in plan development and implementation is critical so that the document is a shared vision among many parties.

It is difficult to realize the vision of community without adequate resources. Areas across the nation are facing serious challenges to their financial health. Reductions in funding or the elimination of programs from various levels of government have contributed to the financial plight of cities. Some cannot afford to maintain services or keep employees. Some have gone bankrupt. Many cannot afford to make bond payments. This lack of financial resources may represent a difficult challenge facing cities and their abilities to engage in and implement Comprehensive Plans and associated actions and programs.

Looking outside the jurisdictional boundaries of a community is also important when developing a Comprehensive Plan. The actions or activities of a local government can cause minor or major repercussions in surrounding municipalities and regions. It is imperative that cities consider the regional implications of our local actions.

Further reading

Individuals interested in learning about state requirements for General Plans, Master Plans, and

Comprehensive Plan should consult the websites of the state agency responsible for the development of such documents.

Most localities have their General Plans on their official websites. Check on the site directory for the agency responsible for developing the General Plan.

For information on General or Comprehensive Plans, readers should consult the Governor's Office of Planning and Research (2003) *General Plan Guidelines*; Kelly (2009) *Community Planning: An Introduction to the Comprehensive Plan*; Fulton and Shigley (2012) *Guide to California Planning*, 4th edition; and Hoch *et al.* (2000) *The Practice of Local Government Planning*.

Questions to discuss

1 **What is a Comprehensive Plan?**

2 **What types of elements are included in a Comprehensive Plan?**

3 **Can localities include additional elements in a Comprehensive Plan?**

4 **Why is it important for text and diagrams to be consistent throughout the Comprehensive Plan and within individual elements of a Comprehensive Plan?**

5 **Describe the role of the public in developing a Comprehensive Plan.**

6 **How is a Comprehensive Plan adopted and amended?**

7 **Why is a city concerned with the environmental impacts of its Comprehensive Plan?**

8 **Why is it important to monitor and update a Comprehensive Plan?**

The techniques of zoning and subdivision regulations

A quiet place where roads are wide, people few, and motor vehicles restricted are legitimate guidelines in a land use project addressed to family needs . . . The police power is not confined to the elimination of filth, stench, and unhealthy places. It is ample to lay out zones where family values, youth values, and the blessings of quiet seclusion, and clean air make the area a sanctuary for people.

Belle Terre v. *Boraas*, 1974

The traditional techniques of zoning

Zoning is the division of an area into zones within which uses are permitted as set out in the zoning ordinance. The ordinance also details the restrictions and conditions which apply in each zone. Types of restrictions found in a zoning ordinance typically include allowed uses, restrictions on the height and size of a structure, minimum lot size, and setbacks and sideyards. Thus, the ordinance for the City of Newark, Delaware, a city with a 2006 population of approximately 30,060 residents, has eighteen classes of districts including residential, business, and industrial. There are eight classes of residential districts which are distinguished by house type and density. For example, one classification provides for districts with single-family, detached houses having a minimum lot area of a half-acre, a minimum lot width of 100 feet, a building setback of 40 feet, a rear yard of 50 feet, and two side yards with an individual width of at least 15 feet (and a combined width of 35 feet). Two other one-family detached residential districts have somewhat lower standards; similarly with one-family, semi-detached residential districts. In the three detached districts, the taking of boarders is restricted to not more than three in any one-family dwelling. For a one-family

dwelling in which the owner is non-resident, the limit is reduced from three to two.

Other residential districts are garden apartments up to three stories in height, high-rise apartments of more than three stories with an elevator, and row or town houses. Certain uses are permissible by 'special use permit'. These include police and fire stations, golf courses, professional offices in a residential dwelling, 'customary home occupations', day-care centers, and private non-profit swimming clubs. The zoning ordinance also provides for a Board of Adjustment to which appeals can be made against the decision of the building inspector in enforcing the ordinance, or for a variance from the provisions of the ordinance where a literal enforcement would result in unnecessary hardship.

Zoning is unique to every city. The zoning classes in Newark, Delaware, are different from the zoning classes in Des Moines, Iowa. The 2005 population of Des Moines is approximately 196,917 residents. It is divided into twenty-seven zoning district classifications, including: agricultural district, one-family residential district, one-family low-density residential district, general residential district, large-lot one-family residential district, multiple-family residential district, planned residential development district, residential historic district, commercial–residential district, neighborhood pedestrian commercial district, shopping center commercial district, heavy industrial district, light industrial district, floodplain district, and floodway district. According to the 2000 Des Moines City Code, Sec. 134-306, the A-1 agricultural district is 'intended and designed to preserve or encourage the continuation of agricultural uses, to ensure urban development occurs contiguous to existing urbanized or urbanizing areas and to prevent premature urban development in areas which are not adequately served by public facilities and/or services'. Under Sec. 134-446, the one-family low-density residential district is 'intended and designed to provide for certain low-density residential areas of the city developed primarily with one-family detached dwellings and areas where similar residential development seems likely to occur'. Des Moines also has a Zoning Board of Adjustment that is empowered 'to hear requests for variances and exceptions from the regulations in the Zoning Ordinance and appeals from the decision of the staff in the administration of the Zoning Ordinance, and to make decisions in such matters'.

Not every city in the United States employs zoning. The Department of Planning and Development regulates land development in Houston through various ordinances and policies. It also makes sure that development requests are in compliance with the city's land development ordinance. A planning commission reviews subdivision proposals and variances. According to the Department of Planning and Development website (http://www.houstontx.gov/planning/Develop Regs/dev_regs_links.html), deed restrictions are the only tool that restricts land use. A deed restriction is further defined on the website as:

> written agreements that restrict, or limit, the use or activities that may take place on property within a subdivision. These restrictions appear in the real property records of the county in which the property is located. They are private agreements that are binding upon every owner in the subdivision. All future owners become a party to these agreements when they purchase property in deed restricted areas.

These restrictions cover the types of uses allowed, the number of units allowed per parcel, setbacks, rules for use, and aesthetics.

The City of Houston also provides home owners with a letter they send to their lender indicating that the city does not have a zoning ordinance. The form letter also indicates that it does not address any separately filed restrictions on the property.

The single-family zone: what is a family?

Since the protection of the single-family home is a major reason for (and a major objective of) zoning, it is clearly necessary to define 'family'. Without a definition it would be possible for a group of unrelated students to live 'as a family' and introduce discordant elements into a single-family zone! But definitions can raise as many problems as they solve; and so it is in this case. The first difficulty is whether it is constitutional to 'penetrate so deeply . . . into the internal composition of a single housekeeping unit'. The answer seems to be in the negative except in a few states. In the notorious 1974 *Belle Terre* case, the US Supreme Court upheld a definition that required a family to consist of persons related by blood, adoption or marriage, or a maximum of two unrelated people. The court held that 'the regimes of boarding houses, fraternity houses, and the like present urban problems. More people occupy a given space; more cars rather continuously pass by; more cars are parked; noise travels with crowds.' It could have been objected that the same would result from a family with four teenage children, but the court was carried away by its respect for judicial deference, its overwhelming concern for the archetypical suburban family – and the poetry of its own words, a further oft-quoted sample of which is given at the head of this chapter.

The matter did not end there, however, since a later case (*Moore* v. *City of East Cleveland*) concerned an embarrassingly nonsensical outcome. The City of East Cleveland, Ohio, had a complex definition of a family which had the result of making one owner's occupancy of her house illegal. The zoning ordinance allowed only traditional nuclear families. The oddity was that all the occupants were related by blood, but the degree of relationship was insufficient to satisfy the ordinance: the family consisted of Mrs Moore, her son, and two grandsons who were first cousins rather than brothers. Mrs Moore received a notice of violation from the City stating that one of the grandsons was an 'illegal occupant'. Mrs Moore refused to remove him, and the City filed a criminal charge. Upon conviction she was sentenced to five days in jail and a $25 fine. The City argued before the Supreme Court that its decision in *Belle Terre* required it to sustain the ordinance. The usual case was made about the need to prevent over-crowding, to minimize traffic and parking congestion, and to avoid an undue financial burden on East Cleveland's school system.

Surprisingly, at least to those who are unfamiliar with the element of unpredictability which is to be found in the workings of the Supreme Court, the justices had great difficulty with this case. However, the majority concluded that the ordinance was an 'intrusive regulation of the family', and that it was distinguishable from *Belle Terre* in that the latter case dealt with *unrelated* persons. The court recognized that it cannot intrude on choices concerning family living arrangements.

The issue is not, however, settled; and perhaps, like many other zoning matters, it may never be. In fact, generally there seems to be a trend to liberalize the meaning of the term 'family' to take into account the freer modes of conjugality that are now more common (see Box 8.1). The test appears to be whether there is 'a legitimate aim of maintaining a family style of living' (Wright and Gitelman 2000: 223).

Group homes

A similar test has been applied to group homes for foster children, the mentally retarded, and other groups to which neighbors may object. The rationale here is that the essential purpose of a group home is to provide a family-like environment (in contrast to the custodial character of an institution). The situation is clear where a foster home consists of a married couple and their children, plus foster children. The issue is more difficult when professionals staff a home, and court decisions in such cases are conflicting.

BOX 8.1 REPRESENTATIVE DEFINITIONS OF 'FAMILY'

Gaylord, Minnesota Zoning Ordinance Section 4 Rules and Definition:

Family means any number of persons living together in a room or rooms comprising a single housekeeping unit and related by birth, marriage, adoption or any unrelated person who resides therein as though a member of the family including the domestic employees thereof. Any group of persons not so related but inhabiting a single house shall, for the purpose of this Ordinance, be considered to constitute one family for each five (5) persons, exclusive of domestic employees, contained in each such group.

San Diego Municipal Code, Chapter 11: Land Development Procedures, Article 3: Land Development Terms, Division 1: Definitions, Section 113.0103:

Family means two or more persons related through blood, marriage, or legal adoption or joined through a judicial or administrative order of placement of guardianship; or unrelated persons who jointly occupy and have equal access to all areas of a dwelling unit and who function together as an integrated economic unit.

In recent years, many states have passed legislation to prevent the exclusionary zoning of group homes. The nature of the legislation varies: some measures are restricted to certain types of home, while others are much broader. Some designate group homes as a 'special exception' under the zoning ordinance; others classify group homes as a separate use to which special standards apply.

In 1985, the US Supreme Court issued a much-awaited decision on group homes in *City of Cleburne, Texas* v. *Cleburne Living Center, Inc. (CLC)*. The CLC wanted to lease a building for the operation of a group home for the mentally disabled. A city ordinance required a special use permit for a group home of mentally retarded individuals but not for other care and multiple-dwelling facilities. Other multiple-dwelling facilities were freely admitted and allowed without permits. CLC applied for the permit and the city council denied it. The reason for the denial was that under Cleburne's zoning, a group home for the mentally retarded was classified as a 'hospital for the feeble-minded' – a use not allowed in the applicable residential zoning zone.

The court ruled that the ordinance requiring a special use permit for group homes for the mentally retarded but not for other care and multiple-dwelling facilities violated the equal protection clause. Moreover, there was no rational reason advanced by the City of Cleburne as to why such group home facilities would pose any threat to the area.

The single-family house: should there be a minimum size?

Photographs of unsanitary, tiny, crowded tenements leave one in no doubt that there are standards below which society will not, in all conscience, wish families to live. These standards vary over time and space. What is considered intolerable in the early 2000s is very different from what was so considered in the 1790s. Similarly, contemporary standards in the United States are very different from those in Bangladesh. Every society has to define for itself the standards at which it expects (and will assist) its people to live. There is

nothing scientific about this: it is a matter for judgment and political decision. Yet the zoning system frequently brings these matters before the courts for adjudication; for example, is the minimum lot size or the minimum floor area prescribed by a zoning ordinance acceptable? Unfortunately, the question is more narrowly conceived than this since the courts commonly operate on 'the presumption of validity': that an act of a legislative body cannot be challenged unless it is blatantly unfair. (This was discussed in the previous chapter.) This makes it difficult to challenge minimum area requirements because the onus of proof is on the plaintiff to show that the provision could not have had a valid purpose. As a result, the argument before (and of) the court is usually couched in terms of a dispute between an individual developer and the local inhabitants: wider issues of exclusion and of regional housing needs tend to be pushed into the background, even if they surface at all.

Two cases illustrate the issue. In a 1953 New Jersey case (*Lionshead Lake*), an ordinance provided that residential areas should have a minimum square footage of 768 for a one-story dwelling, 1,000 for a two-story dwelling having an attached garage, and 1,200 for a two-story dwelling not having an attached garage. Despite so-called 'expert' testimony which maintained that there was scientific evidence on the effect of living space on mental and emotional health, the trial court concluded that the requirements 'were not reasonably related to the public health, were arbitrary and unreasonable, and not within the police powers' of the township. The New Jersey Supreme Court disagreed, and held that

> it is the prevailing view in municipalities throughout the state that such minimum floor area standards are necessary to protect the character of the community . . . In the light of the constitution and of the enabling statutes, the right of a municipality to impose minimum floor area requirements is beyond controversy.

The court soon had cause to regret these words: its decision gave rise to extensive academic discussion which caused it to rethink its position in later cases.

In a 1979 case, it gave prominence to the issue of 'economic segregation'. It noted that, in the quarter-century since *Lionshead Lake*, changes had taken place which were reflected in legislative and judicial attitudes: 'once it is demonstrated that the ordinance excludes people on an economic basis without on its face relating the minimum floor area to one or more appropriate variables, the burden of proof shifts to the municipality to show a proper purpose is being served'.

Large-lot zoning: maintaining community character

Large-lot zoning has the ostensible purpose of safeguarding the public welfare; for example by ensuring that there is good access for fire engines, that roads do not become unbearably congested, or that there is adequate open space. These and similar worthy objectives appear frequently in zoning cases, as does an alternative formulation: to keep out undesirable (that is different) people, and to maintain the social and economic exclusiveness of an area. A leading case arose in the Boston suburb of Needham. To control the amount of development in the area, the town passed an ordinance which provided for a minimum lot size of one acre over much of the area. Though declaring that insular interests must give way to the wider good, the court held that the zoning was valid and reasonable. It was swayed by the fact that 'many other communities when faced with an apparently similar problem have determined that the public interest was best served by the adoption of a restriction in some instances identical and in others nearly identical with that imposed' by Needham.

Other apparently similar cases have been decided differently, but no selection of decisions is necessarily representative. On the contrary, as a leading legal digest expresses the matter: 'the validity of large lot zoning is likely to vary depending on the size of the lot, the circumstances of the community or area involved, and the hostility or lack of it to large lot zoning in a particular jurisdiction' (Wright and Gitelman 2000: 227).

While the Needham example mentioned a one acre minimum lot size, the definition of what constitutes a large lot varies dramatically by city. One community may define it as between 1 and 10 acres, while another community may define it a one house for every 20–40 acres. Large-lot zoning has been used as a growth control technique by a number of communities. In their so doing, it consumes land and could also result in what might be labeled 'rural sprawl'.

Floor area ratio

The floor area ratio (FAR) regulates building bulk while providing the developer some latitude in determining the height of a building and its placement on the lot. It can be expressed as the total floor area divided by the total lot area. Another way to view it would be to take the FAR and multiply it by the lot area. This would equal the total amount of allowable floor space. It simply represents the maximum amount of floor space that can be built on a given lot. The FAR is usually expressed as a decimal fraction.

The following examples will illustrate the FAR concept. If the zoning ordinance specifies that the FAR in a given district is 1.0 and the total lot area is 10,000 square feet, a developer would be able to cover the entire 10,000 square foot site with a one-story high building. It would also allow the developer to construct a two-story building that covers 50 percent of the lot area. It could also allow the developer to construct a four-story building that covers 25 percent of the lot area. In single-family residential districts it is not uncommon to see a FAR of anywhere from 0.30 to 0.60. Thus, if the single-family residential district FAR is 0.60 and the lot area is 5,000 square feet, the maximum permitted amount of floor space that can be built on the lot would be 3,000 square feet (5,000 × 0.60). It is important to remember that the various zoning categories will have different FARs. For example, while the single-family residential district FAR might be 0.60, a commercial office zone might have an FAR of 2.0. Moreover, multiple-family zones could have FARs ranging from 0.75 to 7.0.

Apartments and mobile homes

The reader who has come this far will not be surprised to find that apartments and mobile homes (often far from 'mobile') are the targets of particularly explicit exclusionary practices. However, courts differ in their attitudes to these. Some have gone so far as to approve the restriction throughout an entire jurisdiction of all uses except single-family dwellings. By contrast, other cases have ruled that municipalities must allow all types of dwellings in their area. A leading case is *Girsh*; this invalidated a zoning ordinance that totally excluded apartments from the Philadelphia suburb of Nether Providence, Delaware County. The Supreme Court of Pennsylvania ruled that: 'Nether Providence Township may not permissibly choose to only take as many people as can live in single family housing, in effect freezing the population at near present levels.' Unfortunately, the court did not indicate what rights the owners of the Girsh property had as a result of its decision. Nether Providence subsequently zoned several pieces of land – but not the Girsh property – for apartments, claiming that it had thereby complied with the court's decision. The Girsh property owners disagreed, and after two years won a clarifying order from the court which directed the township to grant the permits required for the development of their site. In the meantime, the township had begun procedures to condemn the Girsh property for a public park. This is a typical example of the way in which the drama of land use disputes is played out.

One obvious question which arises with mobile homes is a definitional one: is not a mobile home a single-family dwelling? Certainly, modern, well-equipped mobile homes in an attractive park may be difficult to distinguish from the stereotypical single-family home which, in fact, nowadays can be largely factory produced. The point becomes one of particular significance with 'manufactured housing' intended for a permanent siting. This type of housing has been built since 1976 under a national code of health and safety requirements. An observer might have thought that locating an immobile manufactured house on a permanent site would have translated a 'mobile home' into a 'single-family dwelling'. Not so, for example,

the Illinois Supreme Court in *Village of Cahokia v. Wright (1974)*, where the zoning ordinance not only restricted manufactured housing to mobile housing parks, but also prohibited such housing from being permanently fixed in such a way as would prevent its removal. The Illinois Supreme Court upheld the ordinance on the grounds that a mobile home might be detrimental to the value of adjacent conventional single-family homes, stifle development in the area, or create potential hazards to public health.

There are innumerable such cases. Some reveal remarkable ingenuity on the part of municipal governments in devising methods for excluding mobile homes: a minimum width for all dwellings; a three acre minimum lot size; a minimum of 'core living space' for all dwellings of 20 feet by 20 feet.

Prior to the Fair Housing Amendment Act of 1988, it was common for local governments to restrict mobile homes to adults and seniors only. This Act makes it unlawful to discriminate against families in the sale, rental, or financing of housing (with some exceptions in the case of housing communities for senior citizens). With changes in design and layout, mobile housing (now more commonly termed 'manufactured housing') has become more acceptable in recent years. Indeed, it is often difficult to identify what is and what is not 'manufactured'.

Conditional uses

There are some uses which, though permissible (and necessary), require review to ensure that they do not have an undesirable impact on an area. Hospitals, schools, day-care centers, and clubs, for example, are needed in a community, but their specific location may give rise to traffic congestion and dangers, or to severe parking difficulties. Similarly with gas stations in commercial districts, and multi-family dwellings in a single-family district. Zoning ordinances typically make specific provision for such developments which require special restrictions. Though terminology varies among municipalities, these are appropriately termed 'conditional uses'. Box 8.2 illustrates the safeguards and criteria to be met for a conditional use permit.

> ## BOX 8.2 CONDITIONAL USE PERMIT SAFEGUARDS AND CRITERIA TO BE MET
>
> 1 The establishment, maintenance, or operation of the conditional use will not be detrimental to endanger the public health, safety, comfort or general welfare.
> 2 The conditional use will not be injurious to the use and enjoyment of other property in the immediate vicinity for the purposes already permitted nor substantially diminish and impair property values within the neighborhood.
> 3 The establishment of the conditional use will not impede the normal and orderly development and improvement of surrounding property for uses permitted in the zoning district.
> 4 Adequate facilities, access roads, drainage and/or other necessary facilities will be provided.
> 5 Adequate measures will be taken to provide ingress and egress designed to minimize traffic congestion in the public streets.
>
> *Source*: St Charles County, Missouri

In order to obtain a conditional use permit, applicants must follow the procedures dictated by the community. First, they must file an application for the conditional use permit. The permit would contain information on such items as a legal description of the land, a parcel map outlining the land in question and all adjoining properties, a deed showing ownership of the land, a plan showing what you want to do to the property, a signature by the owner of the property, and the necessary application fee. The application is then processed by the appropriate governing body. Notices are placed in various locations to alert other individuals and entities that a conditional use permit is being sought by a property owner.

Variances

While a conditional use is one which is permissible under the conditions of the zoning ordinance, a variance involves a relaxation of the provisions of the ordinance. The Standard State Zoning Enabling Act confers on the board of adjustment the power 'to authorize upon appeal in specific cases such variance from the terms of the ordinance as will not be contrary to the public interest, where, owing to special conditions, a literal enforcement of the provisions of the ordinance will

result in unnecessary hardship, and so that the spirit of the ordinance shall be observed and substantial justice done'.

Variances are of two types: 'area' (or 'bulk') and 'use'. The former involves a departure from the requirements of the ordinance in relation to such matters as lot width, lot area, setback, and the like. It recognizes that not all property is created alike. It allows unique circumstances to be considered by a planning commission, zoning board of appeals, or some similar body. By contrast, a use variance allows the establishment (or continuation) of a use which is prohibited by the ordinance. Allowing a house to be built closer to the lot line laid down in the variance would be an area variance; allowing a multi-family house in a single-family district would be a use variance. In many states, the distinction is of no consequence since the same conditions have to be met (as is the case with the SSZEA provisions). In others, the distinction is crucial since use variances are totally prohibited.

The hardship theoretically has to be one which applies to a particular property, not to the personal circumstances of the owner. The rationale for this is that the matter for consideration is the relationship between the particular plot and the wider area. Any effect which a variance has on this wider area will persist after a change of ownership, or even if the hardship ceases. In fact, many variances are given precisely because of personal hardship. One board had an explicit policy of allowing any use variance requested by a disabled veteran – including automotive repair and body work at homes in a residential area, and the sale of groceries in the front room of a residence. This may be unusual, but there is plenty of evidence to show that boards frequently do consider personal circumstances. One board permitted home occupations in cases of personal hardship on the ground that the harm to the particular neighborhood was far outweighed by the economic hardship to the applicant.

The tests set out in the *Otto* case (Box 8.3) have been widely, though certainly not universally, adopted. In particular, the requirement that there be an inability to make a reasonable return has become a standard requirement for variances – though 'reasonable' should not be interpreted to mean 'maximum'. Many cases could be quoted, but even a long list would be misleading since the differences among (and even within) the states on the issue are great.

The types of variances and the processes used to obtain a variance differ by jurisdiction. For example, in Multnomah County, Oregon, variances apply to setbacks. A variance is needed to build or improve a property in a way that is different from the 'dimensional requirements' of the county zoning ordinance. The County recognizes three types of variances: a minor variance (if the proposed land use varies from the dimensional requirements by 25 percent or less); a major variance (if it varies by more than 25 percent); and a residential hillside variance (when a development has an average grade of more than 5 percent from the front of the real property line).

The decision to obtain a variance may be reached through an administrative process (with the planning director making the decision on whether or not to grant a variance) or a hearing officer. The process commences with the property owner getting the proper application and paying the required fee. The property owner is then required to notify neighboring property owners of the proposed site plan and to get their consent. If the consent of the property owners is obtained, a decision is reached within a given number of days as specified in the ordinance. If the proposal fails to get the consent

of the neighboring property owners, a hearing is held. Notice is given to the surrounding property owners and a hearing on the variance request is held. The likelihood that the proposal will damage the public welfare or neighboring property is one criterion that can be used to reach a decision. Whether or not the proposal complies with the General or Comprehensive Plan represents another criterion. Moreover, whether or not a circumstance exists that prevents the property from being used in the manner it is zoned represents another criterion that should be considered. For example, in the City of Greenbelt, Maryland, a variance is granted if the City finds that:

1 A specific parcel of land has exceptional narrowness, shallowness, or shape, exceptional topographic conditions, or other extraordinary situations or conditions;
2 The strict application of this Subtitle 27-230 will result in peculiar and unusual practical difficulties to, or exceptional or undue hardship upon, the owner of the property; and
3 The variance will not substantially impair the intent, purpose or integrity of the general plan or master plan.

The decisions of the planning director or hearings officer can be appealed to the appropriate bodies, as identified in the zoning ordinance.

It was the original intention that variances would be exceptional. It has not worked out that way. The variance is a popular tool of the boards of appeal, who see themselves as a broker for the hard-pressed citizen against the harshness of the law. One writer has suggested that the board of appeals operates as a kind of jury, dispensing rough justice in its hearings of variance applications, resulting in decisions which 'are very apt to reflect the conscience of the community – a close approximation of what most people in the community would think the proper course of action'. Various studies have convincingly shown that boards of adjustment commonly operate according to their own sense of what is right, with little regard for the law or even their local planning department. Most applications are in fact approved.

BOX 8.3 VARIANCES – THE HARDSHIP TEST

The classic statement of the hardship test appears in the 1939 New York case of *Otto* v. *Steinhilber*:

Before the board may . . . grant a variance upon the ground of unnecessary hardship, the record must show that (1) the land in question cannot yield a reasonable return if used only for a purpose allowed in that zone; (2) that the plight of the owner is due to unique circumstances and not to the general conditions of the neighborhood which may reflect the unreasonableness of the zoning ordinance itself; and (3) that the use to be authorized by the variance will not alter the essential nature of the locality.

Since the evidence suggests that illegal use of variances is widespread, it has been proposed that variances should be abolished. It has been argued that this would lead to better and more carefully drafted zoning ordinances. It has also been suggested that variances should be subject to review by a higher authority such as a state review board or specialized courts having metropolitan jurisdiction, or that the power to grant variances be taken away from boards of appeal. But, however much lawyers attack the legal deficiencies of the variance, its popularity at the local level assures its continuance as a major feature of the zoning system. There is more to planning than law.

Spot zoning

'Spot zoning' is the unjustifiable singling out of a piece of property for preferential treatment. It is not a statutory term: it is a judicial epithet signifying legal invalidity. In a Connecticut case, the court warned that an amendment to the zoning map 'which gives to a single lot or a small area privileges which are not extended to other land in the vicinity is in general against sound public policy and obnoxious to the law'. Such spot zoning is frowned upon by the courts but, if a planning commission decides 'on facts affording a sufficient basis and in the exercise of a proper discretion, that it would serve the best interests of the community as a whole to permit a use of a single lot or small area in a different way than was allowed in surrounding territory, it would not be guilty of spot zoning in any sense obnoxious to the law'.

In this particular case, a landowner requested a rezoning of a small piece of land near (but not adjacent to) some new development. The existing zoning was for residential use, but the owner saw a need for some shops, and he proposed to erect a drug store, a hardware store, a grocery store, a bakeshop, and a beauty parlor. He requested an appropriate change of zoning, which the planning commission granted. On appeal, the trial court concluded that the requested change amounted to spot zoning, but the appeal court reversed. Its argument was that, on the facts of the case, there was by no means unanimous opposition from surrounding owners to the proposed development and that, even had there been, 'it was the duty of the commission to look beyond the effect of the change upon them to the general welfare of the community'. This the planning commission had done in deciding to support the rezoning: there was a need for additional stores in the area; and it was the policy of the commission 'to encourage decentralization of business in order to relieve traffic congestion and that, as part of that policy, it was considered desirable to permit neighborhood stores in outlying districts'.

Moreover, a Nevada court has ruled 'the test of spot zoning is whether the amendment was made with the purpose of furthering a comprehensive zoning scheme or whether it was designed merely to relieve the land of a restriction which was particularly harsh upon that particular land'. For example, if a property is designated for industrial use in an area zoned residential in the existing community plan, it would be considered illegal since it is incompatible with the existing plan. In a 2001 ruling, the Montana Supreme

Court ruled that a zoning change that had been approved by a county commission constituted spot zoning because it benefited only a single landowner and infringed upon the property rights of neighboring landowners.

Floating zones

There are a number of potential land uses that local governments anticipate will occur. Cities will create and define a zoning category, along with standards and criteria, but reserve the decision about its location for the future. For example, they might anticipate the need for a shopping center, airport, hospital, or school. Standards are set and described in the zoning ordinance but no location has been designated. As the Ames, Iowa, Municipal Code, Chapter 29, Article 12, Section 29.1200 suggests, 'it "floats" above the zoning map and is dropped or "mapped" on the zoning map upon compliance with standards and the application process provided for in this Article'. In other words, it is not a pre-mapped zoning district depicted on a city zoning map. An example might be the development of garden apartments in an area zoned for single-family residential use if certain criteria and requirements found in the zoning ordinance are met.

As with all zoning-related topics, it is important to consult each state's enabling statute to determine how flexible courts will be in allowing municipalities to use the floating zone concept. One state may view it as essentially spot zoning. Another state may allow its use as long as the proposed use does not conflict with the Master Plan.

Downzoning

While an upzoning may well raise the wrath of the neighborhood, an amendment to rezone to a use of lower intensity – a 'downzoning' – is often the result of neighborhood pressure. Since a downzoning is likely to reduce the value of undeveloped land and limit what can be done on the land, an objection is likely on the part of the landowner.

A good illustration is a 1983 Iowa case, where a city downzoned some six acres of land on which the owner was intending to build a federally subsidized housing project. The downzoning took place after a public outcry, though ostensibly on the ground that the City's electrical, water, sewer, and road systems were inadequate for a concentration of multi-family dwellings in the area. Not surprisingly, the owner claimed that the reasons given were mere pretext, and that the downzoning was racially motivated. The court, however, held that the City's decision had been taken for valid reasons, i.e. the inadequacy of the utility systems. Furthermore, there was no evidence that the City had a discriminatory purpose. Thus, applying the 'fairly debatable' rule, the downzoning was upheld. The court added: 'zoning is not static. A city's comprehensive plan is always subject to reasonable revisions designed to meet the ever-changing needs and conditions of a community. We conclude that the council rationally decided to rezone this section of the city to further the public welfare in accordance with a comprehensive plan.'

By contrast, a Connecticut case was decided the opposite way. The court rejected a downzoning which affected the whole of one of the two districts into which the town was divided. It noted that the downzoning was 'made in demand of the people to keep Warren a rural community with open spaces and keep undesirable businesses out'.

Cases on downzoning abound but, because of their great variety and lack of consistency, it is difficult to make any general sense of them. Ultimately, the cases examine issues on whether the downzoning is consistent with the General Plan, whether it represents a valid use of the police power, whether it is a taking without payment of just compensation, whether any reasonable economic use of the property remains, how much the property value has been diminished, and whether the landowner has any vested rights. However, one point can be made with a moderate degree of certainty: piecemeal downzonings are likely to be examined much more carefully by the courts, without the usual assumption of validity.

Contract zoning and site plan review

Zoning theoretically requires uniform conditions within districts. Uniformity, however, can lead to undesirable rigidity, and it may be to the benefit of both the property owner and the community to depart from a uniform regulation. It offers a bit of flexibility to the perceived inflexible nature of zoning. It is a deviation from the zoning ordinance. It is here that contract zoning can be useful. The jurisdiction has the power to negotiate with the property owner.

Essentially, contract zoning is, as the term suggests, the rezoning of a property subject to the terms of a contract. Box 8.4 illustrates the purpose of contract zoning, as practiced in the Town of Saco, Maine. The State of Maine, in 30A MSRA Section 4352 (8), authorizes its use by requiring that any rezoning be consistent with the State Growth Management Plan; that the rezoned property be consistent with existing and permitted uses within the original zone; and that conditions and restrictions must be related to the physical development or operation of the land. It is also, on occasion, referred to as 'conditional zoning'. The process appears to be one in which the city informs a property owner that it does not have to rezone the property in question. The next step would be for the city to say it might rezone the property if the property owner agrees to do something in return – certain conditions must be met. Typically, the terms of any agreement are negotiated between the owner and the local government following a specific proposal by the owner. Cities that allow contract zoning encourage property owners to meet with neighboring citizens to discuss the proposal.

There is much learned discourse on the validity and the desirability of contract zoning. The argument in favor holds that conditions can render acceptable a use which otherwise would be unacceptable. The contrary argument is that the police power cannot be subject to bargaining, that conditional rezoning is illegal spot zoning, and that local governments have no power to enact contract zoning amendments. Those arguing against contract zoning view it as 'spot zoning'. States and courts differ widely in their attitudes, and overall the position is confused, to say the least. Individuals must consult state enabling legislation to see if contract zoning is allowed. In some states, it is not allowed.

Many of the conditions that have been imposed are now normally included in 'site plan review'. This is the preparation of a site plan for approval by the planning board. Such a review can be a normal zoning requirement, or a special requirement for particular types of development such as cluster zones and planned unit developments.

Site plan reviews are needed to make sure the proposed development is in compliance with local zoning and other municipal ordinances. Plan review represents the first step prior to the issuance of a building permit. They are needed for different types of activities. For example, in North Myrtle Beach, South Carolina, site plan review is required for 'all new development (with the exception of single family homes on existing lots of record), redevelopment, additions, alterations, or changes in the use that necessitate a change in the parking area'. Other areas might require a site plan review for all public and semi-public buildings or for construction in any floodplain area.

Plans to be reviewed generally take the form of a preliminary plan and a final plan. However, it is possible in some areas that a developer might choose to submit a final plan for review. It is a process ripe for negotiation between developer and public officials. The items to be reviewed might include a dimensional site plan, consistency with applicable zoning ordinances, landscaping, drainage, and compatibility with neighboring structures and the surrounding environment. Box 8.5 provides a listing of contract conditions in the City of Moraga, California. Among the personnel that might review the plan are planners, zoning administrators, public works officials, engineers, police and fire departments, building officials, and street officials. These individuals will determine whether the proposed site plan is in accordance with municipal zoning and the General Plan, creates any public facility or traffic problems, complies with all other municipal requirements, and contributes to the protection of public health, safety, and welfare.

BOX 8.4 PURPOSE OF CONTRACT ZONING

Occasionally, competing and incompatible land uses conflict; and traditional zoning methods and procedures such as variance, conditional use permits, and alterations to the zone boundaries are inadequate to promote desirable growth. In these special situations, more flexible and adaptable zoning methods are needed to permit differing land uses in both developed and undeveloped areas, and at the same time recognize the effects of change. In consideration of a change in zoning classification for a particular property or group of properties, it may be determined that public necessity, convenience, or the general welfare require that provisions be made to impose certain limitations or restrictions on the use or development of the property. Such conditions are deemed necessary to protect the best interests of the property owner, the surrounding property owners and the neighborhood, all other property owners and citizens of the City, and to secure appropriate development consistent with the City's Comprehensive Plan.

Source: City of Saco, Maine, *Zoning Ordinance*, Section 1403-1, 2003

BOX 8.5 PROVISIONS FOR CONTRACT ZONING

A.   Compliance with the plans and specifications submitted by the developer;
B.   Time within which the development must be started or completed and controlling the sequence of development, including when it must be commenced and completed;
C.   Protective measures that a developer must undertake for the benefit of neighboring property, such as the construction of fencing or the establishment of buffer areas;
D.   Minimizing adverse impact of the development upon other land, including the hours of use and operation and the type and intensity of activities that may be conducted;
E.   Controlling the duration of use of development and the time within which structures must be removed;
F.   Assuring that development is maintained properly in the future;
G.   Provision for streets, other rights of way, utilities, parks, and other open space;
H.   Payment of an amount of money into a fund for the provision of streets, other utilities, parks or other open space;
I.   Creation or conveyance of interests in lands;
J.   Provision for the grant of a transferable development right (Chapter 8.104) (Prior code section 8-4504).

Source: City of Moraga, California, *Municipal Code* 8.100.040, 2006

Cluster zoning and planned unit development

Traditional zoning is based on the assumption that residential development will take the form of single-family houses on individual lots. New patterns of development emerged in the postwar years which require much more flexibility. This is provided by cluster zoning, which involves the clustering of development on one part of a site, leaving the remainder for open space, recreation, amenity or preservation (see Figure 8.1). Homes are thus on smaller parcels of land.

Standard zoning   Cluster zoning

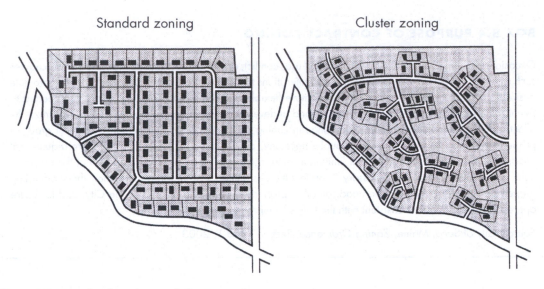

Figure 8.1 Standard zoning and cluster zoning

The overall density of the site is unchanged but, of course, the density of the developed part is increased. This has a number of advantages: the cost of paving and of supplying utilities is reduced; attractive landscape features (or wetlands) can be protected; open space can be provided for recreation (active or passive); and housing can be provided of a type suitable for 'non-traditional' households who do not want the bother of maintaining a large lot.

The basic idea, of course, is not a new one. It goes back to Clarence Stein, Ebenezer Howard, Clarence Perry, and Frederick Law Olmsted. Perhaps its most famous prototype is Radburn, New Jersey: a 149-acre development with a strict separation of road systems, traffic-free residential culs-de-sac, and a continuous inner park.

A refinement of the cluster concept is the 'planned unit development', affectionately known as a PUD. This differs from cluster zoning in that it is more than a design and planning concept: it also provides a legal framework for the review and approval of development. It also can incorporate (or even be confined to) commercial, mixed use, and industrial development. Instead of adhering to preset regulations, the PUD system gives developers the freedom to design developments which

satisfy market demands. In place of elaborate lists of potential uses, a PUD simply sets out the criteria which have to be met (in relation, for example, to noise, vibration, smoke, odors, heat, glare, and traffic generation). Of course, this approach necessitates negotiations between the developer and the municipality: this is the mechanism by which flexibility is achieved. This is a far cry from traditional zoning.

The availability of flexibility in design is a key to the PUD concept. Flexibility can take the shape of reducing building setbacks, altering parking requirements, increasing building heights, modifying lot size, etc. A great deal of negotiation can take place between the developer and the appropriate local officials. Ultimately, the flexibility means that the developer must assure the local officials of certain things. For example, in Milpitas, California, in exchange for the design flexibility of the PUD, the applicant must demonstrate that the development proposal 'does not result in an intensity of land higher than otherwise allowed, provides public benefits that would not be achievable through the normally required zoning standards, does not create unmitigated traffic impacts, is compatible with the surrounding development, and is consistent with the Milpitas General Plan'.

One other feature of PUDs needs to be mentioned: the role of home owners' associations in managing commonly held property. Membership of such an association can be mandatory for the owners of dwellings in a PUD. Whether the result is a happy, democratic way of managing the local environment, or a financial, administrative, and political nightmare depends on the particular circumstances of the development – and the accidents of time, place, and neighbors.

The word 'flexibility' should suggest that negotiations between the developer, adjacent property owners, and the local government are common. Developers often come to the property owners and appropriate government agencies to see if they can negotiate items such as lot size, setbacks, etc. in return for providing something extra to the community. For example, developers have negotiated different lot sizes and setbacks in return to providing more land for a park site. Negotiations between developer, property owners, and various government agencies are quite common in most jurisdictions today.

Performance zoning

Performance zoning offers another opportunity to provide some flexibility into the traditional application of zoning. While traditional zoning identifies the specific land uses allowable in each zoning category, performance zoning identifies the intensity of land use that is acceptable. It is more concerned with the impacts or effects rather than the use of the property. The impacts could be on the environment or on public services. The city identifies the performance criteria or standards and the developer has to determine how to achieve them. Decisions on development proposals are made on a case-by-case basis. For example, a city using performance zoning and a rating/point system might allow the grouping of homes as long as the development meets standards associated with the protection of natural resources, the provision of open space, density, noise abatement, traffic congestion, visual quality, etc. If the standards are not met and the impacts are considered negative, then the development proposal would not be approved.

An example might illustrate how performance standards could work. The City of Davis, California, in Municipal Code 40.05.060, has attached the following performance standards to the issuance of a conditional use permit:

(a) No use shall be the source of such noise, smoke, light, odor or vibrations as to damage or constitute a nuisance to the surrounding properties.
(b) No use shall create a hazard to surrounding uses from explosion, contamination or fire.
(c) No use shall generate pedestrian or automobile travel incompatible in type or amount with the original character of the district.
(d) The planning commission may deny a conditional use permit to any use which does not further the core area plan, or which does not promote the public health, safety, or welfare.

Mixed-use zoning

Allowing two or more additional uses on a parcel is a classic definition of mixed-use zoning (see Plate 7). For example, allowing a commercial use on the first floor and having apartments on the second floor of a building would be an example of mixed-use zoning. Mixed-use zoning is making a comeback. Cities have recognized that mixed use encourages greater density, helps reduce auto-dependency, promotes walking and bicycle use, and helps promote the idea of a community.

The City of Colorado Springs, Colorado, is employing mixed-use zoning to vary neighborhood patterns. Through its Comprehensive Plan, the City is promoting the use of mixed-use developments by identifying and supporting new and existing neighborhood opportunities for mixed-use development. Box 8.6 shows the primary purposes of the mixed-use concept in Colorado Springs.

Non-conforming uses

The introduction of a zoning scheme presents obvious problems with regard to existing uses which thereby

Plate 7 Mixed-use development in San Diego

Photo by author

BOX 8.6 PURPOSES OF MIXED-USE ZONING

The primary purposes of this part are to:

A. Provide appropriate areas for and facilitate quality mixed use development in activity centers that are consistent with the Comprehensive plan's land use and transportation goals, objectives, policies and strategies;
B. Accommodate intensities and patterns of development that can support multiple modes of transportation, including public transit and walking;
C. Group and link places used for living, working, shopping, schooling, and recreation, thereby reducing vehicle trips, relieving traffic congestion, and improving air quality in the city;
D. Provide a variety of residential housing types and densities to assure activity in the district to support a mix of uses and enhance the housing choices of city residents; and
E. Integrate new mixed use development with its surroundings by encouraging connections for pedestrians and vehicles and by assuring sensitive, compatible use, scale, and operational transitions to neighboring uses.

Source: City of Colorado Springs, Colorado, Municipal Code, 7.3.701

become non-conforming. A number of uses pre-date current zoning regulations in many areas around the United States. It is impracticable to have these uses removed; indeed, any such threat would have been sufficient to kill off any idea of zoning. The general approach taken has been to hope that, in time, the non-conforming uses would pass away. This has typically proved not to be the case, and municipalities have striven to find ways to speed up the process. They have had little success.

The courts have been unsympathetic to municipalities which attempt to 'zone retroactively', though some of the early landmark decisions (such as *Hadacheck*, which is discussed in Chapter 5) apparently provided the constitutional basis where the offending use became a nuisance. The most common method of applying a control over non-conforming uses (limited though it is) forbids 'expansion' or 'alteration'. Sometimes a restriction is imposed on rebuilding if a non-conforming use is 'destroyed', for example by fire. Moreover, a use which is 'abandoned' may be refused permission for resuscitation (after a certain number of years). Box 8.7 illustrates how one jurisdiction handles restoring a non-conforming use.

Unfortunately, all these terms have been, and continue to be, subject to intense debate and judicial differences. Perhaps the clearest case in which a non-conforming use may be eliminated is where a billboard is amortized over a number of years. Amortization is sometimes seen as the most painless way of ridding an area of an undesirable use, and the courts have been sympathetic, particularly where the non-conformer is given a reasonable amount of time. However, the political problems remain, as is dramatically illustrated by the success of the billboard lobby in preventing the use of amortization in connection with the federal highway advertising program. Not only was amortization prevented, but the Act actually requires the payment of compensation for the removal of billboards (see Chapter 17). There could be no clearer example of the force of politics in land use planning.

Zoning amendments

A zoning amendment (or 'rezoning', or 'map amendment') is similar to a use variance in that it permits a use which is not allowed by the provisions of the zoning ordinance. However, while a use variance grants the owner an exemption (and leaves the ordinance intact), an amendment changes the ordinance itself. An amendment should be of greater consequence than a variance but practice does not always conform to theory, or even legality.

Zoning is not static. It will need to be changed just as a city's Comprehensive or General Plan needs to be changed. Amendments can be made to the zoning ordinance or to the map. The former deals with the written provisions of the ordinance, the latter with its detailed designation on a map of the area. An amendment can be proposed by a property owner, the city council, or the planning commission. Those individuals or entities that have the power to propose an amendment will vary by jurisdiction. The most

---

BOX 8.7 RESTORING NON-CONFORMING USES

Any non-conforming building or structure damaged more than fifty percent (50%) of the County Assessor's Market Value, exclusive of foundations at the time of damage by fire, collapse, explosion or Acts of God or public enemy, shall not be restored or reconstructed and used as before such happening; but, if less than fifty percent (50%) damaged above the foundation, it may be restored, reconstructed or used as before provided that it is done within twelve (12) months of such happening and that it is built of like or similar materials, or the architectural design and building materials are approved by the Planning Commission.

Source: Gaylord, Minnesota, Zoning Ordinance Section 5, Subdivision 5 (3) Nonconforming Uses

common is a map amendment which allows a more intensive use of a particular area. Such an 'upzoning' is usually to a more profitable use, and it is typically made in response to a request by the landowner. It is also regularly opposed by nearby residents: a more profitable use for an owner (for example, an increase in the permitted density of development) can arouse fears of unwelcome neighbors and a fall in property values – commonly expressed in terms of 'a change in the character of the area'. There are, however, circumstances in which a local government might 'upzone' on its own initiative, as, for example, where it is seeking to attract development to its area.

Every zoning ordinance lays out the procedures for considering a zoning amendment. A proposed amendment should be filed on proper forms by a property owner, planning commission, or local legislative body. For example, after an amendment is proposed, staff studies the issue and makes a report to the appropriate body, in most cases, the planning commission. A written public hearing notice is then required by law. This informs the public of the proposed amendment and gives them an opportunity to speak for or against the proposal.

The planning commission considers the merits of the proposed amendment. It must consider a number of factors in deciding whether or not to approve a text or map amendment. Among the factors to be considered in granting a text amendment are consistency with the General Plan, need for the amendment, consistency with the zoning ordinance, reasons for opposing the proposal, likelihood of prompting additional amendments, and the likelihood that it will alleviate any problems. There are similar factors to be considered in approving a map amendment. For instance, the planning commission needs to know how appropriate the proposed use change is to the surrounding area, the effect of the change on public services, the effect of the change on property values, the reasons for any opposition, and how consistent the change is with existing zoning.

Ultimately, the final decision on a text or map amendment must be made in the public interest and not solely in the interest of any applicant. After the planning commission has advised the city council on a particular text or map amendment, the city council can decide in various ways. It can adopt the proposal that was recommended by the planning commission. It can reject the planning commission's recommendation. It could also modify or alter the planning commission's recommendation and adopt another alternative that has been identified for consideration in the public notice. Finally, should the planning commission fail to make a recommendation, it could change the proposed text or map amendment, adopt it, or adopt another alternative that was identified for consideration in the public notice. Each zoning code will identify the proper procedures to be followed when considering a text or map amendment.

Special district zoning

The term 'special district' is confusing since it has more than one meaning (see Box 8.8). Traditionally, special districts are governmental units established to perform specific functions which, for one reason or another, cannot be performed by the existing general purpose local governments. Examples from the nineteenth century are the toll road and canal corporations. Today there are special districts for education, social services, sewerage, water supply, and natural resources. They can be single-purpose or multi-purpose. (Perhaps the most famous is the Port Authority of New York and New Jersey.)

Special districts that are so designated for zoning purposes, however, are very different. These are areas to which an amendment of the zoning ordinance applies: they thereby become subject to 'special' zoning controls (see Box 8.9). The areas possess some type of 'unique' characteristics that contribute to an area of a city.

In an interesting and illuminating monograph *Special Districts: The Ultimate in Neighborhood Zoning*, Babcock and Larsen (1990) examine their contemporary use in a variety of contexts including, in New York, the Theater District (designed to preserve the area as such by forcibly bribing developers to build new theaters); the Special Fifth Avenue District (designed to stop the influx of banks and airline offices, and to

BOX 8.8 DEFINITION OF A SPECIAL DISTRICT

A special district is a zoning district which imposes special supplemental and zoning regulations for the use and development of land within such districts where there are unique cultural, historic, and physical characteristics that positively contribute to the city's diversity and livability. These supplemental zoning regulations are intended to reduce conflicts between new construction and existing development. They apply in addition to existing zoning regulations.

Source: City of Chicago Zoning Ordinance, Article 5, Section 5.15-1

BOX 8.9 WHAT IS SPECIAL ABOUT A SPECIAL DISTRICT?

Many in the Planning Commission believe that standard, generic zoning could be used to deal with local problems without the need to create a new special district for each neighborhood. The residents get a psychological lift from residing in an area that has a tag to it . . . They know the special regulations; some of them know the twists and bends of the provisions of their districts as well as the lawyers do and probably better than most of the administrators of the ordinance. They become, as Norman Marcus put it, 'zoning freaks'. Their zoning is the one part of the hopelessly complex myriad of municipal laws and policies that city residents believe they can understand . . . They can immediately spot a sign that violates the regulations of their special district or quickly detect a commercial establishment that operates in a way that is in violation of the labyrinthine district regulations.

Thus it appears that the professionals are losing a zoning conflict to the amateurs, a not unheard of event in the zoning arena.

Source: Babcock and Larsen 1990: 97

encourage profitable residential uses above the stores); the Special Garment Center District (designed to prevent the conversion of manufacturing space to office uses and to safeguard the garment industry); and the ill-conceived Special Little Italy District (designed to preserve the Italian character of the community, in disregard of the Chinese residents). San Francisco has sixteen special districts, several of which are Neighborhood Commercial Special Districts. This designation is applied to relatively small commercial corridors in residential areas, with the objective of 'preserving upper-floor residential units in commercial buildings, and keeping fast-food restaurants from taking over the street'. Cambridge, Massachusetts, has a number of special districts with regulations tailored to certain limited areas within the city. For example, a Cambridge Center Mixed Use Development District was created to guide development in the Kendall Square Urban Renewal Area. Chicago has a generic special district: the Planned Manufacturing District, designed to prevent the loss of industrial and manufacturing land to residential and commercial uses. This can be applied wherever it is needed, i.e. wherever the local electors pressure their alderman for one.

In all cases, the intention is to shield the area from market forces. There is nothing 'special' in this: much zoning is essentially of this protectionist and exclusionary nature. The curiosity of special districts is that most of them have little that is special about them. The residents complain about unwelcome changes, or the threat of changes, and the zoning authority responds by giving them a special status. What appears to be special is the large degree of citizen involvement, not only in the designation of the area but also in enforcement.

It seems clear that special districts are being used in areas where they have no justification; and, once established, their popularity with the citizenry makes abolition extremely difficult.

Overlay zones

On occasion, supplemental zoning requirements are placed on an area because it possesses some unique feature. This generally means that an area requires more or special protection. When this occurs, we say that an overlay zone has been placed on the area. Conversely, the underlying zone represents the generic classification that is placed on the property. It could be open space, residential, commercial, or industrial.

A number of areas around the country use overlay zones. Ames, Iowa, has five overlay zones in its Municipal Code, including a single-family conservation overlay zone. This overlay zone was created to help conserve the single-family character of the identified areas. Additional standards dealing with such topics as

garages, driveways, parking, and trees were required to be met by any development in this overlay zone. San Diego, California's Municipal Code identifies a number of overlay zones, including an airport approach overlay zone (providing supplemental regulations for the property surrounding the approach path to San Diego International Airport), a coastal overlay zone (protecting and enhancing the quality of public access and coastal resources), a mobile home park overlay zone (preserving existing mobile home park sites), and an urban village overlay zone (see Box 8.10).

Exclusionary zoning

Exclusionary zoning is said to occur when a zoning requirement or land use regulation excludes certain groups or classes of people from living in a community. For example, a community may attempt to prevent the development of certain types of housing such as apartments or manufactured homes. Preventing these types of housing is essentially telling certain income groups that they are not welcome in the community. Concomitantly, creating a minimum lot size may cause the price of land and the housing unit to be beyond the affordable price range for lower-income individuals and families. Ultimately, exclusionary housing devices could be either overt or covert, or direct or indirect.

BOX 8.10 PURPOSE OF THE URBAN VILLAGE OVERLAY ZONE

The purpose of the Urban Village Overlay Zone is to provide regulations that will allow for greater variety of uses, flexibility in site planning and development regulations, and intensity of land use than is generally permitted in other Citywide zones. The intent of these regulations is to create a mix of land uses in a compact pattern that will reduce dependency on the automobile, improve air quality, and promote high quality, interactive neighborhoods. Urban Villages are characterized by interconnected streets, building entries along the street, and architectural features and outdoor activities that encourage pedestrian activity and transit accessibility. The regulations of this division are intended to be used in conjunction with the Transit-Oriented Development Guidelines of the Land Development Manual and the applicable land use plan.

Source: San Diego Municipal Code, Chapter 13, Article 2, Division 11, Section 132.1101, 2000

The practice of exclusionary zoning effectively limits the range of residential choices for people. Many individuals may have to travel many miles to find affordable housing. Some individuals may find affordable housing but no job opportunities. The lack of public transportation exacerbates the problem. The result becomes the existence of racially or income-divided communities.

The courts have examined numerous zoning regulations that are allegedly exclusionary in nature. They have tended not to focus on whether certain groups have been excluded from a community, but rather to examine the intent of the zoning regulation in question. Their question is whether the intent of the regulation is to exclude certain groups. If so, the community has abused the power it was given to regulate land uses. There is no legitimate government interest being served by a community when it excludes certain groups from living in the area. As McCarthy (1995: 238) has suggested,

> although all zoning ordinances are in many senses exclusionary, the term has come to characterize ordinances challenged as unreasonable and invalid in that they serve to erect walls on the municipality's boundary, according to local selfishness for socially improper goals, beyond the legitimate purposes of zoning.

This is certainly a difficult area for the courts. It is a function of the legislative bodies to design proper zoning ordinances. The courts determine the constitutionality of such ordinances. The remedy would be to design general plans, land use regulations, and housing programs that effectively offer the possibility of providing a variety of housing opportunities for all people. This can be accomplished in a number of ways. Inclusionary housing programs are examined in Chapter 15.

Linkages

The inconstant way in which planning terms are used is illustrated by the use of 'inclusionary zoning' to refer to housing (and other facilities) which developers of major downtown projects are required to make before permission to develop is given. This 'inclusionary housing downtown' is also, and better, termed 'linkage': it is 'included' in a scheme only in the sense that it is linked to it. However, there is some doubt as to how real this link is. Certainly, it is linked in the minds of the municipal officials, who can point to an observed relationship between, say, major downtown development, increases in employment, and rises in housing prices. It is also argued that such development increases the demand for central city housing which, in turn, leads to gentrification and the displacement of poorer residents.

The case is far from universally accepted. Developers, as might be expected, see the matter in a different light. Among their many arguments is that downtown development *follows* and accommodates demand: it does not create it. As one critic nicely put it: 'additions to the supply of office space do not create office employment any more than cribs make babies' (Gruen, 1985: 34). Moreover, there is a danger that linkage fees will kill the golden goose of downtown development, particularly if they are set at a level which will finance significant amounts of housing.

There is merit in both sets of arguments and, as usual, which of them is valid will depend upon the particular circumstances of the time and place. Linkage programs have typically been introduced during a major real estate boom, and usually in connection with downtown commercial development (and mostly with density bonuses as an inducement). The end of the boom drastically changed the economics of the programs, and they were severely cut back. Most cities eschewed them: even if there was concern about the impact of large-scale development, there was a greater fear – that of scaring away private investment.

Linkage schemes are, in general, limited. There is less activity in the real world than the planning literature might suggest. (It is the exceptional which makes good news.) However, where these schemes carry an incentive or bonus, there is a danger that a municipality's desire to obtain contributions from the developer might overwhelm the requirements of good planning in the area. From this odd point of view it is an advantage that municipalities have so little in the way of plans: their absence means that they cannot be sabotaged. But where there is an effective plan, bonusing can destroy it. Seattle provides a good example of what can emerge as a result of an assembly of bonuses.

Purposes of bonusing

All too often bonuses are seen to be self-evidently beneficial. That this is not always so is apparent from the previous discussion. Here we note some of the useful purposes which bonusing can achieve.

Ideally, a bonus should be an incentive for a developer to provide an amenity or facility which is of public benefit, and which the developer would not provide voluntarily (such as a day-care center). There is an immediate difficulty with this: even if there is an agreement about which benefits are desirable, how can it be determined that they will be provided only if an incentive is offered? It has already been noted that New York, in 1982, abandoned many bonuses for mandatory requirements. San Francisco did likewise with matters of downtown design (Getzels and Jaffe 1988: 2). However, these two cities are hardly representative: many cities are extremely anxious to attract downtown development and are therefore far more inclined to provide incentives rather than disincentive conditions.

A favorite objective of bonusing is the promotion of lively street-level retailing in downtown areas – in contrast to the dead, blank walls which so seriously diminish the attractiveness of a street. A good statement of purpose is provided in the Seattle ordinance (see Box 8.11).

Seattle's downtown code provides brief 'statements of intent' for each bonusable amenity. Shopping corridors, for example, are 'intended to provide weather-protected through-block pedestrian connections and retail frontage where retail activity and pedestrian traffic are most concentrated downtown. Shopping corridors create additional "streets" in the most intensive area of shopping activity, and are intended to complement streetfront retail activity.'

Lassar (1989) comments that bonus activities 'run the gamut' and can be clustered around several general categories: building amenities such as urban spaces and day-care centers; pedestrian amenities such as sidewalk canopies and landscaping; housing and human services such as job training; transportation improvements such as station access and parking; cultural amenities such as art galleries and live theaters; and preservation of historic structures. There is, it seems, no end to the ingenuity which can be employed in this area. (Lassar's book covers all these in useful detail.)

Seattle's bonus incentive program was established in the 1960s. It allowed developers the opportunity for additional floor area in two downtown zones if the building had public plazas or arcades, or by agreeing to design building setbacks. Additional bonuses were phased in during the 1970s. In 1985, a new Land Use and Transportation Plan for Downtown Seattle was adopted. It increased the range of bonuses available to developers. It also divided the downtown into eleven functional districts. The Plan incorporated a floor area bonus system that enabled developers to gain additional floor area by providing certain amenities that were deemed beneficial to the public – public benefit features (PBF). The housing bonus and transfer of development rights program were also introduced at this time.

The Washington Mutual Tower gained twenty-eight of its fifty-five stories on account of the amenities

---

BOX 8.11 SEATTLE'S RETAIL SHOPPING BONUS

The intent of the retail shopping bonus is to generate a high level of pedestrian activity on major downtown pedestrian routes and on bonused public open spaces. While retail shopping uses ensure that major pedestrian streets are active and vital, a limit to the amount eligible is set in each zone in order to maintain the dominance of the retail core as the center of downtown shopping activity.

Source: Getzels and Jaffe 1988: 3

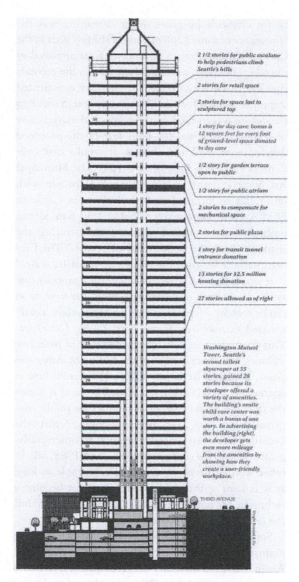

2 1/2 stories for public escalator to help pedestrians climb Seattle's hills

2 stories for retail space

2 stories for space lost to sculptured top

1 story for day care; bonus in 12 square feet for every foot of ground-level space donated to day care

1/2 story for garden terrace open to public

1/2 story for public atrium

2 stories to compensate for mechanical space

2 stories for public plaza

1 story for transit tunnel entrance donation

13 stories for $2.5 million housing donation

27 stories allowed as of right

Washington Mutual Tower, Seattle's second tallest skyscraper at 55 stories, gained 28 stories because its developer offered a variety of amenities. The building's onsite child care center was worth a bonus of one story. In advertising the building (right), the developer gets even more mileage from the amenities by showing how they create a user-friendly workplace.

THIRD AVENUE

Plate 8 Washington Mutual Tower, Seattle: twenty-eight extra stories for public benefits

Courtesy Wright Runstad & Co

compensate for mechanical space, a half-story for a public atrium, a half-story for a garden terrace open to the public, one story for a day-care facility, two stories for space lost to a sculptured top to the building, two stories for the provision of retail space, and two and a half stories for a public escalator to help pedestrians climb Seattle's hills. This was a truly remarkable example of private munificence! (See Plate 8.)

Over the next couple of years, downtown development intensified as a public issue. Streets were being torn up. Increased traffic congestion became commonplace. New construction resulted in a loss of affordable housing in the downtown. These and other problems caused a number of people to question whether limits needed to be placed on downtown building.

The result of this citizen frustration was the development of the 1989 Citizens' Alternative Plan (CAP) – Initiative 31 (Caves 1992). This ballot measure sought to amend Seattle's Land Use Code as it related to downtown zoning by reducing permitted building height, reducing bulk by reducing development bonuses for public and development benefits, limiting the development of new office space with certain exceptions, and requiring that the city commission a study regarding the future management of downtown growth.

The CAP initiative was passed by the voters on May 16, 1989. Although there was a low voter turnout, the voice of the public had been heard. The city would now limit the height, density, and number of skyscrapers in downtown Seattle. The measure could not be amended for two years. Some years later, in 2001, the Seattle City Council adopted ordinance amendments to revise the Denny Triangle TDR (transfer of development rights) area.

The dangers of bonusing

The lesson is clear: once introduced, incentive zoning is difficult to control. In the absence of any overall official plan policy framework, the process 'can engender considerable uncertainty respecting the city's intentions and can give the impression that the underlying basis of the plan is being subverted' (Toronto City

offered by the developer. As of right, the developer was allowed twenty-seven stories. In addition to this, he obtained thirteen stories for a $2.5 million housing donation, one story for a transit tunnel entrance donation, two stories for a public plaza, two stories to

1988: 9). It can also be difficult to ensure that all land-owners are being treated equitably and consistently. Moreover, the absence of basic ground rules results in a process which 'can be extremely time consuming (and costly) and require extensive professional involvement as each application is negotiated' (Toronto City, 1988). See Box 8.12 for a critique of density bonuses.

A fundamental problem with individually negotiated bonusing is that it leads to a situation in which 'it becomes almost politically impossible for a municipality to approach a density increase without demanding the contribution of some amenity' (Bermingham, 1988: 8). Indeed, since it would be difficult for a municipality to grant an increase in density to one owner on more favorable grounds than preceding owners, each grant of a bonus is likely to involve a demand for a higher contribution: 'In other words, from the municipality's perspective, each deal becomes the starting point for the next deal' (Bermingham, 1988: 19). The argument does not have to be accepted in its entirety for its force to be felt.

A different argument contends that 'physical planning standards undermined by [incentive zoning] are not the only interests important to communities. Other values, including those represented by social amenities, contribute to the quality of life, and a city might reasonably resolve that it will tolerate taller buildings and greater congestion in return for more low income housing or daycare facilities' (Kayden 1990: 101). A now classic case of the unhappy times on which New York City bonusing fell is that of the Columbus Circle project, which eventually came before a New York trial court at the instigation of the Municipal Art Society of New York. The agreement reached in this case provided for the acquisition by the developer of the city-owned site, a 20 percent increase in density, and the payment to the City of $455 million, plus another $40 million for improvements to the nearby Columbus Center subway station. The City would also have realized about $100 million in taxes each year from the 2.7 million square feet development. Such riches were tempting indeed, and the City did not resist the temptation. But it fell foul of legal hurdles. The crucial point at issue was the fact that a substantial part of the payment to the City was to

be for citywide purposes. Most damaging was the appearance of some $266 million of the proceeds in the City's 1988 budget, in advance of final approval of the sale. The court invalidated the sale on the grounds that the incentive provided by the City constituted improper 'zoning for sale'. However, though much of the debate was focused on these financial matters, the underlying issue was the huge size of the proposed development, and the shadow it would cast over Central Park. (On October 18, 1987, the Municipal Art Society gathered more than 800 people with umbrellas to form a line from Columbus Circle to Fifth Avenue. On a given signal at 1.30 p.m. all the umbrellas were opened – thus demonstrating the shadow which the building would cause.) The City eventually redesigned the project on a smaller scale.

It would be quite wrong to regard this notorious case as the death-knell of incentive zoning. Far from it, as can be seen by reference to Lassar's 1989 study, neatly entitled *Carrots and Sticks: New Zoning Downtown.* This documents in detail the wide range of incentive schemes which are being operated in many American cities. And, so far as the Columbus Circle case is concerned, New York transgressed because it 'upset the delicate balance between competing public and private interests . . . It made economic return the deciding factor, with scant attention to other public goals and land use considerations' (Lassar 1989: 38). In other words, incentive zoning is acceptable as long as it is kept within bounds and does not become a technique for raising additional municipal funds; but the temptation is difficult to resist!

Some of the initiatives of which this is a striking example are examined further in the following pages. Again, there is a difficulty over terminology. When is an incentive a linkage? When is a bonus a charge? When is an impost a development fee? There may be core elements in each of the terms which distinguish one from another, but these are obscured in the real world by local usage and also by the way in which the elements are combined to suit particular objectives. A good case in point is New York's 'incentive zoning' scheme. This is a tale with a moral.

BOX 8.12 A CRITIQUE OF DENSITY BONUSES

If our city planning theories about appropriate development, servicing and transportation have any validity, the extra density created in one place must either be denied somewhere else or paid for over decades in the expansion of services, utilities and transportation corridors. The costs are not as direct or quantifiable, but the taxpayers will bear them nonetheless.

. . . [what is created is] a circumstance in which one municipal goal (housing, for example) is traded off against another municipal goal (consistent planning) with no necessary relationship between them. If a municipal statement with regard to maximum densities is defensible by planning rationale why should the municipal need for a public swimming pool alter that rationale? Is there not a danger that the planning theory itself will come to be treated as arbitrary and unprincipled – simply one more chip to throw into the urban development poker game?

Source: Bucknall 1988

Incentive zoning in New York

As noted in Chapter 5, the 1916 New York Zoning Resolution employed height, lot coverage, and use restrictions. Unfortunately, to some individuals, allowing extreme densities became commonplace. As the city underwent changes, a number of people claimed that the existing zoning protected existing buildings and limited the size of any new buildings.

The Zoning Resolution of 1961 replaced the earlier zoning. It created the incentive zoning regulations that offered incentives/bonuses to developers in exchange for the provision of various amenities. These amenities could range from such items as open space, day-care centers, parks, arcades to mass transit accessibility points. The incentive system has resulted in the provision of over three million square feet of open space.

The incentive zoning scheme started during one of the city's development booms, and was initially concerned with the provision of urban amenities such as plazas. In the words of William H. Whyte, in his highly enjoyable *City: Rediscovering the Center* (1988: 229):

It seemed a splendid idea. Developers wanted to put up buildings as big as they could. Why not harness their avarice? Planners saw a way. First, they would downzone. They would lower the limit on the amount of bulk a developer could put up. Then they would upzone, with strings. The builders could build over the limit if they provided a public plaza, or an arcade, or a comparable amenity.

At first, the scheme was across the board: bonuses were given as of right to developers who met the requirements set out in an ordinance. There was, therefore, no negotiation: thus for every square foot of plaza space provided, the developer could claim an extra ten square feet of office space. The scheme was a great success — in terms of the number of plazas provided. Indeed, it was really too successful: between 1961 and 1973, over a million square feet of new open space was created in this manner. The incentives greatly exacerbated the overbuilding boom of the 1960s, which led to the high vacancy rates and lost real estate taxes in the 1970s. Moreover, developers found loopholes in the scheme which allowed them to make windfall profits. Even the amenities provided did not escape some criticism: 'A lot of the places were awful: sterile, empty spaces not used for much of anything except walking across. (Whyte, 1988: 234)'.

During this time, the New York City Planning Commission became increasingly more powerful. Its members favored certain design features and looked for those features in the plans for office and apartment towers that were submitted to the commission. Some critics claimed the regulations were too liberal. Starr (1998) argued that this was a problem in that planning commission members tended to not get involved in the design of buildings for safety reasons. Instead, they wanted to see developers meet the building criteria they desired.

The criticisms over New York's zoning continue. In a speech on April 20, 1999, Joseph B. Rose, Chairman of the New York City Planning Commission and Director of the New York City Planning Department, claimed zoning was in 'crisis'. Within the speech, Rose observed:

> We are in crisis because this crucial document has become a hodgepodge of conflicting visions and objectives . . .
>
> We are in crisis because of the many instances our zoning promotes an architectural vision that does violence to our urban fabric.
>
> We are in crisis because we try to micromanage the world's most vital and varied urban economy with regulations that were drafted 40 years ago . . .
>
> And finally, we are in crisis because there are those who would use these evident deficiencies in the zoning resolution as an excuse for discarding the essential character of New York City: the openness to change and the welcoming of new people, buildings, businesses, and ideas.

To him, the resolution needed to keep up with the times. It was time to make some changes. It was now time to provide some flexibility in design and it was necessary to clarify what was allowed and what was not allowed.

The negotiation syndrome

There were two ways of improving the situation. The first was to elaborate the guidelines. This was done, despite fears that it would be too inflexible and would unduly constrain architectural design. The new (1975) guidelines spelled out the rules of the game in considerable detail, and included the maximum height of the plaza, the amount of seating, the minimum number of trees, and so on. The second was to replace the mechanical as-of-right scheme with a special permit process (later designated by the ungainly term Uniform Land Use Review Process – unaffectionately known by its acronym ULURP). Essentially this was a negotiated agreement. By 1980, it was clear that zoning in New York City was in real difficulty. Anticipation of bonuses fed back into higher land prices (though developers sought some measure of protection by signing contingency agreements with landowners, with the higher price to be paid only if the anticipated bonus was granted), and buildings became larger and larger. Promised (negotiated) amenities were sometimes not provided. Citizen groups became increasingly loud in their complaints. Finally, in 1982, midtown zoning was subjected to a sweeping revision. Densities were reduced, and bonuses were largely dropped except for plazas and urban parks. Amenities which had formerly been obtained by way of bonusing now became mandatory. The cumulative effect of these provisions, it was anticipated, 'would go a long way toward eliminating negotiated zoning. They would permit development to proceed on a more predictable and as-of-right basis' (City of New York, 1982: 11).

Form-based codes

There are a number of people who think the strict separation of different land uses and restrictions of use found in a zoning code have outlived their usefulness. They feel focusing solely on the use of the property ignores what is important to the surrounding area. They are tired of the 'legalese' contained in the zoning code and the restrictions placed on the various zoning categories. They are tired of seeing jurisdictions react to individual project proposals. People are looking for alternatives to conventional zoning. Unfortunately, many cities have relied on zoning so long that they do not want to change to something new.

Form-based codes (FBC) offer one alternative to conventional zoning. Cities are developing a proactive approach to development instead of relying on the reactive nature of zoning. The end result is getting in front of an area's development by trying to promote the idea of walkable and mixed-use communities. Ultimately, form and scale of development is of vital importance, especially in how the various parts of the development relate to each other. The Form-Based Codes Institute offers the following definition of an FBC:

> Form-based codes foster predictable built results in a high-quality public realm by using physical form (rather than separation of uses) as the organizing principle for the code. They are regulations, not mere guidelines, adopted into city or county law. Form-based codes offer a powerful alternative to conventional zoning.
>
> Form-based codes address the relationship between building facades and the public realm, the form and mass of buildings in relation to one another, and the scale and types of streets and blocks. The regulations and standards in form-based codes are presented in both words and clearly drawn diagrams and other visuals. They are keyed to a *regulating plan* that designates the appropriate form and scale (and therefore, character) of development, rather than only distinctions in land-use types.
>
> (http://www.formbasedcodes.org)

An FBC that contains graphic descriptions (diagrams are much more powerful than words) is concerned with the desired structure on the property and its relationship to the surrounding environment and streetscape. The code respects community character and the ultimate vision a community wants to achieve. It focuses more on the physical design, form, and visual aspects of development and how the various uses interact with each other, and encourages mixed uses. It does not dwell on the particular use of the property or setbacks, heights, density, floor area ratios, etc. as stated in the zoning ordinance. Market economic forces determine the use of the property, not the zoning code. Box 8.13 offers eight advantages of an FBC over conventional zoning.

According to the Form-Based Codes Institute (2008: 1), an FBC includes the following elements:

- *Regulating Plan.* A plan or map of the regulated area designating the locations where different building form standards apply, based on clear community intentions regarding the physical character of the area being coded.
- *Public Space/Street Standards.* Specifications for the elements within the public realm (e.g., sidewalks, travel lanes, street trees, street furniture).
- *Building Form Standards.* Regulations controlling the configuration, features, and functions of buildings that define and shape the public realm.
- *Administration.* A clearly defined application and project review process.
- *Definitions.* A glossary to ensure the precise use of technical terms.

FBCs can be developed at various scales. There is nothing precluding a city from abandoning traditional zoning and using the FBC on a community-wide basis. However, another city might choose to employ an FBC at a specific sub-area of the community – the neighborhood level. The choice of scale is up to the particular jurisdiction.

Citizen input is critical in developing an FBC. In order to achieve a community vision of what they want, citizens must be actively involved in the process. This participation would occur in public charrettes or public hearings/meetings. Task forces have been created in many cities to get citizens involved in planning. Citizens would be able to make recommendations, voice their support or opposition on potential directions of the area, etc. We should be planning with the people's ideas in mind. They know what they want and do not want.

A key component of the FBC is the 'transect'. Originally coming from the field of ecology, the 'transect' orders land use on a rural to urban continuum. As Duany and Talen (2002: 246) suggest:

> A transect is a geographical cross-section of a region used to reveal a sequence of environments. For human environments, this cross-section can be used to identify a set of habitats that vary by their level and intensity of urban character, a continuum that ranges from rural to urban. In transect planning,

this range of environments is the basis for organizing the components of the built world: building, lot, land use, street, and all of the other physical elements of the human habitat.

The transect zones in the continuum go from 1) rural preserve, 2) rural reserve, 3) sub-urban, 4) general urban, 5) urban center, to 6) urban core. The continuum covers what is appropriate in each zone. As the use progresses from rural to urban, the intensity of use increases. It should be noted that cities do not have to have each of the aforementioned zones. Areas will vary in regards to the strict adherence to the form-based rural–urban transect model. It is place specific in that one community may have fewer transect zones while another community may actually have more transect zones. Figure 8.2 illustrates a generic transect model.

Realizing that strict, traditional zoning was no longer the answer to land use development proposals, the City of Denver, Colorado, updated its Zoning Code in 2010. Instead of employing uniform rules of development throughout the city, Denver decided that rules should vary by neighborhood and crafted a form-based code. The new code offers individuals, architects, and developers more flexibility in maintaining the context and values of specific neighborhoods.

The City of Miami, Florida, also embarked on a rewrite of its zoning code in 2005 after determining that it wanted to focus on form and function instead of the strict regulation of uses found in traditional Euclidian zoning. The city decided to become a pedestrian-friendly community instead of a community that is auto-centric. After various political battles to get it approved, Miami Commissioners approved it in October 2009. The new zoning code was known as 'Miami 21'. It became effective on May 20, 2010.

As noted previously, each city determines the specific transects that it will use. In Miami 21, the following transect zones were employed: natural transect zones; rural transect zones; sub-urban transect zones; general urban transect zones; urban center transect zones; urban core transect zones; civic space zones; civic institution zones; district zones; and waterfront industrial district zones. Within each of these transect zones, the following areas are regulated: building disposition; building configuration; function and intensity; landscape standards; parking standards; standards that integrate individual property with the public realm; and a successful (sequential) relationship between transects.

The City of Flagstaff, Arizona, also chose to employ a rural–urban transect as the organizing principle when it revised the Traditional Neighborhood District from its previous 1991 Land Development Code in 2011. Just as Denver and Miami had determined, the old zoning code was not meeting the needs of the public and the vision of the community. Flagstaff sought to develop walkable, mixed-use environments that would

Figure 8.2 Generic transect model

Courtesy of Center for Applied Transect Studies

BOX 8.13 EIGHT ADVANTAGES TO FORM-BASED CODES

1   Because they are prescriptive (they state what you want), rather than proscriptive (what you don't want), form-based codes (FBCs) can achieve a more predictable physical result. The elements controlled by FBCs are those that are important to the shaping of a high quality built environment.

2   FBCs encourage public participation because they allow citizens to see what will happen and where – leading to a higher comfort level about greater density, for instance.

3   Because they can regulate development at the scale of an individual building or lot, FBCs encourage independent development by multiple property owners. This obviates the need for large land assemblies and the mega-projects that are frequently proposed for such parcels.

4   The built results of FBCs often reflect a diversity of architecture, materials, uses, and ownership that can only come from the actions of many independent players operating within a community agreed-upon vision and legal framework.

5   FBCs work well in established communities because they effectively define and codify a neighborhood's existing 'DNA.' Vernacular building types can be easily replicated, promoting infill that is compatible with surrounding structures.

6   Non-professionals find FBCs easier to use than conventional zoning documents because they are much shorter, more concise, and organized for visual access and readability. This feature makes it easier for non-planners to determine whether compliance has been achieved.

7   FBCs obviate the need for design guidelines, which are difficult to apply consistently, offer too much room for subjective interpretation, and can be difficult to enforce. They also require less oversight by discretionary review bodies, fostering a less politicized planning process that could deliver huge savings in time and money and reduce the risk of takings challenges.

8   FBCs may prove to be more enforceable than design guidelines. The stated purpose of FBCs is the shaping of a high quality public realm, a presumed public good that promotes healthy civic interaction. For that reason compliance with the codes can be enforced, not on the basis of aesthetics but because a failure to comply would diminish the good that is sought. While enforceability of private regulations has not been a problem in new growth areas controlled by private covenants, such matters can be problematic in already-urbanized areas due to legal conflicts with first amendment rights.

Source: Peter Katz, President, Form-Based Codes Institute, http://www.formbasedcodes.org/advantages.html

---

further the character of the community and physical form of the zone.

Subdivision regulations

While zoning is concerned with the use of land, subdivision regulations relate to the division of land for sale. Originally designed to keep track of the legal ownership of land and to facilitate the establishment of clear titles (thus simplifying transactions), they have grown into a formidable tool of land use planning. Published in 1928 by the US Department of Commerce, the Standard City Planning Enabling Act (SCPEA) defines subdivision as 'the division of a lot, tract, or parcel of land into two or more lots, plats, sites, or other divisions of land for the purpose, whether immediate or future, of sale or of building development' (see Chapter 5 for a discussion of SCPEA). SCPEA is not to be confused with the Standard State

Zoning Enabling Act (SSZEA) that, as its title indicates, is concerned with zoning and is discussed in Chapter 5. States soon followed and adopted subdivision enabling legislation.

Though subdivision and zoning are quite distinct in origin, they have come to share some important control features. With zoning, these are built into the zoning ordinance or imposed in the administration of the ordinance. Subdivision has acquired similar features (though it is usually applied only to residential development). The first controls were restricted to matters relating to roads. These ensured, for instance, that any streets built in a subdivision would be aligned with existing streets. These controls were extended to deal with the width of streets and sidewalks, setbacks and such-like. This enabled local governments to prevent the creation of lots that were unacceptably small or badly configured. But it also gave them the scope to impose conditions relating to 'improvements'. It was not a big step, politically, to move from requiring that roads be a certain width to making the actual provision of the roads a condition of subdivision. Many subdivision enabling acts provided for the dedication of these roads – and also sewers, water mains, and other public facilities. As a result of these extensions to subdivision control, 'the subdivision ordinance was well on its way to becoming a development code by the 1950s' (Callies and Grant 1991).

The term 'subdivision' can be defined differently by state and can be broken into various types of subdivisions. For example, the Arizona Revised Statutes (ARS) Title 32, Chapter 20, Article 1 and the Cochise County, Arizona, Subdivision Regulations, Article 1, Section 204, define 'subdivision' as 'improved or unimproved land or lands divided or proposed to be divided for the purpose of sale or lease, whether immediate or future, into six or more lots, parcels or fractional interests'. The State of California, in Government Code Section 66424, has a different definition: 'the division, by any subdivider, of any unit or units of improved or unimproved land, or any portion thereof, shown on the latest equalized county assessment roll as a unit or as contiguous units, for the purpose of sale, lease, or financing, whether immediate or future'. Other variations on these two definitions can be found in the remaining states.

Jurisdictions can also choose to distinguish between different types of subdivisions. For example, it is common for subdivisions to be divided into two categories: major and minor. The Ithaca, New York, City Code defines a major subdivision as 'any subdivision of land resulting in the creation of two or more additional buildable lots'. It goes on to define a minor subdivision as 'any subdivision of land resulting in the creation of a maximum of one additional buildable lot'. Variations in names and definitions can be found throughout the United States.

Jurisdictions will differ as to the goals and purposes of subdivision regulations. Some jurisdictions will have something as simple as assuring reasonable development and to promote the public health, safety, and general welfare. The California Subdivision Map Act provides mandates for cities to follow when processing subdivisions. It gives cities the power to regulate the design and improvement of subdivisions. Curtin and Talbert (2005: 75), in noting two Opinions of the California Attorney General, indicate the following primary goals of the California Subdivision Map Act:

- To encourage orderly community development by providing for the regulation and control of the design and improvement of the subdivision, with a proper consideration of its relation to adjoining areas,
- To ensure that the areas within the jurisdiction that are dedicated for public purposes will be properly improved by the subdivider so that they will not become an undue burden on the community, and
- To protect the public and individual transferees from fraud and exploitation.

Charles County, Maryland, in Article 1, Section 1 of its subdivision regulations, notes that the purpose of the regulations is 'to assure the reasonable and consistent development of land within Charles County, in order to promote the public health, safety, and general welfare and to provide for the creation of development sites suitable for building purposes and human habitation, and to provide for open space in a harmonious environment'. Box 8.14 provides a detailed

BOX 8.14 PURPOSES OF SUBDIVISION REGULATIONS

a. Protect the health, safety, convenience and general welfare of the citizens of the County.

b. Provide for the orderly growth and harmonious development of the County.

c. Require that land be conveyed with an accurate legal description.

d. Establish procedures and standards for all subdivisions.

e. Provide adequate traffic circulation, streets, utilities, waste water treatment, drainage, fire and flood protection, schools, recreation areas and other facilities and services needed or desired by the community in the most cost-effective manner, with the cost being borne by those benefited.

f. Result in individual lots of reasonable utility and livability and to promote neighborhood stability and protection of property values.

g. Promote conservation of those areas with unique natural features and scenic qualities and provide residents with access to these areas.

h. Promote water recharge and clean air.

i. Provide greater design flexibility and efficiency for services and infrastructure including design methods that reduce the length of streets, thus reducing the amount of improved surface and length of utility runs.

j. Encourage well-planned subdivisions by establishing environmentally adequate standards for design and improvement.

k. Provide viable, innovative, cost-effective, voluntary development alternatives.

Source: Cochise County, Arizona, Subdivision Regulations, Article 1, Section 102, 2004

listing of what one jurisdiction hopes to achieve with its subdivision regulations.

State codes and statutes indicate what powers the jurisdiction has over subdivisions. A major power bestowed on the jurisdictions is the ability to control the design and improvement of a subdivision. Design and improvement can be defined in various manners. Box 8.15 shows how the State of California defines the two terms.

Extensions of control continued in later years: to the provision and dedication of schools, police and fire stations, parks, and similar on-site facilities. Later, on the logical argument that new development had 'impacts' beyond the site being developed, conditions were imposed relating to 'off-site' facilities or payments in lieu. Thus, controls originally designed to secure orderly development have been transformed into a complex system of dedication, exactions, and impact fees (see Chapter 9). There is no significant difference in principle between such impositions and those levied under the umbrella of zoning.

As with zoning changes, General Plans, etc., subdivision requests are subject to environmental reviews to determine if any potential environmental impacts or consequences of the proposed action exist. This is a common occurrence with projects that are deemed to be discretionary in nature. For example, the New York State Environmental Review Act requires agencies to consider environmental impacts. In California, the California Environmental Quality Act requires certain subdivision applications to conduct environmental reviews. These reviews would question whether the proposed project would result in a physical change to the project site, its effect on any existing waterway, its impact on trees, vegetation, birds, animals, fish, endangered species, air quality, scenic views, etc. The information contained in the environmental reviews will vary by state.

The relationship between subdivision and zoning is thus somewhat blurred; but there are important distinctions. Both are designed to protect the health, safety, and welfare of the public – the police power.

BOX 8.15 DESIGN AND IMPROVEMENT DEFINITIONS

Section 66418 Design means

1 street alignments, grades, and widths;
2 drainage and sanitary facilities, including alignments, and grades thereof;
3 location and size of all required easements and rights-of-way;
4 fire roads and firebreaks;
5 lot size and configuration;
6 traffic access;
7 grading;
8 land to be dedicated for park or recreational purposes; and
9 such other specific physical requirements in the plan and configuration of the entire subdivision as may be necessary to ensure consistency with, or implementation of, the general plan or any other applicable specific plan.

Section 66419 Improvement means

(a) 'Improvement' refers to any street work or utilities to be installed, or agreed to be installed, by the subdivider on the land to be used for public or private streets, highways, and easements, as are necessary for the general use of the lot owners in the subdivision and local neighborhood traffic and drainage needs as a condition precedent to the approval and acceptance of the final map thereof.
(b) 'Improvement' also refers to any other specific improvements, the installation of which, either by the subdivider, by public agencies, by private utility, by any other entity approved by the local agency, or by a combination thereof, is necessary to ensure consistency with, or implementation of, the general plan or any applicable specific plan.

Source: California Government Codes, Sections 66418 and 66419

They both regulate development. They both are tools used to implement a General Plan. Zoning controls the use of the property. Subdivision regulations control the pattern of development and the quality of development. Subdivision regulations must comply with the zoning ordinance. They cannot be used to amend the zoning ordinance. Zoning is more permanent than subdivision regulations.

Development agreements

The land development process is an intriguing process. Developers want to make sure that local governments will not change 'the rules of the game' on them, while local governments want to get as many concessions from developers as they possibly can. These concerns set the stage for the creation of a development agreement. A development agreement represents a formal statement or agreement between the two parties respecting how the land is to be used. It provides a degree of certainty to the developer that the local government will not change the land use rules and regulations that apply to an ongoing project. In other words, the developer can rely on the rules and regulations in effect at the time a development agreement is signed and executed. It offers the developer protection against later zoning and General Plan changes or amendments.

States give local governments the authorization to enter into a development agreement. The specific procedures to do so could be by either resolution or ordinance. For example, in California, the California Government Code, Section 65865, states 'any city, county, or city and county, may enter into a development agreement with any person having a legal or equitable interest in real property for the development of the property as provided in this article'. In addition, Section 65865(c) requires every city, by resolution or ordinance, to establish procedures and requirements for the consideration of development agreements.

The City of Suisun City, California, provides an example of how cities in California have responded to the state legislation. Suisun City's Zoning Ordinance (Chapter 18.62) details its requirements for a development agreement between the City and a land developer. The ordinance begins by noting that the state authorizes the City to enter into a development agreement and to develop the procedures and requirements for obtaining a development agreement. For example, in Section 18.62.079, the City describes the general contents of a development agreement application:

A. A description of all real property subject to the agreement;
B. The term of the agreement;
C. Identification of parties to the agreement;
D. Conditions precedent to change of parties to the agreement;
E. A development plan;
F. Developmental regulations and uses, including but not limited to the permitted uses of the property, the density or intensity of use and the maximum height and size of proposed buildings;
G. A development program including timing, permits and other authorizations and procedures;
H. Provisions for reservations or dedications of land for public purposes;
I. Conflicts with other laws and ordinances;
J. Defaults, remedies and termination procedures;
K. Cooperative agreement in the event of legal challenge;
L. Reimbursement of City costs;
M. Performance recitals;
N. Waivers and amendments;
O. Severability.

Procedures for review, hearing notice, recommendation of the planning commission, city council consideration, periodic review, amendment or cancellation, are also covered in this section of the Suisun City Zoning Ordinance.

As is the case with any land use regulations, the legality of development agreements has been called into question. Generally, the courts will look to such issues as: does the state allow development agreements, do local development agreement ordinances adhere to the state legislation, is it an exercise of the police power or has government bargained away its authority, and is the development agreement consistent with the general plan? These issues will need to be decided on a state-by-state basis and on a case-by-case basis.

Conclusion

There is no doubt that zoning is not the rigid, simple system of land use regulation that it is sometimes assumed to be. It can display remarkable flexibility. It is not simple: it is increasingly complex. Along with subdivision regulations, it is also like any instrument of public policy, capable of good use and of misuse. Some local governments operate zoning in a highly responsible manner, with a careful balancing of private and public interests. At the other extreme are those who use it, sometimes blatantly, as an exclusionary technique. Of course, in a sense all zoning has inherent exclusionary features: the public policy problem arises when these are used as a means not of regulating land, but of regulating people. This issue has been made clear in this chapter. As times change, cities have turned to new mechanisms to determine how a city looks and grows. Many cities have turned to form-based codes that focus on form and function instead of regulating strict uses of individual properties.

Further reading

For a good explanation of land uses, the Municipality of Anchorage, Alaska, gives real-life examples online (http://www.muni.org/zoning/landuses.cfm). The website provides definitions of uses as well as links to forms and government agencies for approval of conditional uses.

The City of Salem, Oregon, publishes the purpose and application process for a variance on its website (http://pacweb.cityofsalem.net/permits/variances.htm).

The Uniform Land Use Review Procedure (ULURP) can be found online at New York City's Planning Department (http://www.nyc.gov/html/dcp/html/luproc/ulpro.shtml). The website lists the 'evolution of ULURP' as well as the current review process.

Zoning is undergoing major changes in many local governments. A strong advocate of form-based codes is the Form-Based Codes Institute (http://www.formbased codes.org). Different from Euclidian zoning, form-based codes focus on the structure of the buildings rather than their uses. This website offers basic information about form-based codes such as definitions and the availability of educational materials for professionals, including certification courses.

The Planning Department of Clermont County, Ohio, lays out the principles of cluster development on its website (http://www.co.clermont.oh.us/planning/clustering.htm). Also called 'open space zoning', cluster zoning is illustrated in different design patterns.

The City of Phoenix posts a list of frequently asked questions about zoning online (http://phoenix.gov/PLANNING/zonfaqs.html). The detailed answers provide a real account of city zoning and what residents need to know to live in or change the zone.

The Washington State Human Rights Commission provides history and current news for fair housing in the United States (http://www.hum.wa.gov/FairHousing). Its website has a chronology of fair housing legislation, Supreme Court cases, and outreach materials for the public.

Local governments have made subdivisions interactive. For example, Clark County in Washington created a mapping website using GIS that allows visitors to view subdivisions by searching for specific properties or zooming into the map. Visit it online (http://gis.clark.wa.gov/applications/gishome/subdivisions) and click on GIS Digital Atlas for a preview.

Whyte (1988) *City: Rediscovering the Center* gives a fascinating account of incentive zoning (and much else about planning) in New York City. Bressi (1993) also provides a number of articles on the evolution of zoning in New York City in *Planning and Zoning New York City*.

For a discussion of the New York City incentive zoning program and a profile of 503 spaces at the 320 buildings that participated in the incentive program, see Kayden, the New York City Department of Planning, and the Municipal Art Society of New York (2000) *Privately Owned Public Space: The New York City Experience*.

For a discussion of the Seattle Bonus Program and the Citizens' Alternative Plan (CAP) Initiative, see Caves (1992) *Land Use Planning: The Ballot Box Revolution*.

Any standard legal text discusses the court cases dealing with the various instruments of zoning, e.g., Callies *et al.* (1994) *Land Use: Cases and Materials*; Mandelker (1997) *Land Use Law*; Juergensmeyer and Roberts (1998) *Land Use Planning and Control Law*; Salsich and Tryniecki (1998) *Land Use Regulation: A Legal Analysis and Practical Application of Land Use Law*; Selmi and Kushner (1999) *Land Use Regulation: Cases and Materials*; and Salkin and Freilich (2000) *Hot Topics in Land Use Law: From the Comprehensive Plan to Del Monte Dunes*. A concise summary is given in Wright and Gitelman (2000) *Land Use in a Nutshell*. There are remarkably few zoning books that are not legal in character, but Richard Babcock was a lighthearted lawyer with a knack of telling a good story. His books are insightful as well as enjoyable. See his early (1966) *The Zoning Game* and the later *The Zoning Game Revisited* (Babcock and Siemon 1985). A further volume is Weaver and Babcock (1979) *City Zoning: The Once and Future Frontier*. Bair (1984) *The Zoning Board Manual* is a practitioner's guide which discusses the day-to-day work of the zoning board.

For easy reference there are two APA publications: Meshenberg's *The Language of Zoning* (1976), and Burrows' *A Survey of Zoning Definitions* (1989).

Some issues relating to group homes are discussed in Steinman (1988) *The Impact of Zoning on Group Homes for the Mentally Disabled: A National Survey*; Gordon and Gordon (1990) 'Neighborhood responses to stigmatized urban facilities'; and Jacobson *et al.* (1992) *Community Living for People with Development and Psychiatric Disabilities*. See also Dear and Wolch (1987) *Landscapes of Despair: From Deinstitutionalization to Homelessness*.

On mobile homes, see Wallis (1991) *Wheel Estate: The Rise and Decline of Mobile Homes*.

For a statement of the law relating to variances, see Mandelker (1993) *Land Use Law*, pp. 640–52.

On cluster zoning and PUDs, see Tomioka and Tomioka (1984) *Planned Unit Developments: Design and Regional Impact*; and Moore and Siskin (1985) *PUDs in Practice*. Home owners' associations are discussed in McKenzie (1994) *Privatopia: HomeOwner Associations and the Rise of Residential Private Government*.

Special districts are analyzed in depth (and in a particularly interesting manner) by Babcock and Larsen (1990) *Special Districts: The Ultimate in Neighborhood Zoning*. An interesting study of Times Square can be found in Reichl (1999) *Reconstructing Times Square: Politics and Culture in Urban Development*.

Information on form-based codes can be found in Parolek *et al.* (2008) *Form Based Codes: A Guide to Planners, Urban Designers, Municipalities, and Developers*; and Walters (2007) *Designing Community: Charrettes, Masterplans and Form-based Codes*.

Discussions on the 'transect' can be found in Duany and Brain (2005) 'Regulating as if humans matter: the transect and post-suburban planning'; and Duany and Talen (2002) 'Transect planning'.

Subdivisions are discussed within a legal framework in Mandelker (1993) *Land Use Law* (Chapter 9). Callies and Grant (1991) provide a comprehensive picture in 'Paying for growth and planning gain'. See also (for California), Fulton (1991) *Guide to California Planning*; Curtin and Talbert (2005) *Curtin's California Land Use and Planning Law*; and Curtin and Merritt (2002) *Subdivision Map Act Manual*.

Questions to discuss

1 Is zoning inherently exclusionary?

2 Discuss the role of zoning in the protection of property values.

3 Why are 'variances' so called? Do you think that they should be subject to greater control, for example by a state review board?

4 What are the problems that arise over zoning for the single-family home?

5 In what ways has flexibility been introduced into zoning?

6 'Zoning is more about law than policy.' Discuss.

7 What is covered in a city's subdivision regulations?

8 What is the difference between zoning and subdivision regulations?

9 What is a form-based code and why are cities using them?

10 What is a transect?

9

City financing and planning for development

These newcomers bring with them all their fondest hopes of the future. They bring dreams that are the same as ours – dreams of a better life and a better future. What they don't bring with them are the roads, the bridges, the schools, the hospitals, the libraries, the parks, the utilities, the sewers, the waterlines, and all the vast and varied human services that will be needed to realize our dreams.

Florida State Comprehensive Plan Committee 1987: 6

Budgets

Due to poor economic conditions, tax revenues in many areas have steadily decreased in recent years. There is simply less funding available to pays the bills. Many areas are already facing this dilemma while other areas not currently suffering are expecting budget cuts in a host of agencies. For example, Phoenix, Arizona, had to eliminate a number of staff positions over a three-year period from 2007 to 2010. Cutting staff positions is a common reaction to the economic problems. Other areas have had to cut funding for departments or have reduced the number of library hours or eliminated some branches of the local library system. While no one knows how long the current fiscal blight of our cities will last, these budget problems affect service delivery and the residents' quality of life. Nevertheless, some areas continue to construct new buildings during hard economic times. The City of North Las Vegas built a new city hall, much to the chagrin of some of its residents (see Plate 9).

There are a myriad of ways that cities are trying to handle the economic problems facing them. Of course, with all levels of government experiencing financial problems, finding more revenue is problematic. So, what can jurisdictions do? A number of municipalities

have resorted to the unpleasant tool of raising property taxes. This assumes that the state will allow them to do so. Some municipalities start with looking at their employees. Employees are asked to take pay cuts or furloughs. Municipalities may reduce how much they contribute to items such as health care benefits. The end result is a reduction in employees' take-home pay. Some areas are also giving employees incentives to retire earlier.

To weather the budget cuts without layoffs or firings, a number of areas have gone in different directions. Some areas are selling off or leasing surplus sites to raise revenue. Others are eliminating subsidies to non-profit organizations, eliminating park and recreation spaces, and eliminating trash collection from apartment complexes and commercial properties.

The financial plight of the City of Detroit, Michigan, has been widely covered in all types of media. It is plagued by high unemployment rates, housing fore-closures, abandoned properties, worker layoffs, loss of state and federal funds, and a massive amount of debt and unfunded long-term obligations. These and other problems are not new. They have taken years to mate-rialize. As Mayor Dave Bing acknowledged in an address on November 16, 2011, 'simply put, our city is in a financial crisis and city government is broken'.

Plate 9 City Hall, City of North Las Vegas
Photo by author

In April 2012, the long process of restructuring Detroit's fiscal situation began.

Detroit has been adamant about not having an emergency manager or financial overseer appointed by the Governor. Instead, a Financial Stability Agreement between the State of Michigan and Detroit was reached. This Agreement sought to restructure the city's finances. A Financial Advisory Board of nine members (three appointed by the Governor, two appointed by the Mayor, two appointed by the City Council, one appointed jointly by the Governor and Mayor with City Council confirmation, and one appointed by the State Treasurer) would review, monitor, and advise the City on budget compliance, debt issuance, City reorganization, best management practices, and general financial matters. New positions were also created to assist Detroit to get its financial house in order and to better prepare for the future.

Unfunded mandates

A factor that has contributed to the financial plight of numerous cities is the continued imposition of unfunded mandates by the federal government and state governments on lower levels of governments. According to Lovell, *et al.* (1979: 32) an unfunded mandate is 'any responsibility, action, procedure, or anything else that is imposed by constitutional, administrative, executive, or judicial actions as a direct order or that is required as a condition of aid'. These unfunded mandates require lower levels of government to comply with regulations from a higher level of government. They represent a financial burden to the lower levels of government.

The Federal Unfunded Mandate Reform Act of 1995 (P.L. 104-4) represented an attempt to curb the practice of imposing unfunded federal mandates on states and local governments by requiring agencies to issue written cost estimates for the mandates if they equal or exceed $50 million (adjusted annually for inflation) in the fiscal year a mandate is first effective. The act was designed to provide information to decision-makers to address growing concerns on how much it will cost the lower levels of government to comply with the new mandate.

The increasing deficit of the federal government has exacerbated matters. The federal government has reduced its spending, as have the state governments. Programs are being reduced or eliminated at all levels of government. Cities continue to search for relief or remedies to reduce the financial burdens imposed on them by the federal government.

Bankruptcy

With the US currently suffering from a period of recession, fiscal problems plague many cities. Many are facing severe budget shortfalls. Payrolls cannot be met. Some cities have had to sell off city assets. Pension costs are rising. Some cities have sought to merge with a neighboring city or county. Others face costly labor agreements with unions. One city could not afford to pay a breach-of-contract judgment won by a developer that was two times more than its budget. The processes of seeking bankruptcy protection vary by state. Many states require cities to exhaust all possibilities before seeking bankruptcy, including the possibility of putting the city in the hands of a state receiver.

Although no city wants to seek Chapter 9 federal bankruptcy protection, some areas seek it as a last resort. One of the more publicized cities undergoing

bankruptcy is Stockton, California. It is the most populous city in the US to file for bankruptcy protection. With a 2012 population of around 300,000, Stockton represents a prime example of an area that saw prosperous times and is now in its darkest economic period. It is dealing with past bad decisions, bookkeeping errors, high foreclosure rates, dwindling revenues, retired employees with large pensions and retirement benefits, and high unemployment rates. The problems continue to get worse.

Cities in California must follow the procedures set forth in California Government Code Section 53760 prior to seeking bankruptcy protection. This section requires that a local public entity in this state may file a petition and exercise powers pursuant to applicable bankruptcy law if it has participated in a neutral evaluation process or if it declares a fiscal emergency and adopts a resolution by a majority vote of the governing board. In 2010, Stockton declared a fiscal emergency with a deficit of over $20 million.

Stockton did not want to declare for bankruptcy protection. It resisted for some time. However, as the fiscal problems worsened, it saw no other option. It participated in a 60-day mediation or evaluation process to develop a plan to avoid bankruptcy. About half of the states have such a process. Negotiations to restructure debt with creditors failed to materialize. They couldn't close the City's deficit. After three months of working with creditors and failing to reach any resolution, Stockton filed for bankruptcy on June 28, 2012. In doing so, the City is able to break existing contracts without the threat of lawsuits.

Paying for the costs of development

The costs of development include not only the construction costs of buildings (houses, shops, offices, etc.) but also the costs of the services and facilities which are needed to serve these. Sewage disposal, water supply, and other utilities are the most immediately obvious, but the full list ranges much more widely – highways, schools, day-care centers, hospitals and other social services, public transit, the provision of housing (or transportation) for low-income workers needed to service the development, and so on. As downtowns age and buildings are demolished, you pay for the cost of development. Downtown renewal is occurring throughout the United States (see Plate 10).

Who is paying for these? And how? The existing property owners through their property taxes, the developers through exactions, the new residents through special assessments? The possibilities are theoretically almost endless and, not surprisingly, the whole subject bristles with difficulty and controversy.

The story of development charges (or exactions or imposts – there is no standard terminology) in recent decades is one of an ever-expanding net, bringing more and more services within its grasp. The simplest, and

oldest, is the development charge levied to pay for the provision of basic utilities on the site. These charges arose in connection with subdivision control, and were legitimated in the Standard Planning Enabling Act of 1928, which explicitly included a requirement for the provision of infrastructure internal to the development. Such services were normally limited to streets, sidewalks, street lighting, and local water and sewage lines. Services external to the development were paid for by the appropriate suppliers.

This system worked satisfactorily until the housing boom of the post-World War II period, which placed a great strain on the budgets of the municipalities and school districts (and on the tolerance of property tax

Plate 10 Downtown redevelopment demolition in Kansas City, Missouri
Photo by author

payers). Existing property owners were unhappy (sometimes vociferously so) at having to pay increased taxes for the benefit of newcomers and, increasingly, municipalities required developers to make contributions (dedications) of land for such purposes as schools and playgrounds (or, particularly in small developments, cash payments in lieu).

The next step was to extend these contributions to other services which are necessary to serve the development. Typically, these are off-site, such as sewerage and water supply systems, and arterial roads. These 'impact fees' have become increasingly popular for two reasons. First, the reluctance of existing property owners to pay for the servicing of new development grew substantially as federal aid to localities was reduced. At the extreme (as with California's Proposition 13), taxpayer 'revolts' brought matters to a head. Proposition 13 limited the basic property tax rate to 1 percent of the property's assessed value. More broadly, there has been the expansion of popular concern for the environment, which has eroded the traditional belief in the benefits of never-ending growth: this culminated in an articulate and sometimes blinkered no-growth ethic.

States are not blind to these concerns. They have witnessed the financial plight of cities trying to provide the public services needed by new development. Cities can no longer rely on property owners and the property tax to finance infrastructure for new housing. They realize that local governments have suffered from dwindling federal and state funding. One state response has been to provide municipalities with the authorization to charge impact fees as part of the financing for public facilities needed for new development. For example, in Section 82.02.050 (1) of the Revised Washington Code (RCW), the State of Washington sought to do the following:

(a) to ensure that adequate facilities are available to serve new growth and development;
(b) to promote orderly growth and development by establishing standards by which counties, cities, and towns may require, by ordinance, that new growth and development pay a proportionate share of the cost of new facilities needed to serve new growth and development; and

(c) to ensure that impact fees are imposed through established procedures and criteria so that specific developments do not pay arbitrary fees or duplicative fees for the same impact.

The state also requires that the fees for system improvements be reasonably related to the new development and that any fee does not exceed a proportionate share of the costs of improvements that are reasonably related to the new development. As we will see in the next section, the word 'reasonable' is subject to a great deal of interpretation. It is also the subject of many legal disputes.

In some areas, municipalities have long required developers to provide or finance infrastructure which benefits not just a particular development but a wider area, or even the public at large. This has been particularly so in California, where the state courts have taken an unusually relaxed view on the matter. As we shall see, this view was significantly affected by later Supreme Court decisions.

Impact fees

An impact fee is a sophisticated mechanism for shifting from a municipality a part of the cost of the capital investment necessitated by new development. (The question of who bears the cost is discussed later in this chapter.) In tune with the spirit of the age, impact fees are much more complicated than the earlier charges. On the one hand, they can be far more wide-ranging, extending to any municipal capital expenditure required to meet the needs of the inhabitants of the new development, e.g., capital expenditures for public improvements in schools, fire protection facilities, park facilities, water and sewer facilities, and transportation improvements. On the other hand, they are subject to the restraints of a new calculus which attempts to calibrate the marginal impact of the new development upon a municipality. Box 9.1 provides the content of a local impact fee ordinance in the State of Washington. This is a field which has been extensively dealt with in the courts and in a newly developed area of planning expertise. Box 9.2 illustrates the basis for calculating impact fees.

BOX 9.1 CONTENTS OF A LOCAL IMPACT FEE ORDINANCE

The state of Washington also provides guidance as to what needs to be contained in a local ordinance on impact fees. For example, Section 82.02.060 of the RCW indicates that a local ordinance:

1 Shall include a schedule of impact fees which shall be adopted for each type of development activity that is subject to impact fees, specifying the amount of the impact to be imposed for each type of system improvement. The schedule shall be based upon a formula or other method of calculating such impact fees;
2 May provide an exemption for low-income housing, and other development activities with broad public purposes, from these impact fees, provided that the impact fees for such development activity shall be paid from public funds and other impact fee accounts;
3 Shall provide a credit for the value of any dedication of land for, improvement to, or new construction of any system improvements provided by the developer, to facilities that are identified in the capital facilities plan and are required by the county, city, or town as a condition of approving the development activity;
4 Shall allow the county, city, or town imposing the impact fees to adjust the standard impact fee at the time the fee is imposed to consider unusual circumstances in specific cases to ensure that impact fees are imposed fairly;
5 Shall include a provision for calculating the amount of the fee to be imposed on a particular development that permits consideration of studies and data submitted by the developer to adjust the amount of the fee;
6 Shall establish one or more reasonable service areas within which it shall calculate and impose impact fees for various land use categories per unit of development;
7 May provide for the imposition of an impact fee for system improvement costs previously incurred by a county, city, or town to the extent that new growth and development will be served by the previously constructed improvements provided such fee shall not be imposed to make up for any system improvement deficiencies.

Source: Revised Code of Washington, Section 82.02.060

The crux of the matter is the determination of the 'rational nexus' – or, more simply (and therefore less appealing to lawyers) the 'connection' – between the charge levied upon a developer and the burden placed on the municipality by the development. Thus a new development might necessitate the building of a major arterial highway in an adjacent area, but it would be wrong to relate the whole cost of the road to the new development if, as is likely, the highway were required not merely for the new development, but for the area as a whole. There is a useful analogy in the last straw that broke the camel's back. The new development is the last straw: but the main burden on the camel is the straw that is already there.

The rational nexus has become increasingly popular as a broadly acceptable concept for debating the division of costs between the developer and the local authority. Basically, it uses cost accounting methods for calculating what share a new development has in creating the need for facilities. That proportionate share then becomes the basis for a charge.

The rational nexus

The classic statement of the rational nexus concept was set out in the 1987 case of *Nollan* v. *California Coastal Commission*, 483 US 825 (1987). This was the first

---

BOX 9.2 THE BASIS FOR CALCULATING IMPACT FEES

1 The cost of existing facilities;
2 The means by which existing facilities have been financed;
3 The extent to which new development has already contributed, through tax assessments, to the cost of providing existing excess capacity;
4 The extent to which new development will, in the future, contribute to the cost of constructing currently existing facilities used by everyone in the community or by people who do not occupy the new development (by paying taxes in the future to pay off bonds used to build those facilities in the past);
5 The extent to which new development should receive credit for providing common facilities that communities have provided in the past without charge to other developments in the service area;
6 Extraordinary costs incurred in serving the new development;
7 The time-price differential in fair comparisons of amounts paid at different times.

Source: Nicholas *et al.* 1991: 9

---

exactions case heard by the Supreme Court, and its decision has been followed by an avalanche of writings reflecting a range of differing opinions (not all of which can be said to have clarified matters).

The case concerned an application by the Nollans to the California Coastal Commission (CCC) for a permit to demolish their dilapidated beachfront bungalow and replace it with a new and larger three-bedroom house. The commission, which has a policy of increasing access to and along the beach, gave permission conditional on the Nollans providing public access between the sea and their seawall. The CCC claimed the house would obstruct the view of the ocean. Nollan challenged the condition on the grounds that it violated the takings clause of the Fifth Amendment.

The dispute eventually found its way to the US Supreme Court, where it was held that the commission's requirement was an unconstitutional taking of property. Any government action must be proportionally linked to the restriction imposed on the landowner. The essence of the argument was that there was no essential nexus between the permit (which related to the building of a large house in replacement of a small bungalow) and the condition imposed (an easement for public access across part of the Nollans' land). Government had overstepped its boundaries. It

did not show the relationship between the harm being done and the conditions placed on the landowner.

There have been many later cases, one of which was of particular importance. The *Dolan* v. *City of Tigard*, 114 S.Ct 2309 (1994), case related to the requirements imposed on the rebuilding of a commercial property in the commercial area of the Oregon town of Tigard. Dolan owned a 1.67 acre parcel of land that contained a hardware store and parking lot. The parcel also had a creek flowing through part of it. Dolan sought a permit to double the size of the store and to pave the existing parking lot. Tigard subjected approval to two conditions: first that Dolan dedicate land for a stormwater drainage system improvement; and second that Dolan dedicate a strip of land for a pedestrian/ bicycle pathway. In total, Dolan would be required to dedicate some 10 percent of the land. The US Supreme Court had to determine the degree of connection required between a land dedication and the impact of the proposed development on infrastructure and public facilities. Dolan claimed that Tigard had no basis for taking the property and that Tigard did not justify the conditions it imposed on her proposal.

The US Supreme Court accepted the city's argument that the drainage requirement would mitigate the increase in stormwater run-off from Dolan's property, and that the pathway would relieve additional traffic congestion on nearby streets. Thus there was a legitimate state interest in requiring the conditions, and a nexus; but were the conditions imposed by the city reasonable? The court thought not: they were in excess of a 'reasonable relationship' to the proposed development: a 'rough proportionality' was required between the impact of the development and the conditions which could fairly be imposed.

Both *Nollan* and *Dolan* dealt with exactions, not impact fees. In these cases, the court did not question whether the exaction would substantially advance a legitimate government interest. However, the cases are equally applicable to impact fees. Thus there are now two important tests: nexus and rough proportionality. No doubt later cases will develop (or at least debate) these concepts further: some fun should be had deciding whether there is any difference between the familiar 'reasonable relationship' and the new 'rough proportionality'! These tests give a logical framework for deciding what conditions can reasonably be imposed on developers. They also provide the basis for the design of impact fees.

In 1999, the US Supreme Court decided *City of Monterey* v. *Del Monte Dunes at Monterey, Ltd.* Del Monte owned a 37.6 acre piece of ocean-front property that had previously been used as a terminal and tank farm for a petroleum company. It was purchased from Ponderosa Homes, who had previously sought a development permit in 1981. The permit was denied. Del Monte purchased the site in 1986 and cleaned up the abandoned industrial site. It then submitted an application to develop the site as a residential development in conformance with the city's zoning and General Plan requirements. Monterey rejected the proposal but left the door open if Del Monte would develop the site on a smaller scale. The City indicated that the smaller-scale plan 'would be received favorably'. Del Monte scaled down the project but to no avail. The City rejected the development proposal again. Over a five-year period, after submitting five formal proposals, and developing nineteen different site plans, Del Monte concluded that the City of Monterey would never approve its proposed residential development. Each time the site plan was rejected, the City imposed more rigorous demands on Del Monte than in the previous rejection.

Del Monte filed suit in the US District Court for the Northern District of California, charging that the denial of its final development proposal was a regulatory taking without payment of compensation and violated the Due Process and Equal Protection provisions of the Fourteenth Amendment. When the case reached the US Supreme Court, it examined the *Dolan* 'rough-proportionality test' – whether the conditions of land dedications were proportional to the development's anticipated impacts. It held that the *Dolan* 'test' applied only to exactions and would not extend to the denial of the development site plan. In noting the City of Monterey's failure to show good faith in its dealings with Del Monte, it affirmed the earlier judgment of the Court of Appeals and awarded compensation for a regulatory taking of property.

In 2005, the US Supreme Court decided another case involving regulatory takings, *Lingle* v. *Chevron*, 544 US 528. This case failed to garner as much attention as a case decided later that year – *Kelo* v. *New London*, 545 US 469 (2005). *Lingle* v. *Chevron* involved a piece of legislation adopted by the Hawaii Legislature limiting the amount of rent an oil company could charge a lessee – a rent cap. The oil company sued on the grounds that it suffered a taking of property that violated the Fifth and Fourteenth Amendments in that the rent cap failed to substantially advance any legitimate government interest. It should be noted that the oil company did not seek compensation. It sought an injunction to enforce the regulation that it found constitutionally faulty. The US District Court for the District of Hawaii agreed with the oil company, holding that the legislation did, in fact, fail to substantially advance a legitimate state interest.

The court invoked several earlier cases in its decision. In 1980, the court held in *Agins* v. *Tiburon*, 447 US 255 (1980) that a government regulation constitutes or effects a taking if the regulation in question does not substantially advance a legitimate state interest. Two years later, in *Loretto* v. *Teleprompter Manhattan CATV Corp*, 458 US 419 (1982), the court held that if government requires a property owner to suffer a permanent physical invasion of property, however minor, it must provide just compensation. In *Lucas* v. *South Carolina Coastal Council*, 505 US 1003 (1992), the court ruled that government must pay just compensation for a 'total regulatory takings' except to the extent that principles of nuisance and property law independently restrict the owner's intended use of the property.

In its *Lingle* decision, the court held that the 'substantially advances' formula does not constitute a valid takings test. The formula has no proper place in takings law. Since the oil company's argument centered on 'substantially advances' formula theory to support its case, it was not entitled to summary judgment on its claim.

The incidence of charges

It is frequently assumed that charges imposed on developers are passed on to housing consumers. A classic case in support of this is the huge (100 percent) increase in house prices in the Washington metropolitan area which took place in the early 1970s following the simultaneous takeover of the counties of Fairfax, Montgomery, and Prince Georges by anti-development politicians. However, a closer examination of this illustration shows its falsity. There are two important points. First, the *simultaneous* action of these three large counties created a regional land shortage. Second, more typically, a single small municipality (or even a group of municipalities) around, say, Philadelphia or Chicago, could not exert such a market influence: house builders would simply move to more accommodating areas. (Suburbs are often essentially identical.)

The main issue is an elementary one: the incidence of a charge will be determined by market forces. These vary over time and space. The demand for housing in a particularly attractive area may be highly price-inelastic, and thus charges could readily be passed on to buyers. In an area with a plentiful supply of land of similar amenity, the tendency will be for charges to be passed 'backwards' to landowners: developers will pay less for land than they would have done in the absence of charges. At a time of rapid house price inflation, home buyers are tempted to pay high prices in the expectation that prices will rise even further: here the developer should have no difficulty in simply passing on a charge to the buyers. On the other hand,

high mortgage interest rates may make buyers resistant to prices which are increased on account of charges. In short, there is no single answer to the question of who pays.

Existing vs. new home owners

One often hears the argument from residents of a growing community: 'Why should we pay for the expansion of public facilities? Let the developers or the newcomers pay for them. We have already paid our fair share.' Why indeed? But again on reflection, the obvious is not so. It is equally easy – and persuasive – to argue that since growth benefits the community as a whole (not just those involved in the new growth), no extra charges should be imposed on newcomers. In any case, is it not unfair for established households to change the rules for newcomers? Existing residents did not have to pay charges when *they* moved in: why should those who follow them be penalized? Box 9.3 discusses who pays for the infrastructure.

These questions of 'intertemporal fairness' or 'intergenerational equity' are not easy to deal with, and there is no single answer which will fit the situation of widely differing municipalities. Much depends on the historical development of the particular community and the methods employed over time to finance infrastructure. For example, if annual growth is constant, and capital requirements are incremental, the debt (both for replacement of facilities for existing residents and for expansion of facilities for newcomers) is spread over an ever-increasing population. However, some of the investment is for future residents who will not pay their 'share' until further growth takes place. The net effect is problematic. Moreover, continued growth encounters 'thresholds' which require 'lumpy' investments (sewage treatment plants, for example – which cannot be expanded incrementally). The capital cost of these has to be carried for a lengthy period before the full complement of users (and therefore taxpayers) has arrived – by which time, of course, investment in the next 'lump' is necessary. Clearly, it is no easy matter to unravel all these (and similar) matters. The analyst's difficulties are compounded by the impact of inflation,

---

BOX 9.3 WHO PAYS FOR INFRASTRUCTURE?

With time, residents who were established in the community and those who arrived during that year will pay a decreasing share of the cost of facilities that were built for their use. Furthermore, the share of the cost they will bear declines even more as growth rates and financing periods increase.

On the other hand, growth requires new capital outlays for future residents, and established residents must help pay the debt for these capital expansions if they are to be publicly financed. Higher growth rates mean higher rates of facility expansion and, therefore, higher costs to established residents. This impact works as a counteracting balance against the dilution effect described above. The net effect of these forces is not easily determined. It depends upon the magnitude of the growth rate, the interest on borrowed funds, and the length of the financing period.

Source: Snyder and Stegman 1986: 42

---

which has a habit of destroying any notion of equity. There remains plenty of scope for argument at both the academic and the community levels.

Municipal or state bonds

Local and state projects can be financed in a variety of ways. One such way is through the issuance of bonds. By definition, a bond is a debt security where an investor loans money to an entity (city or state) that borrows it for a certain period of time and at a certain interest rate. The borrowed funds are returned, with interest, on a certain date – maturity date.

General obligation (GO) bonds are bonds that are backed by the taxing and borrowing power of the city or state issuing the bond. They are commonly found on local and state ballots. They are generally considered low-interest and low-risk bonds. Some could be tax free. A two-thirds voter approval is generally required for the bonds to be approved.

GO bonds generally serve the entire community or state, depending on who issues the bonds. At the city level, they have been used to finance infrastructure improvements (street and traffic, fire, police, transit, drainage facilities), the construction of public buildings and parks, and the development of affordable rental housing. At the state level, GO bonds could be used for any capital project as defined by the state. For example, the GO bonds could be used to construct school and college facilities, emergency shelters, water and wetland projects, etc.

Revenue bonds can also be used to fund municipal and state projects. They differ from GO bonds in that the bonds are not backed by the full faith and credit obligation of the municipality or state that issues them. They are designed to produce a revenue-producing property. They are paid back by the revenue generated by a specific project, generally user charges.

Revenue bonds can be used for a variety of purposes. Mortgage revenue bonds have been issued to help fund below-market interest and market-rate mortgages for first-time home buyers. Industrial revenue bonds have been issued to finance job creation and business growth in a city or state. Housing revenue bonds have been issued to profit and non-profit developers to purchase, construct, or rehabilitate multi-family housing. In addition, revenue bonds have been used to finance the expansion of a local civic center, construct a tunnel, and build a variety of sports facilities. In fact, financing the construction of baseball and football stadiums and basketball arenas with revenue bonds is a hotly debated topic in the United States.

One issue surrounding bonds is whether or not certain types of bonds are subject to voter approval before they can be issued. The answer to this may vary from state to state. Voter approval may become problematic. In California, due to the infamous Proposition

13, GO bonds are repaid by property tax increases. Voter approval is required. Voter approval is also required in Washington. In the Montana Code Annotated, Title 7, Chapter 15, Article 15, voter approval of urban renewal bonds is required when GO bonds are to be used. Oregon requires voter approval before the jurisdiction can issue the bonds.

The situation differs with revenue bonds. These bonds are self-liquidating. This is the case in Georgia and California. In Oregon, they are generally exempt from voter approval except when the bonds are issued in accordance with the Oregon Uniform Bond Act, if enough signatures are obtained after the bond notice is published.

Another issue facing potential investors is the credit-worthiness of the borrower (the city). The three major credit rating agencies are Fitch, Moody's and Standard and Poor's (see Box 9.4.) These agencies use such criteria as the economy, financial condition of the agencies, management practices of the agencies, and the status of the economy, etc. to determine the rating for a GO bond. The ratings could range from an AAA to a D. The higher the credit rating, the lower the risk

is for investment. The lower the rating can be taken to mean an increase in risk in purchasing the bond. Just because a city's credit rating is high does not mean the bond will not default. Furthermore, credit rating can change over time, due to fluctuations in the above-mentioned criteria.

Special Assessment Districts

Public projects (road improvements, landscaping, street lighting, sidewalks, parks, road construction, etc.) that provide specific benefits to particular properties in a specific area can also be financed through the use of a Special Assessment District. Individual properties in specific areas are levied by a special assessment which is added to their tax bill. The costs of providing the public services are thus shared proportionately by other property owners in the area according to the amount of benefits it receives.

The creation of a Special Assessment District goes through a public project. A letter requesting the creation of one must be submitted to the city, an ordinance

BOX 9.4 COMPARABABLE INVESTMENT RATINGS OF THE THREE MAJOR RATINGS AGENCIES

	Moody's	Standard & Poor's	Fitch
Best Quality	Aaa	AAA	AAA
High Quality	Aa1	AA+	AA+
	Aa2	AA	AA
	Aa3	AA-	AA-
Upper Medium Grade	A1	A+	A+
	A2	A	A
	A3	A-	A-
Medium Grade	Baa1	BBB+	BBB+
	Baa2	BBB	BBB
	Baa3	BBB-	BBB-

Source: http://www.munibondadvisor.com/rating.htm, accessed July 7, 2012

to create the district is prepared, a public hearing notice is publicized, the public hearing is held, a recommendation is made to the local governing body by the appropriate committee or commission, and a decision by the local governing body is made. An election of the affected property owners is then conducted to see if 51 percent of them are in favor of an annual assessment being added to their property tax.

A tax lien is placed on every property in the district. This lien is based on the proportionate share of specific public improvements that benefit their respective properties. Failure to pay the property tax can result in foreclosure on the property.

In 1978, California voters passed Proposition 13 – the People's Initiative to Limit Property Taxation. Passage of this ballot measure capped the property tax by indicating that the *ad valorem* tax on real property could not exceed 1 percent of the assessed value of the property. The virtues and vices of Proposition 13 have been debated for years. Its passage meant other ways of financing public facilities and services had to be found within California.

The Mello–Roos Community Facilities Act of 1982 represented a new method used in California to fund public facilities and services. The Act, through the California Government Code, allows cities, counties, and other public entities to establish community facilities districts, which must be approved by a two-thirds vote of the property owners in the district. Tax-exempt bonds (for large projects) are then sold to fund the public facilities or services. Examples of the facilities and services that could be funded under Mello–Roos are parks, basic infrastructure, police and fire protection, ambulance service, etc. Each district levies taxes in the district to help fund the facilities and services required by a specific plan or subdivision. These districts exist for as long as it takes to repay the bonds. A formula is used to determine the amount each property owner must pay. Although the formula may vary by community, the tax is generally based on the square footage of the property or home. This 'special' tax represents a lien on the property that is added to the property tax. Failure to pay the tax could result in foreclosure.

People buying property in each community facilities district must be told about the existence of the district and the special tax they will be paying. A 'Notice of Special Tax' from the body which levies the tax on the property pursuant to Mello–Roos must be given to the prospective buyer before the property can be purchased.

Capital Improvement Program (CIP)

A CIP represents a city's multi-year long-range plan for improving infrastructure, parks, and other community facilities (see Box 9.5). It links a city's fiscal budget and annual budget to a long-range comprehensive plan and covers different time frames. Most CIPs cover a four- to six-year frame. Other jurisdictions may opt for a longer time period – the CIP for Champaign, Illinois, is a ten-year plan and Chapel Hill, North Carolina's CIP is a fifteen-year plan. CIPs are generally divided into two parts: 1) a capital budget, which is the next fiscal year's budget for capital items, and 2) the capital program, which contains the plan for public expenditures that go beyond the capital budget.

Planning and programming an area's capital improvement needs by projects is a difficult proposition in these days of scarce resources. A wide range of projects could qualify to be in the CIP, including utility projects (flood mitigation, sewer rehabilitation, storm drain replacement, water system improvements), transportation projects, public safety projects, city building projects, airport facility improvements, and parks and recreation projects. The criteria used to determine a capital improvement project vary by jurisdiction. For example, the City of Portland, Oregon, uses a simple definition of a CIP – 'a project that helps maintain or improve a city asset'. Other cities use other criteria to determine if something is a capital project. Fort Lauderdale, Florida, uses the following criteria:

- represent a physical improvement;
- have an anticipated life of not less than ten years;
- cost $5,000 or more.

Box 9.6 provides a more detailed definition of 'capital cost' used by the City of Hagerstown, Maryland.

The goals or purposes of a CIP are fairly simple. One goal is to maintain or repair the capital assets (infrastructure, etc.) that are already in place. Another goal is to provide new capital assets to meet the needs of the public and to enhance the quality of life for the residents. Ultimately, the goal is to prioritize the city's capital expenditures and to make sure the CIP and its fiscal expenditures are consistent with the city's long-range general plan. The Town of Pelham, New Hampshire (2005: 1), indicates that its CIP serves the following purposes:

- To provide the Town of Pelham with a guide to be used by the Budget Committee, Board of Selectmen, and School Board for their annual budgeting process (RSA 674: 5–8);
- To provide a forward looking planning tool for the purpose of contributing to the creation of a stable real property tax rate;
- To aid the Town's elected officials, appointed committees, and department heads in the prioritization, coordination, and sequencing of various municipal and school improvements;
- To inform residents, business owners, and developers of needed and planned improvements;
- To provide a necessary legal basis for the development and proper administration of the Town's impact fee system (RSA 674: 21.V(b)).

How is a project chosen to be included in a city's CIP? Departments do not have the luxury of getting everything that they want. They make 'wish lists' of capital projects. Many cities have forms that are completed for each proposed project that contain the name of the project, the location and description of the project, the justification of the project, the schedule of funds to be expended, the source or sources of funding, and the annual operating budget impact of the project. The projects are then compiled by the agency or department and rated. Each project could be rated on whether the project is legally required, whether the project is considered a maintenance project, whether the project is designed to replace a facility or improve an existing facility, or whether the project is considered to provide a new public service. Other criteria used to prioritize

projects that will be incorporated in the city's CIP might include: availability of financing, the amount of public support for the project, the life expectancy of the project, the percentage of people to be served by the project, the impact on the city's operating and maintenance budgets, and the benefit/cost ratio of the project.

As is often the case in planning, funding sources for projects vary from jurisdictions to states. One factor is certain: there is no one funding source that is used for the CIP. Funding will come from a variety of sources. For example, cities could use the following sources to fund the CIP: surplus funds from the general fund, general obligation bonds, revenue bonds, Community Development Block Grant funds, Transit Occupancy Taxes, State Gas Tax funds, current funds from city operating fund, state grants, gifts, development charges, property owner assessments, etc. The key is to make intelligent and efficient use of all of the funding sources.

CIPs are reviewed by a number of different bodies. Citizens participate in the review process at various times. Some communities have a committee of representatives from the various city departments. Some cities use a citizen committee to examine and prioritize capital projects. A planning commission or some associated body will hear about the capital projects and make a recommendation to the mayor and city council. Finally, the city council will evaluate the various projects, recommendations, etc. and adopt the program as the CIP.

The case of Salt Lake City might better illustrate the steps a city takes for adopting a CIP. Salt Lake City receives capital improvement applications from citizens and city departments. A so-called 'CIP Team' of representatives from various departments reviews and evaluates the capital improvement applications. A citizen board reviews applications and hears presentations on each of the applications. This board also prioritizes the list and presents it to the mayor. The mayor then reviews the list, prioritizes the potential projects, and makes a recommendation to the city council. The city council then holds a public hearing to hear testimony on the various projects, reviews the priority rankings, develops its own rankings of the projects, and makes a recommendation to approve the

BOX 9.5 WHAT IS A CAPITAL IMPROVEMENT PROGRAM?

A Capital Improvement Program (CIP) is a roadmap that provides direction and guidance for the City of Mesa on carefully planning and managing its capital and infrastructure assets. Identifying capital projects and their anticipated funding sources assists in the planning and scheduling of finances for projects and the manpower needed to plan, design, and construct the projects. The CIP promotes coordination of capital projects that are from different program areas but are similar in scope or in the same geographical area of improvement. Examples of projects in Mesa's 5-Year CIP include street construction, water treatment plants, wastewater facilities, park improvements, libraries, mass transit, airport improvements, gas lines, fire stations, police precincts, and public building construction. Land purchases are also listed in the 5-Year CIP since it is considered a capital asset. These projects are long-term in nature (over one year) to complete and are usually financed over a period of time. Typically, a CIP project has a dollar amount over $10,000. The first year of the 5-Year CIP is referred to as the capital budget of a project while the remaining four years are referred to as the programmed amount for a project.

Source: City of Mesa, Arizona, 2006: ii

BOX 9.6 DEFINITION OF CAPITAL COSTS

Capital costs for purposes of the Capital Improvement Program are non-recurring, have a useful life of more than three years, and exceed $5,000 ($10,000 for Enterprise Funds). Capital budget costs include both capital 'projects' and major capital 'outlays.' Project expenditures are for the construction, purchase or major renovation of buildings, utility systems, or other physical structures. Outlay expenditures are for the acquisition of furniture, equipment, or fixed assets, such as trucks, land, or buildings, which also meet the definition of 'capital.'

Source: City of Hagerstown, Maryland, 2006

CIP. The CIP then becomes part of the City's operating and capital budget. Public notices are required before any public meeting is held. It should be remembered that another city could insert its planning commission or some other advisory body into the process.

Tax Increment Financing

Attracting businesses to a blighted, deteriorating, substandard, economically distressed, or decaying for redevelopment purposes area represents a formidable challenge to cities. Box 9.7 illustrates how one state has defined the terms 'blighted' and 'economically distressed'. Who will pay for project costs such as cleaning up the site, financing needed infrastructure, acquiring land, organizational costs, providing utilities, demolishing substandard structures, landscaping, professional services, relocation costs, providing sidewalks, providing street lighting, etc.? These improvements will take a substantial amount of money. Relying on federal and state program funding is no longer a viable option. Federal and state funding for redevelopment activities have been greatly reduced or eliminated. Cities must find new sources of funds for their development activities.

Tax Increment Financing (TIF) represents an option for cities to use for stimulating private investment in blighted or decaying areas without raising taxes. State governments enable cities to create TIF districts. Weber and Goddeeris (2007: 1) indicate that TIF is now used in forty-nine states and the District of Columbia. It works in the following manner. A project area to be redeveloped is identified. The property tax on the property is determined. This represents the base rate on the property before any improvements are made. The base rate is set or frozen at this level for up to twenty-three years, depending on the individual state law. Improvements are then made on the property. These improvements will cause neighboring property values to increase. The rise in property values creates more taxable revenue to be invested in the specific project area. The increased tax revenues represent the tax increments. In other words, the tax increments equal the property taxes accrued after improvements have been made minus the base property rate level that was established at the beginning of the project. The increase in tax revenues is the tax increments used to pay off some of the development costs that would have originally been paid by the parties.

A plan must be written after an area has been designated as a redevelopment area. While the content of such a plan may differ by state, the District of Columbia (Tax Increment Financing Statute Section 2-1217.03) offers the following as the content of a development plan:

1 A delineation of the proposed TIF area;
2 A description of the proposed land uses of the project;
3 The use of the financing proceeds made available pursuant to this subchapter;
4 A pro forma projection of the revenues and expenses of the project;
5 An assessment of the financial feasibility of the project;
6 A description of the timing and phasing of the project;
7 A general description of the compatibility of the project with the Comprehensive Plan;
8 A description of the plan's compliance with the zoning regulations of the District; and

9 An analysis of the projected tax revenue and benefits to be generated by the project.

Once a development plan has been written, it must go through an adoption process. Again, the steps may differ by state. However, a generic first step in the process might be for the governing body to hold a hearing on the development plan. This means that proper notice of the time and place of the meeting must be published in a publication such as a local newspaper. It may also be required to be located in various public places in the affected district. Property owners within a certain radius of the proposed project area must also be notified, generally through certified mail. The hearing will provide citizens the opportunity to voice their support for or opposition to the proposed plan. After the hearing is conducted, the governing body shall vote to adopt or reject the proposed plan. It is then placed in the public record. Any amendments to the plan must also go through a public hearing process.

The original intent of TIF was redeveloping blighted areas. There is no doubt that TIF can be a strong and powerful tool for financing local redevelopment in targeted areas by using increased tax revenues generated by private development in a specific area to pay for public improvements that are needed to enable the development. Some of the benefits to be achieved by using TIF are increased property rates, more jobs, and a stronger tax base. However, any strong tool can be misused or abused for various reasons. Complaints have been heard over how a word such as 'blighted' is defined. Some people claim the word is defined in vague terms. In one case, property owners in an area claimed a proposed project area site was not 'blighted' and did not qualify under the 'spirit of the law'. Legally speaking, however, the area did fit under the definition of 'blight' as written. In another case, a community may have declared productive agricultural lands 'blighted' in the hope of attracting commercial development to the area. Is it fair to take productive lands out of a lower intensive use and then use TIF to put the property into a higher tax producing category? Recommendations to tighten up or strengthen the existing broad or vague definitions have been voiced around the country. TIF is a controversial economic development tool.

There have also been local discussions on who benefits from TIF. Some people feel TIF is being abused when large chain stores are offered tax breaks or other incentives to lure them into an area. These stores do not need the tax breaks or public subsidy. They may move into the area regardless of any offers of tax breaks or incentives. Is this simply a tool being used to benefit private developers or companies at the taxpayers' expense? Why should government take the risks?

Shopping centers or shopping malls offer another example of how TIF might be misused or abused. For example, there could be a case where a shopping mall wants to expand onto adjacent property and the community wants to use TIF to promote local redevelopment. Some people wonder if the adjacent property would be considered 'blighted'. In addition, offering incentives or other public subsidies to expand the shopping mall may be taking away revenue from another shopping mall in the same city. It could also take revenue away from a shopping mall in a neighboring community. In other words, using TIF may cause consequences that were not considered by the appropriate public officials.

BOX 9.7 DEFINITION OF BLIGHTED OR ECONOMICALLY DISTRESSED AREA

a. An area in which the structures, buildings, or improvements, by reason of dilapidation, deterioration, age, or obsolescence, inadequate provision for ventilation, light, air, sanitation, or open spaces, high density of population and overcrowding, or the existence of conditions which endangers life or property by fire and other causes, or any combination of such factors, are conducive to ill health, transmission of disease, infant mortality, juvenile delinquency, or crime, and are detrimental to the public health, safety, morals, or welfare, or

b. Any area which by reason of the presence of a substantial number of substandard, slum, deteriorated, or deteriorating structures, predominance of defective or inadequate street layout, faulty lot layout in relation to size, adequacy, accessibility, or usefulness, unsanitary or unsafe conditions, deterioration of site or other improvements, diversity of ownership, tax or other special assessment delinquencies exceeding the fair value of the land, defective or unusual conditions of title, or the existence of conditions which endanger life or property by fire and other causes, or any combinations of the foregoing, substantially impairs or arrests the sound economic growth of an area, retards the provision of housing accommodations, or constitutes an economic or social liability and is a detriment to the public health, safety, morals, or welfare in its present condition and use, or

c. Any area which is predominantly open and which becomes of obsolete platting, diversity of ownership, deterioration of structures or of site improvements, or otherwise, substantially impairs or arrests the sound economic growth of an area, or

d. Any area which the local government body certifies is in need of redevelopment or rehabilitation as a result of a flood, fire, hurricane, tornado, earthquake, storm, or other catastrophe respecting which the Governor of the state has certified the need for disaster assistance under federal law, or

e. Any area containing excessive vacant land on which structures were previously located, or on which are located abandoned or vacant buildings, or where excessive vacancies exist in buildings, or which contains substandard structures, or with respect to which there exist delinquencies in payment of real property taxes.

Source: Alabama Revised Statutes, 11-99-2(1), 2006

Tax credit programs

A variety of tax credit programs are available to assist in a number of different program areas. Tax credits have been used to help produce and preserve affordable housing. The Low Income Housing Tax Credit (LIHTC) program was created by the Tax Reform Act of 1986 to help acquire, rehabilitate, or construct new affordable rental housing. Private and non-profit developers were eligible to apply for the tax credits. Developers would sell credits to investors to help raise capital for the project. In turn, the credits would reduce the debt the developer would have to borrow from the lending institution.

Rehabilitation tax credits are also available to rehabilitate historic structures and create low- and moderate-income housing opportunities in historic buildings. The tax credits decrease the amount of tax owed on a structure. To qualify for a 20 percent rehabilitation tax credit, a building must be on the National Register of Historic Places or in a registered historic district that houses a business or some other income-producing purpose. Buildings failing to meet those requirements are eligible for 10 percent rehabilitation tax credit.

Minority and low-income communities throughout the US have long struggled to get access to community economic development investments, credit, and financial services. At the end of the Clinton administration, in an attempt to assist these areas, Congress established the New Market Tax Credits (NMTC) Program as part of the Community Renewal Tax Relief Act of 2000. This Program would serve as a catalyst to facilitate investments in low-income communities. By definition, a low-income community is:

> any population census tract where the poverty rate for such tract is at least 20% or in the case of a tract not located within a metropolitan area, median family income for such tract does not exceed 80% of statewide median family income, or in the case of a tract located within a metropolitan area, the median family income for such tract does not exceed 80% of the greater of statewide median income or the metropolitan area median family income.

By making investments in low-income communities through a 'Community Development Entity' (CDE), investors receive credit against their federal income taxes. According to the Internal Revenue Code, Section 45D (c) (1), a CDE is any domestic corporation or partnership 1) whose primary mission is serving or providing investment capital for low-income communities or low-income persons, 2) that maintains accountability to residents of low-income communities through their representation on any governing board or advisory board of the CDE, and 3) has been certified by a CDE as part of a Community Development Financial Institution (CDFI).

The tax credit is 39 percent of the actual capital investment made to the CDE. It is claimed during a seven-year period. For the first three credit allowance dates (years), the applicable percentage is 5 percent. For the four remaining credit allowance dates, the applicable percentage is 6 percent.

In conclusion

There is no universal way to finance urban development. In our current economic climate, many areas are not able to finance development projects in their entirety. They are looking for partners. Multiple players are often evident in project financing. This is common in large state or local projects.

Public-private partnerships have become a popular mode of financing various activities. The partnerships provide a source of capital for projects that facilitate community and economic development. For example, numerous states have enacted legislation authorizing public-private partnerships in operating toll roads and bridges, acquiring or constructing public buildings, building or acquiring a site for a stadium, redeveloping an area.

It is interesting to note that the trend toward greater private 'participation' in the financing of infrastructure is not restricted to the United States. In a comparative study of the United States and Britain, it was noted that in both countries 'the external costs of private land development have, over the past fifteen years, been increasingly borne by private land developers rather than public agencies' (Callies and Grant 1991: 221). The root cause is the inability of local governments to shoulder the increasing demands being made on them. They are therefore searching for new sources of revenue. Imposing levies on new development is a politically painless way of obtaining extra funds.

Further reading

For additional information on unfunded mandates, see: Lovell, *et al.* (1979), 'Federal and state mandating on local governments: report to the National Science Foundation'; Dilger and Beth (2011) 'Unfunded Mandates Reform Act: history, impact, and issues'; Fantone (2011) 'Federal mandates: few rules trigger Unfunded Mandates Reform Act'.

An impact fee consultant, Duncan Associates, has constructed an online informational and educational resource on impact fees (http://www.impactfees.com). Along with links to headlines of news articles reporting on impact fees, the website provides an extensive frequently asked questions section, links to case law and links to the impact fee costs and requirements of almost every county or city in the United States.

There is an enormous, and continually growing, literature on developer contributions to infrastructure and other public benefits. Two good sources (both of information and of references) are Nicholas *et al.* (1991) *A Practitioner's Guide to Development Impact Fees*; and Lassar (1989) *Carrots and Sticks: New Zoning Downtown*. See also Getzels and Jaffe (1988) *Zoning Bonuses in Central Cities*.

Changes in the law and practice of charges can be monitored in the quarterly *Urban Lawyer*. A number of papers previously published in this journal have been collected in Freilich and Bushek (1995) *Exactions, Impact Fees and Dedications*. A very full treatment of the situation in California is given in Abbott *et al.* (1993) *Public Needs and Private Dollars: A Guide to Dedications and Development Fees*. This also has a 1995 supplement.

For discussion of Tax Increment Financing, see: Johnson and Man (2001) *Tax Increment Financing and Economic Development: Uses, Structures, and Impact*; Giles and Blakely (2001) *Fundamentals of Economic Development Finance*; Netzer (2003) *Property Tax, Land Use and Land Use Regulation*; White, Bingham, and Hill (2003) *Financing Economic Development in the 21st Century*.

Interesting discussions of Proposition 13 can be found in Stocker (1991) *Proposition 13: A Ten Year Retrospective*; Shires, Ellwood, and Sprague (1998) *Has Proposition 13 Delivered? The Changing Tax Burden in California*; Fox (2003) *The Legend of Proposition 13*.

Additional information on municipal bonds can be found in: Zipf (1995) *How Municipal Bonds Work*; Bond Market Association and Temel (2001) *The Fundamentals of Municipal Bonds*; Feldstein and Fabozzi (2008) *The Handbook of Municipal Bonds*.

The 'rational nexus' rule is a cost-accounting method which was first elaborated by Heyman and Gilhool (1964) 'The constitutionality of imposing increased community costs on new subdivision residents through subdivision exactions'. See also Ellickson (1977) 'Suburban growth controls: an economic and legal analysis'.

Intertemporal fairness is discussed in Beatley (1988) 'Ethical issues in the use of impact fees to finance community growth'. On 'intergenerational equity' see Snyder and Stegman (1986) *Paying for Growth: Using Development Fees to Finance Infrastructure*. For discussions on who actually pays for development fees and on the impact of impact fees on the cost of housing, see Baden and Coursey (1999) *Effects of Impact Fees on the Suburban Chicago Housing Market*; and Dresch and Sheffrin (1997) *Who Pays for Development Fees and Exactions?*

An interesting comparative study by Callies and Grant (1991) is 'Paying for growth and planning gain: an Anglo-American comparison of development conditions, impact fees, and development agreements'.

For additional information on New Market Tax Credits, see Pappas (2001) 'A new approach to a familiar problem: the New Markets Tax Credit'; Internal Revenue Service (2010) *New Markets Tax Credit*; and the New Markets Tax Credit Coalition, http://nmtccoalition.org.

For information on Community Development Financial Institutions, see Benjamin, Rubin, and Zielenbach (2004) 'Community Development Financial Institutions: current issues and future prospects'; and an entire issue of the Federal Reserve Bank of San Francisco's *Community Development Investment Review* (2005).

Questions to discuss

1 List the costs which are involved in residential development. Discuss who should bear these.

2 Argue the case for and against impact fees.

3 Describe 'the rational nexus'. Do you think that it is a useful concept?

4 'It is the purchaser who bears the cost of charges imposed on developers.' Discuss.

5 Discuss the uses and problems of incentive zoning and bonusing.

6 'Much of planning is in fact negotiation.' Discuss.

7 Describe a development agreement. Why would a developer want one?

8 How do municipalities finance the provision of public services and facilities?

9 What is Tax Increment Financing?

10 What is the difference between a general obligation bond and a revenue bond?

11 What effect did the passage of Proposition 13 in California have on the financing and provision of public services?

12 What is a Capital Improvement Program and why is it important?

13 How can a tax credit program assist in financing economic development?

GROWTH MANAGEMENT

Growth management continues to be one of the foremost issues in land use planning. It remains one of the most hotly contested issues facing areas around the country. At the local government level (discussed in Chapter 10), it has had a surprisingly long history, dating back to the late 1960s and early 1970s. (Involvement by the states – discussed in the following chapter – came later.) The most famous legal cases are *Ramapo* (New York) and *Petaluma* (California), which, in different ways, added the concept of timing to the two traditional planning dimensions of location and use. The idea is a simple, persuasive one: that development should proceed in parallel, or concurrent, with the requisite infrastructure. In addition, the need to refocus growth by infill development, revitalizing existing areas, and centering development around transit centers has emerged in the Smart Growth movement.

Local governments are severely limited in their ability to manage urban growth. The problems associated with growth fail to recognize jurisdictional boundaries. The issues are essentially regional in character. Restraints in one area may simply result in development pressures moving elsewhere in the region.

This represents the rationale for state action. A number of states have assumed responsibilities for growth management, though there are marked differences in the extent to which they have been willing and able to shoulder these. There are also significant variations in the techniques adopted. In Chapter 11, the policies of seven states are examined. Hawaii introduced statewide zoning; Oregon set up a comprehensive system of local planning which had to conform to a long list of state goals; Vermont set up a system of citizen district commissions to administer a development plan system; Maryland pioneered 'Smart Growth'; Florida introduced state controls in 'areas of critical concern' and 'developments of regional impact'; California established a coastal planning system; and New Jersey battled with the introduction of a state plan intended to guide growth and conservation throughout the state. Updates of the various states have been added to the chapter. States continue to promote and pursue activities associated with principles of smart growth. The effectiveness of these and similar policies is a matter of considerable debate, but it is clear that they have proved to be very difficult to implement.

Growth management and local government

The slow-growth movement has proved that it can win elections. What it has not proved, however, is that it can stop growth.

Fulton 1990

Attitudes to growth

American zoning proceeds largely on the basis of decisions regarding individual lots. What is typically ignored is the cumulative effect of an enormous number of 'lot decisions'. This is partly because the zoning machine usually operates without the advantage of a guiding plan; partly because zoning has traditionally been unconcerned with the timing of development (or its relationship to the provision of infrastructure); and partly because the normal presumption of municipalities is in favor of development – the more, the better. The last point goes deep: instead of asking 'is the proposed development desirable in the public interest at this place at this point in time?', the typical municipality starts from the presumption that any development is good and, in any case, it is unfair to penalize a particular owner with a refusal: if one farmer's land has been approved for development, why shouldn't his neighbor get equal treatment?

This traditionally positive attitude to growth is now reversed in some areas, particularly where development has been rapid. Suburban localities have for a long time put up formidable barriers to development which might attract low-income households, but this anti-growth stance has more recently spread, in some areas, to any development which might increase tax burdens or add to levels of traffic congestion which are already considered to be severe. As a result, restrictions have

increased; but, as long as the siting of a development does not detract from the amenities enjoyed by an articulate minority, it is likely to go ahead.

Whether a municipality is for or against growth, however, it is unusual for it to embody its ideas in a formal land use plan. Zoning has generally remained the standard system of land use control. Yet zoning usually operates without concern for wider questions of planning: it is essentially reactive and timeless. The difficulties to which this may be expected to give rise are exacerbated by the fact that zoning maps usually have a similar timeless quality. They show the use to which individual lots of land may – in isolation – reasonably be put, but they do not take into account the effect of the timing of development applications or the effect of a number (and certainly not all) of the proposals emerging at a particular time. The availability of public services (from sewers to roads to schools) does not enter into the political calculus. Development patterns can therefore be haphazard, inefficient and wasteful, costly to service, and cumulatively disastrous – with inadequate public services, gridlock and the like.

Added to the political predispositions are a number of other complicating factors. The dictates of the Constitution are one – particularly the requirement for equal treatment (how does a political body, normally consisting of a very small number of members, defend unequal treatment to landowners on some fuzzy basis of the public interest?). Another is the division of responsibility between different agencies. Transportation planning is frequently the responsibility of an agency different from the one concerned with zoning; schools always are. As a result, zoning is the major discretionary function of municipalities – and sometimes the dominating issue at local elections.

By a curious twist of the tale, action to promote coordinated planning may be interpreted (not always unjustly) as an underhand means of excluding minority groups from an area – what Bosselman *et al.* (1973: 249) have characterized as 'the wolf of exclusionary zoning under the environmental sheepskin worn by the stop-growth movement'. All these considerations help to explain the widespread popularity of large-lot zoning: it results in development which makes the minimum demands on public services (and on the demand for an expansion of them); it pays for itself in the narrow terms of a municipal budget; and it excludes minorities from the area. It also enables a municipality to operate a primitive form of growth management by holding back development pressures. But it can be very inefficient. Large-lot zoning can lead to development which is scattered and expensive to service.

These ideas are explored more deeply in other chapters: they are mentioned here to provide a reference point for the ensuing discussion of the interesting, and largely unsuccessful, attempts to plan the location and timing of urbanization.

Considerable ingenuity has been displayed in devising techniques of growth management. They range from restrictive subdivision and zoning regulations, permits to begin development, caps on the number of new dwellings (either annually or over a period of years), phasing development along with the provision of infrastructure, urban growth limit lines, and the preservation of land for agricultural or other highly restricted uses. A complete list of all the possible measures would be a very long one. Indeed, most planning techniques can be utilized for growth management purposes. Some of the discussion of the subject is therefore scattered among the various chapters of this book.

Any account of this subject must include two machinations in mathematical probity which assumed fame in the early 1970s – the growth control programs of *Ramapo* and *Petaluma*.

The Ramapo growth control program

Ramapo is a town in Rockland County, New York, about thirty-five miles from downtown Manhattan. At the end of the 1960s it had a population of around 76,000 and was growing rapidly. As a result, there was an increasing strain on public services and infrastructure. A master plan had been adopted in 1966, followed by a comprehensive zoning ordinance, and then by a capital improvements program and a phased growth plan (see Box 10.1). The latter provided for the control of residential development in phase with the provision of adequate municipal facilities and services. The various plans covered a period of eighteen years.

The timed growth plan did not rezone any land: the restraint on property use was regarded as being of a temporary nature. This restraint took the form of a requirement that a special permit be obtained for suburban residential development. Thus, where the required municipal services were readily available a special permit would be granted, but where a proposed development was located further away, development could not begin until the programmed services reached the location – unless the developer installed the services. However, a landowner could still erect a single-family dwelling.

The court held that where it is clear that the existing physical and financial resources of the community are inadequate to furnish the essential services and facilities which a substantial increase in population requires, there is a rational basis for phased growth and, hence, the challenged ordinance is not violative of the federal and state constitutions.

Moreover, the ordinance was in compliance with the town's comprehensive plan. It did not seek to do anything contrary to the plan. This decision is of great significance: for the first time land use regulation became legally 'three-dimensional'. To the traditional dimensions of location and use was added the new

BOX 10.1 THE RAMAPO TIMED GROWTH PLAN

The standards for the issuance of special permits are framed in terms of the availability to the proposed subdivision plat of five essential facilities or services: specifically (a) public sanitary sewers or approved substitutes; (b) drainage facilities; (c) improved public parks or recreation facilities, including public schools; (d) state, county or town roads – major, secondary or collector; and (e) firehouses. No special permit shall issue unless the proposed residential development has accumulated fifteen development points, to be computed on a sliding scale of values assigned to the specified improvements under the statute.

one of time. As of 2010, Ramapo had an estimated population of approximately 126,600 inhabitants.

The need for infrastructure to be concurrent with development has become a critical debate around the United States. Too many cases can be cited where the infrastructure needed for a development has lagged behind the development of housing units. This issue is discussed further in Chapter 11.

The Petaluma quota plan

The Ramapo plan implied a limit to the annual number of dwellings that could be built in the area. This was not a predetermined figure: the actual number of dwellings built was dependent upon capital improvements and the ability of developers to acquire points under the point system. The Petaluma plan, by contrast, operated by way of a fixed quota.

Petaluma lies some forty miles north of San Francisco. In the 1950s and 1960s it experienced a steady population growth, from 10,000 in 1950 to 25,000 in 1970. By the latter date, however, this self-sufficient town had been drawn into the Bay Area metropolitan housing market, and development boomed in the form of single-family dwellings. Whereas only 358 dwellings had been built in 1969, the number rose to 591 in 1970, and 891 in 1971. The opening of a highway between San Francisco and Petaluma contributed greatly to the community's growth. Alarmed at this rate of growth, the city introduced a temporary freeze on development. This provided a breathing space during which a growth management plan could be prepared. The plan, adopted

in 1972, fixed development at a maximum rate of 500 dwellings a year (excluding projects of four units or fewer) and established a greenbelt surrounding the city that would serve as an urban expansion boundary for a period of at least five years. To give effect to this control mechanism it was necessary to have a system (Residential Development Evaluation System) which would choose between competing claimants. The instrument devised for this purpose was an annual competition among rival plans in which points were awarded for access to existing services which had spare capacity, for excellence of design, for the provision of open space, for the inclusion of low-cost housing, and for the provision of needed public services. (The policy allocated between 8 and 12 percent of the annual quota to low- and moderate-income housing.) Box 10.2 describes the allocation procedure.

The Residential Development Evaluation System awarded developers zero to five points for various factors. In terms of public facilities, points were awarded for such factors as capacity of the water system, capacity of the sanitary sewers, capacity of the drainage facilities, ability to provide fire protection, capacity of the appropriate schools to handle the increased demand, and the capacity of the major street system to meet the needs of the proposed development. Points were also awarded on site and architectural design quality – this would include usable open space, contributions and extensions of existing foot and bicycle paths, trails, needed public facilities, the extent to which the proposed project contributed to orderly development and provision of units to meet the city's annual low- and moderate-income dwelling unit goal.

The system certainly had its share of critics. Some developers felt the system was too time-consuming and too complex. They didn't understand the system. In order to 'play the game', developers even had to include features that potential residents didn't need. In adding unnecessary features, many developers felt the system requirements simply added to the cost of a home.

Not surprisingly, the development interests in the area were highly alarmed, and a case was brought against the city. The district court declared the plan to be unconstitutional, but the court of appeals reversed. Though it accepted the view that the plan was to some extent exclusionary, it noted that 'practically all zoning restrictions have as a purpose and effect the exclusion of some activity or type of structure or a certain density of inhabitants'. The court's review did not cease upon a finding that there was an exclusionary purpose: what was important was to determine whether the exclusion bore any relationship to a legitimate state interest. The court held that the Petaluma plan did in fact serve such an interest. Moreover, the court held that the plan was certainly not exclusionary in the sense of keeping out low-income households. On the contrary, it was 'inclusionary to the extent that it offers new opportunities, previously unavailable, to minorities and low- and moderate-income persons'. Moreover, the court ruled that jurisdictions can provide for orderly growth. As of 2010, Petaluma's population was pushing an estimated 57,941 residents. It represented the first city in the United States to protect its quality of life through a residential growth management plan.

Other growth control programs

Ramapo and Petaluma are only two of many growth management schemes which have been introduced since the late 1960s. They are particularly notable because of the blessing bestowed upon them by the courts (and their prominence in standard texts). However, not all schemes have been approved by the courts. For instance, the attempt by Boca Raton, Florida, in 1972, to place a cap (of 40,000) on the number of dwellings ultimately to be built in the city (agreed by a public referendum after a superficial review of urbanization trends) was declared unconstitutional. The referendum was designed to reduce the amount of multi-family units in Boca Raton. Among the means to accomplish this were to reduce multi-family densities by 50 percent, rezone multi-family lands to single-family use and to rezone some residential acreage to commercial and industrial uses. Perhaps Boca Raton was just unlucky, though its action was not backed up by the supportive planning studies which courts like to see. Yet – a point that needs to be constantly borne in mind – most schemes are not in fact challenged. For example, there was no challenge to the Californian City of Napa's *residential urban limit line* that was intended to limit the city's population to 75,000 by

---

BOX 10.2 ALLOCATION PROCEDURES USED IN PETULAMA

At the heart of the allocation procedure is an intricate point system, whereby a builder accumulates points for conformity by his projects with the City's general plan and environmental design plans, for good architectural design, and for providing low and moderate income dwelling units and various recreational facilities. The Plan further directs that allocations of building permits are to be divided as evenly as feasible between the west and east sections of the City and between single-family dwellings and multiple residential units (including rental units), that the sections of the city closest to the center are to be developed first in order to cause 'infilling' of vacant area, and that 8 to 12 percent of the housing units approved be for low and moderate income persons.

Source: Construction Industry Association v. City of Petaluma, 522 F.2d 897, 901 (9th Cir. 1975)

the year 2000. This *line* represented the boundary beyond which essential public services would not be provided. It was accompanied by the Napa Residential Development Management Plan, which imposed an annual ceiling on new residential construction.

The reasons for introducing growth management policies vary. Thus, while Napa's *residential urban limit line* was intended to cap the population growth of the area, nearby Santa Rosa had an urban boundary designed to permit all the development which was envisaged for the foreseeable future. If it transpired that further land was required, the boundary could be extended: the objective was not to prevent growth but to ensure that it took place in a desirable manner. It was aimed at the problems of 'scatteration' and the 'unnecessary use' of agricultural land. It also attempted to preserve environmental quality and enhance the aesthetic quality of new housing. Other California cities (such as San Rafael and Novato) have also been concerned essentially with the preservation of open space and the establishment of green belts. When viewing San Rafael for the first time, an individual may get the impression that there remains a great deal of land to be developed. However, this is not the case. Vacant land is a scarce commodity because much of the land has been dedicated as permanent open space. Over a thirteen-year period (1986–1999) the amount of vacant land in San Rafael decreased from 6,000 acres to approximately 250 acres. Consequently, the city is currently focusing a great deal of attention on infill development, attached units, and apartment construction. Similarly, the comprehensive planning program in Montgomery County, Maryland, was 'intended to accommodate growth, and to manage it only to the extent needed to moderate its ill effect' (Porter 1986: 82).

There are numerous means available to local governments to control or manage their growth. While one city may employ a greenbelt to control its expansion, another city may employ an urban growth boundary or a temporary moratorium on development as growth control devices. To be successful, multiple tools need to be employed. The old adage 'one size doesn't fit all' is appropriate. What may work for San Diego may not work for Phoenix, and what works for Phoenix may not work for Las Vegas.

Growth management and infrastructure

The use of infrastructure planning as a major element in land use controls has become popular. A common method of coordinating urban growth and the provision of infrastructure is to require developers to hold back until the necessary provision can be made (as in Petaluma). This also allows developers to proceed if they themselves provide the infrastructure, or the finance for it. (Impact fees, discussed in Chapter 9, may form an element of such a scheme, whether or not the overall intention is the limitation or the management of growth.)

Another permutation of this school of controls is *impact zoning*. This has been particularly popular in the towns of Massachusetts. 'Drawing on NEPA and the lawyer's continuing faith in procedural solutions, these towns have amended their bylaws to require a statement of the impact of proposed subdivisions on town services and the local environment. (Harr, 1977)' The form such an amendment to the zoning bylaw may take is illustrated by the impact zoning scheme set out in Box 10.3.

Initiatives in Boulder

A further example of the use of infrastructure planning in land use controls is provided by the experience of Boulder, Colorado. Boulder has had a long history of planning to preserve its dramatic natural surroundings. It has long been recognized for its stance on environmental stewardship. Frederick Law Olmsted, Jr., produced a report in 1910 on *The Improvement of Boulder, Colorado*, and in 1928 the city became one of the first western cities to introduce a zoning ordinance. Not surprisingly, pressures continued on the peripheral areas, and in particular on the mountain foothills. To stem this pressure, a citizen-initiated 1959 charter amendment established an elevation of 5,750 feet along the mountainsides beyond which utility service could not be extended (Porter 1986: 35). This is the so-called 'Blue Line'. In the 1960s, two-fifths of a one-cent city sales tax was earmarked for open space acquisition (the

BOX 10.3 IMPACT ZONING

In order to evaluate the impact of the proposed development on Town services and the welfare of the community, there shall be submitted an Impact Statement which describes the impact of the proposed development on (1) all applicable town services, including but not limited to schools, sewer system, protection; (2) the projected generation of traffic on the roads of and in the vicinity of the proposed development; (3) the subterranean water table, including the effect of proposed septic systems; and (4) the ecology of the vicinity of the proposed development. The Impact Statement shall also indicate the means by which Town or private services required by the proposed development will be provided, such as by private contract, extension of municipal services by a warrant approved at Town Meeting, recorded covenant, or by contract with home owner's association.

Source: Haar and Wolf 1989: 592

balance went to road improvements). A height limitation of 55 feet was established in 1971 to preserve the views of the majestic Rockies. By the end of the 1980s, the city had spent $53 million on the acquisition of 17,500 acres of open space, most of which lay outside the city limits. By 1998, Boulder had raised over $115 million and acquired over 30,000 acres of open space.

The city was able to exercise some control over development beyond its boundaries by virtue of its utility functions: in a part of the country where water is in short supply, Boulder had virtually complete control of the water in its area (Godschalk *et al.* 1979: 258); but it received a setback in 1976 when the courts ruled that it was unconstitutional to use the powers of a public utility for planning purposes. However, by cooperating with the surrounding Boulder County, a comprehensive plan for the larger area was agreed, and this enabled the two governments to coordinate planning and annexations. Boulder uses a system of phased-in development in which the area is divided into three sections. The first consists of the 19 square miles now within the city limits, and has a full range of public services. The second, 7.5 square miles under county jurisdiction, is targeted to be annexed and serviced within three to fifteen years. The third, some 59 square miles, is not projected for servicing until after fifteen years – if ever. Part of the third section represents areas where the city and county want to preserve existing rural land uses and character. The other part

of section three includes lands where the city and county want to maintain the option of future city expansion beyond the fifteen-year planning period. This was critical since Boulder County was projected to grow from 1998 to 2010 by almost 30 percent.

A 2000 major update to the Boulder Valley Comprehensive Plan focused on two major themes: increasing affordable housing opportunities, including different housing types, and identifying the best sites for housing. The key remains community sustainability. The planning process includes a discussion of the aforementioned themes, a refinement of the issues that underwent citizen review and comment, including a citizen survey, a period for staff to analyze the surveys, and finally, the public hearings necessary to adopt the update. The update reiterated the area's commitment on applying the principles of sustainability to all of its actions. It also stressed the importance of promoting a walkable city, developing trails and trail linkages, preserving historical and cultural resources, archeological sites and cultural landscapes, developing neighborhood design guidelines that promote sensitive infill and redevelopment, and implementing growth management tools that control all aspects of new development and redevelopment.

The time frame of the plan is fifteen years with a required update every five years. This is critical in that changing area conditions may dictate potential changes in the plan and its associated policies. The plan also stressed the importance of regional cooperation since many of the problems facing the area transcend political boundaries.

An interesting feature of Boulder's planning policy is its purchase of development rights, which keeps land in agriculturally productive use but prevents development upon it. Boulder is by no means alone in its concern for protecting agricultural land from urbanization, as the following discussion demonstrates.

Safeguarding agricultural land

The safeguarding of agricultural land has a strangely captivating and persistent appeal. It is uncritically accepted that food-producing land is 'under threat', that its loss is irreversible, and that it is folly to reduce national self-sufficiency in food supplies. This seems singularly inappropriate in a country of the vastness of the United States. Nevertheless, the nation has been losing farmland for a number of years. The number of farms has decreased as well as the amount of land being devoted to farming activities. There is considerable controversy on precisely this point. The issue achieved salience in the 1960s with concern about environmental degradation, urban sprawl, and the pressure for national land use policies. The federal reaction was to mount the National Agricultural Lands Study (US Department of Agriculture 1981). The accuracy of the data presented in this study has been subject to intense debate, but little consensus has appeared. Action at the federal level has been minimal. The most significant legislation has been the Farmland Protection Policy Act of 1981, which requires federal agencies to have regard to the effect of their programs on the loss of agricultural land (see Box 10.4).

In 1981, the US Department of Agriculture's Soil Conservation Service (now known as the Natural Resources Conservation Service) designed the agricultural land evaluation and site assessment (LESA) system. This system was designed to determine the quality of the agricultural lands and to assess the land's agricultural economic viability. It rates and combines soil quality with various other factors in order to rank the relative value of an agricultural site. The land evaluation component of the system simply examines soil quality factors. It can range from the most productive soils to those soils with little or no productive value. The site assessment component analyzes such factors as distance to roads, visual/scenic value, cultural value, water availability, and wildlife habitat values. In other words, it examines any identified limitations on agricultural productivity, other than any soil-based qualities.

The regulations for implementing the Act appeared in draft form in 1982, but became the subject of much controversy. When the final version of the regulations appeared in 1984, they did little more than require that federal agencies consider the impact of their activities on the conversion of farmland: there was no requirement that the activities should be changed as a result of the impacts.

While federal action has been less than dramatic, there has been much action at state and local levels. Indeed, the position has not changed since the National Agricultural Lands Study noted that state and local governments are the prime instigators of agricultural preservation.

BOX 10.4 FARMLAND PROTECTION POLICY ACT 1981

The act requires federal agencies . . . to develop criteria for identifying the effects of federal programs on the conversion of land to nonagricultural uses, and to identify and take into account the adverse effects of federal programs on the preservation of farmland; consider alternative actions, as appropriate, that could lessen such adverse effects; and assure that such federal programs to the extent practicable are compatible with state, units of local government, and private programs and policies to protect farmland.

At the state level, the most common program is some type of favorable tax treatment such as assessment at existing use (farming) value rather than market value (which may include potential development value). This can apply to both property and inheritance taxes. However, it is doubtful whether such tax benefits are very effective on their own: owners may simply enjoy reduced taxes until the time comes when they want to sell. They are, of course, popular with farmers, and therefore they enter into the arena of state politics.

Also popular are 'right to farm' laws: these protect farmers from local ordinances (and private nuisance suits) that restrict normal farming operations. There is little analysis of the effectiveness of such laws, though one study concluded that, while they reduced the number of private actions for farm-related nuisance, they had no effect on the loss of farmland to other uses, especially in peri-urban areas (Lapping and Leutwiler 1987). They may be of greater help to farmers, at least in the short run, than the strategy employed by one Delaware farmer who placed a huge notice on the boundary between his mushroom farm and a new housing development warning prospective buyers of the unpleasant environment into which they were being enticed to move. (The reader may wish to be reminded of the *Hadacheck* case, summarized in Chapter 5.)

These 'right to farm' laws came as a direct result of the growth pressures facing many areas. Farmers had been farming their lands for many years prior to the growth pressures. Nevertheless, residents and developers sued farmers for the nuisances created by the various farm noises and smells. Developers couldn't sell properties. Individuals purchased homes knowing that the farms were in existence. Both groups sought legal solutions to their problems. This came in the form of lawsuits.

Some state legislatures came to the 'aid' of the farmers by passing legislation protecting farmers from such lawsuits. In the 1990s, many of these 'right to farm' laws came under legal challenge. In 1998, the Iowa Supreme Court ruled that the state's law was unconstitutional in that farmers had no right to blanket immunity from such lawsuits (*Borman* v. *Board of Supervisors for Kossuth County*, 584 N.W.2d 309). The

court's decision was later upheld by the US Supreme Court.

State programs tend to be rather blunt instruments: the real action is at the local level (though in fact it is rather modest). Here, there are three main approaches: agricultural zoning, the purchase of development rights, and the transfer of development rights. Agricultural zoning, as its name suggests, restricts use in the defined zone to agriculture. It is a simple technique, but it is open to constitutional challenge and, not surprisingly, it faces strong political opposition from farmers. However, it can be useful when coupled with other measures such as tax incentives or the transfer of development rights.

The most effective way of safeguarding agricultural land (other than outright purchase at market value) is by the acquisition of the development rights. All the states have passed legislation enabling such acquisitions (usually at local level), but the costs are so high that few local governments can contemplate a program on any significant scale. There are, however, various devices for overcoming this difficulty by the transfer of development rights (TDR). This is a relative newcomer to the armory of planning techniques. It is simple in concept but complex in its details (see Chapter 18 for a further discussion of TDRs). In essence, it separates the development value of land from its existing use, and 'transfers' that development value to another site. The owners of the land in the area to be preserved can sell their development rights to developers in designated 'receiving' areas that are thereby allowed to build at an increased density reflecting the value of the transferred rights. Unlike traditional zoning techniques, TDR gives farmers an incentive to retain their land in agricultural use.

Few TDR programs have been implemented, though they have attracted considerable interest. The program in Montgomery County, Maryland, is one of the best known, and perhaps the most ambitious (see Box 10.5). Montgomery County grew from a 1980 population of some 573,000 people to an estimated 932,131 in 2006. With this growth came development pressures for more land. Agricultural lands were consumed as the development pressures continued. The TDR program, created in 1980, designates *preservation*

BOX 10.5 TDR IN MONTGOMERY COUNTY, MARYLAND

Receiving areas are the designated sites to which development rights can be transferred. They must be specifically described in an approved and adopted master plan, a process by which areas are screened to assure the adequacy of public facilities to serve them and to assure compatibility with surrounding development. Each receiving area is assigned a base density. Developers can build to this density as a matter of right. To achieve the greater density permitted under the TDR option, the developer must purchase development rights. No rezoning is necessary, but a preliminary subdivision plan, site plan, and record plan must be approved by the Montgomery County Planning Board.

Source: Banach and Canavan 1987: 259

BOX 10.6 TOWN OF DUNN, WISCONSIN, POINT SYSTEM USED TO DETERMINE WHICH PROPERTIES SHOULD BE PROTECTED

1   Quality of farmland – based on soil quality, size of farm and proximity to other agricultural lands.
2   Development pressures – based on such factors as proximity to sewer services and inclusion of the property in the annexation plans of neighboring cities and villages.
3   Other features – important natural areas or archaeological sites.
4   Financial consideration – including whether the landowner is willing to receive payments for the easement in installments and whether matching funding is available from other sources to buy an easement on the property.
5   Proximity to other protected lands.

Source: Town of Dunn, http://town.dunn.wi.us/PurchaseofDevelopmentRights.aspx

areas where downzoning has reduced development density on about a third of the 500 square mile county to one house per 25 acres. In addition, there is a transferable development right of one house per 5 acres – the density which the land had before designation. This can be sold to developers in receiving areas. Since 1980, Montgomery County has protected 40,583 acres using TDR, or 60 percent of the national total (67,707 acres) (American Farmland Trust 2001: 1).

Cities and towns can also purchase the development rights of a piece of property. It is a voluntary action that compensates property owners for the right to develop a piece of property. Once the development rights to a piece of property are sold, a conservation easement is attached to the property owner's deed permanently restricting the use of the property to agricultural use or open space. This restriction remains on the property owner's deed in perpetuity. The Town of Dunn, Wisconsin, with a 2005 population of 5,307, has developed a successful Purchase of Development Rights (PDR) program that was funded by a property tax increase. Created in 1996, the Dunn PDR program seeks to preserve farmland, protect open space and environmentally sensitive lands, maintain the area's quality of life and rural character, and protect Dunn from encroaching cities and villages. Recognizing that all farmland and open space are not equal and cannot be protected, Dunn employs a point system to determine which properties should be protected (see Box 10.6). As of July 3, 2007, the Town of Dunn has permanently protected some 2,729.48 acres of farm-land and open space, with another 1,869 acres pending.

All the evidence shows that the preservation of agricultural land can involve very large costs. In assessing whether these are justifiable, it is necessary to ask not only what the objectives are, but also who actually benefits. In a review of agricultural land protection policies in New England, it was concluded that there are four interrelated concerns behind the adoption and implementation of such policies in this region (see Box 10.7). These are the difficult-to-define but easily recognizable quality of the rural landscape; environmental degradation (pollution in all its forms); the quality and regional availability of food products; and the various economic benefits of agriculture (such as its beneficial impact on the economy, and the avoidance of the problems of land speculation and rising land prices). Interestingly, it is the 'aesthetic' concerns that predominate. The term is used here in a very wide sense to mean the general quality of the environment (Schnidman *et al.* 1990).

It is noteworthy that these issues are not only inter-related but also somewhat elusive; clearly the major issue is a vague unease and concern about the way in which a familiar and friendly environment is changing. This, of course, is a common feature of the operation of the planning and zoning system. Fischel (1982: 257) argues that 'the real beneficiaries . . . and the real force behind the farmland preservation movement, are local antidevelopment interests'. By contrast, the American Farmland Trust (1988), in its survey of schemes for the purchase of development rights in Massachusetts and Connecticut, underlines the benefits to individual farmers and to the farming industry generally. A more balanced viewpoint is expressed in a monograph emanating from the long-term research program on farmland protection programs carried out by the Florida Joint Center for Environmental and Urban Problems (Hiemstra and Bushwick 1989: xi). As with all good research projects, the conclusions raise as many questions as are answered:

the Center has concluded that urban conversion of agricultural lands, while not posing an immediate threat to America's food supply or its strategic position in international affairs, does warrant concern on other grounds. Certainly, other things being equal, it makes little sense for a society to shift agriculture from better to worse lands if planning and management would allow more efficient uses of land resources. Nor is it prudent to convert agricultural land to urban uses if the urban development in question is itself wasteful and socially expensive. The challenge is to develop farmland protection programs that distinguish inefficient from efficient land uses and promote objectives more complicated than simply indiscriminately saving all agricultural land.

Smart growth

Debates over population growth will never end. People are again questioning the virtues and benefits associated with population growth. They are no longer interested

---

BOX 10.7 FARMLAND PROTECTION IN NEW ENGLAND

State-by-state review of New England farmland protection efforts reveals that in every state one of the most important concerns was the desire to preserve certain aesthetic qualities which agricultural lands provide. The determination to protect open space and local community character evolved primarily from intangible motivations such as the value of farming as a lifestyle that is pleasing to the eye. The traditional Yankee farm, with its small fields surrounded by stone walls, woodlands, and rural architecture, has given the landscape a unique visual character that a majority of New Englanders, both urban and rural, want to protect.

Source: Schnidman *et al.* 1990: 322

in simply having growth at any cost. They want growth to occur. However, they also want to enhance their quality of life. They want to encourage growth while, at the same time, protecting the environment.

To some people, smart growth is a new way of thinking. (See Box 10.8 for the principles of smart growth.) To other people, the techniques used in smart growth have been around for a number of years. They see it as implementing the techniques that have been on the books. As Burchell *et al.* (2000: 823) note:

> Smart growth is an effort, through the use of public and private subsidies, to create a supportive environment for refocusing a share of regional growth within central cities and inner suburbs. At the same time, a share of growth is taken away from the rural and undeveloped portions of the metropolitan areas. This is accomplished by revitalizing existing central cities and inner suburbs so they can participate in the region's future growth. While this is happening, the regional economy is strengthened, residents' quality of life is enhanced, and outer-area natural resource systems are protected and restored. In effect, smart growth encompasses and extends the growth management efforts of the previous decades. One exception is that it is much more progrowth and much less proconservation than earlier growth management efforts.

Ultimately, smart growth advocates a number of things. It sees the need to do away with piecemeal planning and to recognize the fact that one decision can cause repercussions on another issue. Smart growth advocates investing in areas where the infrastructure already exists. There is no need to expend scarce public funds in areas where the funds are not needed. Smart growth also encourages infill development, development bypassed by development. It also encourages development of transportation corridors.

A number of jurisdictions around the United States have embraced smart growth. The city of Austin, Texas, launched a smart growth initiative in early 1998. Its initiative advocates, among a number of things, infill development and revitalization of existing neighborhoods, mixed-use development, designing for people instead of vehicles, environmental preservation, and integrating transportation and land use planning on a regional scale. When the various tools, techniques, policies, and programs are combined, Austin will be determining how and where it grows, how it can improve the area's quality of life, and how it can improve or enhance its tax base.

King County, Washington, also introduced a smart growth initiative, called Shaping Tomorrow, in June 1998. Unexpected high growth rates prompted King County executive Ron Sims to call for the initiative. Adopted by the county council in February 2001, the

---

BOX 10.8 PRINCIPLES OF SMART GROWTH

1   Mix land uses
2   Take advantage of compact building design
3   Create a range of housing opportunities and choices
4   Create walkable neighborhoods
5   Foster distinctive, attractive communities with a strong sense of place
6   Preserve open space, farmland, natural beauty, and critical environmental areas
7   Strengthen and direct development towards existing communities
8   Provide a variety of transportation choices
9   Make development decisions predictable, fair, and cost effective
10 Encourage community and stakeholder collaboration in development decisions.

Source: http://www.smartgrowth.org

comprehensive smart growth plan tries to balance a transportation system, affordable housing, and livable communities with environmental protection. It seeks to support growth while protecting precious natural resources. Some of the key components of the King County Comprehensive Plan are that it:

- allows townhouses, duplexes and apartments in all urban communities;
- further curtails growth in the rural area by requiring development on Vashon Island to be on larger lots and by restricting the size and scale of non-residential buildings into the residential rural communities;
- further protects the commercial forest lands by limiting the amount of land that can be cleared for residences and by requiring an approved forest management plan to accompany all residential permit applications;
- recognizes the evolving role of agriculture in our communities by allowing greater flexibility for the farmer to sell produce on site;
- includes policy support for implementation of measures to address protection and recovery of salmon under the Endangered Species Act listing.

(Sims 2001: 2)

San Diego, called by its residents 'America's Finest City', has been experiencing growth pains for many years. Growth is debated daily among its residents, businesses, and governments. As in other areas around the country, some people see growth as good while others see it as something chipping away at the area's quality of life. Nevertheless, San Diego continues to grow.

Smart growth has become a topic at meetings and forums around the region. A regional plan, currently in its development stage, embraces the ideas behind smart growth. However, San Diegans might be able to go back to the City of San Diego's 1967 Progress Guide and General Plan and see when the seeds of today's smart growth movement were planted. The objectives of the 1967 Plan were:

- creation of a strong central core;
- development of a more compact city;

- prevention of sprawl;
- encouragement of a greater variety and choice in the living environment;
- promotion of a more handsome environment;
- recognition of the importance of San Diego's harbor;
- preservation of the open space system.

Two things remain certain about growth in San Diego: first, growth is going to continue; second, debates over how to control, manage, limit, or accommodate growth will continue to be heard. It is an issue that will not disappear from the public policy agenda.

Various types of incentives are offered by states and localities to further the cause of smart growth. Some cities will waive development fees for smart growth projects. Other cities offer expedited processing of development applications as an incentive or inducement to practice smart growth. Some areas offer funds to build developments near transit stations, loan packages for land assemblage, site demolition, and debris removal, and long-term tax exemption for redevelopment projects. States may also offer localities preferential treatment in qualifying for state funds in the areas of transportation, housing, economic development, and environmental protection. While incentives may induce smart growth activities, failing to offer the above could be considered disincentives.

Direct democracy devices and growth management

Citizens are afforded the opportunity to participate in planning through various mechanisms, such as public meetings, city planning commission meetings, and city council debates over an issue. However, some states give citizens an even larger voice in ultimately deciding a matter of public policy. For example, some states give citizens the opportunity to develop state or local legislative policy, get enough registered voters to sign a petition to get a legislative matter on the ballot, and then allow citizens to vote for the measure. This is called an 'initiative'. Citizens can approve or disapprove a measure. If the measure passes, it becomes municipal law. Citizens can also call for a 'referendum' to see if they approve the measure or if they reject it. Some areas might use the referendum but not the initiative for planning matters because the state has reserved the power to plan to the local legislative body, not to the citizens. The use of direct democracy devices to resolve planning issues must be examined on a state-by-state basis (Caves 1992).

Residents of California have turned to the ballot box on many occasions. The power is reserved to them under the California Constitution, Article II, Section 8(a), which reserves 'the power of the electors to propose statutes and amendments to the Constitution and to adopt or reject them'. Their reasons are no different from the reasons of citizens in other states (Caves 1992). They might be angry at an individual decision made by the city council. They might also be frustrated over the general direction the city council has taken on an issue over the years. It is also possible that they simply want to play a more active role in deciding matters that will directly or indirectly affect them.

California is unlike many states in that the right to use the power of the initiative or referendum on matters of local legislative importance is guaranteed to them by the California State Constitution. It was during the populist movement, in 1911, that this power was added to the California State Constitution. The California Constitution, Article II, Section 9(a) allows 'the power of the electors to approve or reject statutes or parts of statutes'. At the local level, this ability represents the power of the voters to reject an ordinance or resolution which the local legislative body recently passed. This became a way of making sure that the wishes of the citizens were not ignored in favor of the powerful and well-funded lobbyists and business interests.

Ballot-box planning results in a complex, diverse, locally controlled mosaic of planning policies. Policies which have been introduced in this way include caps on the amount of residential development (often based on the Petaluma model discussed earlier in the chapter); density restrictions; infrastructure limits; minimum lot sizes for new residential building; moratoria on development; and designation of areas for conservation or development. Of increased popularity in recent years has been the reservation of certain planning decisions

for future voter approval (such as any change to the local plan). For instance, an initiative of the City of Lodi effectively prohibited further development in the city's peripheral areas without specific approval of the voters – which was repeatedly denied (Fulton 1991:145).

Over the years since 1911, Californians have practiced direct democracy to the point where it is not uncommon to see multiple measures on the ballot relating to planning issues. For example, it is not uncommon to see measures dealing with the following topics on a local ballot: adopting or rejecting a general plan, adopting or rejecting a growth control plan, allowing the construction of apartments, limiting the number of building permits to a specific number each year, creating a buffer zone or greenbelt, extending the time period for an urban growth boundary, limiting the height of building in an area, and requiring voter approval of any rezoning of a tract of land (Caves 1992; Orman 1984).

The power of *initiative* and *referendum* is by no means unique to California, but nowhere else is it used to such a great extent. Between 1971 and 1989, there were 357 ballot box measures concerned with land use planning: on average over two-thirds of these succeed. Hundreds of measures dealing with a wide range of planning issues have appeared on local ballots since 1990. Perhaps the most famous of California's ballot box measures was concerned, not with land use, but with taxes: Proposition 13, passed in 1978, reduced property taxes and limited future increases. In fact, taxation and land use control are closely interrelated; and, in California, the connection is so close as to give rise to what is termed the 'fiscalization of land use'. Its impact is still being felt by municipalities in California.

In November, 1999, voters also rejected ballot measures that required voter approval for housing projects within urban growth boundaries in Livermore, Pleasanton, and San Ramon. More recently, in November 2000, voters in Sonoma County rejected a 'Rural Heritage Initiative' that would have required the approval of Sonoma County voters for thirty years of any change to the county general plan that would lead to more housing or commercial development in areas zoned for agriculture, open space, rural lands, or rural residential units.

In 2012, outside of California we continue to witness a number of ballot measures for the electorate to decide. We saw measures seeking to consolidate separate planning and environmental commissions appointed by the city council into a joint commission. In Oregon, voters were asked if an area could be partially removed from an urban growth boundary, to annex territory to a city and to create an urban renewal district in an unincorporated part of a county. In Virginia, voters in one area were asked to support the financing and development of a light rail system. Voters in a jurisdiction in Maine were asked to decide the fate of redeveloping two elementary schools into a community center and public library.

Critics have advanced a number of reasons for opposing the use of direct democracy devices. Some individuals claim that using direct democracy undermines our system of representative government. Citizens elect representatives to make public decisions on their behalf. Unfortunately, many elected officials have abdicated this responsibility and turned the power to make decisions on some matters back to the citizens themselves. The fear of not being re-elected by the voters may be a factor in letting the voters decide some issues.

A common complaint voiced in using the initiative or referendum is the actual wording of the ballot measure. To put it plainly, many measures are written in words that are hard to understand by the general public. Technical terms are not translated so that the voter can get a good grasp on the meaning and intent of the measure. Moreover, some measures are written with little or no intent over the measure's implementation should it be approved by the voters. On other occasions, the wording appears to be if you vote yes on a measure, you are actually voting against something.

Some researchers are concerned with voters not being knowledgeable enough on the issues to understand ballot measures. They claim that voters don't study the issues enough to make an informed decision. In fact, some individuals don't vote on the issue. Their decision on whether or not to vote for or against the measure depends upon who is speaking on behalf of the measure or who is speaking against the measure. For example, a number of years ago California had several competing insurance measures on the state ballot. Ralph Nader, a famous consumer advocate, backed one of the measures. Some individuals claimed that they voted the way they did because Ralph Nader would not give them bad or incorrect information.

Opposition to the use of direct democracy can also stem from the fact that, in some areas, companies are hired to get the signatures needed to qualify a measure for the ballot. Direct democracy has created an industry to fill this need to obtain signatures. Individuals are paid on a per signature basis. The more signatures they obtain, the more money they make. To compound this problem, in some people's minds, is the problem that the individuals getting the signatures may not actually know anything about the ballot measure. Their job is simply to get signatures and nothing more. All they are doing is trying to convince individuals that the measure deserves a chance to be put on the ballot, not trying to get the signer to commit to a certain vote.

The sheer number of measures on a ballot is another complaint against direct democracy. Any number of measures on the same issue could qualify for a place on the ballot. The differences between the various measures could range from very minor changes to drastic changes. Many voters get confused over the alternative measures and occasionally simply don't vote for any of them.

The actual number of people voting on the measures is also a complaint when we are discussing direct democracy. The number of people voting in many municipal and state elections is not that great. With small turnouts in elections, a small number/percentage of voters will be making the decisions for an entire city. To some individuals, this further complicates issues between the majority and other groups of citizens.

A final concern over the use of direct democracy is that interest groups, development concerns, and some individuals are using their financial resources essentially to buy the outcome of the election. In some areas of the country that have witnessed growth control measures on the ballot, development groups have spent exorbitant amounts of money on billboards and other advertisements to defeat a measure. This should not be taken to mean that these groups always get their way. Many voters have seen through some advertisements and campaigns and defeated the development interests.

Conclusion

In this chapter a brief account has been given of the urban growth controls operated by a number of local governments, together with a summary of some policies relating to the safeguarding of agricultural land. The policies discussed are interesting, and they are certainly popular with local residents – who, of course, have not themselves been prevented by the controls from living in the area. There is no lack of critics, some of whom are very sure of themselves. Ellickson (1977), for example, describes growth controls as a type of 'home owner cartel', while Frieden (1979) slates 'the defense of privilege'. However, the evidence is variable in reliability, and equivocal or contradictory in its results. Thus, while one study concluded that the Petaluma policy resulted in higher house prices, another found that (in both Petaluma and Ramapo) there was little demonstrable effect on subsequent development.

However, it may sometimes be that, though policy is expressed in terms of urban growth management, the real purpose is to secure leverage in the planning process to obtain benefits for the locality. Stiff restrictions on development may prompt developers to offer 'amenities' on a scale or of such a character as could not be legally required by the local government.

Moreover, what happens to growth pressures which are successfully stemmed in one area? Do they necessarily move to another area where development is in the public interest? How can any rational assessment be made of such matters without a proper land use plan? Interestingly, Judge Choy made a similar point in the *Petaluma* case, where he noted that:

> If the present system of delegated zoning power does not effectively serve the state interest in furthering the general welfare of the region or entire state, it is the state legislature's and not the federal courts' role to intervene and adjust the system . . . the federal court is not a super zoning board and should not be

called on to mark the point at which legitimate local interests in promoting the welfare of the community are outweighed by legitimate regional interests.

It is possible to debate the relative successes or failures of local or regional growth management programs. There is no doubt that problems associated with growth do not respect political boundaries. The problems have extraterritorial impacts. There is jurisdictional fragmentation among areas and special districts. The development of a growth management program is one thing; the implementation of the program is an entirely different beast. The program may be very hard to understand and implementation gaps may occur between the various parties. Some programs may offer exceptions or loopholes that could effectively hinder the successful implementation of a program. Another program might allow certain land use development activities that were previously prevented, if the party can show a community need that is being filled, for example, the need for affordable housing. If this occurs on a regular basis, some people will question whether the program is actually being implemented.

Some areas might decide not to pursue a growth management program in times of our current economic woes. These jurisdictions see growth management programs of impeding possible economic growth. In time, many cities have relaxed regulations in hopes of attracting economic growth. They are essentially attempting to reduce the regulatory barriers that stand in the path of economic growth and job creation.

In short, policies relating to growth management cannot be adequately designed and implemented on a local basis: a regional or state outlook is required. Many of the problems localities face are not confined to municipal boundaries. They transcend boundaries. For example, the decision of a locality to use land near its borders for a landfill will impact upon its neighbors. The smells associated with a landfill can travel for miles. Moreover, the development of housing in one area could lead to a premature demand for infrastructure in another area. Localities can no longer think that their individual actions don't affect neighboring municipalities. Many of their problems are regional in nature and must be handled on a regional basis. A few states have realized this, and are making attempts to create a new intergovernmental system of land use control. This is the subject of the next chapter.

Further reading

The State of Washington has helped to manage growth with its Growth Management Services division (http://www.cted.wa.gov/growth). This division provides technical, educational, and financial resources to assist local governments and other planning organizations meet the requirements of the Growth Management Act of 1990. The website outlines the goals of growth management and the role of the Growth Management Services division.

Volusia County, Florida, combines growth management with resource management into the same department (http://volusia.org/growth). Focusing on quality of life issues, the Growth and Resource Management department provides links to smart growth updates in the county, as well as other growth measures, such as zoning and land conservation.

The Center for Green Space Design (http://www.greenspacedesign.org) is a website that promotes the inclusion of green spaces in the plan of a community. By providing a list of books, online resources and a guide to community involvement, the Center walks the visitor through the assessment, design, facilitation, and implementation of green space.

American Farmland Trust (http://www.farmland.org) is a non-profit organization created by farmers and conservationists to protect the best farmland, change US farm policy, plan for agriculture and promote healthy stewardship of the land. The Trust provides farmland reports, information about the farm policy campaign it advocates in Washington, DC, and information about other state farming programs.

The Smart Communities Network offers a detailed description of the process of transferring development rights (http://www.smartcommunities.ncat.org/landuse/transfer.shtml). After providing a description of land use strategy, the Network offers links to programs throughout the United States, such as the National Association of Realtors 'Field Guide to Transfer of Development Rights' and many individual states' programs.

The Smart Growth Network is the result of a partnership between the US EPA, non-profit organizations, and local government organizations. The website, called Smart Growth Online (http://www.smartgrowth.org), is an offshoot of the Network and supplies online news, resources, and techniques for smart growth.

The US EPA engages in the dialogue of smart growth with its website (http://www.epa.gov/smartgrowth). On the website visitors can submit proposals for smart growth developments, apply for grants, and learn more about smart growth concepts.

For readings on direct democracy see: Goebel (2002) *Direct Democracy in America: A Government by the People*; Fossedal (2005) *Direct Democracy in Switzerland*; and Altman (2011) *Direct Democracy Worldwide*.

By far the greater amount of writing on growth management is concerned with state (not local government) policies. The legal texts on land use typically have a substantial discussion of the main cases. Schiffman (1990) *Alternative Techniques for Managing Growth* is a modest but very useful book which outlines the various growth control measures available to local government. Porter (1986) *Growth Management: Keeping on Target?* details many of the measures in operation in the early 1960s, as does Godschalk *et al.* (1979) *Constitutional Issues of Growth Management*.

Nelson and Duncan (1995) examine growth control principles and practices and how they can be integrated into a comprehensive system in *Growth Management Principles and Practices*. Porter (1996) *Performance Standards for Growth Management* covers how areas are using performance standards to measure and control the effects of proposed development. Porter (1997) also offers a useful source in describing various strategies, programs, and techniques for managing growth in *Managing Growth in America's Communities*.

Critiques of growth management policies abound. See, for example, Fischel (1990) *Do Growth Controls Matter? A Review of Empirical Evidence*; and Landis (1992) 'Do growth controls work?' See also Stanilov *et al.* (1993) *A Literature Review of Community Impacts and Costs of Urban Sprawl*.

There is a useful set of articles reviewing growth management programs and their achievements in the Autumn 1992 issue of the *Journal of the American Planning Association* (58: 425–508). Also well worth studying is Chinitz (1990) 'Growth management: good for the town, bad for the nation?'

On agricultural land, see Brower and Carol (1987), especially the paper by Banach and Canavan on the Montgomery County program (and its TDR scheme); and Schnidman, *et al.* (1990) *Retention of Land for Agriculture: Policy, Practice and Potential in New England*. An article which combines theory, a literature review, and some empirical data from Oregon is Nelson (1992b) 'Preserving prime farmland in the face of urbanization'. See also Daniels (1991) 'The purchase of development rights: preserving agricultural land and open space'. Daniels (1999) *When City and Country Collide: Managing Growth in the Metropolitan Fringe* provides an overview of growth management and urban planning techniques for fringe metropolitan areas.

There is, of course, the irony that while some public policies are geared to keeping land in agricultural use, others (particularly federal agricultural subsidies) are aimed at limiting output. For an absorbing account of *How the US Got into Agriculture and Why It Can't Get Out*, see Rapp (1988).

A smart growth network has been formed by a number of organizations concerned with how US communities were growing. Its members are devoted to encouraging development that serves the economy, the community, and the environment. Members of the smart growth network include such bodies as the US EPA, the National Association of Counties, the Congress for the New Urbanism, the International City/County Management Association, the American Farmland Trust, the National Association of Realtors, the National Wildlife Federation, the Natural Resources Defense Council, the Urban Land Institute, and the National Trust for Historic Preservation.

There is a growing amount of literature devoted to various aspects of smart growth. The American Planning Association (2002) published *Growing Smart (SM) Legislative Guidelines*. See also: Planning Advisory Services (1998) *Principles of Smart Development*; Burchell *et al.* (2000) 'Smart growth: more than a ghost of urban policy past, less than a bold new approach'; Porter (2002) *Making Smart Growth Work*; Szold and Carbonell (2002) *Smart Growth: Form and Consequences*; Duany, Speck, and Lydon (2009) *The Smart Growth Manual*; O'Connell (2009) 'The impact of local supporters of smart growth policy adoption'; Freilich, Sitkowski, and Mennillo (2010) *From Sprawl to Sustainability: Smart Growth, New Urbanism, Green Development and Renewable Energy*; and Green Leigh and Hoelzel (2012) 'Smart growth's blind side.'

Questions to discuss

1 Describe the various methods of managing local urban growth. Which do you think is the most effective?

2 Discuss the case for and against growth management policies.

3 What is involved in the transfer of development rights (TDR)? For what planning purposes can this be used?

4 Evaluate the reasons given for farmland protection.

5 Is there a regional dimension missing from municipal growth management plans?

6 Consider the arguments of Chinitz (1990) on whether growth management is 'good for the town, bad for the nation' (*Journal of the American Planning Association* 56: 3–8); and the response by Fischel (1991) and Neuman (1991).

7 What are the reasons behind smart growth?

8 Why do some jurisdictions decide not to control growth?

Growth management and the states

This country is in the midst of a revolution in the way we regulate the use of our land. It is a peaceful revolution, conducted entirely within the law. It is a quiet revolution, and its supporters include both conservatives and liberals. It is a disorganized revolution, with no central cadre of leaders, but it is a revolution nonetheless.

The tools of the revolution are new laws taking a wide variety of forms but each sharing a common theme – the need to provide some degree of state or regional participation in the major decisions that affect the use of our increasingly limited supply of land.

Bosselman and Callies 1972

Urban growth problems

This now famous quotation from Bosselman and Callies was written in the heady days of the 1970s, and its promise has not been fulfilled. Only a few states were initially involved in land use planning (as distinct from environmental planning); there are marked differences among them in the purpose and scope of their involvement; successes have been limited, and typically modest. Since then, more states have become involved in the realm of land use planning and have adopted various forms of statewide controls.

In this chapter, seven illustrative types of state land use planning are discussed: Hawaii, Oregon, Vermont, Florida, Maryland, California, and New Jersey. Each of these states faced one or more problems seen as requiring state action. These problems continue to plague the states. Hawaii was troubled by the rapid urbanization of its valuable agricultural land; Oregon experienced growth pressures along its coastline, and also problems of urban development and speculation; Vermont faced a sudden large increase in development pressures; Florida experienced phenomenal growth; Maryland was growing and facing the problems

associated with sprawling development; California was witnessing a large loss in public access to the coast; and New Jersey (the most urbanized state in the nation) was facing massive urbanization. There are clearly some similarities among these states: all have had to devise ways of dealing with growth. But, as will become apparent in the following account, each has its distinctive set of problems, goals, constraints, and plans. Thus Hawaii introduced statewide zoning, Oregon set up a comprehensive system of local planning which had to conform to a long list of state goals, Vermont set up a system of citizen district commissions to administer a development plan system, Florida introduced state controls in 'areas of critical concern' and 'developments of regional impact', Maryland was an early leader in developing smart growth policy at the state level, California established a coastal planning system, and New Jersey battled with the introduction of a state plan intended to guide growth and conservation throughout the state. Many of the initial provisions have been revised, for a variety of reasons ranging from a recognition that the early provisions were inadequate to the impact of changes in political control. The stories continue to unfold, of course.

Hawaii

It all began in Hawaii.
(Bosselman and Callies 1972)

Hawaii's approach to land use control is, as elsewhere, a product of its history; but this history is very different from that of the other forty-nine states. Indeed, the land use planning system which has emerged is unique but, since it was the first, it is of some significance, as well as being of intrinsic interest. The discussion is, however, quite brief.

The particular history of Hawaii led to a concentration of land ownership which lasted until a combination of events brought into being the 1961 Land Use Law. This law gave Hawaii the nation's first system of state land use planning. The concern was not simply with land ownership, but with the effects of the policies being operated by the landowners. In short, an increasing pace of development (particularly by way of premature subdivisions) threatened Hawaii's agriculture-based prosperity. Thus, unlike the situation which is typical of other states, controls were introduced, not to retard or control growth but to safeguard and promote economic development.

Another distinctive feature of Hawaii is its highly centralized governmental structure, with the state having responsibility for major services such as education, welfare, and housing. There is no tradition of autonomous local government: its four main islands represent four separate counties with no lower levels of local government. The state is the general-purpose level of government.

The 1961 Act established a Land Use Commission which was charged with designating all land in four 'districts': urban (4 percent), agriculture (48 percent), conservation (47 percent), and 'rural' (less than 1 percent) (Hawaii Department of Business, Economic Development and Tourism 1999). The urban districts cover land which is in urban use or which will be required as a reserve area for urban purposes in the foreseeable future. The administration of zoning in these districts is the responsibility of the counties. The designation provides no rights to urban development: it merely signifies that the county *may* zone the land

for urban development under its zoning code. Thus the counties can impose more restrictive conditions, but they cannot relax the commission's regulations.

Agricultural districts cover land used not only for agricultural purposes but also for a range of other uses including 'open area recreational facilities'. The state and county governments share responsibilities in the agricultural districts. In establishing these districts, the commission is required to give the 'greatest possible protection' to land which has a high capacity for intensive cultivation. Conservation districts are primarily forest and water reserve zones, but also include scenic and historic sites, mountains, and offshore outlying islands. The administration of planning in the conservation districts lies directly with the state government, operating through the Land Board of the Department of Land and Natural Resources. The final small category of 'rural districts' was added to permit low-density residential lots. These are principally small farms and rural subdivisions which are inappropriate for either the agricultural or urban designations. Administration lies with the Land Commission, which operates within the framework of the State Plan, which was approved by the legislature in 1978.

Hawaii was the first state to enact a comprehensive plan. It is a short document developed by the Office of State Planning which sets out a series of 'themes' and policies for the state covering a wide range of issues including health, culture, education, and public safety, as well as land use, population, and the environment. The provisions of the State Plan relate to a number of important matters concerning population growth and distribution. These include the carrying capacity of each geographical area; the direction of urban growth primarily to existing urban areas; and the preservation of greenbelts and critical environmental areas.

Hawaii is the only state to operate a centralized statewide system of land use controls. The particular history and governmental system of Hawaii accounts for this system of land use controls. Previously, in 1978, Hawaii's four counties had only rudimentary land use and zoning schemes. However, matters have changed greatly since then, and they now have well-developed planning capabilities. In addition to having the authority to zone in accordance with a general plan, each county must designate an existing agency to maintain and update laws regarding regulatory powers over planning. Moreover, they prepare and revise the general plan. This raises the question as to whether the statewide system of controls is still appropriate or whether it is still evolving. There are also wider questions now being raised about the adequacy of the bureaucratic Hawaiian system to meet the needs of the islands (Callies 1994).

Hawaii's population continues to grow. Its estimated 2011 population of 1,374,810 is an increase of 6.9 percent since 2006. This continued growth exacerbates the state's problems of a lack of affordable housing, a reliance on a service-based economy, and the need to improve a deteriorated infrastructure.

Hawaii's latest activity involving state land policy is reviewing the mid-1970s State Plan and the state's comprehensive planning system. The last time the State Plan was reviewed and revised was in the mid-1980s. Its twelve functional plans (agriculture, conservation lands, employment, energy, health, higher education, historic preservation, housing, recreation, tourism, transportation, and water resources development) were last updated in 1989 and 1991. The Hawaii 2050 Task Force, comprised of twenty-five members, is charged with developing a broad-based Hawaii 2050 Sustainability Plan.

The evolution of the Task Force, which was created by Hawaii Senate Bill, later called Act 8, is interesting. The Senate Bill was passed by the state legislature on May 3, 2005. It was returned by Governor Linda Lingle on July 11, 2005, unsigned for several reasons. First, she felt that charging the Office of the State Auditor to develop the new Hawaii State Plan was not the proper place for it to be developed. To her, it would be an inappropriate function for that office. Second, requiring other state departments to provide assistance to the project without providing any funding was also inappropriate. The Governor's veto was overridden in a Special Session of the Hawaii Legislature on July 12, 2005.

The Task Force will seek guidance from a wide variety of constituents on ways to achieve a sustainable Hawaii. It is an ambitious program that will develop a new planning strategy for Hawaii. In a September 2007 Draft Plan, the Task Force identified the foundation of Hawaii 2050, the so-called 'triple bottom line' – where the economy, environment, and community goals are in balance (p. 9).

In 2011, Governor Neil Abercrombie signed Act 55, which created the Public Land Development Corporation (PLDC). Governed by a five-member board of directors, it serves as the development arm of the Hawaii Department of Land and Natural Resources. It is designed to create and facilitate partnerships between various parties that are designed to improve Hawaii's communities and to make wise use of the state's natural resources. Although it is still devising its administrative rules, the PLDC has met with opposition. Some residents and organizations claim it is all about money and that the Department of Land and Natural Resources budget has been decimated. The sad shape of the Department's budget puts its ability to fulfill its stated mission into question. Some have even called for the repeal of Act 55.

Oregon

> The form of planning in Oregon is not so much different from that in other states, but the substance is. In most states, the cities and counties may plan and zone; in Oregon they must. In most states, standards for local planning are not uniform from one jurisdiction to another, are not particularly high, and are not enforced by any state agency; in Oregon, general planning standards (the goals) are the same throughout the state, they are high, and they are administered by an agency with clout.
>
> (Rohse 1987)

Oregon has had a good, lengthy track record for state planning initiatives and has been held as an example for other states. For example, between 1969 and 1971 five laws were passed (the so-called 'B' laws) which provided for public access to the beaches; issued bonds for pollution abatement; banned billboards; earmarked funds for bicycle paths; and mandated returnable bottles. Other laws established the Oregon Coastal

Conservation and Development Commission, and mandated local governments to prepare comprehensive land use plans and develop land use controls. In 1973, after much negotiation and compromise, came the Land Conservation and Development Act, which greatly increased the powers and responsibilities of (mandatory) local planning and provided for a set of state planning guidelines which local plans are required to follow.

As usual, there is no single factor which explains why Oregon acted as and when it did. Certainly, a catalyst was political – in the form of Governor Tom McCall and Senator Hector MacPherson, who, in promoting new legislation in 1971, started with strong forces on both sides of the debate, making compromises to get the bill passed. But there was a popular base on which this blitz was waged. Oregonians have a particular pride in the beauty of their state: they see it as a precious heritage which demands to be preserved. Bolstering this is a strong and vocal conviction that 'Oregon must not become another California'. (This conviction has been fueled in recent years by the continued influx of Californians seeking an environment similar to that of California, but with much lower house prices.) The concern emanates from visible pressures on the land: urban encroachment in the Willamette Valley, land speculation in the fragile landscape of the eastern part of the state, and degradation of the marvelous shoreline.

The Act established a number of state planning goals that would be achieved through comprehensive planning at the city and county levels. These goals, and guidelines on how to achieve the various goals, dealt with such issues as citizen involvement; land use planning; agricultural lands; forestlands; recreational needs; housing; urbanization; estuarine resources; coastal shorelands; and ocean resources. Not everyone supported the legislation and its planning goals. For the next six years, many attempts were made to repeal the legislation. In 1995 alone, more than seventy bills were considered by the legislature that would weaken the legislation. The various attempts were either defeated in the legislature or vetoed by the governor.

Oregon's land use law is comprehensive only in the sense that the planning guidelines apply to the whole of the state. The actual preparation and implementation of local plans is the responsibility of the 240 cities and 36 counties. Thus there is in no real sense a 'state plan': there are 276 local plans that have been developed in accordance with state standards and have been reviewed and approved by the state. Nevertheless, the importance of the state requirements should not be underestimated: plans have to conform to a range of specific state land use policies. These include the containment of urban growth: each municipality's plan has to delineate an *urban growth boundary* (UGB) which defines the limit of urban development and its separation from rural land. Growth is strongly discouraged outside of the boundary. It represents a flexible tool in that a UGB can be changed depending upon the need for more land.

A state planning agency, the Land Conservation and Development Commission, was established to ensure that state policy is implemented. Its seven members are appointed by the governor and confirmed by the senate. Members serve four-year terms and can serve no more than two consecutive full terms. Its first tasks were to adopt the statewide planning goals and to review and approve (technically termed the 'acknowledgment' of) local plans. The review of amendments (of which there are several thousand every year) is dealt with by the commission's administrative arm, the Department of Land Conservation and Development. However, the department has no power to prevent a municipality from adopting an amendment to which it objects. In such cases, it would normally appeal to a body known as the Land Use Board of Appeals (LUBA). This board is comprised of three members appointed by the governor and confirmed by the Oregon Senate. It was established in 1979 to review all governmental 'land use' decisions and all 'limited land use' decisions and to provide a simple means for settling land use disputes without the need to go through the state circuit courts. Examples of 'land use' decisions, as stated in Oregon Revised Statutes, 197.015 (10) include comprehensive plans, zone changes, conditional use permits, variances, and rural land divisions. 'Limited land use' decisions, as indicated in Oregon Revised Statutes, 197.015 (12), include an urban partition, urban subdivision, urban site review decision, and an urban design review decision.

The important lubricant in this system is the wide provision for citizen involvement. This is in the political tradition of Oregon: it is common for local governments to establish citizen advisory committees (CACs) for every city neighborhood and county district.

> The groups typically meet monthly. Their advice and concerns are given to the planning commission or governing body. CAC meetings are quite informal, and are open to all, without dues or formalities of membership. CAC meetings are often attended by members of the planning department, who can answer technical questions or keep a record of comments.
>
> (Rohse 1987: 57)

According to Oregon Revised Statutes 197.160(1), local governments are also required to establish and support 'an officially recognized citizen advisory committee or committees broadly representative of geographic areas and interests related to land use and land use decisions'. At the state level, there is the Citizen Involvement Advisory Committee, which has several functions: to advise the commission on matters of citizen involvement, to promote 'public participation in the adoption and amendment of the goals and guidelines', and 'to assure widespread citizen involvement in all phases of the planning process'. However, it is important to note that it has no authority over any state agency or local government agency.

Intergovernmental relations are often characterized by a heavy measure of bluff. A state law may mandate a local government to do something, but if there is no machinery for ensuring compliance or no financial incentive, nothing may happen. The Oregon system has teeth in it. It provides tangible incentives to local governments to prepare plans and obtain state 'acknowledgment' of them. Certain state contributions to local budgets are dependent on this. In addition to such financial considerations is an incentive which is even more effective: once a local government's plan and land use regulations have been approved by the commission, the state's role in local planning is greatly reduced. There is no longer any need for it: the goals have been incorporated into the plan.

Urban growth boundaries (UGBs) are, understand-ably, a source of contention: in more senses than one they are the cutting edge of planning implementation; and wherever they are drawn someone will be upset. There is a lot at stake and powerful proponents and opponents often engage in protracted heated debates. Property owners can accrue substantial profits when their land is brought inside of the UGB. Others are more concerned about the loss of prime farmland and the loss of environmental amenities. Nevertheless, they have worked out much better than might have been expected. Each of Oregon's 240 cities has adopted an urban growth boundary which has been reviewed and approved by the commission. Difficulties have gener-ally been overcome by a judicious degree of flexibility; for instance, some of the areas just beyond the urban growth boundaries have been designated for eventual urban expansion on a comprehensively planned basis.

In 1998, Portland's Metropolitan Council decided to expand the Portland Metropolitan UGB. The exact amount of expansion was a compromise between competing forces. The UGB system has been effective in controlling urban growth in the Willamette Valley. This is a significant achievement; but development pressures continue to mount. The plans were drawn up for the period ending in the year 2000, and need to be reviewed for the new century. There is a lot of discussion as to whether UGBs have lived up to the legislature's expectations (Staley *et al.* 1999). To some, they have done a good job in restricting growth and protecting the environment. To others, they have resulted in restricting the production of needed hous-ing, therefore causing the price of housing to increase out of the range of many individuals. Challenges are still facing Oregon and its growth. It is recognized that it is critical to maintain an effective partnership between all levels of government, the private sector, non-profit groups, and the general public. Ultimately, in the judgment of Arthur C. Nelson: 'The challenge facing Oregon now is how to properly recognize the urban form it has created through UGB policies, and its implications, in what manner it should be reassessed, and how best to consciously facilitate that urban form' (Abbott *et al.* 1994: 45).

Much activity has occurred in Oregon since 2003. In a May 29, 2003 speech, Governor Kulongoski spoke of the need to reassess Oregon's system of land use regulation. Two years later, Senate Bill 82 was passed, which created a Task Force on Land Use Planning. The purpose of this Task Force, more commonly referred to as the 'Big Look Task Force', as found in Section 1 (2)(a–c), was to study and make recommendations on the following areas:

1　The effectiveness of Oregon's land use planning program in meeting current and future needs of Oregonians in all parts of the state;
2　The respective roles and responsibilities of state and local governments in land use planning;
3　Land use issues specific to areas inside and outside urban growth boundaries and the interface between areas inside and outside urban growth boundaries.

The Task Force was divided into work groups to evaluate the different areas. The Task Force was to make recommendations to the state legislature in 2009. Funding has been an issue facing the Task Force. Funding was removed by the state legislature in June 2007. Governor Kulongoski wrote to the Task Force on June 28, 2007 indicating that he would try to get the funding restored.

The Task Force issued some preliminary findings in a July 2007 report. Within the report, the Task Force noted a number of conclusions, including that Oregon had fared better in containing sprawl than had most states; the state land use system had protected farms and forestlands; citizens strongly supported private property rights; the state's land use system was viewed by the public as a regulatory program; future population and employment growth were expected in high growth areas; and Oregon would be hard pressed to finance and maintain the infrastructure needed to accommodate the expected growth in the state.

Oregon has certainly gained notoriety over its land use planning program over the past thirty years. More recently, the state system has come under scrutiny by a number of parties. Many individuals felt that the state's land use regulations had reduced the value of properties for many Oregonians. Others felt that government had no right regulating away the use and value of their properties. Ultimately, cries for reducing

government regulations and repairing the system were heard loud and clear.

Two statewide ballot measures have influenced land use in Oregon in 2004 and 2007. In 2004, Measure 37 was drafted in response to government land use regulations. Drafted by Oregonians in Action, a non-profit organization, Measure 37 sought to remedy the perceived unfairness of Oregon's land use planning regulations. The group felt that Oregon had failed to consider the ramifications of its land use regulations. It would be a model for regulatory reform. It would provide relief, in the form of just compensation, to landowners for government regulations that reduced the value of their properties. The Measure would also result in less litigation since cities could now purchase land from landowners instead of engaging in lengthy court battles. Reduced to its simplest form, Measure 37 said that if government takes your property, it should pay for it.

Whenever ballot measures are used to decide an issue of public importance, you can be sure that an opposing side will surface. When the two sides start campaigning, the old slogan, 'truth in advertising', goes out the window. Allegations will be thrown out by one side and countered with allegations from the other.

The arguments against Measure 37 came quickly. It was claimed that Measure 37 was a poorly written, too complex, and flawed ballot measure. To some, it would represent a step backward in Oregon's land use regulatory system. Too much progress had been made to plan for orderly development. Measure 37 was perceived to be a threat to the progress Oregon had made in land use planning. In addition, many people and groups thought the main result of Measure 37 was to add more government red tape to the process. Moreover, there was no money to pay the landowners compensation for their property. Neighbors would be pitted against other neighbors to receive just compensation. The specter of people being treated differently by Measure 37 concerned many individuals and groups.

After a somewhat testy campaign involving numerous individuals and groups, Measure 37 was passed on December 2, 2004 by a margin of 61 percent to 39 percent. It was passed in thirty-five out the thirty-six counties in Oregon. The measure went into effect on December 2, 2004.

Two years after its passage, the Oregon State Supreme Court ruled on the constitutionality of Measure 37 in *MacPherson* v. *Department of Administrative Services*, 340 OR 117, 130 P.3d 208 (2006). A trial court had found Measure 37 unconstitutional for violating the equal privileges and immunities guarantee of the Oregon Constitution, the legislative plenary power, separation of power constraints, and for violating the Fourteenth Amendment to the US Constitution. The Oregon Supreme Court did not concur with the trial court's arguments against Measure 37. It upheld the constitutionality of Measure 37 on February 21, 2006.

Measure 37 allowed property owners to file claims asking for just compensation for the taking of their property due to the land use regulations. The number of claims probably surprised government officials. As of March 12, 2007, a staggering 7,500 claims had been filed. With the government being unable to process the claims in a timely fashion, the governor signed a bill on May 10, 2007 extending the time frame to process the claims that had been filed.

Frustration continued to build over Measure 37. The slowness in processing claims continued to anger property owners. The confusion and uncertainty over the intent and implementation of Measure 37 prompted the creation of a new ballot measure – Measure 49. Box 11.1 provides the wording of both measures.

Proponents of Measure 49 claimed that it was sorely needed because Measure 37 was not working and needed to be fixed. It was simply allowing too much development that was threatening Oregon's quality of life. Proponents also claimed that Measure 49 would correct some of the unintended consequences of Measure 37. These unintended consequences included destroying farmland and forestlands and then developing subdivisions on the properties. They felt too much development was being allowed. These problems led to many individuals feeling that Measure 49 was needed because Measure 37 was being abused by speculative developers.

Opponents to Measure 49 countered these arguments. Oregonians in Action, a statewide non-profit organization dedicated to protecting property rights

and excessive land use regulations and the developer of Measure 37, claimed that Measure 49 was a blatant attempt to repeal Measure 37. It claimed that Measure 49 was poorly written and that it was misleading. To others, these problems would result in more litigation by property owners because the Measure would wipe out property rights and allow government to take private property without payment of just compensation.

Measure 49 was passed by Oregon's voters on November 6, 2007 by a margin of 61 percent to 39 percent. The voters apparently felt that Measure 37 went too far and voted to modify or clarify the right to build a reasonable number of houses on property; to limit large-scale development and to protect farms and forests; and to close the loopholes and exceptions found in Measure 37. Nevertheless, the battles are not over in Oregon. Court action will undoubtedly follow this latest attempt to regulate land use in Oregon.

UGBs for cities and counties are required to be periodically reviewed and updated. Reviews are done to see if the current UGB will meet the needs of the area for the next twenty years. If it is deemed not to be able to meet future needs, one option available is to expand the UGB. In 2011, Portland's Metropolitan Council voted to expand the area's UGB to increase four areas for industrial development and residential development. In June 2012, the Oregon Land Conservation and Development Commission verbally approved the UGB expansion.

Vermont

It is the traditional settlement pattern (village, town, and countryside) that reflects the essence of Vermont. In order to maintain the essential character and ethic of Vermont's built environment, there should be a clear delineation between town and countryside through effective planning and supportive land development.

(Governor's Council on Vermont's Future 1988)

Vermont is a largely rural state in which the pressures for development come mainly from outside in the forms

BOX 11.1 OREGON MEASURES 37 AND 49

Measure 37

The following measures are added to and made part of ORS chapter 197:

(1) If a public entity enacts or enforces a new land use regulation or enforces a land use regulation enacted prior to the effective date of this amendment that restricts the use of private real property or any interest therein and has the effect of reducing the fair market value of the property, or any interest therein, then the owner of the property shall be paid just compensation.

(2) Just compensation shall be equal to the reduction in the fair market value of the affected property interest resulting from enactment or enforcement of the land use regulation as of the date the owner makes written demand for compensation under this act.

(3) (8) . . . in lieu of payment of just compensation under this act, the governing body responsible for enacting the land use regulation may modify, remove, or not to apply the land use regulation or land use regulations to allow the owner to use the property for a permitted use at the time the owner acquired the property.

Measure 49

Modifies Measure 37 (2004) to give landowners with Measure 37 claims the right to build homes as compensation for land use restrictions imposed as they acquired their properties. Claimants may build up to three homes if previously allowed when they acquired their properties, four to ten homes if they can document reductions in property values that justify additional homes, but may not build more than three homes on high-value farmlands, forestlands and groundwater-restricted lands. Allows claimants to transfer homebuilding rights upon sale or transfer of properties; extends rights to surviving spouses. Authorizes future claims based on regulations that restrict residential uses of property on farm, forest practices. Disallows claims for strip malls, mines, other commercial, industrial uses.

Sources: State of Oregon, Secretary of State, 'Voters' guide for November 2, 2004 General Election'; State of Oregon, Secretary of State, 'Voters' pamphlet, November 6, 2007 Special Election'

of meeting tourism needs and the development of vacation homes. In terms of population (some 563,000) it ranked forty-eighth in the 1990 census. Its estimated 2011 population is 626,431 – a minute increase from 2006. It has a highly decentralized local government system of small New England communities, with nine cities, 237 'organized towns', five unorganized towns (in sparsely populated areas) and fifty-seven incorporated villages which are urban in character. Fourteen counties largely exist only on paper, and there is therefore a void between state and local government which has been filled by regional state administration. Despite this, Vermont had no tradition of state planning; yet in 1970 it passed a growth management measure which introduced a statewide planning system.

The major reason for this was a transformation of Vermont from a state whose young people traditionally left for better opportunities elsewhere into a state beset by the problems of unprecedented growth. This was caused by a number of factors. The extension of the interstate highway system brought Vermont within easy traveling distance of the 40 million inhabitants of the urbanized areas to the south. Several economic changes also took place within the state, some of which were related to this new accessibility, including the expansion of the ski industry and the growth of second homes. Farms became endangered. Almost suddenly, Vermont changed from a remote area to an easily accessible vacation, second-home, and commuters' haven. It is, of course, a beautiful state, and Vermonters are proud of their quality of life and environmental stewardship.

The resultant growth in population led to development pressures and increased land costs (and taxes) which were alarming to the conservative Vermonters. A 14 percent growth in population during the 1960s, though modest by the standards of California or Florida, was greater than the increase over the previous half-century. It was, moreover, concentrated in particular areas, and therefore its impact was striking. The local governments of Vermont were quite unable to deal with this unprecedented situation. None of the towns had a capital budget program, and few had a zoning ordinance. They were literally at the mercy of developers. Something had to be done.

In a remarkably short space of time, the state government acted, and legislation (Act 250) was passed in 1970. It was in response to the continued growth pressures facing Vermont: the state's way of guiding growth while protecting and conserving the lands and environment of the state. The legislation introduced a development permit system administered by an appointed environmental board and district environmental commissions. It also provided for the preparation of three statewide plans: an interim land capability plan (an inventory of physical data); a land capability plan (to guide 'a coordinated, efficient, and economic development of the state'); and a final land use plan. The development permit system has worked reasonably well, but the plans have given rise to a number of difficulties, and the land use plan never emerged.

Establishing new agencies of government is always problematic. In particular, there is the perennial issue of decentralized versus centralized control. In Vermont, it was clear that the local government system could not

administer the development permit scheme, but there was little enthusiasm for giving more power to the state. The solution adopted placed the major responsibility for administering the development permit system on lay citizen district commissions – with the right of appeal to a lay state board. Thus the process is decentralized, but in a way which bypasses the established local governments. Ideally, of course, a plan should have preceded the introduction of this system, but there was insufficient time: the development pressures were too intense.

In the absence of a plan, Act 250 provided a list of criteria against which development applications are to be judged. These include a wide range of environmental, aesthetic, and land use issues. For example, development proposals are not to be approved if they would cause 'unreasonable congestion or unsafe conditions on highways', or create an 'unreasonable burden' on the ability of the municipality to provide services, or have an 'adverse effect' on the scenic beauty of the area. Such criteria clearly give a considerable range of discretion. However, they apply explicitly only to large developments (see Box 11.2), and decisions can be appealed to the Vermont Environmental Board. The burden of proof rests with the applicant with respect to a subdivision or to the party opposing the applicant with respect to the subdivision.

Vermonters were in favor of controlling unwanted development, but they were dubious about plans, particularly if these were to be drawn up by state bureaucrats. A land capability plan was accepted, but plans which would limit local discretion were strongly opposed. In the words of a local planning consultant, 'The idea had never been to limit the options for local folks, but rather to stem the destructive tide of "flatlanders" bent on citifying Vermont' (Squires 1992: 14). The anxieties here were so strong that the statutory provision requiring a state land use plan was repealed.

Of course, plans typically have multiple objectives, some of which may be difficult to harmonize, while others may be contradictory. The Vermont planning system was directed to improving the quality of large-scale developments; there is in reality little of a 'plan' in it. The approach is essentially 'reactive': it evaluates planning proposals which are submitted for approval;

BOX 11.2 DEVELOPMENTS REQUIRING A PERMIT IN VERMONT

1   Housing developments of ten or more units by the same applicant within a five-mile radius.
2   Developments involving the construction of improvements for commercial or industrial purposes on a tract of more than one acre in towns without permanent zoning and subdivision bylaws and on a tract of more than ten acres in towns with such controls.
3   Developments involving the construction of improvements for state or municipal purposes of a size of more than ten acres.
4   All developments above an elevation of 2,500 feet.

it does not direct growth to areas which are considered by planners to be suitable for growth. In short, as a growth management system, a lot remains to be desired. A 1988 report by the Commission on Vermont's Future appointed by Governor Madelaine Kunin underlined the perceived weaknesses, and stressed the fact that 'a consequence of the failure to adopt comprehensive local and regional plans is that basic planning decisions are left to the regulatory process'. As a result of the local nature of most of the land use controls, suburban and resort developments were continuing at a rapid rate: there was an urgent need 'to introduce planning into the regulatory process'.

Legislation was passed in the same year in the form of the Vermont Growth Management Act of 1988. This (Act 200) specified the minimum contents of local and regional plans (including land use, housing, transportation, utilities, education, and natural resources). It authorized impact fee ordinances for local governments that had adopted plans and Capital Improvement Programs. It retained the existing regional planning commissions with wider powers and subject to a requirement that they cooperate with other agencies and levels of government. All regional commissions and state agencies are required to ensure that their planning is consistent with twelve broad state planning goals. In this revised system, the regional commissions become the vital force in growth management: they are assigned the responsibility for reviewing and approving local plans, and for commenting on state agency plans.

A new agency, the Council of Regional Commissions, was created to review regional and state agency plans for compatibility with state goals. It was also to serve as an impartial mediator to decide disputes among municipalities and regional planning commissions and between regional planning commissions and state agencies, 24 VSA Section 4305 (b). Additionally, a Municipal and Regional Planning Fund was established to assist municipal and regional planning commissions. A geographic information system (to which all commissions and agencies contribute data) is financed from this fund.

Passed at a time of economic prosperity (and major governmental initiatives in environmental planning), the new legislation seemed to promise a major improvement in Vermont planning. But there was much that it did not do (partly because of opposition during its passage), and further opposition quickly followed. This increased as the economy deteriorated, but its origins were deep. A well-funded and organized Citizens for Property Rights group has attracted much support, and the controversy continues.

The requirements of Act 200 for consistency with the state's planning goals provide a substantive framework for plan preparation and implementation. (Plans have to demonstrate consistency with the goals, or good reasons for departing from them.) The goals are therefore a unifying element in the planning system. This appears to have had good effect in coordinating the plans of state agencies. The Act does not, however, mandate local planning; and municipalities can elect not to submit a plan for review by the regional planning commissions. Nevertheless, more planning is being undertaken than ever before.

The revised Vermont system is a neat balancing act between the requirements of area-wide planning and the strong proclivity of Vermonters for local control. It is also balancing private property rights with local control. But, in essence the planning process works from the 'bottom up', though within the framework of state policies. There has thus been little change in the Vermont allegiance to local control.

In 2006, Act 183 was passed. This Act implements Act 200's goal for the need for compact settlements. It required the Department of Housing and Community Affairs and the Natural Resources Board to work together on developing a growth centers program. A growth center represents a compact area with mixed-use development. It is defined in 24 VSA Section 2791 (12)(A) as 'an area of land that is in or adjacent to a designated downtown, village center, or new town center'. The characteristics of a growth center can be found in Box 11.3. The issue of coordination between state, regional, and local levels of government remains important.

There are currently six communities that have been approved as designated growth centers: Williston (2007); Bennington (2008); Colchester (2009); Montpelier (2009); Hartford (2010); and St Albans City (2010). As noted in 24 VSA Section 2793c(e), each growth center designation is for a period of twenty years. They are reviewed on a five-year basis. Designation can be modified, suspended, or revoked if the area has not achieved the required regulatory changes specified in its growth center plan.

BOX 11.3 CHARACTERISTICS OF A DESIGNATED GROWTH CENTER IN VERMONT

1   It incorporates a mix of uses that typically include or have the potential to include the following: retail, office, services, and other commercial, civic, recreational, industrial, and residential uses, including affordable housing and new residential neighborhoods, within a densely developed, compact area.

2   It incorporates existing or planned open public spaces that promote social interaction, such as public parks, civic buildings (e.g., post office, municipal offices), community gardens, and other formal and informal places to gather.

3   It is organized around one or more central places or focal points, such as prominent buildings of civic, cultural, or spiritual significance or a village green, common, or square.

4   It promotes densities of land development that are significantly greater than existing and allowable densities in parts of the municipality that are outside a designated downtown, village center, growth center, or new town center, or, in the case of municipalities characterized predominantly by areas of existing dense urban settlement, it encourages in-fill development and redevelopment of historically developed land.

5   It is supported by existing or planned investments in infrastructure and encompasses a circulation system that is conducive to pedestrian and other non-vehicular traffic and that incorporates, accommodates, and supports the use of public transit systems.

6   It results in compact concentrated areas of land development that are served by existing or planned infrastructure and are separated by rural countryside or working landscape.

7   It is planned in accordance with the planning and development goals under section 4302 of this title, and to conform to smart growth principles.

8   It is planned to reinforce the purposes of 10 VSA Chapter 151.

Source: 24 VSA Section 2791 (12)(B)

Florida

> Deep-rooted love affairs are always difficult to terminate, and Florida's love affair with growth has been no exception.
>
> (DeGrove 1984)

Florida's growth in post-World War II years has been phenomenal – a result of its attractive environment, its warm climate, and its low taxes. In 1950 the state had a population of 2.8 million. This increased to 5.0 million in 1960, 6.8 million in 1970, 9.7 million in 1980, 12.9 million in 1990, 15.1 million in 1999, and 16.0 million in 2000. Since 2000, the state's population had increased by 19.2 percent to an estimated 19,057,542 in 2011. It remains the fourth most populous state in the Union. Such a growth would have presented problems in any state, but the problems in Florida are compounded by its unique, fragile, and complex natural environment. These are the most difficult in precisely the areas of the greatest growth – in the southern part of the state. If ever a situation cried out for strong planning measures, this is it.

It took some time for Floridians to appreciate and acknowledge this, but concerns about rapid growth finally resulted in a legislative response after the serious drought of 1971. Of particular importance was the Environmental Land and Water Management Act of 1972, which provided for the designation of *areas of critical state concern* (ACSC) and for special measures for dealing with *developments of regional impact* (DRI). The appeal of these instruments was that they furnished a nice balance between state, regional, and local interests. Development normally remains the responsibility of the municipalities, but in the case of an ACSC or DRI, higher levels of government are involved. As such, private property owners felt the government had infringed upon their property rights, while local governments objected to the state questioning their land use authority.

ACSCs are recommended by the state planning agency. Four areas have been designated: the Big Cypress Swamp Area (Miami-Dade, Monroe and Collier counties), the Green Swamp, the City of Key West-Florida Keys, Monroe County, and the City of Apalachicola. While ACSCs are designated by the state, DRIs are a matter for local governments, subject to review by the regional planning council and the state. A DRI is designated only when a development is proposed. The system is therefore a reactive one, and it was made more difficult initially because of the absence of a comprehensive state plan. It was, however, an improvement on the previous system in that it brought into the development approval procedure the regional level of government. All of the state is now covered by eleven regional planning agencies (which are essentially multi-county councils of government).

The system was characterized by persuasion: persuasion of one level of government by another, and persuasion of developers by the municipalities. As a result, the great majority of developments were approved (though with conditions attached). The obvious weaknesses in the system (particularly the absence of a state plan, lack of funding for local planning, and the inadequacy of state review of plans) led eventually to the introduction of major changes in the planning system.

The turning point in Florida planning came in the mid-1980s, when the state overhauled its planning system at state, regional, and local levels. The revised system is in essence one of growth management. A hierarchy of plans features a comprehensive state plan with which the plans of state agencies ('functional plans') and regions (regional plans) must be consistent; similarly with local plans. The state plan adopts twenty-five goals and policies, ranging from education to housing, from health to natural resources, and from air quality to property rights and plan implementation. (See Box 11.4 for the land use goal and its associated policies.)

Florida planning has thus been transformed. In place of the 'bottom up' character of the earlier legislation, it is now unequivocally 'top down'. Under the Local Government Comprehensive Planning and Land Development Regulation Act (Chapter 163, Part II, Florida Statutes), all municipalities and counties are obliged to prepare and adopt comprehensive plans. The requirement that these be consistent with the state plan is not mere rhetoric. Regional review teams review the comprehensive plans and, if the plan is not 'compatible with the goals' of the state plan, the state can impose some severe sanctions, particularly the withholding of funds. Some 467 counties and municipalities have adopted plans determined to be in compliance with state law (Florida, Department of Community Affairs 2000: 10).

A remarkable provision requires local governments to coordinate the provision of infrastructure with urban growth. Development can be permitted only to the extent that the infrastructure can support it: 'public facilities and services needed to support development shall be available concurrent with the impact of the development'. Local governments are required to design adequate and realistic 'level of service' (LOS) standards for roads, sewers, drainage, water, recreation, and (if applicable) mass transit. Development which would fail to maintain LOS standards cannot be permitted unless the deficiency will be made good by the provisions of the capital investment plan.

These *concurrency* provisions, together with a strong emphasis on compact urban development, have given Florida a powerful tool of growth management. Implementation has been made feasible by the preparation of mutually consistent plans. (Local and regional plans are statutorily required to conform with the state plan.) It has, however, been weakened by inadequate state funding of infrastructure. This problem has been dealt with, to a limited extent, by increased local taxation and by the imposition of impact fees on developers. Nevertheless, the long-term viability of the system is dependent upon a stable solution to the infrastructure financing issue. Such a solution is not yet in sight. Until this intransigent hurdle is overcome, Florida's impressive planning system will have more unfilled promise than achievement.

Florida's system of planning for growth is still evolving, as is the case for all of the states described in this chapter. There continues to be constant legislative

BOX 11.4 FLORIDA LAND USE GOAL AND POLICIES

Goal

In recognition of the importance of preserving the natural resources and enhancing the quality of life of the state, development shall be directed to those areas which have in place, or have agreements to provide, the land and water resources, fiscal abilities, and the service capacity to accommodate growth in an environmentally acceptable manner.

Policies

1 Promote state programs, investments, and development and redevelopment activities which encourage efficient development and occur in areas which will have the capacity to service new population and commerce.
2 Develop a system of incentives and disincentives which encourages a separation of urban and rural land uses while protecting water supplies, resources development, and fish and wildlife habitats.
3 Enhance the livability and character of urban areas through the encouragement of an attractive and functional mix of living, working, shopping, and recreational activities.
4 Develop a system of intergovernmental negotiation for siting locally unpopular public and private land uses which considers the area of population served, the impact on development patterns or important natural resources, and the cost-effectiveness of service delivery.

debate on growth. In response to this debate, Governor Jeb Bush created, through Executive Order 200-196 (July 3, 2000), a twenty-three-member Growth Management Study Commission to review what Florida has done regarding growth management, what Florida is currently doing in the area of growth management, and to issue a report with recommendations for addressing Florida's growth in the twenty-first century.

The Commission's report concluded that Florida's earlier growth management policies and programs were too rigid and that the state must make some changes. For example, it recommended that the old DRI program be replaced. There was also a call for the state to create incentives for revitalization to encourage municipalities to participate in urban revitalization and rural development. Governor Bush apparently recognized that controlling growth requires a long-term commitment on behalf of all parties.

In 2005, the Florida Legislature passed legislation that amounted to a 'pay as you grow' philosophy. This legislation was designed to make sure roads, schools, and water were available to meet the needs of a growing population. In the following year, the Florida Impact Fee Act of 2006 was passed. It held that a developer could demand to see how impact fees were used and calculated. The legislation also created a Florida Impact Fee Review Task Force, a fifteen-member advisory body that would review the use of impact fees to finance local infrastructure in Florida. The Task Force has found that the use of impact fees has grown in Florida because local governments do not necessarily generate revenue resources to keep pace with the growing infrastructure demands. Additional legislation urged the use of infill development to help curb urban sprawl. Most recently, in 2007 Governor Crist signed legislation privatizing toll roads and legislation that removed airports from concurrency requirements and certain urban areas from the oversight of the Florida Department of Community Affairs. Opponents felt these pieces of legislation promoted unmanaged growth and failed to deal with problems of congestion on Florida's roads.

Concerns over growth continued to be in the news in 2008. A statewide ballot measure was proposed that would amend the Florida Constitution by granting voters the right to decide whether or not to adopt or amend local land use plans. The specific wording of the ballot measure, Amendment 4, was:

> Establishes that before a local government may adopt a new comprehensive land use plan, or amend a comprehensive land use plan, the proposed plan or amendment shall be subject to vote of the electors of the local government by referendum, following preparation by the local planning agency, consideration by the governing body and notice. Provides examples.

The measure failed to get enough signatures to qualify for the 2008 ballot. It did, however, later qualify for the 2010 ballot, where it was defeated.

In October 2011, the Florida Department of Community Affairs (DCA), which since 1969 had monitored and reviewed city and county growth management decisions, was abolished. Governor Rick Scott felt the state had too many regulations that were impeding job creation. Job creation and economic growth were high on his agenda. He vowed that the state would become more job friendly. The DCA would be merged with several other agencies. A new Department of Economic Opportunity would oversee and coordinate economic development, housing, growth management, and community development programs. It would also serve as the home of a Division of Community Development. This office would now be the State Land Use Planning Agency and serve to guide Florida's development by offering assistance to city and county governments for disaster planning, community planning, and revitalization. It would also review DRIs.

Maryland

According to the US Census, Maryland's population in 1990 was 4,781,461. In 2000, its population had increased to almost 5.3 million residents – an increase of 10.8 percent from the 1990 figures. The population continues to grow, with an estimated 2011 figure of 5.8 million residents – a 10 percent growth from 2000.

BOX 11.5 VISIONS TO BE INCORPORATED INTO MARYLAND COUNTY AND MUNICIPAL PLANS UNDER 1992 ACT

1 Development concentrated in suitable areas.
2 Sensitive areas are protected.
3 In rural areas, growth directed to existing population centers and resource areas are protected.
4 Stewardship of the Chesapeake Bay and the land is a universal ethic.
5 Conservation of resources, including a reduction in resource consumption, is practiced.
6 To assure the achievement of the above, economic growth is encouraged and regulatory mechanisms are streamlined.
7 Funding mechanisms are addressed to achieve these visions.

Source: Economic Growth, Resource Protection, and Planning Act of 1992

Accompanying this growth are concerns over sprawl and its effects on the environment.

Maryland has been proclaimed to be a leader among the states in developing comprehensive state planning and growth management policies. Former Governor Parris Glendenning (1995–2003) has been called a leader of the 'smart growth' movement and continues to be active in this area on a national basis. A number of early activities prompted the state's concern over growth and its associated impacts. In the 1980s, there was a renewed interest in examining growth on a statewide basis. Individuals were reminded of the connections between population growth, development patterns, and environmental degradation. Governor Walter Schaefer (1987–95) appointed a Year 2020 Panel in 1987 to evaluate alternative strategies to deal with growth patterns in Maryland. He later appointed a Growth Commission to examine the visions of the 2020 Panel. The Growth Commission recommended legislation that would increase the state's role in managing growth. The legislation was opposed by local governments and failed to be passed in 1991. The state legislature passed legislation the following year.

The Economic Growth, Resource Protection and Planning Act of 1992 was enacted on October 1, 1992. This Act recognized the important role the state played in guiding growth and development in Maryland. Within the Act, a number of visions were to be incorporated into county and municipal plans (see Box 11.5). An eighth vision was added to the list by the Maryland General Assembly in 2000. Working through the municipal and county plans seemed to make sense because of their impact on growth and conservation. The Act also required state agencies to make sure their activities were consistent with state policies on growth and conservation. Concentrating development, protecting sensitive areas, conserving resources, and directing growth to existing population centers were key themes of the Act.

In 1997, Maryland passed smart growth legislation to curb the problems of encroaching sprawl. It recognized that government had to share the responsibility in causing the problems. The Smart Growth Priority Funding Areas Act of 1997 called for the development of existing areas where local governments want the state to direct investment in future growth. There would be no state funding to support infrastructure outside of these areas. Funding inefficient and sprawling development would no longer be tolerated. The legislation also authorized counties to designate additional smart growth areas. A Rural Legacy Program represented a counterpart to the growth areas by targeting funding for land preservation efforts. These programs would comprise what is known as Maryland's smart growth efforts. A Smart Growth Subcabinet was created that would assist in the implementation of Maryland's Smart Growth Policy, make recommendations to the governor on smart growth issues, and provide a forum for intergovernmental issues affecting smart growth. Box 11.6 provides Maryland's Smart Growth and Neighborhood Conservation Policy.

Three years later, the Maryland Legislature directed the Maryland Department of Planning to develop guidelines for creating 'smart neighborhoods'. 'Smart neighborhoods' were defined as 'relative self-contained new communities with a compact mix of residential, commercial, employment office, and civic land uses and range of housing choices with a design that fosters pedestrian and bicycle activity, public safety, environmental protection, long-term investment, efficient use of infrastructure, and efficient provision of public services'. Incentives were offered to communities that encouraged the development of smart neighborhoods.

An Office of Smart Growth, housed in the Governor's Office, was created in 2001 to coordinate the state's smart growth efforts. Governor Robert Ehrlich, Jr. was elected in 2002 and pledged his commitment to smart growth by promoting the idea of walkable communities. Two years later, the office was moved to the Maryland Department of Planning.

Ehrlich proclaimed his support for Maryland's smart growth programs. However, critics claim that during his tenure, he weakened environmental laws and cut funding for land preservation and new environmental programs. In 2005, he announced his program of 'Priority Places', which sought to clarify state policies for land use and smart growth. This program sought to protect Maryland's quality of life by promoting economic development in existing communities. The Governor's critics felt that he had harmed key programs and had undermined the original goals of smart growth.

Governor Martin O'Malley assumed office in January 17, 2007. As a believer in smart growth, Governor O'Malley has embarked on a program designed to make Maryland, once again, a leader in the smart growth movement.

In 2010 the Maryland Sustainable Growth Commission was established. This enabled a permanent body to advise Governor O'Malley on growth and development issues facing Maryland. Among the responsibilities of the Commission were to monitor progress on planning efforts regarding sustainable growth, to promote interjurisdictional planning cooperation and coordination, investigate the relationship between various state plans and local land use plans, engage the public through programs associated with smart growth, and make recommendations regarding state laws, regulations, etc. that are needed to achieve state policies on smart and sustainable growth.

BOX 11.6 MARYLAND SMART GROWTH AND NEIGHBORHOOD CONSERVATION POLICY

1   State agencies give priority to central business districts, downtown cores, empowerment zones, and revitalization areas where funding infrastructure projects or locating new facilities;
2   State agencies review, evaluate and coordinate programs, services and activities in Priority Funding Areas to enhance and support Community revitalization;
3   State agencies work with local jurisdictions to ensure that programs and activities in a rural area will sustain the character of villages in the area;
4   State agencies encourage locating workshops, conferences, and other meetings in Priority Funding Areas and support available business in these areas when planning such activities; and
5   State agencies encourage federal agencies to adopt flexible regulations and standards which are more responsive to state and local policies and can be used to support Smart Growth Policies.

Source: Executive Order of Governor Parris Glendenning, July 1, 1998

The next year saw the adoption of PlanMaryland – Maryland's official State Development Plan. Governor O'Malley signed an Executive Order on January 1, 2011, recognizing it as the leading guide to the economic and physical development of Maryland. While individuals and organizations debate the merits of PlanMaryland, the Maryland Department of Planning's (2011: np) Executive Summary of Plan-Maryland indicates:

> PlanMaryland will not remove local planning and zoning authority. It seeks to improve coordination between state agencies and local governments on smart growth because too often the actions of the State have been at cross-purposes to achieve a common goal of making existing communities stronger, healthier, cleaner and safer. Goals for planning, development, conservation and sustainable quality of life are interdependent; not the work of several agencies occasionally coordinating. Independent initiatives by the State or local governments won't achieve these goals.

Nevertheless, some individuals feel it intrudes on a local jurisdiction's authority of land use. Others are concerned that the document ignores the needs of rural communities seeking economic growth.

In 2012, the Maryland Sustainable Growth and Preservation Act was passed. This legislation sought to minimize any land use or pollution problems associated with septic systems. It also helped local governments control the use of septic systems.

California

> Probably no state employs more planners or produces more plans, and probably nowhere else in the country does planning and development engender more discussion at the community level. But, for all that, California has proven, over the past decade, incapable of managing its growth.
>
> (Fulton 1991)

California has had a long and checkered history of planning endeavors. It has seen a prodigious number of plans, which continue to flow – though their destination is more often planning libraries than implementation. There is, however, no machinery for state intervention in local land use planning. At first sight this is curious, since California is the leading state in environmental planning, and its elaborate system of environmental review has had a profound impact on environmental considerations in local planning. Nevertheless, no state agency exists to review or approve land use plans.

Instead of state intervention in local planning, California has enabled citizens to take an active role in deciding land use and environmental matters through the ballot box (Caves 1992). As noted in Chapter 9, the California Constitution, Article II, Section 8(a) reserves 'the power of the electors to propose statutes and amendments to the Constitution and to adopt or reject them'.

It is against this background that state involvement in land use planning is restricted. Curiously, however, there is a major exception which itself came about as a result of an initiative. After failing to pass the state legislature, an initiative promoted by environmental groups led to the enactment of the Coastal Act in 1972 (despite a well-funded aggressive counter-campaign by developers, oil companies, and the like – and the opposition of Governor Ronald Reagan). Thus California has a full-scale coastal planning program, and this has survived several hundred bills to kill or cripple it (Fischer 1985). It is this program which is of relevance to this chapter.

Strictly speaking, California's coastal program is not a statewide comprehensive planning endeavor: as its name suggests, it is concerned only with the coast. But that coast is 1,100 miles long, from Oregon to the border with Mexico (and the coastal planning area is up to five miles wide in rural areas). The coastal area is comprised of some 1.5 million acres of land and reaches from three miles at sea to an inland boundary that varies from a few blocks in the more urban areas of the state to about five miles in less developed regions. It is therefore very much akin to a statewide planning area. It is administered by the California Coastal Commission.

The coastal plan is not concerned solely with environmental protection: it seeks to ensure that the coastline is used intelligently and sensitively, with due regard to both the environment and the needs of coastal-related development. However, the plan is highly restrictive in respect to the preservation of wetlands, historic, scenic, agricultural, and forest lands. The basic goals given in the legislation are set out in Box 11.7.

In the early years of the program, administration was by interim planning commissions. These were independent of local government, and they operated that way: they made very little effort to develop any collaborative relationships with their constituent municipalities. This changed dramatically after 1976, when, subject to conditions, plan-making and regulatory responsibility was returned to local government. The conditions were several. The statute mandated development of local coastal plans, with regulatory authority over most development to be transferred back to local government only after the Commission had certified that the plan was in conformity with the policies of the Coastal Act. Further, the Commission retains some important planning and regulatory responsibilities, including permanent jurisdiction in some areas such as tidelands, submerged lands and trust lands; reviewing and acting upon appeals from local permit decisions; reviewing and authorizing amendments to plans; implementing public access programs; and periodically reviewing the implementation of certified plans to determine if the plans are being implemented in conformity with the provisions of the Coastal Act, and making recommendations to local governments or the legislature. Thus the legislation clearly establishes a shared responsibility between the Commission and local governments.

The California Coastal Commission is a regulatory body. To complement its operations, a State Coastal Conservancy was established. Among its many functions, this helps to carry out coastal improvement and restoration projects to implement policy established through the plans and regulations of the Commission and local governments. It is empowered to buy land, and to restore or resubdivide it, or sell or transfer it to others (whether at a profit or a loss). It carries out a wide range of functions in furtherance of the policies and regulations of the Commission and the local governments.

One priority is the maximization of public access to and along the shoreline. The Commission requires, as a condition for the granting of a permit, that public access be provided. (This is a condition which achieved national publicity in the planning world with the 1987 case of *Nollan* v. *California Coastal Commission*, which is briefly discussed in Chapter 9.) This involves the dedication of an easement to a public agency that is willing and able to accept responsibility for maintenance and liability. Since huge numbers of conditional permits have been issued (1,800 in the twelve years up to 1985: a potential of more than fifty

---

BOX 11.7 CALIFORNIA COASTAL PLAN GOALS

1 Protect, maintain, and, where feasible, enhance and restore the overall quality of the coastal zone environment and its natural and artificial resources.
2 Assure orderly, balanced utilization and conservation of coastal zone resources taking into account the social and economic needs of the people of the state.
3 Maximize public access to and along the coast and maximize public recreational opportunities in the coastal zone consistent with sound resources conservation principles and constitutionally protected rights of private property owners.
4 Assure priority for coastal-dependent and coastal-related development over other development on the coast.
5 Encourage state and local initiatives and cooperation in preparing procedures to implement coordinated planning and development for mutually beneficial uses, including educational uses, in the coastal zone.

miles of additional shoreline access), this is no small task; and it is one which financially hard-pressed local governments are none too happy to accept. The future of the coastal program is uncertain. It has aroused a great deal of continuing opposition, and its budget is under constant attack. More directly concerned with growth management are the attempts at regional planning which have been made in some of the major urban areas of the state. Here we look at two metropolitan agencies (for the San Francisco Bay Area and for the Los Angeles and Southern California region), both of which are struggling with widely conflicting views of the future regional planning of their areas.

The Bay Area 'Bay Vision Commission', created in 1989, has stressed the diversity of views which were represented on the Commission and the difficulty of reaching agreement for its *Bay Vision 2020* report (see Box 11.8). It was proposed that a regional commission be set up combining, for a start, the functions of the Bay Area Air Quality Management District, the Metropolitan Transportation Commission, and the Association of Bay Area Governments (and later the Regional Water Quality Control Board and the Conservation and Development Commission), but this still leaves sixty-two other agencies! There is, moreover, disagreement on the constitution of the proposed regional commission, its role in equalizing tax burdens, and its power to control developments of regional importance. But 'we strongly believe in maintaining the integrity of existing local governments and their autonomy over local decision-making' (Bay Vision 2020 Commission 1991: 38). As Joseph E. Bodovitz, the commission's project manager, has commented, 'there is no ground swell of readiness to plug into a regional political system. Indeed, there is great antipathy' (Stanfield 1991: 2330). The Commission's 1991 report and proposed legislation two years later failed to garner enough support in the State Senate. However, many of the groups are involved in collaborative projects that seek to achieve some of the Bay Vision 2020 goals. For example, groups such as the Association of Bay Area Governments, the Air Quality Management District, the Metropolitan Transportation Commission, the Regional Water Quality Board, and the Bay Conservation and Development Commission joined forces in a project called the Smart Growth Initiative. This project will ultimately produce a smart growth strategy for the region that will show how the Bay Area region could grow over the next twenty years.

There is a clear parallel with the 1990 report on Los Angeles and Southern California by 'The 2000 Partnership'. Though existing governmental agencies 'cannot adequately plan for and manage growth on a regional level', no new planning authority is proposed; instead a new council would consolidate the current planning powers of existing agencies, and subregional councils could be formed on the basis of cooperation between local governments. It is suggested that the new regional council 'would have the authority to make and implement policies when a city, county, or special district was determined to have failed to meet regional objectives within a specified time limit'. This sounds as if the regional council would have some teeth, but these are quickly drawn: regional and subregional plans would be subject to the agreement of the constituent authorities who would have ample opportunity for 'consultation' and 'bargaining'. At the most, the proposal amounts only to control over 'limited areas of regional impact'. It remains to be seen what the effect will be of the federal planning requirements for transportation planning introduced by the Intermodal Surface Transportation Efficiency Act. This is discussed in the following chapter.

The California Environmental Quality Act (CEQA) that was passed in 1970 could be considered as a state growth management tool. An Environmental Impact Review (EIR) is the informational document required by CEQA. According to Title 14 California Code of Regulations, Chapter 3 Guidelines for Implementation of the California Environmental Quality Act, Article 1, Section 15002, the basic purposes of CEQA are to:

1 inform governmental decision-makers and the public about the potential, significant environmental effects of proposed activities;
2 identify the ways that environmental damage can be avoided or significantly reduced;
3 prevent significant, avoidable damage to the environment by requiring changes in projects through

BOX 11.8 BAY AREA REGIONAL PLANNING DEADLOCK

We have noted that current forecasts predict an increase in the Bay Area's population from the current six million to well over seven million by the year 2000. Some of us have concluded that there is a point beyond which the Bay Area's population must not be allowed to grow if the natural resources of the Bay Area are to be protected adequately. Others of us believe that such a population limit is neither desirable nor possible to achieve. Still others believe that the issue is not population growth itself, but the need to manage development so that natural resources are not degraded as population increases. All of us agree, however, that the environmental impacts of an increasing population and an expanding economy will require a new, more comprehensive ability to plan and make regional decisions for the Bay Area.

Source: Bay Vision 2020 1991: 4

the use of alternatives or mitigation measures when the government agency finds the changes to be feasible;

4 disclose to the public the reasons why a government agency approved the project in the manner the agency chose if significant environmental effects are involved.

Section 15126.2(d) goes on to indicate that an EIR must 'discuss the ways in which the proposed project could foster economic or population growth, or the construction of additional housing, either directly or indirectly, in the surrounding environment' – so-called 'growth-inducing impacts'. For example, if additional infrastructure capacity is added, population growth may follow. In addition, a new residential or commercial development will generate the need for other, secondary services such as gas stations, restaurants, shopping centers, etc. It must then be determined if the proposed project's growth-inducing impacts will cause significant environmental impacts.

The State of California continues to undergo tremendous population growth. Hanak (2005: v) has indicated that between 2000 and 2030, California is expected to add 14 million residents, to reach a total of 48 million residents. The estimated 2011 population of California is 37,691,912 – an 11.2 percent increase over the 2000 population of 33,871,648. Current problems of unaffordable housing, congestion, concerns over the availability of water, and deteriorating infrastructure

will undoubtedly worsen. In a December 26, 2007 press release, Governor Arnold Schwarzenegger noted that the California Department of Finance had estimated that California needed $500 billion worth of infrastructure over the next two decades. Among the current activities of the State of California are legislation making infill development and transit-oriented development easier to develop, a call to develop public-private partnerships to build needed infrastructure, a call to develop a comprehensive water infrastructure plan that ties economic growth and environmental protection to infrastructure, and requiring certain cities and counties to submit an Annual Progress Report indicating how land use decisions relate to the adopted goals and policies of their General Plan and how the jurisdiction is monitoring long-term growth.

New Jersey

Statewide comprehensive planning is no longer simply desirable, it is a necessity.

(*Mount Laurel I* 1975)

New Jersey has had a series of important regional planning initiatives. The most famous is the Pinelands Commission, established in 1979, which is responsible for the planning, development, and adoption of a comprehensive management plan for some one million acres in the southern part of the state. Earlier, the

Coastal Area Facility Review Act of 1973 created a regional commission to develop an environmental inventory of the New Jersey coastal area and to regulate large developments in the coastal area by requiring each permit application to include an environmental impact statement.

Former US Secretary of the Interior Bruce Babbitt has called the New Jersey Pinelands 'the best broad-scale regional planning in America' (New Jersey Pinelands Commission 2006: 1). Its area covers 1.1 million acres of land in seven counties and fifty-six municipalities. The area contains many rare, threatened, and endangered plant and animal species and many other scenic, cultural, and recreational resources. Continued rates of haphazard development threatened the area. As stated in Title 13, Section 18A-2 of the Pinelands Protection Act:

> the continued viability of such area and resources is threatened by pressures for residential, commercial and industrial development;
>
> . . . a certain portion of the pinelands area is especially vulnerable to the environmental degradation of surface and ground waters which would be occasioned by the improper development of use thereof; and
>
> . . . the current pace of random and uncoordinated development and construction in the pinelands area poses an immediate threat to the resources thereof, expecially in the survival of rare, threatened and endangered plant and animal species and the habitat thereof, and to the maintenance of the existing high quality of surface and ground waters.

Legislation was passed in 1978 in the US Congress (through the Federal National Parks and Recreation Act) and in 1979 by the State of New Jersey (through the Pinelands Protection Act) to protect the Pinelands National Reserve. The Pinelands Commission was established within the New Jersey Department of Environmental Protection to preserve and protect the Pinelands National Reserve. The Commission is comprised of fifteen members; seven appointed by the governor of New Jersey, one member appointed by each

of the seven county boards, and one member appointed by the US secretary of the interior.

A Comprehensive Management Plan (CMP), prepared and adopted by the Pinelands Commission, protects the area and its future development. The Commission implements the CMP in cooperation with the various units of local governments with the Pinelands areas, the State of New Jersey, as well as appropriate federal offices, agencies, and departments. A sampling of the Commission's functions include: reviewing zoning, land use ordinances and master plans for consistency with the CMP, protecting cultural and historic resources, supporting land protection efforts, and reviewing water quality plans. Local governments are primarily responsible for implementing the CMP.

In 1980, the Democratic administration of Governor Brendan Byrne created a State Development Guide Plan. Though this was short lived (Republican Governor Thomas Kean abolished it in the following year), the courts continued to use it in the implementation of the Mount Laurel policy (discussed further in Chapter 14) to identify growth areas where municipalities were required to set aside some 20 percent of their new housing for lower-income families. During the early to mid-1980s, the New Jersey economy boomed, migration (of both people and jobs) into the state grew, and political pressures for more effective planning increased. In response, a State Planning Act was passed in 1985, establishing a State Planning Commission and its staff arm, the Office of State Planning.

The Commission was charged with preparing the primary instrument for coordinating planning and growth management in the state – the State Development and Redevelopment Plan. The statute provides that the plan shall protect the natural resources and qualities of the state, while promoting development in locations where infrastructure can be provided. It also establishes statewide objectives in a variety of areas including land use, housing, and economic development. The plan is intended to be used to guide the state's capital expenditure.

The plan was prepared according to procedures spelled out in the Act. First, a preliminary plan was approved by the Commission. This was then used in an interactive planning process called *cross acceptance* which was intended to integrate municipal, county, regional, and state land use plans as well as the capital facility plans needed to assure efficient services. This process was the crucial mechanism for obtaining support for the plan from the local governments whose cooperation is essential for its implementation. It represented a type of dialogue between the various levels of government, the private sector, and the public. New Jersey has a strong tradition of home rule, and the 567 municipalities are very suspicious of state action in the land use field. It was therefore essential that the plan preparation process should involve the active participation of the municipalities (see Box 11.9).

The process was an involved one; there was even a *Cross Acceptance Manual* prepared by the Office of State Planning. However, in essence the idea was simple: the authorities that needed to coordinate their activities were given a mechanism by which they could talk until agreement or compromise was reached. Obtaining consistency between the various local, county, and regional plans with the state plan was important to the realization of the state policies. However, as was expected, not all was plain sailing. Disagreements were bound to occur. Nevertheless, a sufficient level of agreement was reached to permit the plan to be finalized, vague though it is in important respects.

The New Jersey State Development and Redevelopment Plan was approved in 1992, after a long period of debate and public hearings. Its intent was to coordinate public and private actions and to help guide growth into compact forms. It establishes statewide goals and objectives for a wide range of policies including land use, housing, infrastructure investments, energy resources, air quality, water resources, economic development, transportation, recreation, and historic preservation. The plan embraces the concept of growth areas, though it is coy in identifying these (except in the case of the older cities). Several hundred other locations are identified as areas where development, redevelopment, and economic growth are considered to be in the public interest, but these are not actually designated, and no growth targets are established. There is thus a high degree of uncertainty in the plan – a result of the acute political difficulty in obtaining agreement among conflicting interests.

Much of the recent growth in New Jersey has been along transportation corridors, and this pattern is likely to continue in the future. The plan takes this fact as a basis for a major strategy of developing centers in the prosperous corridors. These centers are not envisaged as an elongation of the corridors: on the contrary, they are to be high density consolidations around existing development. Their attraction is that of good transportation (which an elongation of a corridor would jeopardize). The development of centers provides the opportunity for enhancing the transportation advantages.

Major features of the plan are its emphasis on mixed-use centers and on the expansion of existing urban areas. These are considered to have sufficient capacity to meet the anticipated population growth in the state up to the year 2010. An impact assessment study, undertaken by Robert Burchell of the Rutgers

---

BOX 11.9 CROSS ACCEPTANCE IN NEW JERSEY

The term 'cross acceptance' means a process of comparison of planning policies among governmental levels with the purpose of attaining compatibility between local, county, and state plans. The process is designed to result in a written statement specifying areas of agreement or disagreement and areas requiring modification by parties to the cross acceptance.

Source: N.J.S.A. 52: 184–202 (6)

University Center for Urban Policy Research, concludes that the implementation of the plan would save some 130,000 acres of land at no appreciable increase in the cost of development. However, the plan is not self-implementing and though there are procedures for certification of the consistency of local plans with the state plan, it is unclear how this will work out. The mechanisms for implementation are as uncertain as the provisions of the plan. Indeed, it is not at all clear what the long process of plan preparation has actually achieved. Certainly, there is nothing equivalent to the implementation provisions to be found in Oregon.

In 1999, Governor Christine Whitman signed legislation creating an open space preservation program that allocated monies for farmland protection, state land acquisition, and offering matching grants to local governments and non-profit organizations that advance the goals of the state legislation. This land acquisition would act as a catalyst to slow down the continuing sprawling development. In the following year, Governor Whitman signed a transportation bill providing monies for reducing congestion. New Jersey continues to encourage development where the needed infrastructure is already in place. Governor Whitman left office to become Administrator of the US Environmental Protection Agency on January 31, 2001.

After a series of acting governors, James McGreevey assumed the office of governor on January 15, 2002. One of his first acts was to issue Executive Order #4 on January 31, 2002. Within this Executive Order, Governor McGreevey acknowledged the state's policy of promoting smart growth and reducing the negative effects of sprawl and disinvestment in older communities throughout New Jersey. To make sure state agencies adhere to the principles of smart growth and to the State Plan, he created a Smart Growth Policy Council within the Office of the Governor. Among the responsibilities of the council were to make sure state agency functional plans, programs, and projects were consistent with the principles of smart growth and the State Plan; recommend any legislative and administrative changes to promote smart growth and the State Plan; and develop initiatives to assist local governments

and communities to achieve smart growth objectives. A lack of leadership and disagreements within the Council impeded the case of promoting smart growth in New Jersey.

In his 2003 State of the State Address, Governor McGreevey reiterated his belief that unrestrained and uncontrolled development threatened the residents' quality of life. This type of development was wasteful, jeopardized the state's water supplies, contributed to school overcrowding, and contributed to the increased congestion of the state's roads. To him, regional solutions were the answers to these and other problems.

The position of Smart Growth Ombudsman was created under the Smart Growth Act, Public Law 2004, Chapter 89. The Ombudsman reviews administrative rules to make sure they are consistent with the State Plan and works with various parties to provide balance between protecting the environment, protecting public health, and providing for growth.

A new round of cross acceptance was initiated in 2004 after the State Planning Commission approved the release of the Preliminary State Development and Redevelopment Plan. Cross acceptance allows localities to compare their Master Plans with the State Plan to determine if changes are needed to achieve more consistency between the Master Plans and the State Plan.

Governor McGreevey resigned his position in 2004 and was replacing by an acting governor until 2006. John Corzine was elected to the position in 2006. Unfortunately, planning and smart growth were not on his agenda. As such, a lack of leadership contributed to the lack of implementation of the State Plan.

The State Planning Act, NJ Statute Section 52:18A-196 (2006) acknowledged the importance of coordinating statewide planning with local and regional planning. The New Jersey State Legislature also acknowledged, in Section 52:18A-196(c), that 'it is of urgent importance that the State Development Guide Plan be replaced by a State Development and Redevelopment Plan designed for use as a toll for assessing suitable locations for infrastructure, housing, economic growth and conservation'. This Plan would balance development and conservation objectives to meet the needs of the state. The objectives of the plan can be found in Box 11.10.

One noteworthy effort is still under way in New Jersey. The New Jersey Department of Transportation's Transit Village Initiative, originally started in 1999, is designed to reward revitalized and redeveloped communities around mass transit stations. These areas are taking advantage of the existing infrastructure and public transit. As of 2012, there were twenty-six designated Transit Villages in New Jersey. Designated areas receive priority status for state funding and technical planning assistance as well as being eligible for grants from the New Jersey Department of Transportation.

Chris Christie was elected Governor in 2010. He has pledged to continue discouraging urban sprawl.

In 2011, he released a new State Strategic Plan which proposed a plan for how the state would invest in sustainable economic growth. Within the document, four state goals were identified: targeting economic growth, effective planning for vibrant regions, preserving and enhancing critical state resources, and aligning state government agency plans with the new State Strategic Plan. The problem of the ongoing loss of open space continues to plague New Jersey.

Residents of New Jersey continued to be interested in controlling growth. In a 2011 survey, residents continued to support statewide proposals for smart growth. They want to repair the existing public infrastructure and not build new roads, protect the state's drinking water, and preserve open space and farmlands.

BOX 11.10 NEW JERSEY STATE DEVELOPMENT AND REDEVELOPMENT PLAN

The State Development and Redevelopment Plan shall be designed to represent a balance of development and conservation objectives best suited to meet the needs of the State. The plan shall:

a. Protect the natural resources and qualities of the State, including, but not limited to, agricultural development areas, fresh and saltwater wetlands, flood plains, stream corridors, aquifer recharge areas, steep slopes, areas of unique flora and fauna, and areas with scenic, historic, cultural and recreational values;

b. Promote development and redevelopment in a manner consistent with sound planning and where infrastructure can be provided at private expense or with reasonable expenditures of public funds. This should not be construed to give preferential treatment to new construction;

c. Consider input from State, regional, county, and municipal entities concerning their land use, environmental, capital, and economic development plans, including to the extent practicable any State and regional plans concerning natural resources or infrastructure elements;

d. Identify areas for growth, limited growth, agriculture, open space conservation and other appropriate designations that the commission may deem necessary;

e. Incorporate a reference guide of technical planning standards and guidelines used in the preparation of the plan; and

f. Coordinate planning activities and establish Statewide planning objectives in the following areas: land use, housing, economic development, transportation, natural resource conservation, agriculture and farmland retention, recreation, urban and suburban redevelopment, historic preservation, public facilities and services, and intergovernmental coordination.

Source: NJ Statute Section 52:18A-200

Conclusion

> The status of planning . . . has been substantially altered by the adoption of a state land and growth management system.
>
> (DeGrove 1984: 389)

State involvement in growth management has increased significantly as the needs to overcome the inherent problems of local land use planning have become apparent. Though it is premature to declare the 'revolution' suggested by Callies and Bosselman, this involvement and the ways in which it has been sustained, developed, and imitated do constitute a remarkable change in the attitudes of states to land use planning.

In addition to the seven states discussed here, several more have passed or proposed legislation. For example, Rhode Island passed a Comprehensive Planning and Land Use Regulation Act in 1988 (DeGrove and Miness 1992). This requires consistency between every local government's comprehensive plan and the state's comprehensive plan. Local plans are reviewed by the Department of Administration for consistency with the Act. Any disagreements are decided by a Comprehensive Appeals Board, which can if necessary substitute a plan of its own (a unique provision). In 2001, Governor Lincoln Almond, by Executive Order, created a Growth Council which would provide guidance to local communities on land use matters. Low-density development continued to plague the state. The need to create more compact development was evident to voters. In 2006, the state adopted Land Use 2025. This plan recognized that the state had gotten away from developing dense urban centers. The plan also identifies an Urban Service Boundary. Development is to be concentrated inside the boundary. This will serve to increase density and get away from the low-density, scattered large-lot development that has caused taxpayers a lot of money to service. The key words are to create compact and mixed-use development.

Utah passed a Quality Growth Act in 1999 which gave incentives to areas establishing quality growth areas and encouraging growth in these areas instead of the fringes of urban areas. The legislation also created a Quality Growth Commission which stressed the needs of promoting intelligent growth and for developing intergovernmental cooperation on matters of growth. On January 1, 2004, Governor Olene Walker, through an Executive Order, launched the Utah Quality Growth Communities Program. This program would designate so-called 'Quality Growth Communities' which would receive various incentives for planning and developing quality growth programs. These incentives included preferential access to state resources and better rates on infrastructure loans. Nineteen communities received this designation in 2004. A Twenty-first Century Communities Program was the rural counterpart to the Quality Growth Communities Program.

Maine also passed a Comprehensive Planning and Land Use Regulation Act in 1988. Its goals can be found in Box 11.11. In 1995, the program was moved from the State Legislature to the State Planning Office. Its new emphasis was to focus on smart growth; especially preventing sprawl and directing state capital funding to designated growth areas. An early evaluation of the program in 2003 indicated that all levels of governments needed to work together and coordinate public investment opportunities in transportation. Another evaluation took place in 2006 which examined what had transpired since the enactment of the Act in 1988. In this evaluation, called 'Resolve 73', state officials called for a complete study of state laws, policies, etc. in the areas of land use planning, management, and regulation. The ultimate goal of the evaluation was to improve the statewide planning process and how the various bodies deal with growth and development. A major conclusion of the 2006 evaluation was to focus on issues of regional and statewide significance. As of 2007, 287 towns in Maine had comprehensive plans that were consistent with the 1988 Act. These plans are evaluated every four years to make sure they continue to meet the goals and purposes of the Act. The latest activity in Maine occurred on March 13, 2007 when Governor John Baldacci signed an Executive Order creating the Governor's Council on Maine's Quality of Place. This council is charged with examining how land use decisions are made within Maine and how the state can prevent the continuing problem of sprawl since approximately 70 percent of the state's growth occurs in rural areas. Its goal was to strengthen the state's economic and social vibrancy in its rural and urban places. A 2011 report was issued detailing how it could achieve this goal through integrated bicycle and pedestrian connection.

Washington State followed in 1990 with its Growth Management Act, which requires comprehensive plans for populous and other fast-growing local governments. These comprehensive plans must designate growth areas. There must also be consistency between the local and county plans. Moreover, as in other areas, the needed infrastructure must be concurrent with development. Amendments to the Act have been adopted in most legislative sessions since 1992. In 1991, Growth Management Hearing Boards were created to rule on any disputes that surface from county and city growth-planning policies.

The 1990 Growth Management Act contained Administrative rules and guidelines designed to help localities adopt their comprehensive plans. During the 2007–2009 period, the Washington State Department of Community, Trade and Economic Development will be updating the administrative rules implementing the Act. These new rules will assist jurisdictions to interpret the requirements of the Act.

In 2006, Governor Chris Gregoire established the Smart Communities Award, which acknowledges how areas throughout Washington are developing effective growth management planning strategies. The 2007 winners included projects focusing on public-private partnerships, a neighborhood investment strategy, a capital facilities plan, a downtown sub-area plan, a redevelopment project, and an area action plan. The Smart Communities Award continues to recognize the contribution of local leaders to promoting smart growth in 2012.

Tennessee, in 1998, passed a law (Public Chapter 1101) requiring all cities and counties, through a coordinating committee, to develop and adopt comprehensive growth plans, including designating urban growth boundaries, planned growth areas (where growth is expected outside of incorporated

---

BOX 11.11 MAINE STATE GROWTH MANAGEMENT ACT OF 1988 GOALS

1 To encourage orderly growth and development in appropriate areas of each community, while protecting the state's rural character, making efficient use of public services, and preventing development sprawl;

2 To plan for, finance, and develop an efficient system of public facilities and services to accommodate anticipated growth and economic development;

3 To promote an economic climate, which increases job opportunities and overall economic well-being;

4 To encourage and promote affordable, decent housing opportunities for all Maine citizens;

5 To protect the quality and manage the quantity of the state's water resources, including lakes, aquifers, great ponds, estuaries, rivers, and coastal areas;

6 To protect the state's other critical natural resources, including without limitation, wetlands, wildlife and fisheries habitat, sand dunes, shorelands, scenic vista, and unique natural areas;

7 To protect the state's marine resources, industry, ports, and harbors from incompatible development and to promote access to the shore for commercial fishermen and the public;

8 To safeguard the state's agricultural and forest resources from development which threatens those resources;

9 To preserve the state's historic and archaeological resources; and

10 To promote and protect the availability of outdoor recreation opportunities for all Maine citizens, including access to surface waters.

Source: 30 MRSA Section 4312, sub-3 (1989)

municipalities), and rural areas that are reserved for such activities as farming and forestlands. There were incentives for having approved plans as well as penalties for not completing the growth plans. These plans must be consistent with all provisions of Public Chapter 1101. A main area of activity in Tennessee is the power of cities to annex territory.

There is no formal state land use planning body that is vested with the authority to conduct comprehensive long-range statewide planning. The state continues to experience the problems associated with sprawl. Ultimately, land use decisions impact upon development decisions. They affect local revenues, the costs of services, the availability of alternative transportation options, and possible environmental degradation.

Arizona has passed several pieces of legislation designed to save open space and to help manage growth. Faced with continuing growth, a Growing Smart Act was passed in 1998 to help meet the state's objective of saving open space and managing growth. In 2000, Senate Bill 1001, better known as 'Growing Smarter Plus', was passed. It sought to help cities and counties plan better for growth. Both pieces of legislation gave cities and counties more tools to plan more effectively for growth. In February 2001, Governor Jane Dee Hull appointed a Growing Smarter Oversight Commission/Council which would monitor how Arizona was doing in implementing its smart growth programs and would make recommendations on how to make programs better and to suggest any new programs that might better achieve state objectives. In 2004, Governor Janet Napolitano asked the commission to recommend guiding principles and recommendations to help Arizona develop 'quality growth'. It was hoped that the principles would help in coordinating activities between the state and local governments, developing best practices and performance criteria to determine if state and local planning goals are being met. The commission/council issued guiding principles in 2006 on ways to better manage the explosive growth of Arizona. These principles dealt with the following areas: responsibility and accountability, preservation of community character, stewardship, opportunity, infrastructure, and economic development. Ultimately, the commission/council

called for the removal of any barrier that prohibited local governments from planning and managing growth intelligently.

Colorado has seen a great deal of activity surrounding smart growth. Numerous pieces of legislation have been introduced but failed to gain passage. However, in 2000, Governor Bill Owens signed a number of pieces of legislation that serve as the cornerstone of Colorado's smart growth policy. An Office of Smart Growth was created in 2000 whose purpose was to provide financial and technical assistance to help local governments address growth issues. The office also offers Heritage Planning Grants for communities cooperatively planning to manage growth. As of 2007, over $1.8 million has been awarded to communities since 2000. Due to state budget cuts, funding is not available for this program.

Growth management has become concerned with far more than channeling urban growth in desirable directions. It has necessarily involved a large number of regional policy issues. These range from concerns for the protection of land *against development* (including agricultural land, natural resources, fragile environments, and amenity) to concerns for *the promotion of development*, such as housing, transportation, and economic growth. (Some of these have been discussed in this chapter, and others are discussed in later chapters.) This broadening of interest is not accidental: growth management is inherently a governmental process which involves many interrelated aspects of land use (see Box 11.12 for the elements of growth management). The process is essentially coordinative in character since it deals with reconciling competing demands on land and attempting to maximize locational advantages for the public benefit. This can be done adequately only if all the relevant factors are taken into account. To illustrate: a narrow approach could lead to a worsening of the problem of affordable housing. This was a widespread concern in the early days; for example, Oregon's growth boundaries were initially criticized for being likely to increase land prices and thus housing costs. In fact, the opposite has occurred because the densities within the boundaries were increased: Oregon's goals were intentionally comprehensive, and included issues such as housing which are essential elements of the well-being of the state. Any growth management approach which omitted concerns for such vitally important aspects of the socio-economic life of the state would be not only inadequate, but also unacceptable.

Acceptability across the spectrum of interests is the key characteristic of successful growth management policies. Securing this acceptability is difficult, enormously time consuming, and fraught with political problems. Moreover, it is an ongoing process: the determination of land uses, the timing of development, the coordination of development with the provision of infrastructure all involve continuing debate and planning, the achievement of consensus, and the provision of adequate finance. In short, growth management is a major part of the continuing process of government.

The importance of acceptability stems not simply from the dictates of a democratic system, but also from the necessity for cooperation in implementation on a regional basis. Without the necessary cooperation, the system will not work. It also needs strong public support for both the policies and the taxation required to finance them. Many of the difficulties facing growth management policies have stemmed from a lack of sufficient support, particularly with funding. The continuing support for Oregon policies is in no small part due to the emphasis placed on citizen involvement. (It is significant that Florida has followed Oregon with its '1000 Friends', who perform an active role in monitoring both local and state activities in growth management.)

One particular problem of acceptability has frequently arisen in connection with state agencies (Wickersham 1994: 543). It is curious, but true, that a state often has acute difficulty with its own agencies. Having been established with specific goals to do a specific job, they can be loath to compromise their mission by taking on wider considerations. They are specially designed to carry out their particular functions; they have specialist staff for these purposes; they have their own political supporters; and they often resist 'compromising' their work by taking on extra – and perhaps conflicting – objectives. More apparent is the conflict between state (and regional) goals and the objectives of individual local governments. This is the hub of the growth management machine: in the final analysis, it is the local governments which operate most land control policies. They have to be persuaded to accept not only limitations on their actions, but also a subjugation of these to wider interests – hence the importance of 'acceptance', 'conformity', 'consistency', and similar concepts in the lexicon of growth management. Techniques to make these work are limited: they range from bribery to force; but generally they involve a great deal of debate. All planning requires a lot of talk, but none as much as issues which involve reconciling local interests with wider goals.

It is too early to judge the impact of state involvement in land use planning. It is uncertain whether it is an expanding sphere of government which has established strong roots, or a temporary burst of activity which will not last. However, there is now much more

---

BOX 11.12 ELEMENTS OF GROWTH MANAGEMENT

1 Consistency among governmental units;
2 Concurrency: requiring infrastructure to be provided in advance or concurrent with new development;
3 Containment of urban growth: the substitution of compact development for urban sprawl;
4 Provision of affordable housing;
5 Broadening of growth management to embrace Economic Development (the 'managing to grow' aspect);
6 Protection of natural systems, including land, air and water, and a broadened concern for viability of the rural economy.

Source: Based on DeGrove and Miness 1992: 161

planning activity by states and much more inter-governmental cooperation than even a decade ago; and there are indications that interest in effective growth management is increasing. The effectiveness itself, however, is less clear. Moreover, the states discussed in this chapter are, of course, exceptional: otherwise it would not be interesting to write about them. Overall, there seem to be some grounds for cautious optimism in a limited number of states.

However, even this cautious optimism has to be qualified by an issue of overwhelming importance: the increased social fragmentation of the metropolitan areas. As the flight to the suburbs continues, the problems from which so many are fleeing thereby get worse. These wider issues of growth management demand a higher political profile than they usually receive. In this respect, there are few grounds for any optimism.

Not every state is engaged in developing state land use plans that deal with growth management. State environmental regulations and infrastructure policies may serve as mechanisms to control growth. Some states might leave it to the lower levels of government since state constitutions might reserve the power to plan to these governments. Moreover, in tight economic times, cities might opt to pursue economic development and growth. There is no doubt that some areas have accepted environmental degradation and other problems associated with growth at the expense of creating new jobs.

Further reading

Hawaii's Land Use Commission is still active and illustrates the highly centralized government structure of the island (http://luc.state.hi.us). Maps, pending petitions and an overview of the Commission can be found on its website.

Metro, Oregon's regional government, offers an explanation of many land use issues on its website (http://www.metro-region.org). By using the website search tool, a visitor can review the concepts that are behind the urban growth boundary, its 2040 growth concept and other issues.

Oregon's Department of Land Conservation and Development (http://www.oregon.gov/LCD) has an attractive website that provides visitors with an overview of the department, including state statutes and publications on land use planning in Oregon. Along with information about the Citizen Involvement Advisory Committee, the website enumerates statewide planning goals with internet links to information about each one.

The Governor of California has an Office of Planning and Research (http://www.opr.ca.gov). This office coordinates a state-level review of planning and environmental documents, provides legislative analysis to the governor, researches policy changes, and also advocates for small businesses. It is a virtual warehouse of information and regulations.

The California Coastal Commission (http://www.coastal.ca.gov) offers a live internet webfeed of its meetings. Its website also provides publications and information on programs run by the commission.

The quotation at the head of the chapter is from Bosselman and Callies (1972) *The Quiet Revolution in Land Use Control*. A review up to the end of the 1970s was written by Callies in 1980, 'The quiet revolution revisited'. At about the same time there appeared the full-length study by Healy and Rosenberg (1979) *Land Use and the States*, followed by DeGrove (1984) *Land, Growth and Politics*. DeGrove's original work has been updated in DeGrove and Miness (1992) *The New Frontier for Land Policy: Planning and Growth Management in the States*. A review of state systems is provided by Wickersham (1994) and covers Florida, Georgia, Maine, Maryland, New Jersey, Oregon, Rhode Island, Vermont, and Washington. Zovanyi (1998) covers state legislation regarding growth management in *Growth Management for a Sustainable Future*. On the individual states DeGrove provides the most detailed account up to the beginning of the 1980s, while DeGrove and Miness update this (selectively) to the beginning of the 1990s.

Legal materials are included in Callies *et al.* (1994) *Land Use: Cases and Materials*.

There is a constant stream of books on state growth management policies. The reader should check the latest. Those used in the preparation of this chapter (in date order of publication) are: DeGrove (1984) *Land, Growth and Politics*; Porter (1992) *State and Regional Initiatives for Managing Development*; Buchsbaum and Smith (1993) *State and Regional Comprehensive Planning*; Stein (1993) *Growth Management: The Challenge of the 1990s*; Abbott *et al.* (1994) *Planning the Oregon Way*; Weitz (1999) *Sprawl Busting: State Programs to Guide Growth*; DeGrove (2005) *Planning Policy and Politics*; Ingram, Carbonell, Hong, and Flint (2009) *Smart Growth Policies: An Evaluation of Programs and Outcomes*.

Accounts of the policies of individual states are:

California: DeGrove (1984); Stein (1993); Fischer (1985) 'California's coastal program: larger-than-local interests built into local plans'; Fulton and Shigley (2012) *Guide to California Planning*.

Colorado: DeGrove (1984).

Florida: DeGrove (1984); DeGrove and Miness (1992); Porter (1992); Stein (1993); Audirac *et al.* (1990) 'Ideal urban form and visions of the good life: Florida's growth management dilemma'; Koenig (1990) 'Down to the wire in Florida: concurrency is the byword in the Nation's most elaborate statewide growth management scheme'; Catlin (1997) *Land Use Planning, Environmental Protection and Growth Management: The Florida Experience*. See also, Nicholas and Steiner (2000); Ben-Zadok (2005); Chapin *et al.* (2007) *Growth Management in Florida: Planning for Paradise*.

Georgia: DeGrove and Miness (1992); Buchsbaum and Smith (1993); and Stein (1993).

Hawaii: DeGrove (1984); Callies (1984) *Regulating Paradise: Land Use Controls in Hawaii*; Callies (1994) *Preserving Paradise: Why Regulation Won't Work*.

Maine: DeGrove and Miness (1992).

Maryland: Knapp and Frece (2007); Howland and Sohn (2007); Frece (2009); Lewis and Knapp (2012).

New Jersey: DeGrove and Miness (1992); Buchsbaum and Smith (1993); and Stein (1993).

North Carolina: DeGrove (1984).

Oregon: DeGrove (1984); Buchsbaum and Smith (1993); Stein (1993); Rohse (1987) *Land-Use Planning in Oregon: A No-Nonsense Handbook in Plain English*; Knapp and Nelson (1992) *The Regulated Landscape: Lessons on State Land Use Planning from Oregon*; Oliver (1992) '1000 friends are watching: checking out the record of Oregon's pace-setting public interest group'; Abbott *et al.* (1994) *Planning the Oregon Way*; Hibbard, Seltzer, Weber, and Emshoff (2011) *Toward One Oregon: Rural-Urban Interdependence and the Evolution of a State*; and Adler (2012) *Oregon Plans: The Making of an Unquiet Land-Use Revolution*.

Rhode Island: DeGrove and Miness (1992).

Vermont: DeGrove (1984); DeGrove and Miness (1992); and Porter (1992).

Washington State: DeGrove and Miness (1992).

Questions to discuss

1 What are the objectives of growth management policies?

2 Why are growth management policies so difficult to implement?

3 Outline a theoretically effective growth management policy.

4 Why have growth management policies widened to include such issues as housing and economic development?

5 Discuss the importance of citizen involvement in growth management policy-making.

6 Why have states become involved in growth management?

7 How can ballot measures affect state growth policies?

8 Can state growth policies impede job growth?

PLANNING AND DEVELOPMENT ISSUES

Environmental planning and policy encompasses a huge field, ranging from the disposal of household refuse and toxic waste to the protection of endangered species, from clean air and clean water to the control of vehicle emissions, from soil erosion to desertification to wetlands – to name but a few issues. Legislation abounds at both federal and state levels: the federal environmental legislation alone encompasses over a hundred statutes which have been passed during the last sixty years. More than a dozen federal agencies have major environmental responsibilities, and every state has an administrative organization for environmental protection. Any comprehensive account clearly has to be highly selective. The academic writer has the luxury, denied to the policy-maker, of being able to omit important relevant matters, and to choose those which are thought to illustrate adequately the nature and problems of environmental policy. The state level can also be largely ignored, ostensibly on the ground that state laws mirror or supplement federal provision. The states are, however, vital to the implementation of federal policies. In fact, the implementation of environmental policy operates in an intergovernmental and intragovernmental context, with each level of government playing a role.

Even with major omissions, the discussion here is very long. To make it less daunting, the main discussion of environmental policies (Chapter 12) starts with an outline of the growth of diverse environmental concerns and their culmination, during the 1970s, in the burgeoning of 'environmental policy'. The first of this new generation of policies was, ironically, an act to force government agencies to take environmental issues into account in all fields of public policy.

Following a review of this National Environmental Policy Act, three substantive areas of environmental policy are summarized: air, water, and waste. Current concerns over the continued development of coastal areas, the identification and protection of threatened or endangered species, and wetlands are also discussed in the chapter. This is followed in a separate chapter (Chapter 13) by a discussion of the problems and limits of environmental policy, including the importance of promoting environmental justice. More obviously than in many areas of public policy, the constraints imposed on the environmental policy-maker are very apparent. At the same time, the fundamental importance of underlying values is clear; paradoxically, though this is a field involving much scientific expertise, many issues are too uncertain to be left to experts.

Most land controls deal with readily measured matters such as the height and bulk of buildings, the uses which they are to serve, and perhaps (more problematically) the transportation implications. A very large number of development issues would need to be covered in a comprehensive text. Here a selection is limited to three: transportation, housing, and community and economic development. These represent three of the most important and difficult of today's development issues. Transportation is the essential 'connector' of activities. Upon its adequacy depends the efficiency and convenience of settlements. The importance of transportation in the urbanization process was discussed in Chapter 2. In Chapter 14, the focus is on its centrality in modern society, the problems to which it gives rise, the need to link transportation to land use, and the huge and expensive measures needed to deal with these problems. There is also a caution: our understanding of the complexities of metropolitan areas (within which four-fifths of the population live) is limited: in the current state of knowledge, we are unable to plan transportation systems with confidence about their adequacy or even their effects.

Housing, covered in Chapter 15, is another important development issue, but it is much more: it provides a home (which over two-thirds of households own or are buying); it represents a major item of household expenditure; it has a long life and therefore requires continued maintenance. If it is deemed inadequate, there can be drastic neighborhood effects (and these can themselves lead to lower standards of maintenance). There are many other aspects to housing: its location affects households' access to opportunities, it is a major land use, and an important source of municipal revenue. The list could easily be lengthened. Large parts of this book deal with various aspects of housing: urbanization (Part 1), land use regulation (Part 2), growth management (Part 3), and environmental quality (Conclusion). In this part, the focus is on the working of the housing market, the provision of affordable housing, and (continuing the discussion on housing discrimination) access to affordable housing. A discussion of the growing problem of home foreclosures and ways of assisting home owners to avoid foreclosures is included in this chapter.

Housing is also an issue of – and in – community and economic development, which is covered in Chapter 16. The promotion of community and economic development has for long been a concern of federal and state government. Their involvement has varied over the years, in large part according to the political and economic philosophies of the various administrations. The Community Development Block Grant has been an important measure of support for many years, and was embraced in the measures of 'community empowerment' introduced by President Clinton. Recognizing that all of the various federal and state programs will not solve the nation's community and economic development issues, President Bush encouraged that federal funding opportunities be made available to the many faith-based organizations already working on community and economic development issues. President Obama has reiterated the importance of faith-based organizations in providing needed services to community residents throughout the US. Chapter 16 provides a brief history of some of these endeavors, and highlights the different political philosophies which they reflect.

There are increasing concerns for the quality of development. Chapter 17 discusses aesthetic controls. These began with billboards, for reasons which today seem quaint (they were considered to be 'hiding places and retreats for criminals and all classes of miscreants'). Nowadays, aesthetic controls are explicitly accepted (though not always without protest) in terms of design quality. Billboards continue to attract attention and litigation, but controls now extend to building design and the skyline. Lawyers and architects tend to take a different approach to the issues, raising concerns over freedom of speech, due process and equal protection, and a taking of private property. There is a wide range of opinion throughout the country on the ethics, law, and practicability of controls. Many areas, however, seem to get along nicely, either with an absence of controls or with a system of judging what is acceptable.

The concern for aesthetic values joins with an interest in history, architecture, and culture in a movement for historic preservation: this is the subject of Chapter 18. At first this was entirely voluntary but government has assumed increasing responsibilities. This was inevitable since the issues involved are often political and constitutional. Preserving things of value from the past can raise issues of property rights, of regulatory controls, and conflicts with other governmental policies (as, for example, when a new road is proposed through a historic area). Interest in the past continues to increase and, as with natural areas, there is now a problem of dealing with very large numbers of visitors. At the same time, concerns for historic preservation have taken on a wider interest in the human heritage.

Environmental policy and planning

ENVIRONMENTAL CONCERNS

The first great fact about conservation is that it stands for development.

(Gifford Pinchot 1910)

Environmental awareness

Words change their meaning over time. Nowhere is this clearer than in the environmental field. Gifford Pinchot is sometimes referred to as 'the father of conservation' yet, as the quotation shows, used the term in a way which is quite different from that of today. Pinchot was responsible for the establishment of the US Forest Service in 1905, and had strong views on the need for managing the forests in the interests of long-term commercial development. Forests had to be managed like any other crop. Wanton exploitation was inefficient: good management involved sensible conservation. Pinchot had a strong influence on forestry policy, but his was not the only view being expressed about natural resources and the environment. Then as now, attitudes toward the environment varied widely. The traditional view had been that nature had to be conquered. Nature had to be defeated, or at least tamed. Land and natural resources seemed limitless: why conserve them? – there was always more over the next ridge. The cornucopia of the New World presented a huge market place for exploitation, development, and profit.

Pinchot's was only one of many voices speaking out against this innocent profligacy. His concern was utilitarian. Others provided more romantic, artistic, religious, and transcendental visions. John James Audubon presented the beauty of birds in his paintings (even if he shot them first); Ralph Waldo Emerson warned that 'nature cannot be cheated' (though he believed that nature had a capacity for self-healing); Henry David Thoreau embraced nature's role for the spiritual nourishment of humans (and, though he presented no ideas for implementing his philosophy, it later became the intellectual foundation of the movement for wilderness preservation). Later writers provided the beginning of a basis for action, though it was many years before this could be seen. Remarkable among these was George Perkins Marsh, who, in 1864, published *Man and Nature; or, Physical Geography as Modified by Human Action*. Far from the seducing idea of nature being self-regenerating, Marsh stressed the irreparable damage which human activity could inflict upon the land. His immediate influence was small (though Pinchot used some of his ideas), but he set out the fundamental ideas of what we now know as ecology. He prompted an increasing realization of the interconnectedness of things.

More successful in getting action were those who focused on specifics – such as John Muir, who, in addition to espousing the intrinsic value of the wilderness, campaigned for the Yosemite national park. Though his success had more to do with his close friendship with President Theodore Roosevelt than the force of his arguments, he greatly strengthened the campaign for national parks as well as the promotion of tourism as an economic incentive for preserving such areas. He also helped in the formation of the Sierra Club (1892), which grew into a major force in the campaign for preserving wilderness.

However, 'the economics of superabundance' largely prevailed until the New Deal of the 1930s saw some lurches in a new direction: the Tennessee Valley project in 1933 (a project focusing on integrated resource management through such activities as flood control, generation of hydroelectric power, erosion control, and reforestation efforts), the establishment of the US Soil Conservation Service in 1935 (an organization providing programs and services in soil and water conservation), the expansion of the public domain with new forests, and the abortive attempt to bring all federal land responsibilities together in a Department of Conservation. Though World War II intervened, the conservation ethic was now on firmer ground, and events of the postwar years gave it a strong forward impetus. These events ranged from worrying disasters to a flowering of books and articles on the environment, and from congressional action (tentative at first, but growing in strength) to bureaucratic activism.

Attitudes evolve over time, and it is seldom possible to point to a date when change can be said to have taken place or emerged. By common consent, Earth Day 1970 is the convenient date marking the culmination of a series of changes. This environmental celebration was preceded by a flood of writings critical of the way in which the environment was being maltreated. Among these were the works of Lewis Mumford (on urbanization), René Dubos (on drugs and their effect on microorganisms and 'the chain of life'), Aldo Leopold (on a land ethic), Paul Ehrlich (on overpopulation) and Rachel Carson's *Silent Spring* (on pesticides). There were many more.

Environmental awareness was increasing in other ways. Pressure groups campaigning for change mushroomed at local and national levels. David Brower achieved fame by his aggressive leadership of three bodies. First, he gave the Sierra Club a new lease of life. His style, however, proved too much even for the rejuvenated organization and he was forced to resign. He then established a new body with a name taken from a quotation of John Muir: 'The earth can do all right without friends, but men, if they are to survive, must learn to be friends of the earth.' Friends of the Earth (which bred parallel organizations in several countries) soon achieved a high profile but Brower ran

into further difficulties, and he moved again – this time setting up the Earth Island Initiative. Whatever Brower's shortcomings, there is no doubt that 'he helped rekindle the transcendental flame lit by Thoreau and Muir, and played a major role in pulling the old preservationist movement out of the comfortable leather armchairs of its clubrooms and into the down-and-dirty arena of local and national policy-making' (Shabecoff 1993: 101).

Other groups were established at this time, including the Conservation Foundation, the Natural Resources Defense Council, and the Environmental Defense Fund. By the end of the 1960s, the number of members in organizations such as the National Audubon Society, the National Wildlife Federation, the Sierra Club, and the Wilderness Society was increasing dramatically. Environmental concerns were moving to center stage.

Against such a turmoil of activity, it is unlikely that any one factor can be identified as the most important; but it is generally accepted that the emblem of the time is the unlikely one of a book on pesticides, Rachel Carson's *Silent Spring*. First published in 1962, the book offers an extraordinarily eloquent, moving, and lucid presentation of the environmental dangers of manufactured poisons (such as DDT, which was banned in 1970 under the Clean Air Act). Its dramatic message, expressed in almost poetic terms, gives it a place among the great books of the century.

The first Earth Day

However compelling Carson's arguments may now seem, they did not precipitate rapid action: the forces ranged against environmental policy were too powerful. But the increasing public awareness of environmental problems gradually put the environment on the political agenda. A number of environmental disasters added to the growing concern – and, in turn, made the public sensitive to disasters that previously might have had little publicity beyond their immediate locality: a huge spill off the California coast in January of 1969 sent vast quantities of crude oil onto the beaches of Santa Barbara and neighboring towns; the bursting into flames of Cleveland's Cuyahoga River in June of 1969; fish killed by toxic waste in the Hudson River; beaches fouled by garbage. Reports multiplied as environmental concern grew.

April 22, 1970 saw a remarkable series of activities throughout the nation, ranging from teach-ins to litter collection, and also including the pouring of oil into a reflecting pool belonging to the Standard Oil Company of California as a protest against oil slicks. On March 21, 1970, Earth Day brought together a wide range of supporters for environmental protection calling for the celebration of nature and life on the planet. It signaled an important change in environmental politics. Concern for the environment was no longer restricted to a few: its broad base demanded a political response that Congress was quick to recognize. So many

politicians joined in the Day's activities that Congress was forced to close down.

Congress had already passed the National Environmental Policy Act (NEPA) requiring all federal agencies to take account of environmental factors (see Box 12.1); many more environmental laws followed. This period of frenetic law-making was quite exceptional: it was contrary to the normal incremental approach which distinguishes the political process. Nor was the passing of legislation the only evidence of the new environmental activism. The designated area of wilderness increased from 10 million acres to 23 million acres. Seventy-five parcels of land were added to the National Park Service over the same period; and the National Wildlife Refuge System grew similarly. In such ways were environmental policies pursued.

BOX 12.1 NATIONAL ENVIRONMENTAL POLICY AND RESPONSIBILITIES OF THE FEDERAL GOVERNMENT

Sec. 101–42 USC. Section 4331

(a) It is the continuing policy of the Federal Government, in cooperation with State and local governments, and other concerned public and private organizations, to use all practicable means and measures, including financial and technical assistance, in a manner calculated to foster and promote the general welfare, to create and maintain conditions under which man and nature can exist in productive harmony, and fulfill the social, economic, and other requirements of present and future generations of Americans.

(b) In order to carry out the policy set forth in this Act, it is the responsibility of the Federal Government to use all practicable means, consistent with other essential considerations of national policy, to improve and coordinate Federal plans, functions, programs, and resources to the end that the Nation may:

1 Fulfill the responsibilities of each generation as trustee of the environment for succeeding generations;
2 Assure for all Americans safe, healthful, productive, and aesthetically and culturally pleasing surroundings;
3 Attain the widest range of beneficial uses of the environment without degradation, risk to health or safety, or other undesirable and unintended consequences;
4 Preserve important historic, cultural, and natural aspects of our natural heritage, and maintain, wherever possible, an environment which supports diversity, and variety of individual choice;
5 Achieve a balance between population and resource use which will permit high standards of living and a wide sharing of life's amenities; and
6 Enhance the quality of renewable resources and approach the maximum attainable recycling of depletable resources.

THE NATIONAL ENVIRONMENTAL POLICY ACT

The National Environmental Policy Act has a potentially important role to play in an integrated effort to achieve sustainable development.

National Commission on the Environment 1993

Environmental control of federal programs

Much environmental policy takes the form of controls operated by government over the actions of private bodies. In modern societies, however, a great deal of activity is undertaken by government itself, operating through a profusion of agencies. The range and diversity of this activity is enormous – from the development of natural resources to the building of roads, from the dredging of harbors to the administration of national parks, from the promotion of funding of social programs to the conduct of military operations. Thus the governmental machine which is responsible for protecting the environment from unacceptable private actions is itself responsible for a huge number of activities which can equally affect the environment – and, in some cases (as with nuclear power or military investments) can be particularly hazardous to the environment. In the real world, there is no guarantee that a governmental agency will act in a way that safeguards the environment. On the contrary, there is abundant evidence that, given the choice, a governmental agency will seek to achieve its specific, narrow objectives without regard for wider public considerations. Concern for environmental matters will normally be ranked as subordinate to the objectives for which the agency has been established. Any doubt about the validity of this contention would be settled by examining the Department of Energy's 'gross mismanagement', deliberate deception, and suppression of information on its fourteen military nuclear facilities – including the concealment of major accidents (Rosenbaum 2007: 122).

Thus there is a real problem: how is the environment to be protected from unacceptable actions on the part of government? Or, to put the matter more vividly, even if more loosely: who controls the controllers? There is, unfortunately, no ready solution to this conundrum. The art of government is not akin to driving a machine: it is a highly diffuse process that at best is extremely difficult to manage, and at worst is beyond control. It necessarily operates by dividing its functions into manageable parts with specific responsibilities. An agency established to carry out a particular function cannot be required to give a higher priority to the protection of the environment: this would compromise its very *raison d'être*. All that can be done is to devise a mechanism that obligates governmental bodies to give full and serious attention to environmental factors in the course of carrying out their functions.

This is what NEPA does (and many states have similar provision in relation to state government). It requires all federal agencies 'to the fullest extent possible' to carry out their functions in accordance with environmental policies which are set out in broad terms in the Act. For this purpose, there are a number of 'action-forcing' procedures, of which the most important is the requirement for an environmental impact statement (EIS) in connection with any federal action 'significantly affecting the quality of the human environment'.

The components of an EIS can be found in Section 102 (42 USC. Section 4332) I:

I   the environmental impact of the proposed action,
II any adverse environmental effects which cannot be avoided should the proposal be implemented,
III alternatives to the proposed action,
IV the relationship between local short-term uses of man's environment and the maintenance and enhancement of long-term productivity, and
V   any irreversible and irretrievable commitments of resources which would be involved in the proposed action should it be implemented.

The EIS is a *procedural* mandate or requirement: it is for the agencies themselves to determine the implications of the EIS for the proposed action. At first sight, this may seem to give agencies a remarkable degree of freedom to interpret NEPA as they wish, since it is each agency itself, and not any superior controlling body, that has the responsibility for deciding what action, if any, is required following an EIS. Though this is true, there are several qualifications. Federal agencies have to abide by regulations made by the Council on Environmental Quality (CEQ): a body established by NEPA to oversee the implementation of the Act. These regulations have to be followed rigorously. The procedural rules relating to an EIS (which are outlined below) dictate a process of thorough examination and reporting which the agencies must follow *to the fullest extent possible*. Public involvement plays an important role in this process, and appeals can be made to the courts. NEPA involves a complex process in which power does not rest in any single place: the agencies, the public, the courts, the CEQ and other government departments including the Environment Protection Agency (EPA) all play a role. There is thus a typical system of dispersed power.

Federal organization for NEPA

There are two federal agencies that have the responsibility for the working of NEPA: EPA and CEQ. Though they have different functions, it is simpler to regard the two bodies as sharing responsibility for the development and oversight of national environmental policy.

The EPA is the largest federal regulatory agency, and it has very wide-ranging environmental management responsibilities, including the responsibility for EIS review. It was established by executive order of President Nixon in 1970, and it carries a huge administrative burden. Nixon perceived a need to rethink and reorganize the federal government's role in environmental policy. Instead of various piecemeal initiatives, he called for the creation of a single agency that would consolidate the environmental programs of numerous agencies. Among its many functions, EPA receives all environmental impact statements and checks them for completeness. More significantly, it reviews statements for their adequacy 'from the standpoint of public health or welfare or environmental quality'. Any EIS that is judged to be unsatisfactory is referred to CEQ for resolution.

CEQ is responsible for environmental policy coordination, monitoring and reporting on environmental quality, and the working of NEPA. It was established as a body within the Executive Office of the President. As such, it has a position of power and can act as 'the environmental conscience of the executive branch', though how real this is will depend on the stance of the president. (A current awareness service is required to keep up to date on such matters.) Its chair is appointed by the president and serves as a key senior advisor to the president on environmental policies and initiatives. The CEQ has several functions, including advising the president on environmental matters; producing an annual report; the monitoring of environmental trends; and coordinating and overseeing federal agencies and their compliance with environmental policies. Its regulations define the ways in which NEPA is implemented. These regulations set out the details of the environmental review process.

The procedures have to be taken seriously, and environmental values must be pursued 'to the fullest extent possible'. Thus, to quote from one of the multitude of court cases on the Act (*Calvert Cliffs*), there is no 'escape hatch for footdragging agencies': NEPA imposes a duty on federal agencies that cannot 'be shunted aside in the bureaucratic shuffle'.

The environmental review process

Perhaps the key section of NEPA is its requirement for all agencies of the federal government to integrate environmental considerations into their operations. In essence, it is a tool designed to help decision-makers make better decisions. Any federal action is subject to NEPA if it is qualified as 'major' and can have a significant impact on the environment. This includes the application of policies, new legislative proposals, adoption of plans and programs, and the approval of specific projects. This broad approach covers not only direct action by a federal agency, but also any action taken by other public or private bodies which are funded by, or require the approval of, the agency.

There is no exemption from the NEPA mandate unless Congress has explicitly made an exception (as it did with the closure of certain defense bases) or unless there is a 'clear and unavoidable conflict in statutory authority'; for example, where an agency is statutorily required to take action so rapidly that it is impossible to prepare an EIS within the time frame. The courts have made it clear, however, that they will not allow this to be a loophole; and they have been kept busy dealing with unacceptable claims by agencies!

A special procedure applies to cases in which it is unclear whether or not an EIS is required (see Figure 12.1). These arise where it is not evident whether an action will have a significant effect on the environment, or where identified significant effects can be 'mitigated' satisfactorily. The procedure involves preparing an *environmental assessment* (EA). This must provide sufficient evidence to demonstrate whether the action will not (or will) have a significant environmental impact. If the EA shows that there will be no significant impact, the agency is not required to prepare an EIS. If, on the other hand, the conclusion is that there will be a significant impact which cannot be mitigated, a full EIS follows. Where mitigation is decided upon, the agency must conform to additional requirements for public review. A finding of no significant impact is known in the trade as a FONSI – or, as the case may be, a 'mitigated FONSI'. Such is the way in which the rhetoric of environmental policy is translated into the language of bureaucracy.

According to CEQ regulations, 'the primary purpose of an EIS is to serve as an action-forcing device to insure that the policies and goals defined in the Act are infused into the ongoing programs and actions of the federal government' (Bass and Herson 1993: 122). It is thus not the production of a passive documentary analysis of environmental impacts: it is intended as an important part of the decision-making process. It must not be used to rationalize or justify a decision already taken. It therefore has to be carried out in advance of a decision – and early enough to influence the decision. Indeed, its coverage extends beyond the proposed action to embrace reasonable alternatives. Moreover, a process known as *scoping* is required: this is a public process which should start very soon after a decision

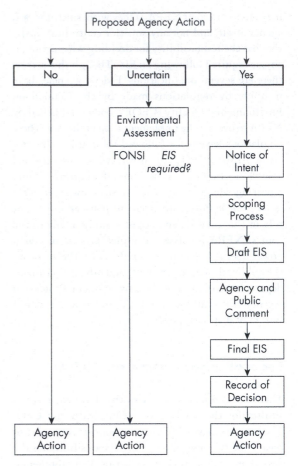

Figure 12.1 The NEPA environmental review process

Source: Based on Bass and Herson 1993: 15

has been taken to carry out an EIS. It seeks the views of the public and other agencies on what is to be covered in the EIS, and the alternatives to be examined. Generally, a draft EIS is required, and public hearings may also be necessary. Whatever the detailed requirements, the overriding purpose is to ensure that there is the widest possible input from other agencies and the public.

It is important to appreciate that at no point do the procedures provide for adjudication on an EIS. The system is essentially one of publicity, inquiry, discussion, and negotiation. When all this is completed, the agency takes its decision (which could be to reject environmentally preferable alternatives because of overriding non-environmental matters). The decision has to be presented in a manner laid down in the regulations: a written *record of decision*, which is available to the public, has to explain the decision, the alternatives that were considered, and any mitigating measures which have been adopted.

Agencies are not forced to accept a view that they think conflicts with their own interpretation of the findings of the EIS; nor is an agency required to adopt an environmentally preferred alternative or measures of mitigation (though explanations are required in the EIS). NEPA deals only with procedural matters: it imposes no duties concerning the protection of the environment. Moreover, there are no statutory powers of enforcement. Such enforcement as exists lies with the courts, and therefore in the last resort with the alertness and resources of the public in general and of interest groups in particular.

Congress has thus clearly relied upon lawsuits as an important way of obtaining agency compliance with NEPA. Against this background, it is not surprising that the courts have had a heavy load of cases dealing with NEPA issues. Thousands of cases have been filed, and almost every federal agency has been involved in litigation. Some have been more involved than others: transportation has been a particularly lively area.

With the legal process forming such a significant part of the NEPA system, the issue of who may bring a case before the courts is an important one. The standing issue has given rise to a great deal of litigation in environmental cases (Findley and Farber 2000: 2). Briefly, to have standing under NEPA, plaintiffs have to be able to assert that there has been an injury to a part of the environment that they use. The injury has to be shown to be causally related to the allegedly illegal action on the part of a federal agency. Thus, when Walt Disney Enterprises wanted to build a $35 million resort in an area of great natural beauty in the Sierra Nevada Mountains, the Sierra Club was able to claim that its members would be injured by the development since they would no longer be able to roam through an unspoilt wilderness.

There are several points to note here. First, there was an environmental interest at stake and, since NEPA protects environmental interests, the plaintiffs had standing. Second, the plaintiffs were asserting their own interests – not those of a third party; they could thus claim direct 'injury'. (Related to this is the rule that it is not acceptable to claim a generalized interest – such as 'humankind' or 'the poor' or 'recreationalists'.) Third, the 'injury' was remediable by a favorable judgment. These are three requirements for a case to be successfully prosecuted.

Generally, the Supreme Court has consistently ruled that an agency decision cannot be set aside because of its effect on the environment. This logically follows from the position that NEPA is purely procedural. It can be argued that the court has been unduly narrow in its approach to this issue: a triumph of form over content. Nevertheless, the EIS process is important not only for facilitating legal challenges, but also for serving broader purposes. It alerts environmentalists to the disclosed implications of administrative issues which otherwise might not have been apparent; it serves as an early warning system for newly arising issues; and it compels agencies to carry out discussions and maintain contact with environmental groups. In these ways, EIS has helped to bring environmentalists into the policy arena: they now have a role that has been legitimated and facilitated. This success of environmental groups was noted by business interests who followed a similar pattern of organization, research, and lobbying. Particularly noteworthy has been the effectiveness of the not-for-profit legal foundations (such as the Pacific States Legal Foundation) that have participated in both negotiation and litigation. The EIS thus has brought about a greater degree of participation in the process of designing environmental regulations. It has thereby subscribed to the democratic principle that all who are affected by public policy should play a part in determining what this should be.

A number of states have enacted state legislation similar to NEPA. In 1970, California enacted the California Environmental Quality Act (CEQA). Government agencies are required to consider the environmental consequences of proposed projects prior to approving plans and policies for the project. California Public Resources Code Section 21065 defines 'project' as:

> An activity which may cause either a direct physical change in the environment, or a reasonably foreseeable indirect change in the environment, and which is any of the following:
>
> a) An activity directly undertaken by any public agency.
> b) An activity undertaken by a person which is supported, in whole or in part, through contracts, grants, subsidies, loans, or other forms of assistance from one or more public agencies.
> c) An activity that involves the issuance to a person of a lease, permit, license, certificate, or other entitlement for use by one or more public agencies.

CEQA applies to discretionary projects – projects that require an exercise in judgment in deciding the fate of a project. It does not apply to ministerial projects – projects that are approved if the action is found to be in conformity with the standards found in applicable ordinances and laws.

Certain types of projects are exempt from CEQA. There are categorical exemptions for projects deemed not to have a significant impact on the environment, such as changes in existing facilities, replacement or reconstruction of existing structures and facilities, new construction or conversion of small structures, and minor alteration to land. There are also statutory exemptions for such items as ongoing projects, feasibility or planning studies, waste discharge requirements, and timberland reserves.

If a proposed project is not exempt from CEQA, an initial study is undertaken to determine if an Environmental Impact Report (EIR) is needed. If it is determined that the proposed project will not have any significant environmental impact, the lead agency prepares a Negative Declaration – a statement indicating that an EIR is not needed because there will be no significant adverse environmental impacts.

Should the initial study indicate the presence of significant adverse environmental impacts, an EIR must be prepared. The lead agency first prepares a notice of preparation of the EIR after it has consulted with the applicant, the appropriate agencies, and the public. A draft EIR is then prepared and is circulated to the public and various public agencies. After a review and comment period for the draft EIR, a final EIR is written. According to the CEQA Guidelines, a final EIR must contain the following information: table of contents, summary of the proposed action and its consequences, project description, environmental setting, and evaluation of environmental impacts. Private sector actions that are either approved or permitted by public agencies are also subject to CEQA.

A number of other states also require agencies to analyze the environmental impacts of their actions. Many of the laws are very similar to NEPA. According to the Code of Virginia, state agencies are required to submit Environmental Impact Reports on 'major' state projects (Section 10.1–1188). 'Major' is defined as 'the acquisition of an interest in land for any state facility construction, or the construction of any facility or expansion of an existing facility . . . hereafter undertaken by any state agency, board, commission, authority or any branch of state government, including state-supported institutions of higher learning, which costs $200,000 or more'. Washington (RCW 43.21 C.030–031) requires an Environmental Impact Statement 'for proposals for legislation and other major actions having a probable significant adverse environmental impact by all branches of the state, including state agencies, municipal and public corporations, and counties'. Other states possess similar environmental policy or environmental quality legislation.

CLEAN AIR

> 'The purity of the air of Los Angeles', an enthusiast wrote in 1874, 'is remarkable. The air when inhaled, gives to the individual a stimulus and vital force which only an atmosphere so pure can even communicate.'
>
> (Boorstin 1973)

Technics and politics

The cleanliness of air might seem, at first sight, to be essentially a technical subject. Nothing could be further from the truth: it is a highly political matter. This is partly because the technicalities are highly problematic, but also because measures to cleanse air (whether by removing contaminants or by preventing them from entering the air in the first place) involve costs and benefits that arise in different areas and therefore affect people differently. These distributional effects arise in all areas of public policy, but they are particularly troublesome with air pollution since the underlying scientific base is weak: there is a degree of ignorance about air pollution that must surprise the layman. The huge area of ignorance means that inadequate facts are capable of widely differing interpretations – the perfect base for lengthy and frequently inconclusive political argument. As will be shown later, these difficulties are by no means confined to clean air, but extend over wide areas of environmental policy.

Early clean air policies

Policies for clean air have a long history. It was as early as 1881 that Chicago and Cincinnati passed laws to control smoke and soot from furnaces and locomotives. A hundred other cities followed in the next thirty years; by 1950 the number had risen to over 250. State action came much later, with Oregon being among the first in 1952; but by 1970 all the states had passed air pollution control legislation (Stern 1982). The federal government came onto the scene in 1955, and several Acts were passed in the following years. At first, the federal role was restricted to providing financial and technical aid to the states. In the mid-1960s, however, a more positive federal role emerged in relation to cross-boundary pollution and the setting of emission standards for motor vehicles. The 1967 Air Quality Act went further, and promoted the planning of air pollution strategies. Air quality regions were to be designated to cover areas (within state borders) of interconnected air pollution problems. The states were to establish air quality standards for these areas and

develop plans for their achievement. Standards were to be based on advice from the Department of Health and Welfare on the health effects of common pollutants.

It soon became apparent that this system would not work effectively. The Department of Health and Welfare was tardy in designating air quality regions and in preparing the advice the states needed; and even where the necessary guidelines were produced, few states developed plans. By 1970, not a single state had devised a complete program for dealing with any pollutant. Moreover, the automobile industry had proved to be adept in circumventing emission controls. Against a background of mounting public concern about the environment (dramatically evidenced by Earth Day), it was clear that stronger federal action was needed.

Pollution and economic development

Much of the problem facing the states stemmed from their concern to safeguard their economic development. Any individual state that took positive action to control air pollution could be at an economic disadvantage, since new industries would naturally select the cheaper locations of the states that had no controls. Whether or not this was in fact a significant factor in location decisions, it was certainly seen as such. There was also some concern, particularly in the case of industries operating in several states, that there would be widely varying standards. Nowhere was this more important than with vehicle emission standards; and manufacturers pressed for national standards.

The problems went much wider than this, however; and they still do. First, of course, is the perennial confrontation between environmental and industrial interests. Environmentalists can evoke powerful images of ecological devastation (such as the *Exxon Valdez* oil disaster) and risk to health. Equally, those opposing environmental controls are able to rouse the specter of declining economies and the loss of jobs. Second, there are major regional conflicts, which also remain as part of the permanent political landscape. Particularly striking is the conflict between the Midwest and the

Northeast, the root of which is the competitive pro-duction of coal for power plants. The Midwest produces low-cost coal which has a high sulfur content while the Northeast produces coal which has a low sulfur content but a high cost. Sulfur is a major cause of acid rain, and it is a target of much clean air policy. The sulfur content of Midwestern coal can be reduced, but this involves a cost which reduces its economic advantage. The two regions have diametrically opposed interests in the control of sulfur. Any measure imposing costs on power plants reduces the economic advantage of the Midwest mines and increases that of the Northeast. Thus the political conundrum arises as to what controls are to be operated and who is to bear the cost.

This is the classic problem with air pollution controls; and it has other dimensions. High chimney stacks for coal-fired power stations using Midwestern coal are not very expensive. They considerably reduce the pollution in the Midwestern states; but they do so by transporting the pollution to other areas, predom-inantly in the East. As a result, a part of the real cost of the use of Midwestern coal is borne in the Northeast. But it is not only the Northeast that receives the pollution: it is spread over a very wide area. It is also mixed with other pollutants produced both locally and far afield. The resulting cocktail is made up of a variety of pollutants, in differing amounts, from different areas: it is impossible to determine the origin of the ingredients. The areas that suffer the impact of the pollutants, whether they be Northeastern cities or Northern lakes and forests, naturally view the problem as one of wide geography which should be equally widely shared. But 'clean' states do not see it this way. The states in the Sunbelt, for example, have a good proportion of modern industrial and utility facilities. Their investment in clean air has already been made: why should they contribute toward the cost of cleaning up in the dirty states?

The issues are further complicated by attitudes to the control of new pollution sources. It would seem common sense to impose stricter standards on new industries than on those already existing, since they can meet them more easily. Introducing pollution control measures when a plant is being built is considerably cheaper than adding them later. There is thus a national

gain at a relatively low cost. Such arguments are attractive to older areas since they see the extra costs as reducing the competitive advantage of clean states. The clean states, on the other hand, argue that the burdens of clean-up should be borne by the areas where the emissions are the greatest. These and similar arguments raged in the debates on the 1970 Clean Air Act, which finally imposed different requirements for new and existing plants.

The structure of clean air controls

Though the legislation has been amended considerably since 1970, this basic division continues. All new plants are required to conform to *new source performance standards* that are devised separately by EPA for each industry to take account of costs and the 'best available technology'. Existing plants are subject to EPA nation-wide emission standards known as *national ambient air quality standards*. These standards constitute acceptable levels of pollution in the ambient (outside) air. They represent national objectives that are to be met at some future date (which is subject to postponement when they prove unattainable). They relate to six major pollutants: carbon monoxide, ozone, particulate mat-ter, sulfur dioxide, nitrogen dioxide, and lead.

There are many other substances that pollute the air, but little is known about the effects that these have on health. The degree of ignorance is quite alarming: even the extent of pollution is very uncertain. Though EPA is required to identify and designate air pollutants and to establish emission standards for them, progress has been slow. This has been largely because of a lack of research. The needed research is laborious, expensive, and slow to produce results; and the 'results' tend to be of sufficient uncertainty as to create lengthy debate, particularly on the part of the firms that are responsible for the pollution and for clean-up costs.

Acid rain

Rain is naturally acid, due to the presence of natural elements, but the degree of acidity can be increased

BOX 12.2 ACIDITY

The major chemicals that produce acid rain are sulfur dioxide, nitric oxide and nitrogen dioxide (known collectively as nitrogen oxides) and hydrocarbons (and also, to a lesser extent, ammonia and carbon dioxide). Acidity is measured on a logarithmic pH scale which goes from 0 to 14. Pure water has a pH of 7; any reading above this is alkaline and any reading below it is acidic. Rainwater has a normal acidity of 5.6; damage to the natural environment is associated with a pH of 5.0 or lower; at the extreme level of 3.0 extensive damage to buildings will occur over time. A very extreme reading of 1.69 was recorded in 1984 south of Los Angeles, in the Californian town of Corona del Mar. To appreciate how severe this was, it can be compared with battery acid, which has a pH rating of 1. (Since the pH scale is logarithmic, a pH of 3.0 is ten times stronger than one of 4.0.)

by pollution to such an extent that it results in harmful environmental effects (see Box 12.2). These effects are caused by acid deposition that may be borne by air, dew, fog, and wind, as well as by rain. The term 'acid rain' is therefore not quite accurate, but it is typically used in preference to acid deposition.

The effect of acid rain varies according to local conditions such as the wetness of the ground, the rate of run-off, and the vegetation cover. In urban areas, erosion effects of acid rain differ according to the types of building materials: limestone is affected much more than granite or even sandstone. High acidity, of course, is a killer: large areas of forest and lakes in the Northeast of the United States and in Canada have suffered greatly. High smoke stacks have been responsible for carrying pollutants away from the source to far-distant areas.

The burning of fossil fuels (mainly coal and oil) is a major factor in the creation of acid rain (and thus, as frequently happens, energy policy and environmental issues and policies intertwine). Conventional power stations are the worst offenders in producing sulfur dioxide and nitric oxide. The United States is second only to the former Soviet Union in sulfur dioxide emissions, the principal offending states (in decreasing order) being Ohio, Pennsylvania, Indiana, Illinois, Missouri, Wisconsin, Kentucky, Florida, West Virginia, and Tennessee (Pickering and Owen 1994).

Ozone

Ozone is a gas that is both enormously beneficial and dangerous. In the stratosphere, it protects the Earth from harmful rays; at ground level, it can form a constituent of hazardous noxious pollutants. Its chemical composition – three atoms of oxygen (O_3), compared with oxygen's two (O_2) – makes it highly reactive. It readily combines with other substances, often thereby forming a most obnoxious mixture. Smog is its public image. Unlike air pollutants such as carbon, nitrogen, and sulfur oxides, ozone is a 'secondary' pollutant formed by reaction between primary pollutants and natural constituents of the air. These 'precursor' pollutants consist mainly of nitrogen oxides and *volatile organic compounds* (VOCs), including hydrocarbons such as benzene. The largest causes of ozone are motor vehicles, 'small stationary sources' such as paint shops and dry cleaners, and large refineries and chemical plants. Indeed, much of modern economic activity creates ozone, but it is also produced as a result of natural processes. In short, the production of ozone is extremely varied and complex, and there is much about it that is not understood. As a result, ozone is particularly difficult to regulate. Virtually every major urban area in the United States fails to meet the ozone quality standard (which many experts consider is too low for the protection of human health). Ozone is a carcinogen; it can cause bronchitis, asthma, and other pulmonary diseases; indeed, it can impair many bodily functions, and even give rise to heart failure. It is also harmful to trees, crops, and aquatic systems.

Improved vehicle emission controls have greatly reduced the impact of individual vehicles, but the improvement has been more than offset by increased traffic. It seems that there is little prospect of significant quick (or even slow) solutions to the problem, though the regulatory mechanisms affecting ozone have been strengthened and widened in the 1990 Act.

State implementation plans

Responsibility for enforcing air quality standards lies with the states, operating through their *state implementation plans* (SIPs). These SIPs show how each state will meet federal air quality standards for six 'criteria' pollutants: ozone (O_3), particulate matter (PM_{10} and $PM_{2.5}$), carbon monoxide (CO), nitrogen dioxide (NO_2), sulfur dioxide (SO_2), and lead (Pb). The individual plans allow states to adjust their controls according to the clean-up costs in their areas (and they operate in conjunction with state provisions). They impose pollution limits for geographical divisions called *air quality control regions*, detail the arrangements for control, and set out measures for clean-up (and for emergencies). The plans are subject to the approval of EPA and have the force of law. States are induced to formulate and operate their implementation plans by the power of the federal purse: grants for highway and sewer works for example can be withheld from recalcitrant states.

Air pollution varies not only among industrial plants but also, of course, among (and within) individual states. The question becomes how to treat areas that already meet national standards. Are they to be required to make their contribution to the national clean-up effort by aiming for higher standards? This may make sense nationally, but it is likely to be opposed locally: why should costs be incurred in an area that is already in conformity with national standards? The additional costs of being required to operate more stringent controls than elsewhere would (so it could be argued) unfairly create barriers to the economic development of the area. There is an alternative: the

area could be allowed to *increase* its pollution as long as this does not bring it below the national standard. The political difficulty here (much stressed by the environmental lobby) is that such a policy for clean air would deliberately and explicitly be promoting dirty air in some areas.

The 1970 Act did provide for some differentiation in regional standards that became a natural target for litigation. A case brought by the Sierra Club was decided in the Club's favor, and EPA was barred from approving any state plan that permitted increased levels of pollution in the clean areas. Though the decision was widely applauded by environmentalists, it aroused violent objection from groups concerned with economic development in the West. The latter claimed, for example, that it 'precluded further development of vast energy resources' (in New Mexico); and 'continued poverty in many rural areas' (of Utah) (Ackerman and Hassler 1981). A Bill to amend the 1970 Act was killed in 1976 by a filibuster mounted by Western senators who argued that it would seriously affect economic development and energy exploitation.

The problem of the clash between clean air policies and local economic development is one that seems likely to remain insoluble. An argument that clean air policies themselves create employment (which they certainly do) carries little weight: even if the new employment were created in the areas which suffer because of pollution controls, it would be a very happy coincidence if it provided jobs for those displaced. The intractability of the problem is evidenced by a provision in the 1990 Act for a program of compensation for workers who lose their jobs as a result of clean air policy.

Areas of severe pollution

In addition to the withdrawal of federal funds from states that failed to produce satisfactory implementation plans, the 1970 Act made provision for the imposition of federal plans. Though EPA was (for strong political reasons) very reluctant to act in default, it was forced to do so in a number of cases because of lawsuits by environmental organizations. (This

Plate 11 Los Angeles smog
Photo courtesy of Public Domain photos (http://www.pdphoto.org)

was done in Phoenix, Chicago, and several places in southern California.) (See Plate 11.)

The 1990 Act took a different approach. It was recognized that some parts of the country would have great difficulty in meeting clean air standards, and that the imposition of national standards would be unworkable (as the failure of the earlier state implementation plans testified). The solution adopted was essentially one of differential standards devised on the basis of the severity of the pollution problem. Areas with acute difficulties are designated *non-attainment areas*. This status is determined on the basis of pollution criteria for ozone, particulate matter, and carbon monoxide. The air pollution levels in these areas exceed national ambient air quality standards. Each non-attainment area has a deadline for meeting the standards, and the programs for this (state implementation plans) are geared to the severity of the problems: the more acute the problem, the more aggressive must be the program. Of particular importance here is the permit system that was introduced by the 1990 Act. This requires *all* major stationary sources to have an operating permit from the state specifying all the conditions being imposed on that source. These include not only the amount of allowable emissions, but also the requirements for monitoring and maintenance.

Vehicle emissions

Vehicle emissions are a major source of air pollution: they account for over a half of all air pollution. Though a great deal of effort has been made to reduce it, real achievements in making vehicles cleaner have been offset by increases in vehicle use. Thus, though cars produced in the mid-1990s emit between 70 and 90 percent *less* pollution than their 1970 counterparts, vehicle travel has more than doubled during this period, partly as a result of urban development patterns. At the same time, concern has grown about other, previously unrecognized, environmental threats such as acid rain and global warming. The political response has been a major strengthening of vehicle emissions control. Its political salience is epitomized by the fact that instead of standards being set in the

usual way by EPA, Congress has actually written them into the legislation.

Controls have been operated by placing most of the responsibility for reducing emissions on manufacturers and, consequently, standards have been nationally operative, irrespective of any differences in local air quality. This has now changed, and increasing responsibility has been placed on vehicle owners through state vehicle inspection programs. State implementation plans can include these and other measures such as discouraging traffic by the control of parking and encouraging carpooling (discussed in Chapter 14). The 1990 Act provides for the phasing out of lead and also for tighter tailpipe (exhaust) standards. EPA is required to study the desirability and feasibility of further changes in standards. Cars produced since 1994 are required to be equipped with 'onboard diagnostic systems' which feature dashboard warning lights that show when emission control equipment is malfunctioning. Stiff penalties face 'backyard mechanics' who tamper with emission controls. In areas with severe pollution problems special regulations apply.

California, which has acute air quality problems, has gone further than the federal government in its efforts to reduce emissions in ozone non-attainment areas. The so-called 'California Pilot Program', established by the EPA administrator, requires manufacturers to produce specified numbers of 'clean' fuel (i.e. electric) cars. These vehicles have to meet severe standards; how they are met will depend on the results of a research and development effort (that so far have been disappointing). A number of states are adopting policies based on those of California, which are sometimes described as 'technology-forcing': the standards involved are stricter than can be met with existing technology. Evidence that such an approach can be effective (even if not as quickly as its protagonists would wish) is provided by the successful development of catalytic convertors; but it seems dangerous to rely on a technological quick-fix to environmental problems – a point which is developed later in this discussion.

Californians continue to have a 'love affair' with the automobile. Nevertheless, California has continued to develop legislation designed to reduce greenhouse gases. On July 22, 2002, Governor Gray Davis signed a bill requiring the California Air Resources Board to develop greenhouse gas (carbon dioxide) standards for vehicles for automobile models starting in 2009. These standards are designed to reduce emissions and to increase fuel efficiency in automobiles sold in California.

The 1990 Act and its future

It remains to be seen how the wide-ranging and complex provisions of the 1990 Act work out in practice. Congress, having labored mightily in producing the Act, has passed responsibility to EPA and the states. Their previous record has been uneven but, given the changes in public attitudes, and the strengthened legislative framework, it may be hoped that implementation will improve. In addition to the provisions outlined above, mention should be made of the increased penalties for emission offences, most of which have been made criminal felonies. Moreover, the courts have increased powers to compel EPA and the states to comply with legal requirements; and there are extended citizen suit provisions. It has also been suggested that the workers' compensation program may bring labor and environmental interests together in a way that has hitherto been impossible because of the specter of job losses (Bryner 1993). On the other hand, Congress would be running true to form if it has set clean goals at an unrealistically high level.

Problems with particulate matter

There have always been concerns over the air we breathe. Of great concern today is the problem of particulate matter, occasionally referred to as particle pollution. The US Environmental Protection Agency (2005b: 1–3) defines particulate matter as 'the generic term for a broad class of chemically and physically diverse substances that exist as discrete particles (liquid droplets or solids) over a wide range of sizes . . . Particles may be emitted directly or formed in the atmosphere by transformations of gaseous emissions such as sulfur oxides, nitrogen oxides, and volatile organic compounds'. Once inhaled, the particles go directly to the lungs. Dirt, soil, dust, mold, and ash from vehicles, stoves, and fireplaces would be considered in this category. These particles are considered harmful, especially to individuals suffering from heart disease, emphysema, or asthma.

As noted previously in this chapter, states are required to develop SIPs to show how they are meeting national air quality standards. On March 29, 2007, the US EPA issued a rule defining requirements for state plans to clean the air (to protect both public health and public welfare) in areas with levels of fine particle pollution that do not meet national air quality standards. For particles 2.5 micrometers in diameter and smaller, states should meet the $PM_{2.5}$ standards by 2010.

The Clean Air Act has been praised and condemned (see Box 12.3). On the one hand, the Act provides an

---

BOX 12.3 PRAISE AND CRITICISM OF THE CLEAN AIR ACT (CAA)

The federal CAA has been both criticized as a cause of sprawl and praised as a useful tool to curb it. Critics contend that by barring increases in air pollution in cities where the air is unhealthy, the law drives businesses and development to outlying areas, thus increasing sprawl and the air pollution from its attendant motor vehicle travel. This is the basis for claims that the Act can have the perverse and unintended effect of increasing air pollution rather than reducing it. However, the Act's defenders argue that it actually can deter sprawl by providing incentives for transit-oriented, compact development, and urban revitalization.

Source: Moore 2001: 3

incentive for compact development that discourages the use of automobiles. On the other hand, air quality requirements are more stringent in developed areas. This may lead businesses and industries to seek out outlying areas where the air is more likely to be clean. Moreover, federal funding could be restricted for transportation projects in areas failing to meet the air quality standards. Again, funding could go to outlying areas that meet air quality standards, thereby increasing potential motor vehicle travel.

Kyoto Protocol

In 1997, a number of industrialized nations entered into an agreement to reduce greenhouse gases (carbon dioxide, methane, nitrous oxide, sulfur hexaflouride, hydrofluorocarbons, and perflurocarbons) that lead to global warming (climate change) by 2012 – the Kyoto Protocol. The Protocol went into force in 2005. As of November 2007, 175 parties have ratifed the Protocol. Each of the signatories determined its own target to reduce emissions by a certain amount. In order to promote sustainable development, each party agreed to implement policies and measures, as identified in Box 12.4. The earlier Montreal Protocol, which entered into force in 1989, sought to phase out substances responsible for ozone depletion.

The United States has not ratified the Kyoto Protocol. It signed the Protocol but never ratified because it wanted developing nations to sign it. President Bush also felt that China and India should not be exempt from the Protocol. Moreover, he felt that it would hurt the US economy. In 2002, he announced plans to cut greenhouse gas emissions by 18 percent by 2012. Nevertheless, a number of states and local governments have agreed to also reduce greenhouse gas emissions and support signing the Protocol.

BOX 12.4 RESPONSIBILITIES UNDER THE KYOTO PROTOCOL

1 Enhancement of energy efficiency in relevant sectors of the national economy;

2 Protection and enhancement of sinks and reservoirs of greenhouse gases not controlled by the Montreal Protocol, taking into account its commitments under relevant international environmental agreements; promotion of sustainable forest management practices, afforestation and reforestation;

3 Promotion of sustainable forms of agriculture in light of climate change considerations;

4 Research on, and promotion, development and increased use of new and renewable forms of energy, of carbon dioxide sequestration technologies and of advanced and innovative environmentally sound technologies;

5 Progressive reduction or phasing out of market imperfections, fiscal incentives, tax and duty exemptions and subsidies in all greenhouse emitting sectors that run counter to the objective of the Convention and application of market instruments;

6 Encouragement of appropriate reforms in relevant sectors aimed at promoting policies and measures which limit or reduce emissions of greenhouse gases not controlled by the Montreal Protocol;

7 Measures to limit and/or reduce emissions of greenhouse gases not controlled by the Montreal Protocol in the transport sector;

8 Limitation and/or reduction of methane emissions through recovery and use in waste management, as well as in the production, transport and distribution of energy.

Source: Kyoto Protocol, Article 2, 1(a)(i–viii)

Global warming

Global warming, or as it is now more commonly referred to 'climate change', continues to be a hotly contested topic of discussion. In 2006, former US Vice President Al Gore's film, *An Inconvenient Truth*, brought the issues into a world spotlight. Gore discussed the problems associated with climate change and the need to reduce the use of fossil fuels – the greenhouse effect. The film was widely praised and criticized at the same time. Some scientists applauded his bringing the topic to the international audience, while other scientists claimed the science used in the film was incorrect. They thought he was being alarmist. Nonetheless, the topic is worthy of discussion.

As previously noted, greenhouse gases (e.g., carbon dioxide, methane, and nitrous oxide) trap heat in the atmosphere. The heat cannot escape, thus, the rise in the Earth's temperature. Some gases occur naturally, others we produce. The burning of oil and coal to generate energy represents the largest source of greenhouse emissions, followed by transportation (the use of fuel in automobiles, trucks, and airplanes), and then our usage of oil and gas to heat our homes. The Earth's temperature will continue to rise as we increase our emission of greenhouse gases.

There are a variety of effects that are linked to climate change. We are witnessing the shrinkage of glaciers, causing a rise in sea level around the world. Coastal areas will be more prone to flooding. Changes in the climate around the world will affect growing seasons as well as a distributional change in animals and plants. A number of questions are being raised on the effects of climate change. Is climate change causing an increase in the number of tornadoes and hurricanes? Is climate change causing an increase in the likelihood of wildfires? Is climate change causing the extinction of some endangered species?

That the United States has not ratified the Kyoto Protocol is a fact that is disturbing to a lot of researchers. However, the United States has noted a desire to reduce greenhouse gas intensity. In his January 23, 2007 State of the Union address, Bush called for the reduction of gas usage over the next ten years. The need to develop new technologies to reduce greenhouse gases and to work with the international arena to reduce climate change has also been noted by the Bush administration.

CLEAN WATER

Ever since chemists began to manufacture substances that nature never invented, the problems of water purification have become complex and the danger to users of water has increased.

(Carson 1962)

The succession of professional approaches

Water policies illustrate rather clearly the role of fashion in the environmental policy field. Unlike clothing, however, these fashions acquire a life of their own, and persist even when it has become clear that a replacement is timely. That is because they have the backing of experts (Tschinkel 1989). The first expertise to hold sway in the development of water policy was that of physicians, for whom the problem of polluted water was solved by washing it away. By the middle of the nineteenth century, the common method for disposing of household wastes was by way of the storm drain. This proved a most effective way of reducing the spread of cholera. Unfortunately, since the raw sewage was dumped into areas from which drinking water was obtained, there were side-effects of some danger to public health. The solution to this was chlorination, which again was effective: typhoid was virtually eliminated. This public health achievement, however, was not without its own problems. Large quantities of diluted sewage cause eutrophication and contamination of bodies of natural surface water, thus wasting enormous quantities of both water and nutrients, and also raising the question as to whether the chlorine itself has harmful effects. Eutrophication is a naturally occurring process when sewage and other materials such as fertilizers are introduced into a body of water. Nitrates and phosphates essentially 'fertilize' the water and stimulate the growth of algae. As the water ages, these materials 'choke' the water and prevent light from reaching various depths of the water, deplete oxygen, kill fish, and may make the water unpalatable.

The problems of the next stage – reducing the quantity of nutrients and pathogens entering water systems – were dealt with by the engineering profession with a huge program of sewage treatment plants. Though this has partly (though not completely) solved one water pollution problem – that of *point pollution* (pollution originating from a single source) – it has been quite inadequate to deal with wider water pollution problems. Despite the enormous cost involved, EPA's Water Quality Inventory reveals that a third of all surface waters do not meet water quality standards. The reason for this is that much water pollution arises not from a particular 'point' such as a sewer or discharge pipe or ditch, but from *non-point sources*. These include agricultural run-off, leaking gasoline storage tanks, landfills, abandoned mines, run-off from irrigation and from salted roads, seepage from septic tanks, and hazardous waste sites. This frightening mixture of pollution sources, which was dramatically brought to public attention by Rachel Carson's *Silent Spring*, presents serious difficulties for scientists. Previous generations of public health and engineering specialists may have had supreme confidence in their remedies: their contemporary counterparts have no grounds for any such confidence. Though biologists may be 'waiting in the wings to solve these problems', it is apparent that there is a high degree of ignorance on both the causes of and the cures for these problems (Tschinkel 1989: 161). (To complete the record of professional succession, it may be noted that, given the degree of scientific ignorance, lawyers are now the dominant professional actor on the environmental stage.)

From this brief account of the fate of successive generations of experts, it is clear that water pollution presents a wide range of difficult analytical, technical, and political problems. As is typical of environmental issues, there is a paradox here: the complex of issues requires breaking down into its constituent elements but also demands a large degree of policy coordination. Without the former, important factors may be missed; without the latter, programs may have serious shortcomings. Regrettably, the current situation is far short of the ideal. To give one example: involved in some way with water policy are twenty-seven federal agencies, over 59,000 water supply utilities, fifty state governments, and thousands of local governments and special districts (Smith 2003: 110).

Federal water policy

Given the general abundance of water in the United States, it is not surprising that there has been no long history of water policy. The first Congressional Act dates from 1948, when federal research and funding were introduced. Later legislation extended federal responsibilities and, in 1965, states were required to establish water quality standards. It was, however, the 1972 Federal Water Pollution Control Act (strengthened by later amendments) that established the current regulatory framework. There is a separate Safe Drinking Water Act, originally passed in 1974, which regulates and sets national standards for public water supplies. It represented the primary federal law designed to protect and ensure the quality of the nation's drinking water. It has been amended on several occasions.

Section 502(14) defines a 'point source' as 'any discernible, confined and discrete conveyance, including but not limited to any pipe, ditch, channel, tunnel, conduit, well, discrete fissure, container, rolling stock, concentrated animal feeding operation, or vessel or other floating craft, from which pollutants are or may be discharged'. Agricultural stormwater discharges and return flows from irrigated agriculture are not included in the definition. Pollution from point sources is controlled by a system of EPA permits: the National Pollution Discharge Elimination System (NPDES) introduced in 1972. These are source-specific: each has to be individually determined. Enforcement is undertaken by the states, whose pollution-control programs (which are grant-aided) have to be approved by EPA. Municipal sewage treatment plants are required to meet EPA standards. Pollution from non-point sources is mainly regulated by the states, as is groundwater.

Water quality standards

The 1972 Act was a prime example of 'technology-forcing' legislation. It embodied the policy (mainly for point source pollution) 'that the discharge of toxic pollutants in toxic amounts be prohibited'. The aim was for 'fishable and swimmable' waters by 1983, and complete elimination of discharges into navigable waters by 1985. The Act set standards relating to water quality and effluent limits. The latter were based on the principle that all pollution is undesirable and should be reduced to the maximum extent that is technologically possible. Imperious deadlines were set to meet the standards based upon concepts of the 'best practicable technology', the 'best available technology economically achievable', and such-like.

With hindsight it is obvious that the standards were far from 'technical'. They had to be interpreted, and this involved negotiation with representatives of polluting industries. Since the character of the pollutants and the polluting process varied among industries, individual standards were required for different types of industry. Contrary to the intention, the realization was essentially political and, not surprisingly, the goals were not met. Nevertheless, the legislation appears to have had some positive effect in the control of water pollution, though it is difficult to assess how much! Evidence on trends in water quality is elusive, adequate data are sparse, and it is impossible to isolate the effects of particular programs. A National Water Quality Inventory Report to Congress is prepared every two years: this summarizes the information collected by the states. The information is incomplete, but it shows that about two-thirds of the waters assessed are of sufficient quality to support uses such as fishing and swimming, and therefore meet the goals of the Clean Water Act. It is clear, however, that there are some serious problems, for example of groundwater pollution (particularly from agricultural run-off) and of discharges from inadequate municipal sewage treatment facilities.

From inadequate information, all that can be said with any degree of certainty is that the nation's waters do not appear to have deteriorated in quality. Given the extent of population and economic growth, this is some degree of achievement even if it is far from the ambitious original hopes. Matters could have been far worse! Unfortunately, there are indications that some of the problems could increase. This will become clear from an examination of several areas of water pollution control policy.

Municipal treatment plants

The improvement of municipal sewage treatment plants has for long been a major target of federal policy; and some impressive progress has been made. Unfortunately, serious problems remain (see Box 12.5).

With high rates of pollution from agricultural run-off, urban streets, large-scale destruction of wetlands, floodplains, coastlines, and such-like, 'we are actually going backwards in our efforts to restore the health of our aquatic ecosystems' (Adler 1994: 19).

Finance for the provision and maintenance of municipal plants has not been adequate, despite a huge expenditure. Part of the problem has been that, though grants were available for capital costs, the states received no support for running costs; the result was the building of 'Cadillac projects without the funds or technically qualified operators to maintain them, and some plants operate substantially below design capacity' (Ingram and Mann 1984). Voting capital funds is always politically more popular than supporting running costs.

Grants were initially at a high level (75 percent of construction cost); but this was reduced to 55 percent in 1981, and replaced by a loan program in 1987. The need for further upgrading of treatment plants remains. The inability of plants to cope with wastewater is a primary cause of pollution. In 1989, EPA reported that over two-thirds of the nation's 15,600 wastewater plants had 'documented water quality or public health problems', and it estimated that $83 billion was needed to bring plants up to the required standard (Smith 2003: 112).

BOX 12.5 PROGRESS WITH CLEAN WATER – SOME INDICATORS

The federal government invested $56 billion in municipal sewage treatment from 1972 to 1989, with total federal, state, and local expenditures of more than $128 billion. By 1988, 58 percent of the US population was served. This improved treatment resulted in an estimated reduction in annual releases of organic waste by 46 percent, despite a large increase in the amount of waste treated. The same measure viewed from the opposite direction, however, shows a glass only half full. In 1988, public sewer systems serving 26.5 million people provided only minimum treatment, and 1.5 million people had no treatment at all, with raw sewage discharge into public waters.

In 1990, the Clean Water Act's 'swimmable' goal was met in about three-quarters of our rivers and estuaries, more than 82 percent of our lakes, and almost 90 percent of our ocean waters . . . But this leaves a large number of water bodies which are unsafe for swimming – one out of ten ocean miles, and one in five lake acres. Closer analysis indicates that many more waters are not really safe for swimming.

Source: Based on Adler 1994: 10–11

Non-point source pollution

Broad definitions can be found throughout the US. Oregon Administrative Rule 340-41-006 (17) defines non-point source pollution as 'diffuse or unconfined sources of pollution where wastes can either enter into or be conveyed by the movement of water to public waters'. By definition, non-point source pollution is problematic: there is no one easily recognizable point at which it can be controlled. Its origins are diffuse and varied. It results from the way in which an industry is structured, managed, and operated: thus improvements may require major changes. Added to this is the fact that the industries concerned are politically powerful: agriculture, construction, forestry, meat packing, shipping, and many others have strong support in Washington and in the state capitals. Bringing influence to bear in these quarters (influence which may spell jobs) is a far cry from diverting a sewer pipe. Coupled with technical difficulties, these problems have retarded progress with the abatement of non-point source pollution. As such, it remains a national concern.

There are numerous sources of non-point source pollution. This pollution does not originate from any one source. Agriculture is the worst non-point source offender: agricultural run-off contains many toxics from pesticides, fertilizers, animal waste, and similar materials. To abate these requires changes in standard agricultural practices. States have been unwilling to take an aggressive approach, and have tended to adhere to policies of gentle persuasion. There is, however, an Agricultural Water Quality Incentives program that provides technical assistance and subsidies for measures that reduce source contaminants. Urban land use practices also contribute to non-point source pollution. Commonly known sources would include: landfills, land development activities, erosion, roads, parking lots, etc. All sources of pollution are exacerbated by wind and water flows.

Groundwater pollution

Groundwater – despite its name – is water which flows under the ground. Most (98 percent) of the global amount of available fresh water is groundwater stored in aquifers – the pores and cavities of rock strata. In the United States, about a quarter of fresh water comes from this source – and much more in the arid West. Pollution of groundwater is a mounting problem of alarming proportions. It comes from numerous sources: agricultural activities (particularly fertilizers and other chemicals used in modern agriculture), discharges from sewage works, urban run-off, oil discharges, waste sites of many types, acid rain; indeed, all the pollutants which are discussed in this chapter (and many more) find their way into groundwater. Thus all measures that go toward reducing pollutants also help to protect groundwater, including the Superfund legislation examined in the next section. (This is an illustration of the interrelation of environmental issues.)

One of the many problems with pollution generally is that chemicals can be in use for many years before their hazardous nature is appreciated, and thus, needed measures of environmental protection are retarded. As a result, pollution builds up and becomes more difficult to tackle. Even when a pollutant is recognized as such, its presence may remain undetected for many years. This is partly because of the unknown dangers in the thousands of abandoned hazardous waste sites, and also because the slow rate of movement of groundwater increases the difficulty of detecting pollution. There are added complications caused by the variability of ground conditions.

All this adds up to an emerging problem of frightening dimensions. This is well recognized by scientists – 'groundwater pollution could become one of the scourges of the age' (Hiscock 1995: 246) – but it does not have the political salience of other types of pollution.

Safe drinking water

The fragmented nature of water pollution control is illustrated by the existence of a separate Safe Drinking Water Act. Originally passed in 1974, following public concern about harmful chemicals in drinking water supplies and the inadequacy of state programs, the Act required EPA to monitor and regulate twenty-two water contaminants. There were additional discretionary powers to extend controls over further contaminants. Between 1974 and 1986, EPA introduced regulations relating to the twenty-two specified contaminants, but progress beyond this was very slow. Frustrated by the tardy rate of progress, Congress eventually adopted a more directive approach. An

amending Act of 1986 required EPA to regulate eighty-three specified contaminants and also an additional twenty-five every three years. At the same time, EPA's responsibilities were increased and strengthened. By 1992, EPA had issued regulations for all but seven of the eighty-three specified contaminants.

The Act was amended again in 1996. Protecting public health was emphasized to a greater degree. States were required to identify contaminants that posed the greatest risks to human health, thereby threatening the nation's drinking water supply. EPA was also able to get better scientific information for developing more cost-effective regulatory decisions. Funding was also available to states to fund upgrading of water systems.

In 1999, there were about 170,000, publicly or privately owned, public water systems regulated under the Safe Drinking Water Act. These served 250 million people (the remainder obtained their water from private wells). A very large number of systems are small and have difficulty in shouldering the financial burdens of compliance. This difficulty is significantly increased by the 'twenty-five every three years' mandate, and is criticized by EPA as adding to the regulatory burden and detracting from the implementation of regulation of priority contaminants. In some cases, 'contaminants have been forced onto regulatory schedules that out-pace EPA's ability to develop needed technical information, some regulations have unquantified benefits, yet impose significant costs'. It thus seems that the forced pace of regulation imposed by Congress has not worked well. The US Environmental Protection Agency (1993) has concluded that 'a fundamental reform' of the legislation is needed which would focus on priority public health threats. There is fragmentary evidence that these threats are real. Since the sources of this evidence include the General Accounting Office and EPA (as well as a Ralph Nader study), this gives rise to some concern (Rosenbaum 2007: 227).

The limits of the regulatory approach

There has for a long time been criticism of the favored regulatory approach. Though some failures in implementation are undoubtedly due to intrinsic difficulties, many argue that 'a major share of the responsibility for the slow rate of progress must be assigned to the inappropriate incentive structures created by the regulatory approach to pollution control' (Freeman 1990: 145). The advantages of alternative approaches are discussed in the last section of this chapter.

It would be foolhardy to include in this discussion policy changes that are under consideration at the time of writing. The political uncertainties are far too great! Suffice it to say that issues being addressed include the control of polluted run-off, watershed management, further restriction on the discharge of toxics, and a strengthening of enforcement procedures. The focus is on the protection of water, rather than treatment after it has been polluted.

WASTE

The sedge is wither'd from the lake
And no birds sing
John Keats, *La Belle Dame Sans Merci*

The nature of waste

Humans are indeed 'wasteful': large quantities of the by-products of economic processes are not used and, in fact, are perceived as having no use. This is either because no one has thought of a use, or because any use that has been considered is judged to be uneconomic. But this is determined at least in part by the way in which the costs of production and their unwanted by-products are calculated and shared. If a firm is free to dump its unwanted by-products without regard to the costs imposed, it has no economic incentive to find uses for them. The cost is passed on to those who suffer from environmental pollution and degradation, or to those who have to shoulder the burden of clean-up. Moreover, these costs can be much higher than those that would have been involved in introducing more efficient (less polluting) methods of manufacture or systems of recycling. One reason

for this is that biological and chemical processes acting upon waste can render it far more harmful than it was in its initial state – and therefore costly to treat. Another is that regulatory systems can be enormously expensive and, as we shall see, not always effective.

Waste mattered little in primitive societies: little of it was produced, and it caused no harm, natural processes being sufficient to perform a self-purifying function. As wealth and populations grew, waste increased, and natural processes became insufficient, especially in urbanizing areas; innovative but more harmful manufacturing systems were introduced; and waste of greater toxicity was produced. The development of new goods involved a widening range of manufacturing processes in which the constituent materials were selected solely for their ability to contribute to producing the wanted good: whether they also produced unwanted by-products was incidental. By-products constituted 'waste' which was to be got rid of in the cheapest way possible. Waste was even to be seen as a sign of wealth: a coal tip growing on the edge on a mining community or a chimney pouring out black smoke from an urban factory were indications of prosperity. The environmental impacts were ignored, or regarded as incidental, or simply (as also with industrial diseases) not understood.

What constitutes waste in one system of economic production is not considered waste in another. The squatter settlements on the edges of third world cities eloquently demonstrate the potential value of urban 'waste': materials jettisoned as being of no use in the city are put to good use in providing shelter and primitive amenities. In affluent cities themselves, simple incentives for recycling can transform something that is 'waste' into a marketable commodity. Well-known examples are returnable 'deposits' on bottles and cans, and policies for preferential use of recycled paper in government departments, which, given their prodigious use of paper, is not to be underestimated. The 'value' of waste paper is nicely illustrated by a report from California which notes that 'a ton of loose office paper can be sold for $30. Bale the paper and the market price rises to $150. Pulp the paper and the market price reaches $570. Convert the pulp to writing paper and the price can climb to $920 a ton' (Schwab 1994: 47).

The opposite also holds: a marketing system which puts a premium on attractive, well-packaged consumer goods creates enormous quantities of waste packing materials. Even a rule introduced for the benefit of the public health – requiring food to be wrapped, for example – can increase waste packaging. Convenient new packaging (from the ubiquitous plastic bag to the polystyrene supports in boxes of consumer durables) creates new and problematic forms of waste. Equally convenient new throw-aways create increasing waste-disposal problems: 1.6 billion pens, 2.6 billion razors, and 16 billion diapers, for instance, are added each year to the mountains of municipal waste in the United States. The diaper has become a particularly large problem, resulting in between 3 and 4 percent of the solid waste collected by municipalities; one study estimates the cost of 'disposal' at $4 billion a year – almost the same as the value of the market (Cairncross 1992: 215). On the other hand, all these modern inventions make life easier, sometimes very obviously so, as with disposable diapers, sometimes less obviously, as with the greater convenience and shelf-life of packaged foods. Indeed, better packaging might lead to less waste of food.

Sometimes, waste can be turned into a wanted good. In addition to the systems of recycling newspapers, bottles, and cans that households have taken to heart, there are a large and increasing number of recycling technologies. Plastics are a case in point. These can now be turned into substitutes for timber, concrete, and other building materials. Perhaps the most striking example of recycling is that of cars. The once-common site of a wretched junkyard of abandoned cars has now largely gone, not as a result of effective environmental programs but because of technological changes in the auto industry. These enabled steel to be profitably made entirely from scrap. Concomitantly, the automobile shredder provided the means of separating the various materials. Unfortunately this bit of magic is in jeopardy since the growing pressure to increase fuel economy has led to a substitution of plastics for metal. These are not only more difficult to deal with: they also reduce the value of car hulks and thus the incentive to recycle.

It is apparent from this short recital that the idea of 'waste' is not a straightforward one. It varies over time, among countries, between industrial processes, and according to the controls operated by governments. The last point is of particular importance: the quantity and character of waste (and methods of waste disposal) can be affected by public policy. Given appropriate mechanisms (whether regulatory or economic), most waste can be disposed of with relative ease. There is one exception: hazardous waste, of which a ton per head of population is produced each year in the United States. This issue continues to be debated and contested at all levels of government.

Hazardous waste

Hazardous waste is, in one sense, easy to define: it is simply waste that is hazardous. But so are smoke, agricultural run-off, and leaking chemicals. Some of these types of pollutant affect air or water more than land, but pollutants can affect any or all of these media. The point does not need to be labored: there are difficulties of defining categories of waste, and these arise in part as a reflection of the approach taken to their mitigation. Some legislation focuses on the source of the waste (as with nuclear waste), some on the medium it affects (air and water), some on its character (toxic). There are also differences in the ways in which different wastes are dealt with: policies for clean air and water, for instance, are focused on making these media clean rather than on disposing of the pollution. Hazardous waste is typically thought of in terms of land pollution, but it can affect all environmental media, and the way in which it is dealt with legislatively and operationally is in part a result of the way in which the problems were initially interpreted and defined. Had the accidents of history been different, hazardous waste might have been viewed differently. It should also be noted that, as interpreted by regulations, the US definition of hazardous waste is by no means all-inclusive: it excludes the wastes produced by households and by agriculture, mining, and drilling operations.

Love Canal

Love Canal, a 49-acre site including a canal located in Niagara Falls, New York, was one of a number of highly publicized and visible disasters that have precipitated major legislative responses (others include Bhopal and the *Exxon Valdez* oil spill). Popular environmental history tells that the discovery of the 'ticking time bomb' of 21,000 tons of chemical waste, including dioxin, halogenated organics, and pesticides, at the Hooker Chemical site in Niagara Falls, New York, revealed a 'public health emergency' of 'great and imminent peril'; quick action by local residents and rapid response by state and federal agencies led to the evacuation of the residents. The Love Canal nightmare and the associated risks to public health caused by chemical dumping led to the enactment of the Comprehensive Environmental Response, Compensation and Liability Act (CERCLA) in 1980. This legislation, better known as the Superfund, was designed to clean up hazardous waste sites whose owners were negligent, denied responsibility, or could not be found. The Superfund paid for the costs of the clean-up. These costs were recovered when the courts determined the party responsible for dumping the hazardous wastes.

The passing of the Superfund legislation aimed at dealing with similar catastrophes throughout the country. The reality is different, and much more complicated. The terms used to describe the nature of the problem (in quotation marks above) have special meanings in their original context that were misinterpreted by the public. This is hardly surprising: who was to know – or believe – that the phrase 'public health emergency' was a jargon term used to ensure that Love Canal would legally qualify for federal emergency relief funds, or that the phrase 'great and imminent peril' was an administrative trigger for the allocation of funds for public health studies? The nuances of the legal and administrative meanings of such terms were not appreciated by the public, or by the residents of Love Canal, or by the press, or by the involved politicians. The result was, indeed, quick action (relatively) and, since it was feared that similar catastrophes might arise elsewhere, national legislation

dealing with toxic sites. (One further unfortunate effect of the 'crisis' was that public health studies were carried out hurriedly and inadequately – and with results which were later discredited.)

The legislation, however, was already in the pipeline. Although Love Canal was not its cause, it certainly contributed to the need for legislation and greatly increased political support for it. Members of Congress were quick to see the public reaction to Love Canal and the specter of thousands of similarly abandoned lethal sites throughout the country. More important in the long run, Love Canal served as a catalyst and had a major impact on the character of the legislation. In particular, the widespread public concern provided EPA with an opportunity to widen its mission and to take on new responsibilities for the public health: it made good use of the opportunity. In doing so, it built its case on the basis of the Love Canal problem (Landy *et al.* 1994: 142).

Superfund legislation

The difficulty of assessing the scale of the hazardous waste problem has complicated the task of devising a sensible regulatory system. There was no way of knowing how many hazardous waste sites there were, what was in them, how dangerous they were, or how much the cost of clean-up would be. Information disclosed by companies charged with dumping hazardous waste was slow, at best. Indeed, the only thing that was certain was that the answers to such questions were unknown. In fact, history shows us that we tend to learn of many problems 'after the fact' – not while the problems are occurring. It was accepted that there was a great deal of ignorance, though no one knew just how much. Yet Congress had to give a lead: it did this by requiring EPA to develop a *National Priority List* (NPL) of sites posing immediate threats to people living or working near the sites that are in greatest need of clean-up. No indication was given as to how these sites were to be selected, but they were to be eligible for Superfund finance. Beginning with 400 sites, the list was to be added to each year on the basis of information obtained by EPA and the states. Sites not included in the list would fall to the responsibility of the states, an issue of great concern to the states.

The regulatory system for dealing with hazardous waste was introduced by the Resources Conservation and Recovery Act (RCRA) in 1976. It was an amendment of the 1965 Solid Waste Disposal Act that focused on the management and disposal of municipal and industrial waste. The legislation established a permit program for disposing of hazardous wastes. It is an extraordinarily complicated piece of legislation, and its complications were increased by later acts, of which the best known is the 1980 Comprehensive Environmental Response, Compensation and Liability Act. (Mercifully, even its acronym CERCLA has given way to the popular term 'Superfund'.) With this legislation, Congress intended to establish a comprehensive 'cradle-to-grave' system for regulating wastes. It prioritized a list of substances that had been found to pose the most significant threats to human health. The 1999 priority list's 'top 10' substances were: arsenic, lead, mercury, vinyl chloride, benzene, polychlorinated biphenyls, cadmium, benzo(a)pyrene, polycyclic aromatic hydrocarbons, and benzo(b)fluoranthene. The list is periodically revised.

The control of hazardous waste operates over the three main participants in the waste production and disposal process: generators, transporters, and operators of treatment, storage, and disposal facilities (known by their acronym as TSDs). A manifest system tracks hazardous waste from its generation to its final disposal: records are kept of each stage of the journey made by the waste from its production to its final disposal. There are heavy fines for violators. This impressive-looking system is less effective than might be expected. In the first place, there is little monitoring. Second, the system applies only to waste that is moved from the site where it is generated. Thus, all the waste that is dealt with by the producers is not covered: this is the majority — estimated to be up to 90 percent. Though some on-site disposal requires a permit, the self-management thus allowed is subject to only very limited monitoring. Inspection by EPA is rare, and state agencies do not have the resources for regular monitoring.

It should be noted that implementation of this system (as with much else in the environmental protection field) depends essentially on the capability (and willingness) of the states. It was assumed by Congress that the widespread public concern about hazardous waste would prompt the states to set about implementing the scheme with enthusiasm. In fact implementation has been very varied. Some states have done little, while a few have shown much initiative and (coupled with additional powers from the state legislature) have reached high standards of effectiveness. New Jersey's Environmental Compensation Responsibility Act requires a hazardous waste site assessment on the sale or transfer of any industrial or commercial property. This 'has provided a tremendous impetus for careful site assessment, completely transforming the local real estate market' (Mazmanian and Morell 1992: 88).

More generally, however, the record of the states is disappointing — partly, in their view, because of inadequate federal funding. Progress was particularly disappointing in the early years, but a change began to take place in the early 1980s and, though still patchy, state implementation improved. Paradoxically, one result was a *reduction* in the number of landfills. This came about because of the unwillingness of many operators to incur the expenses of the new EPA standards. Rather than upgrade, many landfills simply closed. These closures, together with the expenses incurred by operators who did upgrade, and a continuous growth in the amount of waste being produced, resulted in a marked rise in land disposal costs. Increasingly, it began to be realized that there was a far better alternative to disposal, namely waste *treatment*. This was embraced in new legislation – the 1984 Hazardous and Solid Waste Amendments Act. This gave pride of place to waste reduction, followed by recycling, treatment, and – as a last resort – land disposal.

The Act was written in a way that forced implementation: so-called 'hammer provisions' required EPA to introduce new controls by fixed dates, with automatic arrangements. These were designed not only to overcome any tardiness on the part of EPA, but also to circumvent antagonistic action by President Reagan (who correctly saw that stronger regulation of waste disposal would significantly affect a large number of firms).

The American Recovery and Reinvestment Act of 2009 (ARRA) provided some $600 million to help with the clean-up of Superfund sites around the US. These sites had contributed to a number of public health problems. Funding went for such activities as installing new septic systems, removing buried drums containing hazardous materials, completing a water treatment plan, installing a pump-and-treat system, connecting property owners with contaminated drinking water wells to a public water supply, conducting soil excavation, and re-vegetating property.

Liability and compensation

What is particularly frightening about hazardous waste is its 'time bomb' character (to coin the term which quickly stuck to Love Canal). It can be a very long while before the toxic effect begins to appear – by which time it is too late to take preventive measures, and also late for ameliorative action. Illnesses resulting from some contaminants may take years to manifest. Actions may not occur when the individual is first exposed to the contaminant. The sense of apprehension is backed up by other emotions – of confusion, distrust, betrayal, and even treachery. What had happened to the skills and competence of American industry? How could government let this happen? Wasn't some agency monitoring the disposal situation? How could such a successful machinery of production have wrought such a disaster on unsuspecting people?

This sense of outrage emboldened Congress to pass some severe penalties on those responsible for producing hazardous waste. Instead of the traditional legal doctrine of negligence, a far more severe doctrine was invoked: that of strict and several liability. This meant that excuses and mitigating circumstances are irrelevant, and that all who have been involved in the generation of the waste are liable. Thus, the common legal immunities are absent. At the same time, the law was made retroactive. The term 'Superfund' by which the waste regulation system is generally known is a misnomer: it gives the impression that there is a huge federal fund available for clearing up hazardous sites. In fact, the legislation is designed to pass the costs

on to the maximum extent possible. These costs are not cheap. They could be astronomical. There are complex provisions intended to identify the responsible waste-producing parties and make them pay for the cost of disposal. The cost of litigation alone is mindboggling. Only when no *potentially responsible party* (inevitably known as a PRP) can be identified is federal funding available. This Superfund, financed mainly by a tax on chemical manufacture, was established in 1980 with an initial spending limit of $1.6 billion (increased to $8.5 billion in 1986, following the alarm caused by the Bhopal explosion).

The incredible complexities to which this has given rise have proved hugely profitable to lawyers. Though the courts have taken the very sensible approach that efforts to achieve clean-up have priority over the allocation of costs, the subsequent wrangling over costs can take a very long time. It is not unusual for there to be seemingly endless arguments not only from those initially identified as being responsible for the hazardous waste (present and past owners and operators, generators, and transporters) but also from those making counterclaims, cross-claims, and third party claims. (The Act expressly authorizes PRPs to make claims against other PRPs; the court allocates costs 'using such equitable factors as the court determines are appropriate'. In spite of the length and cost of litigation, it can be very much cheaper than the clean-up costs. A study of one Landfill Superfund Site produced estimates of the total cost of clean-up ranging from $50 million to $4.5 billion. As involved attorneys comment, 'with so much at stake, it was very difficult for the various PRPs (site owner, site operator, industrial generators, and transporters) to agree how to proceed with cleanup or to fund clean-up activities' (Muse *et al*. 1995: 135). Just to illustrate one of the difficulties that can arise, generators can be liable for clean-up costs even if they had no knowledge of the site to which their waste was transported.)

The liability provisions are so strict that firms who might have been willing to voluntarily pay their 'fair share' have been deterred from doing so since they could find themselves forced to pay far more than a fair share – at worst, the whole clean-up cost. For the large firms, Superfund has become 'the legal equivalent of a survivor-pays-all game of roulette'. One state official has wryly observed that a Superfund listing 'can actually be counterproductive in achieving clean-up' of a site (Mazmanian and Morell 1992: 37).

The great emphasis on liability is in striking contrast to the way in which compensation is dealt with: there is, in fact, no provision for compensation to those affected by contaminated sites (local residents, former workers). The absence of victim-compensation was deliberate: it was intended to ensure that resources went to clean-up and were not depleted by compensation. But this did not stop victims suing PRPs. The precedent was set by the residents of the area around Three Mile Island (the site of the 1979 near-catastrophic meltdown of a nuclear reactor); levels of compensation reached huge proportions with Love Canal, where the residents finally reached a settlement of $20 million.

In looking at these large issues of industrial hazardous waste, sight should not be lost of the importance of risks that are much closer to home. Many private garages and garden sheds contain an abundance of highly toxic aids to gardening, car upkeep, and house cleaning and maintenance. Many home owners appear to be somewhat ignorant of this problem and do not know which products might contain highly toxic chemicals. This is in spite of labels on the products. Data on these hazards are largely anecdotal: ignorance of the extent of the problems of storage, use, and disposal is profound. They may present a significant potential risk but, like non-point source water pollution and radon, they have less salience than the more dramatic forms of pollution. They therefore arouse little public attention and therefore little action; and, as already noted, household waste is excluded from the definition of hazardous waste. Tragically, they may be awaiting their own disaster, their own type of *Exxon Valdez* or Love Canal incidents.

These domestic hazards do not fit neatly into the structure of pollution controls. But, given the nature of this structure, this is not surprising. Particularly baffling to a newcomer is the distinction between hazardous waste and toxic substances. Surely hazardous waste *is* composed of toxic substances? Had the history of environmental policy taken a different turn, they might have been dealt with in a comprehensive manner. As it is, not only do they have different control systems: toxic waste actually involves two systems of its own – one for pesticides and one for other 'toxic substances'. (In fact this oversimplifies the situation since there are twenty-four acts that deal with toxic substances!)

Toxic substances and pesticides

There is a sense in which the period following World War II can be described as the chemical age. The huge expansion in the production of chemicals, many of them being human-made (synthetic), amounted to a revolution not only in terms of numbers (more than four million between the mid-1940s and the mid-1960s), but also in their impact on agriculture, on electronics, on industrial processes and on everyday life.

New insecticides proved to be highly efficient in controlling pests and insect-borne disease. Herbicides have been similarly successful it controlling weeds. Together with the use of fertilizers, these chemicals have brought about huge increases in agricultural productivity as well as other benefits such as reductions in tillage, labor requirements, and soil erosion.

Regrettably, there were costs involved that were not apparent at first. Thus, it became apparent that chemicals that were effective in killing insects and weeds had serious environmental effects. As Rachel Carson explained, in prose which could be readily understood, synthetic pesticides such as DDT (dichloro-diphenyl-trichloroethane) have extraordinary power. Unfortunately, this power is not confined to killing insects: it can have disastrous effects on all forms of life. DDT is highly persistent and can be readily passed from one organism to another through the food chain. Through this process it can become heavily concentrated, with severe and even fatal effects on humans. Such chemicals are (in Carson's memorable phrase) 'elixirs of death'. There are many other proven carcinogens, including dioxin, asbestos, and polychlorinated biphenyls (PCBs); more than 500 such chemicals have been prohibited or restricted by EPA.

However, not all harmful toxics can be readily identified, and there is a great deal of uncertainty about the health effects of a large number (though a much greater number are harmless). Though a substance may have serious, even fatal, latent effects, the 'latency period' can be long, and since those affected may also be exposed to other toxics, it is a very complicated matter to isolate the effects of individual toxics. It follows that it is also difficult to determine what regulatory controls are appropriate. It took five years of inquiry and debate before Congress was able to decide on an acceptable approach to the control of toxics. The Toxic Substances Control Act (TSCA), passed in 1976, requires EPA to screen and track banned industrial chemicals that impose unreasonable risks on the public health and environment that are either produced or imported into the United States.

The nature of regulation differs between new and existing chemicals. New chemicals gave rise to particularly acrimonious disagreement but, despite opposition from the chemical industry, it was decided that EPA should review all new chemicals before production commences. Manufacturers have to shoulder the burden of proof that the new chemical is safe, and EPA can ban or hold up production until it is satisfied about safety.

By contrast, existing chemicals can continue to be marketed unless EPA invokes a review procedure. It does this where the safety data are deemed to be insufficient and where it is felt that there is an unreasonable safety risk. In such a case, after a lengthy full rule-making process (which includes public notification, time for public comment, and testing), EPA can require testing. When adequate data are available about a challenged chemical, EPA has very broad powers that it can invoke: these range from stricter labeling requirements to an outright ban. This power was used to ban chlorofluorocarbon propellants in aerosols (because of their effect on the ozone layer).

Controls over pesticides have a longer history, though initially this was for the purposes of consumer protection from fraudulent goods. Control for environmental objectives did not arise until the burgeoning of public opinion in the 1960s (the time when Rachel Carson's *Silent Spring* was published).

As with toxic substances, pesticides which are already on the market can be challenged by EPA only by way of a lengthy involved procedure. New pesticides, however, have to be licensed by EPA before they can be marketed. Licenses are given when the manufacturer can show that they pose no unreasonable risk. In determining risk, account is taken of 'the economic, social, and environmental costs and benefits of the use of any pesticide'.

It is inherently difficult to judge the effectiveness of toxic substances policies. The state of scientific knowledge is too inadequate for even a rough judgment to be made. And, of course, to the extent that policies are successful in preventing toxics being introduced to the environment, their effects are not there to be seen!

Nuclear waste

The reader may be surprised that no mention has been made in this account of nuclear waste. The reason that it appears separately, almost as an appendix to the main discussion, is that it is dealt with quite separately from other wastes. As with a number of other government functions (such as coal mining control and reclamation), it does not fall within the responsibility of EPA, but its federal guardian – the Nuclear Regulatory Commission (NRC). This is an independent agency set up in 1974 by the Energy Reorganization Act (taking over the functions of the Atomic Energy Commission) which is responsible for regulating the civilian use of nuclear materials (it covers nuclear reactors, including the use of nuclear materials and the disposal of nuclear waste) and for developing policies and regulations governing or licensing nuclear reactors and materials safety. It also has the responsibility of licensing the building of new nuclear power stations, but there have been none of these for many years. All issues relating to the safety and environmental aspects of nuclear power rest with the Commission.

The history of nuclear waste disposal has been a dismal one, even after the shock of the Three Mile Island incident killed the dream of a nuclear age in which energy would be clean, safe, and cheap. It has proved to be none of these, and the regulatory machine has proved incapable of dealing with the increasingly complex and horrendously dangerous problems that have arisen. Many of these problems were quite unexpected: among the list of unanticipated difficulties have been severe operational problems with nuclear power plants (some of which flowed from basic design faults), rapid deterioration of plants, mismanagement, severe safety problems – all accompanied by escalation of both costs and public anxieties. Some of these problems could probably have been avoided (or at least lessened) by better management and planning, but the early days of nuclear power were characterized by a high degree of optimism and a belief that any teething difficulties would be overcome by technological solutions. But the most troublesome – and unsolved – issues were totally unexpected: above all the question of safely disposing of nuclear waste – which grew increasingly difficult to deal with as public concern (and outright fear) made it impossible to find adequate sites. The original assumption was that spent fuel would simply be reprocessed, with the residue being dealt with safely by advanced technology – an assumption that proved to be false. Added problems arose with the temporary storage of waste. For these and other reasons nuclear waste became a huge liability, and bitter interstate battles raged on site selection.

Eventually, Congress was forced to act: the Nuclear Waste Policy Act of 1982 was intended to solve the long battle over sites by introducing a scrupulously fair and open process which, it was hoped, would satisfy everybody. In fact it satisfied nobody. The initial three sites nominated (Deaf Smith County, Texas; Hanford nuclear military reservation, Washington; and Yucca Mountain, Nevada) were overwhelmed by controversy, and Congress attempted another solution by summarily designating Nevada to be the home of the first site. (The designation was accompanied by a large bribe – $20 million annually.) However, after two years' preparatory work, involving an expenditure of $500 million, the Department of Energy abandoned the project on the grounds of inadequate technical quality. Difficulties arose with other projects such as the Waste Isolation Pilot Plant near Carlsbad, New Mexico.

Public fear about the dangers of nuclear waste sites has been a major factor in this sad story and will probably continue to be a significant factor in the future. This fear is justified since there are so many uncertainties about making nuclear waste safe; and the unfortunate history to date now bedevils the issue. Even if a solution were to be found, it is likely that the news would be met with disbelief. As if this were not bad enough, the salience of the issue has diverted attention from another emerging problem: that of safely decommissioning nuclear facilities at the end of their useful life. Rosenbaum dismally concludes that 'Waste management and nuclear power decommissioning problems will trouble Americans for centuries and remain a reminder of the technological optimism and mission fixation that inspired Washington's approach to nuclear technology development' (Rosenbaum 2007: 278). Even if this should happily prove to be too pessimistic a judgment, it is clear that the story is likely to continue for some time without a happy ending.

OTHER ENVIRONMENTAL CONCERNS

This chapter has examined a number of critical environmental issues. Population and development pressures dictate the need for continuous monitoring. Our environmental resources are precious resources that we cannot afford to abuse. Renewable resources have the ability to clean themselves to some extent. However, their renewability is severely hindered when habitats are cleared and asphalted.

Coastal zone management

Population and economic development forces have continued to place great strains on the nation's coastal waters and adjacent shorelands. Many people started worrying about the future of our coastal areas. The need to protect and regulate coastal development became of paramount importance. In 1972, the Coastal Zone Management Act (CZMA) was enacted. The legislation acknowledged the importance of the many ecological, cultural, historic, and aesthetic benefits of the coastal zone, and called for its effective management (see Box 12.6). Moreover, as noted in Section 1451 (c) of the legislation, the increasing and competing demands upon the lands and waters of our coastal zone occasioned by population growth and development, including requirements for industry, commerce, residential development, recreation, extraction of mineral resources and fossil fuels, transportation and navigation, waste disposal, and harvesting of fish, shellfish, and other living marine resources, have resulted in the loss of living marine resources, wildlife, nutrient-rich areas, permanent and adverse changes to ecological systems, decreasing open spaces for public use, and shoreline erosion.

As such, the development and protection of the nation's coastal resources is a delicate balancing act.

The expansive nature of the policy issues involved in coastal zone planning and management provides an excellent example of the need for effective communication and coordination of a plethora of participants at all levels of government, the private sector, and the non-profit sector. CZMA is administered through the US Department of Commerce.

CZMA provided state grant funding to states to develop state coastal programs. Virginia's Coastal Resources Management Program was established by Executive Order in 1986. This plan covers such issues as coastal lands management, coastal primary sand dunes, fisheries, shoreline sanitation, and tidal and non-tidal wetlands. Maryland's Coastal Zone Management Program was also established by Executive Order and approved in 1978. Michigan's Coastal Management Program was also developed under the CZMA and

BOX 12.6 NATIONAL COASTAL ZONE POLICY

1   To preserve, protect, develop, and where possible, to restore and enhance the resources of the Nation's coastal zone for this and succeeding generations;

2   To encourage and assist the states to exercise effectively their responsibilities in the coastal zone through the development and implementation of management programs to achieve wise use of the land and water resources of the coastal zone, giving full consideration to ecological, cultural, historic, and esthetic values as well as the needs for compatible economic development;

3   To encourage the preparation of special area management plans which provide for increased specificity in protecting significant natural resources, reasonable coastal-dependent economic growth, improved protection of life and property in hazardous areas, including those areas likely to be affected by land subsidence, sea level rise, or fluctuating water levels in the Great Lakes, and improved predictability in governmental decision-making;

4   To encourage the participation and cooperation of the public, state and local governments, and interstate and other regional agencies, as well as of the Federal agencies having programs affecting the coastal zone, in carrying out the purposes of this title;

5   To encourage coordination and cooperation with and among the appropriate Federal, State, and local agencies, and international organizations where appropriate, in collection, analysis, synthesis, and dissemination of coastal management information, research results, and technical assistance, to support State and Federal regulation of land use practices affecting the coastal and ocean resources of the United States; and

6   To respond to changing circumstances affecting the coastal environment and coastal resource management by encouraging States to consider such issues as ocean uses potentially affecting the coastal zone.

Source: CZMA, Section 1452

approved in 1978. Federal actions that are likely to affect resource use in the coastal zone must be consistent with federally approved state coastal management programs.

Endangered species

In 1973, Congress enacted the Endangered Species Act (ESA). This legislation acknowledged the multiple effects that growth and development pressures have had on various species of fish, wildlife, and plants. Many have been lost. Others are now threatened with extinction, while other species will soon follow. As of January 10, 2013, the US Fish and Wildlife Service has identified 321 plant species and 1,115 animal species as threatened or endangered in the United States. The Act defines 'endangered species' as 'any species which is in danger of extinction throughout all or a significant portion of its range other than a species of the Class Insecta determined by the Secretary to continue to constitute a pest whose protection under the provisions of this Act would present an overwhelming and overriding risk to man'. The term 'threatened species' means 'any species which is likely to become an endangered species within the foreseeable future throughout all or a significant portion of its range'. Ultimately, the Act recognizes the multiple aesthetic, ecological, historical, recreational, and scientific values or benefits associated with these resources. It requires federal agencies to ensure that federally authorized and funded projects do not jeopardize any endangered or threatened species or modify or destroy their habitats.

There is an opportunity for a non-federal property owner to undertake a project or activity that may conflict with an ESA-listed plant or animal species. In 1982, the ESA was amended to allow for 'incidental take permits'. The legislation defines 'take' as 'to harass, harm, pursue, hunt, shoot, wound, kill, trap, capture, or collect, or to attempt to engage in any such conduct'. Concerns that lawful activities of property owners may unintentionally harm a listed species contributed to the call for the creation of this incidental take permit (ITP). An ITP may be granted with conditions that allow the 'taking' of the ESA-listed species if the

presence of the species interferes with a legally permitted land use activity. According to Section 10(a)(2)(B), the conditions attached to the issuance of an ITP are:

(i) the taking will be incidental;
(ii) the applicant will, to the maximum extent practicable, minimize and mitigate the impacts of such taking;
(iii) the applicant will ensure that adequate funding for the plan will be provided;
(iv) the taking will not appreciably reduce the likelihood of the survival and recovery of the species in the wild; and
(v) the measures, if any, required under subparagraph (A)(iv) will be met, and he has received such other assurances as he may require that the plan will be implemented, the Secretary shall issue the permit.

In order to obtain the ITP, the property owner must develop a 'Habitat Conservation Plan' (HCP). Section 10 specifies that an HCP must include:

(i) the impact which will likely result from such taking;
(ii) what steps the applicant will take to minimize and mitigate such impacts and the funding that will be available to implement such steps;
(iii) what alternative actions to such taking the applicant considered and the reasons why such alternatives are not being utilized; and
(iv) such other measures that the Secretary may require as being necessary or appropriate for such purposes of the plan.

The plan is designed to minimize any potential harmful effects a proposed project or activity will have on the listed species. It may involve a single party or multiple parties. An increase in the number of parties involved in the HCP will require a great deal of communication, cooperation, and coordination between the various parties.

HCPs can be found throughout the United States. They vary greatly in size from less than 1,000 acres to over 1 million. As of October 2012, more than 430

HCPs have been approved by the US Fish and Wildlife Service and the National Marine Fisheries Service. Additional HCPs are in various stages of the planning process.

There are a number of representative HCPs that could be discussed. In Pima County, Arizona, an HCP was created to allow property owners to engage in a lawful economic use of their property while planning for the survival of an endangered species of pygmy owl. As a mitigating action, the cooperating parties involved in the HCP are moving some trees and saguaros – a favorite habitat of the pygmy owl. The parties are also monitoring the effects of their activities on the affected species. In Garfield County, Utah, an HCP was developed so that property owners could develop their property. In this instance, the parties were required to relocate Utah prairie dogs to other sites prepared expressly for them.

In 1991, California enacted the Natural Community Conservation Planning (NCCP) Act. According to Section 2801(i) of the legislation, the purpose of community conservation planning is 'to sustain and restore those species and habitat identified by the California Department of Fish and Game that are necessary to maintain the continued viability of biological communities that are impacted by growth and development'. The legislation did not deny the development of property but, instead, sought to make any development compatible with the protection of vegetation and wildlife. Developing a natural community conservation plan is similar to the HCP. The first program developed under the legislation was the Southern California Coastal Sage Scrub NCCP program. The program covered some 6,000 square miles and incorporated parts of another five counties.

The San Diego, California region was part of this program. The region represents an intriguing example of conservation planning. Economic development pressures accompany the region's population growth. The San Diego Association of Governments (SANDAG) projects the region to grow in population from a 1995 population of 2,669,300 to 3,853,300 in 2020. This represents a 44 percent increase in population. San Diego County contains over 200 plant and animal species that are federally and/or state listed as endangered, threatened, proposed, or candidates for listing. It has been called a 'hot spot' for biodiversity and species endangerment. The stage is ripe for potential disputes between growth advocates and conservation advocates.

In 1993, an Ongoing Multi-species Planning Agreement was signed by multiple parties, including SANDAG, the City of San Diego, the County of San Diego, the US Fish and Wildlife Service, the California Department of Fish and Game, and the California Resources Agency to create a regional planning management system with the goal of creating a large network, or ecosystem preserve, of habitat and open space within the San Diego region. It consolidated several 'subregional' habitat planning efforts within San Diego County. Each 'subregional' effort would be considered a plan under the 1991 NCCP Act.

The regional habitat planning program covers the unincorporated territory of San Diego County, the eighteen municipalities within San Diego County, and various independent special districts, an area of over 900 square miles. Development is not being denied. Private property rights are being protected while the native vegetation and habitat needs of multiple species are being preserved. A key to the program is that it deals with multiple species, not a single species.

There are a number of complex issues that must be faced. Coordinating a program when property is owned by the national government (military reservations), other levels of government, and the private sector is no easy task. Acquiring properties will be costly and funding must be secured. Developing implementing agreements specifying the roles and responsibilities of the various parties involved in the program will be a delicate balancing act.

Wetlands

Defining what is called a 'wetland' is not an easy task. Definitions of the term abound in the literature. according to agency and need. Cowardin *et al.* (1979: 11), in *Classification of Wetlands and Deepwater Habitats of the United States*, have suggested 'there is no single, correct, indisputable, ecologically sound definition for wetlands, primarily because of the diversity of wetlands and because the demarcation between dry and wet environments lies along a continuum'. Nevertheless, the US Environmental Protection Agency (1995: 1) defines wetlands as 'areas where water covers the soil, or is present either at or near the surface of the soil all year or for varying periods of time during the growing season'. Wetlands can be further broken down into coastal or inland wetlands. Individuals also refer to wetlands as swamps, marshes, and bogs.

For many years, people thought of wetlands as waste-lands. They were subject to much abuse. Development pressures caused vast amounts of wetlands to be filled. Some were drained for agricultural and forestry pur-

suits. Pollution destroyed their levels of productivity. Unfortunately we really do not have good estimates on how much wetland we have lost.

We have come to realize the values of wetlands. They serve as valuable resources for water quality and hydrology (filtering toxins and acting as a natural filter), flood protection, shoreline erosion, fish and wildlife habitat (spawning areas), natural products for our economy (shellfish), and for recreation and aesthetics (birdwatching) (see Plate 12). Box 12.7 shows the multiple benefits we accrue from wetlands.

In 2001, the US Supreme Court, in *Solid Waste Agency of Northern Cook County* v. *US Army Corps of Engineers*, 531 US 159 (2001), removed small

Plate 12 Parker River refuge, Newburyport, Massachusetts
Photo by author

waterways from federal jurisdiction under the Clean Water Act. The ruling prompted environmentalists to become concerned over small wetland areas, especially non-navigable, isolated wetlands, isolated ponds, and waterways. They felt the court's decision weakened the protection of wetlands. In essence the court indicated there was 'no need to protect wetlands unless they are close to navigable rivers, streams, and their tributaries'. A couple of years later, environmental organizations claimed the Bush administration weakened the Clean Water Act by having the field staffs of the Army Corps of Engineers not require permits for the pollution or destruction of wetlands within a single state or that are not associated with any navigable waterway (e.g., lake or river).

In 2004, Bush declared a new national policy to go beyond no net loss. Clinton had supported the same goal in 1993. As noted in *Conserving America's Wetland's 2006* (Council on Environmental Quality 2006: 1), he indicated that instead of simply limiting the loss of wetlands, the country would expand the amount of wetlands. In that same publication, the chairman of the Council on Environmental Quality informed members of Congress that 1,797,000 acres of wetlands had been restored, created, protected, or improved since 2004.

Concerns over the continued loss of wetlands prompted the development of new instruments. Mitigation banking represents one of these instruments designed to preserve or restore wetlands (see Box 12.8). It represents a proactive approach, under Section 404 permits, to curtailing the continued loss of wetlands or to lessening the impact of an activity on a wetland by creating comparable new wetland areas. They can be created by a government agency, a private sector enterprise, or a non-profit organization that wants to engage in wetlands restoration or protection activities. A formal agreement with the US Army Corps of Engineers is required before any activities are started. The agreement 'describes the wetland area's restoration plan and establishes the number of environmental credits the restoration work can potentially generate', (US Environmental Protection Agency 2006: 1). A permit applicant purchases credits from the mitigation

BOX 12.7 BENEFITS ACCRUED FROM WETLANDS

Wetlands are home to wildlife. More than one-third (1/3) of America's threatened and endangered species live only in wetlands, which means they need them to survive. Over 200 species of birds rely on wetlands for feeding, nesting, foraging, and roosting. Wetlands provide areas for recreation, education, and aesthetics. More than 98 million people hunt, fish, birdwatch, or photograph wildlife. Americans spend $59.5 billion annually on these activities.

Wetland plants and soils naturally store and filter nutrients and sediments. Calm wetland waters, with their flat surface and flow characteristics, allow these materials to settle out of the water column, where plants in the wetland take up certain nutrients from the water. As a result, our lakes, rivers and streams are cleaner and our drinking water is safer. Man-made wetlands can even be used to clean wastewater, when properly designed. Wetlands also recharge our groundwater aquifers – over 70% of Indiana residents rely on groundwater for part or all of their drinking water needs.

Wetlands protect our homes from floods. Like sponges, wetlands soak up and slowly release floodwaters. This lowers flood heights and slows the flow of water down rivers and streams. Wetlands also control erosion. Shorelines along rivers, lakes, and streams are protected by wetlands, which hold soil in place, absorb the energy of waves, and buffer strong currents.

Source: Indiana Department of Environment, http://www.in.gov/idem/programs/water/401/wetlandimport.html

BOX 12.8 MITIGATION BANKING

Mitigation banking has been defined as wetland restoration, creating, enhancement, and in exceptional circumstances, preservation undertaken expressly for the purpose of compensating for unavoidable wetland losses in advance of development actions, when such compensation cannot be achieved at the development site or would not be as economically beneficial. It typically involves the consolidation of small, fragmented wetland mitigation projects into one large contiguous site. Units of restored, created, enhanced or preserved wetlands are expressed as 'credits' which may subsequently be withdrawn to offset 'debits' incurred at a project development site.

Source: Federal Register, November 28, 1995: 58606

bank at the site of the impacts (on-site mitigation) or, in some case, at an off-site location. The credit or credits that have been purchased are used to compensate for development activities. Whether constructing comparable wetlands is possible continues to be debated by environmental researchers. The appropriate regulatory bodies from state and federal governments must approve the building of any mitigation bank.

Mitigation banks exist across the United States. The Everglades Mitigation Bank in Florida is a 13,500 acre wetlands restoration project owned and operated by the Florida Power and Light Company. The project site is home to some fifty threatened, endangered, and listed species. The Rancho Jamul Mitigation Bank is located in San Diego County, California and is operated by a private concern. The site consists of 150 acres of riparian and wetland habitats and is designed to restore two creeks and create habitat for an endangered bird of the region. It is owned by a private group and authorized to develop the bank and sell credits to make up for the loss of wetland habitat. The Snohomish Basin Mitigation Bank in the State of Washington was a former wetland that was drained to enable agricultural pursuit. The 230-acre site will restore the former wetland and will create new wetlands. The bank will sell credits to accomplish these tasks.

In-lieu-fee mitigation programs also exist where an applicant pays a fee to a public agency or another organization that has an agreement to use the fees for restoration, creation, enhancement, or preservation activities. These fees are theoretically provided to a

sponsor instead of the project applicant conducting any on-site mitigation activities and should be in or near the same watershed area. The use of in-lieu-fees has generated some controversy in that they are not considered 'to meet the definition of mitigation banking because they do not typically provide compensatory mitigation in advance of project impacts' (US Army Corps of Engineers, US Environmental Protection Agency, US Fish and Wildlife Service, and National Oceanic and Atmospheric Administration, 2000: 3).

The Reagan years

The Reagan years saw a halt to environmental policy initiatives – except those that involved a reversal of previous policies. The agenda had changed: it was now 'regulatory reform'; the shackles imposed on the American economy by previous administrations were to be removed, thus releasing the inherent powers of private enterprise. Environmental deregulation was the overriding objective of those appointed to the senior positions in the Reagan administration, and they set about their tasks with vigor. Their successes were significant, though less far-reaching than they had anticipated. They found that the public was not as enamored of the implications of deregulation as had been thought, and the very achievements of deregulation prompted a resurgence of environmental concern. Moreover, the separation of powers among the branches of American government ensured that moderating influences would be significant. Indeed, some areas of environmental policy, such as clean air and water, were actually strengthened during these years. Before the Reagan administration came to its appointed end, the reaction was abundantly clear, and positive environmental action was at the forefront of domestic policy. Public opinion was seen to triumph over even a popular president's agenda. Though much effort was expended in the battle between Reagan's onslaught and the defenders of environmental policy, these years also witnessed a re-evaluation of the adequacy and viability of the extraordinary range of policies that had been introduced over the preceding two decades. Though largely ignored by the Reagan administration, this period of heart-searching was a useful investment of time (Vig and Kraft 1990: 18). In the first place, it was apparent that the legislation had made remarkably optimistic assumptions about the speed with which the technical problems posed by compliance could be solved. Second, the administrative and compliance costs were also underestimated. Added to these difficulties were the legal challenges used by the affected industries to avoid the costs of compliance, as well as the time and effort involved for the regulatory agencies. In short, the legislation posed problems of implementation that were unanticipated. There was thus a ready-made agenda for a new administration.

Reagan's 'New Federalism' beliefs saw him cutting EPA's budget and personnel levels. He disliked the large public bureaucracy and felt that environmental regulations had gotten out of control. He encouraged returning environmental responsibilities to the states. Many individuals thought his environmental policies were designed to weaken federal enforcement of environmental regulation, which would please the private sector. Moreover, his lack of commitment to implementing existing federal environmental policies was obvious in one of his cabinet appointments – James Watt as secretary of the Department of the Interior.

In 1981, when Reagan appointed Watt as secretary of the Department of the Interior, the appointment generated a great deal of controversy. Watt was considered by many to be strongly anti-environmental. He appeared to be openly hostile to the environment. This was indeed a strange presidential appointment.

He supported developing federal lands for various commercial purposes, including mining, drilling, harvesting, and timber. Some of his proposals threatened a number of scenic areas around the country.

Controversial appointments did not end with Watt. In 1981, Reagan appointed Anne Gorsuch as administrator of EPA. She was a disciple of Reagan's belief in regulatory reform of the federal government. At her May 20, 1981 swearing-in ceremony, she noted that EPA was one of the largest and most important government agencies in the country. She agreed with Reagan that EPA was too big and that the country's environmental regulations were not good for business. She advocated relaxing Clean Air Act standards. People started wondering if her appointment was designed to dismantle the agency or to protect the environment. She resigned under pressure in 1983.

The environment was among the salient issues of the 1988 election, with both Bush and Dukakis vying for the honor of being the true leader in the field. Following his election, Bush quickly moved to establish himself as a real friend of environmentalists and, though his record was patchy, he returned environmental policy to center stage. Despite his support for Reagan's underlying beliefs in the efficacy of the market place, he espoused a number of environmental causes, particularly the strengthening of the clean air policy.

The second Earth Day was celebrated twenty years after the first, in 1990. It seems that environmental protection is firmly established as a central feature of domestic (and, less certainly, foreign) policy. How far this is deeply entrenched, however, is another matter. As Walter Rosenbaum has noted, 'the political ascendance of American environmentalism has occurred during two decades of almost uninterrupted domestic economic growth'. There has been no pitting of job losses against environmental losses, except in a number of localities – and where this has occurred, jobs have typically won (Rosenbaum 2007: 342).

Beyond Reagan

The remainder of this chapter describes a number of important national environmental actions that have been undertaken since the Reagan administration. The Clean Air Act Amendments were passed in 1990 during the Bush administration. A Toxics Release Inventory was released in 1990, alerting the public to pollutants released from specific facilities. During the Clinton administration, an incentive-based acid rain program to reduce emissions was launched. Municipal incinerators were required to reduce toxic emissions from 1990 levels. A program to clean up abandoned sites (brownfields) and then return the sites to a productive use was created. In addition, new regulations for lower emission standards for a variety of vehicular types were developed and implemented.

Bush went on record indicating he was a strong advocate of environmental conservation and for the protection of property rights. His record is yet to be determined. He called for the development of technologies for new sources of energy, so-called 'smart' technologies. In 2002, he announced 'The Clear Skies Initiative' that is designed to improve air quality by reducing power plant emissions for certain pollutants through a market-based approach. In addition, he pledged to reduce the growth of greenhouse gases.

In an April 27, 2011 press release, the Council on Environmental Quality stressed that the Obama administration has emphasized a commitment to protecting America's waters and to restoring rivers and watersheds. To do so would require a partnership and coordination between multiple partners. More information on Obama administration environmental initiatives can be found in Chapter 4.

NOTE: Further Reading and Questions to Discuss for this chapter are located at the end of Chapter 13.

13

The limits of environmental policy

I know of no safe depository of the ultimate powers of society but the people themselves . . . and if we think them not enlightened enough to exercise their control with a wholesome discretion, the remedy is not to take it from them, but to inform their discretion.

Thomas Jefferson

Introduction

In 1988, beaches in New York, New Jersey, and elsewhere on the Atlantic coast had to be closed because of pollution: among the evidence were hypodermic needles, syringes, blood bags, and other repulsive medical waste. Not surprisingly, public alarm was immediate. The alarm was increased in the localities affected by the temporary solution of closing the beaches, where the impact on local economies was sometimes severe. Further political response quickly followed. Several states passed or debated legislation; EPA established a task force to consider the problem; and Congress held hearings that led to the passing of the Medical Waste Tracking Act in the same year. This legislation amended the earlier Solid Waste Disposal Act. The term 'medical waste' included such items as cultures and stocks of infectious agents and associated waste; human blood and blood products; syringes, needles, and surgical blades; laboratory wastes; dialysis wastes; and discarded medical equipment. Facilities generating these and other types of medical waste were required to package and label waste prior to sending them to any treatment or disposal facility. The legislation did not only concern hospitals. Other medical facilities including physician offices, dental practices, and veterinary hospitals were also subject to the legislation.

This was a remarkable demonstration of rapid governmental responsiveness. Unfortunately, the action was far from effective since, despite the apparent obvious evidence, the real culprit of beach pollution was not medical waste: it was municipal sewage. Only a small proportion of the beach closings were due to 'medical-related' waste. The majority were due to high levels of fecal coliform that resulted from sewer overflows in periods of heavy rainfall. The overflowing pollutants were carried by weather and tide conditions down the coast. The solution therefore lay not in the better handling of medical waste, but in hugely expensive investments in municipal waste systems (Fiorino 1995: 155).

EPA has continued its fight against the careless disposal of medical waste. In 1998, it entered into a voluntary partnership with the American Hospital Association and other organizations to advance pollution prevention and reduction by reducing hospital wastes by 33 percent by 2005 and by 50 percent by 2010. Today, it is estimated by the Hospitals for a Healthy Environment that hospitals generate 6,600 tons of waste per day. This represents a 15 percent higher figure from data released in 1992.

This incident is of particular interest for the analyst of environmental policy since it highlights the importance of three interrelated issues: politics, ignorance, and public opinion. Reference has already been made in earlier chapters to these, but they are so important that it is worthwhile examining them more thoroughly.

The role of politics in environmental policy is central. There are several reasons for this. First, there are huge areas where unequivocal solutions to environmental problems simply do not exist: in the final analysis, the decision has to be one of judgment – which is another way of saying that it is a political one. In a democracy, this means that the decision is taken openly with the 'facts' (such as they are) being freely available and subject to public discussion. Even when relevant information is available, questions of interpretation remain. These involve value judgments that will differ according to individual and group beliefs and attitudes. When so much is uncertain, it is important that the political process is as free as possible from undue influence or unjustified restraint. It seems clear that democratic political systems are more attuned to environmental needs than are dictatorships. The rights of access to information, of free protest, of electing governments, and all such features of democracy are effective as well as inherently desirable.

There are only a very few things that are certain. Policy development and implementation are not two of those things. Ignorance (or scientific uncertainty, if the term is preferred) is not only widespread in the environmental field, but it is also not much of an exaggeration to say that it is commonplace. The rate of technological innovation has been so great for so long that the area of uncertainty is now vast. The easy environmental problems are behind us: those that remain are much more difficult to deal with; and they are constantly being joined by new ones opened up by technological advances and by belated discovery of the long-term pollutant effects of earlier innovations. As a result, much policy is based on quicksand rather than on firm scientific ground. This makes it difficult to inform, persuade or force public opinion; and without supportive (or at least tolerating) public opinion, no policy can work.

Scientific uncertainty has occasionally been used as an excuse for environmental inaction. Some researchers may feel that unless they can prove a causal effect between two items, it is better not to engage in any activity. This might be analogous to the phrase 'discretion is the better part of valor'. It could also be argued that this inaction may have exacerbated an environmental problem.

However, public opinion is not a slave to scientific fact; indeed, the gradual realization that there is not a clear 'scientific' solution to all problems has increased public distrust of 'official' views, whether these be expressed by scientists, politicians, or any others who purport to have clear answers. But distrust also acts as a safeguard against bogus science, or the unwarranted promotion of a particular interpretation. The old adage that knowledge is power now has to be qualified, since unshared knowledge may not be politically acceptable: the very authority of science has been dramatically weakened. (Monuments range from Chernobyl in the old Soviet dictatorship to the nuclear waste sites in the United States.) Thus science, politics, and public opinion intertwine and create a new image of the aligned 'expert' who is recognizably associated with a particular viewpoint – an environmentalist, an economic expansionist, or whatever.

The essential argument of this chapter having been summarized in bold terms, the constituent elements can now be examined in detail. The starting point is the nature of current policies.

Technocratic policy

As annual expenditure on pollution control continues to increase and as awareness of the range of environmental problems increases, it has become apparent that there will never be sufficient resources to deal with all of them. It is therefore important that policies should be kept under review, and consideration given to changes in the pattern of expenditure. That changes might be appropriate is suggested by the fact that the present pattern is not the result of a carefully considered strategy. On the contrary, it largely reflects surges in public opinion – from clean air and water in the 1970s, to toxic pollutants in the food chain and on waste sites in the 1980s, to the current concern for global ecological problems. However justified these peaks in public concern may have been, it is at least questionable whether the resultant array of policies is 'optimal'.

Many argue that too much effort is directed at risks that are small but scary-sounding, while larger, more commonplace ones are ignored (Morgan 1993). Some figures look compelling: if the mathematics (and the underlying assumptions) are correct, the United States is spending at a rate of $12 million per potential victim of hazardous waste pollution but only $5,000 per potential victim of indoor radon. This clearly suggests that more lives would be saved by transferring resources from hazardous waste control to radon control. However, public attitudes rate the two dangers quite differently. Indeed, EPA reports have concluded that the agency follows priorities that are often very different from those that its own experts consider to be the largest environmental risks. Frequently, the public and Congress (reflecting public opinion) focus on problems that experts consider to be of relatively small importance.

A bill introduced by Senator Moynihan in 1992 (and again in 1993) responded to this by requiring EPA to seek 'ongoing advice from independent experts in ranking relative environmental risk . . . and to use such information in managing available resources to protect society from the greatest risks to human health, welfare, and ecological resources'. Senator Moynihan's initiative is a good example of the technocratic approach to public policy: what Jonathan Lash, president of the World Resources Institute, has described as 'a nostrum to quell the effects of public ignorance and to prevent the contamination of the domain of experts, with its hard, quantitative, reproducible results, by unscientific values' (Lash 1994: 75).

Values and risks

The inadequacy of this technocratic approach is that it marginalizes the crucial issues of value. The idea that there is an objective, scientific, value-free solution to problems is quite false. Science is not like a piece of arithmetic, where the 'answer' can be found by a feat of intelligence. The shortcomings can be seen by examining the difficulties encountered in the attempts to develop 'risk analysis'. At first sight this seems an eminently sensible approach: risks would be scientifically evaluated, and resources then allocated according to the severity of the risk.

The difficulties of this approach are several. In the first place, the scientific information required is lacking and, though efforts to reduce the area of ignorance and uncertainty will be helpful, it is common experience that research findings often raise new questions that demand further research (Unman 1993). Moreover, 'a growing body of experience seems to suggest that, in fact, more research and better technical information actually exacerbate conflict among experts and in the policy process' (Brooks 1988). Herein lies the insurmountable problem facing risk assessment: it involves issues of value – and these cannot be settled by inquiry. Though research can contribute to the debate by making it better informed, no amount of research can avoid ultimate questions of value. And values are not the preserve of the environmental expert: they are essentially a matter for the public.

A few illustrations of the value-laden issues that arise are eloquent. Is the death of a child worse than the death of an adult? Is a slow death by cancer worse than a swift death in a storm-caused flood? Is equity important? Is it relevant to take account of the benefits that accompany a hazard? Are involuntary risks worse than those voluntarily incurred? Does fear of an event increase the seriousness of a risk? (Lash 1994: 79).

The list can easily be lengthened by encompassing other issues: How should a health risk be weighed against employment benefits? What importance should be attached to wildlife, the rivers and forests, landscape, and ecosystems? Is beauty relevant?

Perceptions of risk

Differing perceptions of risk underlie many issues in environmental policy. A good example arises with the siting of waste disposal facilities (Mazmanian and Morell 1992: ch. 7). Scientists see the risks in terms of statistical likelihood. Officials are particularly concerned with the problem of alternatives and the relative 'risk' of opposition among the residents of different sites (a factor relevant to environmental equity); they are also prone to consider the electoral implications of different sites. Private developers of waste sites see risk in the normal economic sense of the chances of making a profit.

Residents, however, do not perceive risk in statistical, financial, or policy terms; they see risk in stark terms of frightening danger. Moreover, it is being imposed upon them by human forces that they can counteract. This makes the risk different from natural disasters that are acceptable because they cannot be predicted or prevented. In moving to an area with a natural risk, that risk is accepted – even if there is a real belief that it will not come to pass. Furthermore, while there is an accompanying benefit to a natural risk (the benefits of the location), there is no such benefit with an imposed environmental risk. Any benefit goes to the developer! (Risks that have accompanying benefits are much more acceptable – as with those jobless communities which welcome nuclear waste for the economic benefit it brings.)

The problems are made more difficult by feelings of deep distrust. Institutions of government and business that were at one time trusted (or at least accepted) are now subject to what has been termed a 'confidence gap' (Lipset and Schneider 1983). This arises from the inherent complexities of many contemporary issues, the failures to anticipate disasters such as Love Canal and Three Mile Island, and clear evidence of scandals and corruption (such as Watergate and the incredible savings and loan fiasco – not to mention Vietnam!). The NIMBY syndrome is not simple parochialism: it has deep and justifiable roots.

The situation is considerably worsened by the huge area of scientific ignorance surrounding many environmental issues. If the experts do not really know, how can their judgment be accepted? Increased information and provision for public participation is of little or no assistance in the face of antagonism; indeed, it can serve to confirm suspicions rather than allay them. The battle in Congress over nuclear waste reflects the deadlock accurately.

If this were all that could be said on the matter, the outlook would be bleak indeed. There are, however, some rays of hope, weak though they may be. At one time, it seemed that California had the answer with its YIMBY ('yes in my backyard') program, which was developed (after a very lengthy period of discussion and negotiation) on the basis that each county would provide facilities for dealing with the hazardous waste produced within its boundaries. This both stimulated more strenuous efforts of waste prevention, and was accepted by all fifty-eight counties as being fair. Unfortunately, it fell foul of state politics. It might have a better fate elsewhere (Mazmanian and Morell 1992: 192–203).

Another approach might be by way of a negotiated contract between a community and a site operator: this could be effective if the basic fear is not of hazardous waste itself, but of the way in which it is managed (Elliott 1984).

Risk and equity

One of these points – that of equity – can be illustrated by some actual cases (Hornstein 1994). Assessments of the risk involved in eating fish ignore the fact that non-whites eat more carcinogenic fish than the population at large, and that non-whites may prepare their food in ways that increase their exposure to contaminants. Similar issues arise with Native Americans, who are extremely vulnerable to polluted food (how relevant is it that they are few in number?). Assessment of the health risks from toxic chemicals or hazardous waste facilities are made on an 'aggregate' basis which ignores the much higher risks faced by racial groups who disproportionately work with such chemicals or live near waste facilities.

These are not exceptional cases. Indeed, there is extensive documentation on the widespread extent to which environmental hazards are disproportionately located in minority and low-income areas. A long series of reports have consistent findings: 'Blacks make up the majority of the population in three out of four communities where landfills are located' (US General Accounting Office 1983); 'Three out of every five Black and Hispanic Americans lived in communities with uncontrolled toxic waste sites' (United Church of Christ 1987); 'Racial minority and low-income populations experience higher than average exposures to selected air pollutants, hazardous waste facilities, contaminated fish, and agricultural pesticides in the workplace' (US Environmental Protection Agency 1992). It is also clear that there is discrimination in the enforcement of environmental regulations (Lavelle and Loyle 1992). Much of this discrimination is influenced by the effectiveness of articulate wealthier groups – an unacceptable aspect of the healthy role to be played by public involvement in environmental policy (Saleem 1994). This issue is commonly referred to as 'environmental justice'. In a response to a frequently asked question on the US EPA website, EPA defined environmental justice in the following manner:

> Environmental Justice is the fair treatment and meaningful involvement of all people regardless of race, color, national origin, culture, education, or income with respect to the development, implementation, and enforcement of environmental laws, regulations, and policies. Fair Treatment means that no group of people should bear a proportionate share of the negative environmental consequences resulting from industrial, municipal, and commercial operations or execution of federal, state, local, and tribal environmental programs and policies. Meaningful involvement means that: (1) potentially affected community residents have an appropriate opportunity to participate in decisions about a proposed activity that will affect their environment and/or health; (2) the public's contribution can influence the regulatory agency's decision; (3) the concerns of all participants involved will be considered in the decision-making process; and (4) the decision-makers seek out and facilitate the involvement of those potentially affected.

The issue of equity has become increasingly more important and more visible in the area of environmental policy. Concerns over issues associated with environmental justice continue to be debated inside and outside the halls of government. In 1992, the US EPA created an Office of Environmental Justice to address and provide advice to the EPA administrator over growing concerns over this complex issue. Regional EPA offices were to create an environmental

justice coordinator to help facilitate the work of the newly created Office of Environmental Justice.

In response to public concerns, the US EPA began a commitment to environmental justice in 1992. The National Environmental Justice Advisory Council (NEJAC) was created a year later for advising and making recommendations to the US EPA administrator on how to incorporate environmental justice into EPA policies, programs, and activities.

On February 11, 1994, President Clinton signed an Executive Order that made every federal agency examine their policies, programs, and activities. As noted in Section 1-101, 'each Federal agency shall make achieving environmental justice part of its mission by identifying and addressing, as appropriate, disportionately high and adverse human health or environmental effects of its programs, policies, and activities on minority populations and low-income populations in the United States and its territories and possessions, the District of Columbia, the Commonwealth of Puerto Rico, and the Commonwealth of the Mariana Islands'. To help facilitate cooperation and coordination between the various agencies, an Interagency Working Group on Environmental Justice was created. The administrator of EPA or the administrator's designee would convene the Working Group comprised of the heads of the following executive agencies or their designees: Defense, Health and Human Services, HUD, Transportation, Justice, Office of Management and Budget, Office of Science and Technology Policy, Office of the Deputy Assistant to the President for Environmental Policy, Office of the Assistant to the President for Domestic Policy, National Economic Council, Council of Economic Advisors, and other officials designated by the president. NEJAC's Charter was renewed in 2005 and is reviewed annually.

Since that time, issues surrounding environmental justice continue to surface. At various times, federal officials have been both praised and cursed for environmental stances and decisions. Discussions and debates continue to be heard over the development and potential passage of national environmental justice legislation.

There are also wider issues. In ranking environmental risks, how should 'environment' be defined?

Should it include overcrowding or poverty or lack of access to services? In determining policy, should the emphasis be on preventing pollution – and accepting a major upheaval in socio-economic organization? Or is it better to concentrate on a regulatory system that determines 'acceptable' amounts of pollution, thereby minimizing upset to existing economic processes and structure?

The last point may seem to be a radical one though, in fact, some of the most striking environmental improvements have been achieved by prohibition of certain pollutants. Lead was removed from gasoline; DDT and PCBs were banned; phosphate was severely restricted in detergents; and the nuclear test ban treaty led to a major reduction in Strontium 90. Barry Commoner (1994) has argued that this is the route which policy should take. He is prepared to follow his argument through to its logical conclusion: a pollution prevention strategy could be fully implemented 'only by undertaking massive, wholesale transformations of systems of production: energy, transportation, agriculture, and major industries, the petrochemical industry in particular'.

This takes us too far afield for discussion here, though it has to be accepted that there is a severe limit to the amount of environmental improvement that is possible by means of regulating emissions. Waste policy is explicitly moving toward the control of sources, and the mandated electric car is similarly a recognition that emission controls are insufficient. Another straw in the same wind is the verdict of the International Joint Commission on the Great Lakes (1994) that the only safe way of dealing with the pollution of the Great Lakes is to virtually eliminate the toxics at source. This binational US–Canada body was created in 1909. The Commission's Eleventh Biennial Report indicates that progress to restore the chemical and biological integrity of the area has proceeded at a slow pace.

Such changes cannot be brought about quickly, but they are possible – as is demonstrated by the examples already quoted and from changes in public behavior in relation to smoking, recycling, and garbage sorting.

The Pollution Prevention Act 1990

It has been commonly known that within the United States millions of tons of pollution are produced every year. While vast amounts of dollars are spent producing the eventual pollution and waste, vast amounts of dollars are also spent trying to clean up the problems caused by the pollution and waste. Debates surfaced around the need to focus attention on reducing pollution at the source instead of waiting for the problems to materialize. The force of the arguments in favor of focusing policy on reduction of pollution at source is now widely accepted, and the 1990 Pollution Prevention Act promotes an integrated approach. EPA has the responsibility of developing a detailed, coordinated strategy to promote source reduction; to promote research to disseminate experience in this field; to advance better methods of data collection and public access to environmental data; and to report biennially on the implementation of the pollution prevention strategy. Source reduction is defined in the legislation (42 USC. 13102, 5A) as 'any practice which reduces the amount of any hazardous substance, pollutant, or contaminant entering any waste stream or otherwise released into the environment (including fugitive emissions) prior to recycling, treatment, or disposal; and reduces the hazards to public health and the environment associated with the release of such substances, pollutants, or contaminants'. The source reduction could be accomplished through a variety of methods including 'equipment or technology modifications, process or procedure modifications, reformulation or redesign of products, substitution of raw materials, and improvements in housekeeping, maintenance, training, or inventory control'. Manufacturing plants are required to report annually (and publicly) on their releases of toxic chemicals to the environment, and the steps taken for source reduction, recycling, treatment, and disposal.

The strategy is based on the key principle of encouraging initiative in the private sector: voluntary effort rather than slavish adherence to detailed federal codes is considered to be more efficient and cost-effective. EPA has the role of integrating pollution prevention options into strong regulatory and enforcement programs.

It will take some time for this system to show results, particularly since it involves major changes in approach. Instead of mastering and implementing the detailed rules set by EPA, firms will have the scope for determining the best way of meeting objectives. This means that costs and benefits have to be assessed. It will be interesting to see how far the interests of individual firms harmonize with broader socio-economic interests.

Balancing costs and benefits

No policy is cost-free and, at some point, the costs outweigh the benefits. How is that point to be determined? It might be thought that an obvious goal of environmental policy should be to abolish pollution, or at least reduce it to proportions that have negligible effects on health. Unfortunately, this is simply impossible unless all economic activity is brought to a halt; and even low levels of pollution involve some risk. There is *no* level of pollution that is risk-free for everybody. Even if virtually all of the population would be unaffected by a low-level pollutant, there will always be some individuals who are so sensitive that any level would affect them; and there may be others who, while not being made ill by a pollutant, may be discomforted by it. (A final point worth remembering is that not all pollution is human-made: volatile organic compounds are emitted from natural sources such as plants and trees, and these natural emissions can become a significant contributor to ozone.) It follows that, whether explicitly or by default, a decision has to be taken on the levels of pollution that are to be tolerated. This also means taking a decision on which types of pollution are to receive priority treatment.

Answers to such questions can come from experts or from those who are affected by pollution. As earlier discussion showed, these two groups give markedly different answers, and it was argued that a democratic political system required decisions to be taken by the public. However, the public needs to have a factual basis on which the favored option can be decided. In the words of Jefferson (quoted at the head of this chapter), discretion should be informed. Is there any

objective way in which this can be done? The answer to this has two parts. First, there is the contribution that can be made by economists who have the techniques for measuring costs and benefits. Second, there is the American decision-making process that ensures that decisions are taken by the operation of pluralist politics.

Economic analysis

It is an elementary axiom of economics that expenditure on one good precludes expenditure upon an alternative. Thus, if resources are devoted to cleaning up hazardous waste, the same resources cannot be used for protecting endangered species (or for developing hospital services, or for road building). The cost of any environmental measure can be expressed in terms of forgone alternatives. It follows that an important question for environmental policy is the way in which resources are to be allocated among different specific environmental objectives and, when the objectives have been agreed, how these can most economically be met. There are so many worthwhile environmental objectives, and so many different ways of meeting them, that it is typically difficult to choose among them. Economic analysis provides a useful set of techniques to assist in this complex process.

Despite its utility, there has been little use of economic analysis in environmental policy-making. Many laws are 'cost-oblivious': they mandate the achievement of a goal (such as 'clean' water or air) with total disregard for the costs involved; and they often require this to be done by a certain date or in a certain way, again regardless of costs. This disdain for economics cannot be explained simply in terms of politics: it often stems from a worrying feeling that there is something inappropriate, unseemly, or even immoral in using economic tools. On this line of thinking, the use of prices debases environmental benefits and puts them on an equivalent basis with market commodities which are bought and sold; and it denies the inherent value of the environment, which is seen as being of a different order from the value of perfume or television sets. Such views are sometimes passionately held, and are supported by reference to an alleged moral imperative of safeguarding the intrinsic values of the wilderness, of nature, and of an uncontaminated environment. On this line of thought, such environmental legacies cannot be valued as if they were market goods. More widespread is the unease (if not repugnance) over the valuation of life in those cost-benefit analyses that compare the costs of different pollution prevention measures in terms of lives saved. Yet assigning a dollar value to life is basically no more difficult for an economist than determining the market value of transplant organs or motor vehicles or hamburgers. This 'commodification' does not sit easily on the public conscience.

These points have weight, and they are increased by an appreciation of the limitations of cost-benefit analysis and the ways in which it can be subverted. One of the major limitations is the paucity of relevant data. There is a large area of ignorance about the physical, biological, and chemical mechanisms which operate in the environment to allow figures to be calculated with confidence; and benefits are elusive, if only because there is a wide variation in the values which different people place on environmental 'goods'. In addition to these problems, opponents of economic analysis tend to be especially worried about the extent to which it can falsify. The classic example is the use of cost-benefit analysis in water resource development projects. It is now widely known that these analyses used methods that understated costs and exaggerated benefits. As a result of these distortions, a number of schemes went ahead that were costly and environmentally damaging. The height of absurdity was reached when serious consideration was given to the building of a dam in the Grand Canyon (Carlin 1973). Clearly, the extremism of which environmentalists are accused is not unique to them.

Economic incentives in environmental regulation

The use of economic incentives for environmental improvement has been discussed for a number of years. Reagan signed an Executive Order calling for the consideration of market-oriented regulatory alternatives in environmental regulation. The Clean Air Act Amendments of 1990, a result of the Bush administration, supported the use of economic incentives. Later presidents have continued to support their use.

Public opinion is in favor of regulatory standards because of their apparent fairness: all are required to meet the same target. Polluters may also like them because of the certainty they give to the market. In fact, the fairness is illusory. Fixed standards impose quite different costs on different firms. Some can meet standards easily, particularly if their machinery is modern. Older firms, by contrast, may need to invest heavily in new plant to meet a standard; and they will understandably seek to negotiate a less onerous standard. More important in terms of effective environmental improvement, firms will tend not to seek anything beyond the regulatory standard even if they can achieve a higher standard at relatively low cost. They will have no incentive to do so, unless they thereby obtain other benefits.

There are considerable advantages to be derived from designing pollution controls in a way that gives firms economic incentives to reduce pollution to the maximum extent. An incentive can make a firm take a totally different approach to its waste. If, for example, a tax is levied for every ton of waste produced, a firm will not be satisfied with calculating the economics of compliance: it will be motivated to review its processes to reduce its waste to the minimum. It has an inducement to calculate the real cost which its waste involves – a cost which otherwise is borne by the environment.

Of course, this is not to suggest that the interests of individual firms and of environmental policy are now uniformly harmonious: most will simply obey the law and follow their self-interest. 'It is not the job of companies to decide what values ought to be attached to natural resources and what the priorities of environmental policy ought to be' (Cairncross 1992: 299).

The use of regulatory instruments and the absence of an economic incentive to reduce pollution does not mean that firms simply abide by the dictates of the regulation. Far from it, the incentive is to avoid the costs of compliance. Thus the regulatory agency has to demand detailed records, inspect the record-keeping system, carry out site inspections, and undertake other such control functions. If, however, the resources of the regulatory body are not sufficient to enable this to be done, some firms at least will be tempted to circumvent or even ignore some part of a regulation. Since administrative resources are typically inadequate, this is a significant issue. The laxer the day-to-day controls, and the higher the costs of compliance, the greater will this temptation be. Overstretched agencies may well know that some firms are in default, but they may have some difficulty proving it, or they may have to accept a firm's assurance that it is doing the best it can or, given the pressure of work, they may simply leave the relevant file in the pending tray. Particularly bad cases may be prosecuted, but this takes even more time and resources, and the court is frequently unpredictable. There is no need to labor the point further. The incentives to adhere to a regulation are weak for the individual firm; the costs of enforcement are high for the agency.

Congress has a penchant for legislating the detailed manner in which environmental policies are to be carried out. Though there have been political explanations for this, particularly in the Reagan years when the administration was bent on reducing controls over business, detailed congressional provisions have unfortunate long-term effects. They shackle the administering agencies with approaches that are a product of the time; later, more effective approaches, technological innovations, and the lessons to be learned about the weaknesses and strengths of particular types of controls have to be ignored.

The range of economic instruments

The range of economic incentives, according to a 2001 EPA Report, 'The United States experience with economic incentives for protecting the environment', would include:

1 Pollution charges, fees, and taxes;
2 Deposit-refund systems;
3 Trading programs;

4 Subsidies for pollution control;
5 Liability approaches;
6 Information disclosure; and
7 Voluntary programs.

The simplest economic instrument is a tax on pollution, levied at a rate determined, for example, in relation to the damage caused by the pollutants and the costs of clean-up. Such a tax could be levied on lead or carbon content. (Several European countries have such a carbon tax.) The tax provides an immediate incentive to firms to reduce their use of the pollutant – and it is a continuing incentive. The difficulty arises in setting an equitable rate – a problem that also arises with marketable pollution rights.

Economic incentives can be applied to some types of waste with a deposit-refund system. This is essentially the same as the charges on returnable bottles, though rather more complicated. The producer of something which would become a waste after it has been used in a manufacturing process (a solvent for instance) would be required to pay a charge for each unit produced. This would increase its price (thereby introducing an incentive for reduction in its use). A refund of the charge would be payable to anyone who returned the solvent after its use. This system has the advantage of providing a disincentive to illegal tipping. (Clear evidence of its effectiveness with bottles is provided by their disappearance from the countryside after the charge was introduced.) The same system can be applied to motor vehicles.

Emissions trading or cap-and-trade programs

The problems associated with increased greenhouse gas (GHG) emissions have been well documented. These GHGs can occur naturally in the environment (e.g., carbon dioxide, methane, nitrous oxide, and water vapour) or they can be man-made (e.g., chlorofluoro-carbons, hydrocarbons, perfluorocarbons, and sulphur hexafluoride). Regardless of their origins, GHG emissions have been increasing in recent years due to such factors as continued population growth, continued deforestation, and the increased burning of such fossil fuels as coal, gas, and oil.

Several techniques have been established for 'trading' in emissions. One of the earliest was the *offset* technique that was introduced to allow new economic development (and its accompanying new pollution) to take place without increasing the level of pollution in an area. This was achieved by further 'offsetting' reductions in the emissions from existing plants. In effect this creates a trading market in emissions. This not only enables firms to achieve lower compliance costs, but it also results (because of EPA rules) in a net reduction in emissions. An advance on this was the 'bubble' policy. This first arose in the case of large industrial plants that had several smoke stacks that were notionally grouped in a bubble: controls were operated over the bubble as a whole instead of over each stack. This enabled the company concerned to 'trade' between stacks, thus achieving a lower cost of compliance. Again, a net reduction in emissions can be obtained. A variation on this technique allows 'emission banking', whereby firms can earn credits for achieving reductions in emissions greater than required, and apply them elsewhere (either to their own emissions or, by sale, to other firms). Techniques such as these encourage technical innovation and more efficient methods of emissions control. It is also suggested that they may encourage the early retirement of older, dirty facilities.

The 1990 Clean Air Act went further with its outright attack on the coal-fired power plants that contribute so significantly to acid rain. That it was able to do so was a result of a change in the political climate which made it imperative that the new Act should be passed. One result of this was that Midwest politicians found themselves isolated from other parts of the country, which now objected to paying for the cost of clean-up in the Ohio Valley – a good example of the territorial politics which underlies environmental policy (Cohen 1992). The Act mandates a major reduction in sulfur dioxide emissions by the electricity-generating plants of eighteen states (mainly in the Midwest). This is being effected by an emissions trading program that effectively establishes a market in pollution rights. Each plant is given an emissions

permit that sets the limit to its 'allowance'. If this is exceeded, a fine is payable and a reduction made in the following year's allowance. The allowances can be traded, and thus plants that are not able to reduce their emissions by the required amount can buy credits from plants that have made excess reductions. In this way, plants that find it very costly to reduce their emissions can trade with plants that are able to make a reduction more cheaply. The theoretical result is that the maximum reduction in emissions is made at the lowest possible cost. The essential feature of this trade in emissions is that the total is limited; and, if it works successfully, the cap can be reduced.

The Los Angeles Basin has always had severe air quality problems. It is generally considered the smoggiest area in the United States. To help it meet National Ambient Air Quality Standards by 2010 for nitrogen oxides and sulfur oxides, officials decided to turn to the market place. Any business that beat or cut its emissions (70 percent reduction for nitrogen oxides and 60 percent for sulfur oxides) by 2003 could trade its emission credits on the open market. Box 13.1 discusses the flexibility and cost savings for businesses under its Regional Clean Air Incentives Market (RECLAIM).

The State of California has recognized the severity of worsening air quality and global warming and passed the Assembly Bill (AB) 32 Global Warming Solutions Act of 2006. Reducing the amount of GHG emissions became state law. Under AB 32, California would seek to cut GHG emission levels to 1990 levels by 2020. The California Air Resources Board (CARB) was given the authority over the multiple sources of GHG emissions.

Two years later, in 2008, Senate Bill (SB) 375 was passed. This legislation supported AB 32 and required that CARB, in consultation with Metropolitan Planning Organizations (MPOs), develop regional targets for reducing GHG emissions. The legislation also sought to encourage areas to start planning for sustainable communities. Each MPO would be required to prepare a 'Sustainable Communities Strategy' (SCS) which would identify how the region would meet its GHG emissions target from automobiles and light trucks. Meeting this target could be accomplished only by integrating land use, housing, and transportation planning.

Most people felt that the state policy reducing GHG emissions would be challenged in some form or another. In 2010 that challenge took the form of a statewide ballot measure – Proposition 23. This measure sought to suspend the implementation of AB 32 until the state's unemployment rate dropped to 5.5 percent or less for a full year. The ballot measure failed. Voters upheld the law in 2010.

California completed its first carbon auction in November 2012. With its program of reducing carbon emissions, the state places a limit or cap on how much GHG emissions can be released. If a source exceeds its limit, it can pay the state extra through the purchase of carbon allowances. The CARB-administered auction generated $290 million for carbon credits in 2013 and in 2015. The credits cost $10.09 per ton of carbon, a figure lower than was expected. Carbon auctions will be held over the eight years following on from 2012.

The latest attempt at creating an emissions trading program by the federal government was the proposed American Clean Energy and Security Act (ACES) of 2009. This bill sought to create clean energy jobs, achieve energy independence, reduce global warming and pollution, and transition to a clean energy economy. The bill sought to create a cap on the amount of GHGs that could be emitted on a national level. The bill was passed by the US House of Representatives but was ultimately defeated by the US Senate.

While the US failed to enact ACES, there are attempts at regional emissions trading or cap-and-trade programs. Box 13.2 describes what a cap on emissions does. A Regional Greenhouse Gas Initiative (RGGI) was developed by states in the Northeastern part of the US and the eastern part of Canada as a cap-and-trade program. It was specifically focusing on GHG emissions from power plants. The RGGI auctioned off emission allowances in 2008. A second regional program was the Midwest Greenhouse Gas Reduction Accord (MGGRA). Six Midwestern states and the Canadian province of Manitoba sought to reduce GHG emissions through a market-based cap-and-trade program. The Western Climate Initiative (WCI) represents another regional compact to reduce GHG emissions. The WCI is comprised of seven western states and four Canadian provinces.

The European Union (EU) has also developed a cap-and-trade program – EU Emission Trading Scheme (EU ETS). Over twenty-five countries participate in the initiative. The program was designed to assist countries in meeting their Kyoto Protocol commitments. The program has met with mixed success.

Emissions trading has been welcomed by economists and policy analysts who see it as a highly efficient method of pollution control which avoids the problems of a purely regulatory regime. As Ellerman *et al.* (2003: 1) note, 'emissions trading is one of several market-based approaches that theoretically should improve the performance of regulatory regimes designed to improve air quality by giving sources the flexibility to achieve emissions constraints more cheaply than command-and-control alternatives'. It allows a company to determine how, or if, it will reduce its emissions. A company can purchase emission credits from a company that lowered its emissions and has available credits to sell. It certainly has political attractions: for conservatives, it is consistent with market principles; for liberals, it promises environmental improvement at an economic cost. It is akin to a user fee 'that ensures that those who benefit from the use of a natural resource pay for those benefits' (Bryner 1993). However, not all would agree. To the layman, there is something distasteful in such a market. Additionally, there may be concern that the market would distribute costs and benefits unequally. As already indicated, there is considerable evidence that minorities bear a disproportionate burden of pollution (for example, in the location of hazardous waste sites).

Others see emissions trading programs in a negative fashion. They don't believe all companies will invest in new technologies that would reduce GHG emissions. To these individuals, such a program increases the cost of energy, which is ultimately passed on to the consumers. This is essentially an energy tax that penalizes consumers. There is also a belief that such programs may result in a loss of manufacturing jobs. This could occur as companies outsource manufacturing processes to other countries where GHG emissions aren't as severe or regulated as they are in the US.

Economic incentives therefore have an important role to play in environmental policy (though they are not appropriate in all circumstances – in relation to monopolies, for example). They have other real advantages. They involve much less cost for government than do regulatory systems. They require less information – which, as has been repeatedly demonstrated in this discussion, is at a premium. They have an effect on all firms – the small (difficult to regulate) as well as the large (easier to regulate); and they avoid 'negotiation' of deals between regulators and the regulated – as well as suspicion that this is widespread. Above all, they place the costs of pollution where they properly belong: on polluters.

The 'polluter pays principle' is thus a very important one. In practice, individual firms will go beyond what

---

BOX 13.2 DEFINITION OF A CAP

A cap sets a maximum allowable level of pollution that penalizes companies that exceed their emission allowance. No other system can guarantee to lower emissions.

- *The cap is a limit* on the amount of pollution that can be released, measured in billions of tons of carbon dioxide (or equivalent) per year. It is set based on science.
- *It covers all major sources of pollution.* The cap should limit emissions economy-wide, covering electric power generation, natural gas, transportation, and large manufacturers.
- *Emitters can release only limited pollution.* Permits or 'allowances' are distributed or auctioned to polluting entities: one allowance per ton of carbon dioxide, or CO_2 equivalent heat-trapping gases. The total amount of allowances will be equal to the cap. A company or utility may emit only as much carbon as it has allowances for.
- *Industry can plan ahead.* Each year, the cap is ratcheted down on a gradual and predictable schedule. Companies can plan well in advance to be allowed fewer and fewer permits – less global pollution – each year.

Source: http://www.edf.org

BOX 13.3 INFORMATION AS AN ENVIRONMENTAL INCENTIVE

The British subsidiary of Rhone-Poulenc decided to build a computerized waste-accounting system to keep track of the waste each plant generates and the costs of disposing of it. The data go back to each plant every month. 'The first time I did this', the manager in charge of the system told the newsletter ENDS, 'there was quite a sensation. I was besieged by calls saying "Are you absolutely sure?" It was a revelation. They were jolted from blissful ignorance about their true product costs.'

Source: Cairncross 1992: 291

---

is prompted by price-signals. There are at least two reasons for this. First, there is the highly efficacious policy instrument of requiring firms to collect and provide information about their activities. Information is a management tool of crucial importance: it can alert firms to facts of which they were totally unaware (see Box 13.3). This is in addition to its public use and the effect which publicity can have on a firm's public standing – which in turn can precipitate action.

Second, the people who run business are subject to the same shifts in opinion and perception as is the general public. Companies have embraced the environmental ethic as clearly (even if not usually as enthusiastically) as dedicated environmentalists. That this is partly a matter of self-interest cannot be denied, but enlightened self-interest works toward social benefit in the same way as does altruism. But it is difficult to believe that the individuals who run the nation's businesses are unaffected by the environmental awareness and concern that is now so widespread. There is an unmistakable concern for the environment, at least on the part of many large companies.

The right to know

The 1986 federal Emergency Planning and Community Right-to-Know Act requires certain classes of manufacturing companies and federal facilities to report annually on a wide range of toxic emissions (Hadden 1989). (The data are reported to EPA, maintained in a Toxics Release Inventory (TRI), and made available to the public on request.) Begun in 1988, EPA issues

an annual report on the collected toxics data. For example, the 2000 TRI found that a total of approximately 38 billion pounds of TRI chemicals 'in production-related waste was reported as managed during 2000'. This elementary idea is intended not only to assist regulatory bodies and community groups, but also to provide an incentive to firms to keep their emissions low in order to avoid public agitation or disapproval. Though this might be considered wishful thinking, the power of knowledge should not be underestimated. It may even be useful in informing companies (as well as their shareholders) of their environmental behaviour. Portney (1990: 280) recounts the story of the Monsanto Company, whose chairman, on seeing the level of his company's reported emissions, immediately pledged to reduce those emissions by 90 percent over the following five years, even though they were not in violation of any law. Whether such action results from a desire for good public relations, a real concern for environmental quality, or enlightened self-interest does not much matter – though the concern of industrialists for the environment may well be higher than they are often given credit for.

This Act also required industry to change the way in which emissions were reported from parts-per-million to pounds per year. The changeover revealed 'surprisingly big' numbers – 'not only in terms of lost product and substantial cost savings but also in terms of credibility'. The quotation is from a paper by a spokesman for the Chemical Manufacturers Association (continued in Box 13.4). It is seldom that a technique for reducing pollution proves so effective and cheap.

BOX 13.4 EXPLAINING TO THE PUBLIC

It was clear that explaining those numbers to the public would be difficult. No matter how comfortable a company or plant had become with the safety of a parts-per-million concentration at the fence, the emissions of hundreds of thousands, or even millions, of pounds per year was tough to explain – even if it was only a fraction of a percent of the company's throughput. This revelation caused most plant managers – often for the first time – to put themselves in their neighbors' place and consider what other people might be thinking about the perceived safety of a plant's operations. Real confrontation of issues began. Risk communication experts were retained and plant managers learned about the importance of listening to their neighbors' concerns . . . and about how to understand and deal with an outraged public.

Source: Chemical Manufacturers Association

In 2001, the Bush administration issued a new rule regarding poisoning from lead emissions. Recognizing that lead has been a persistent problem in humans and the environment, the reporting threshold for industries would be lowered to 100 pounds for each facility. This represents a drastic change from the previous requirement – facilities would have to report lead and lead compound emissions if the facility manufactured or processed more than 25,000 pounds annually or if the facility used more than 10,000 pounds annually. The rule did not apply to lead contained in stainless steel, brass, and bronze alloys.

Public participation

The ability for members of the public to participate in the formulation of environmental policy has become an integral part of the public policy-making process. Public notice and participation requirements (e.g., public hearings, forums, meetings, advisory committees, etc.) can be found through environmental policies. For example, the CEQ requires agencies to make efforts to involve the public in fulfilling their efforts throughout the NEPA process. In addition, under the 1990 Clean Air Act, citizens are able to participate in meetings designed to determine how state and local governments fulfil their responsibilities of cleaning up air pollution. Ultimately, various procedural requirements exist for public participation to occur in the

formulation of environmental policy. Showing that a particular body has failed to adhere to the appropriate procedural requirements represents a common challenge to any decision that may have been made.

It is striking that many of the major changes in environmental policy have resulted from public pressure, itself roused by environmental disasters. Love Canal provided the political base for the 1980 Superfund legislation, and the 1984 chemical accident at Bhopal, India, which killed more than three thousand people and injured several hundred thousand, led to the Emergency Planning and Community Right-to-Know Act; the *Exxon Valdez* oil spill in 1989 was followed by the Oil Pollution Act of 1990. However, the role of the public in environmental policy is much more than that of alerting politicians. As the discussion on waste disposal demonstrates, some problems demand public involvement. Though such involvement may prove inadequate in itself, there may be no possibility of any solution without it.

The discussion on the shortcomings of risk assessment underlined the importance of values and the limited validity – and viability – of technocratic approaches. Putting these various points together adds up to a powerful case for a high degree of public participation. It needs to be stressed that this is not to suggest that the public's role is simply to supply value judgments to experts. Facts and values are inextricably intertwined – and this, in a democracy, is how it should be. The conclusions of a conference on risk assessment

BOX 13.5 FACTS AND VALUES

The conflict between 'expert' rankings and those of the public-at-large is less one of 'facts versus values' than one of 'values versus values'. We tend to talk about 'non-quantifiable factors' as if they were accessories to risk estimates that might marginally change these estimates . . . But what if the variation contributed by values is equal to or greater than the variation among risk estimates? The 'fact' that indoor radon may cause 100 times the death toll of Superfund sites moves from foreground to background if citizens view preventing each injury from the latter cause as thousands of times more important than the former.

Source: Finkel and Golding 1994: 336

(Finkel and Golding 1994: 335) are particularly telling here (see Box 13.5).

Public involvement in environmental policy is an essential feature of 'bottom-up' regulatory systems as distinct from 'top-down' systems that are run by experts and administrators. To coin a phrase, the issues are too important to be left to experts. However, it is a mistake to think solely in terms of active participation in the actual administration of pollution control. More typically, the involvement is through the normal channels of pluralistic politics. Usually this involves the state, the local governments, and citizen groups. Citizen groups often enter the scene by way of protest, either in a NIMBY way (campaigning against a waste disposal site, for instance) or in protest against the desecration of some natural feature (objecting to the flooding of a valley or to the continued pollution of a river). In the classic case, they start by vociferous opposition, graduate through learning about the problems involved, and eventually finish as powerful participants in a workable solution to the environmental problem that first motivated them.

If the necessary political framework exists, such a process can develop into a wider environmental movement. This framework might actually be brought into existence by the coming together of like-minded citizen groups, as with state environmental coordinating councils. Donald Snow, in his account of *Inside the Environmental Movement* (1992: 29), has noted:

The existence of a long-lived and successful 'coordinating council' with a paid staff and a grassroots board is a good indicator of whether a given state has a rich enough corps of environmental activists. In most states where they occur, these coordinating councils often arose from the need to enhance the representation of many local conservation organizations before state legislatures and administrations.

Such groups operate as a political link between local groups and the state government. The benefit is reciprocal. Not only do the local groups gain from access to the machinery of state environmental policy, but also the state gains a concerned active partner which can inform and advise it of the perspectives of localities and bring new questioning and ideas before it. Since public support can be an important ingredient of state environmental policy, this is no mean input.

The states are at the cutting edge of the implementation of environmental policy. Discussions of 'federal policy' can easily underplay the importance of the states. Though federal powers and funding are obviously important, they are only part of the total public policy operation. Environmental policy is largely carried out by the states; and this is inevitable since there is a definite limit to what can be done by the federal government on its own. Even with environmental issues that clearly fall within federal responsibility (such as transboundary pollution), the states play an important role. Indeed, the general complaint is not that the federal government takes the lead role, but that it gives insufficient resources to the states to enable them to carry out their responsibilities.

States, of course, have constitutional powers; but there is more to federal–state relations than these. A major power of the states lies in their knowledge of their diverse territories. Washington can rarely operate effectively (at least for any length of time) at the state level. It lacks the effective power, knowledge, and 'feel' that are needed for state administration. Moreover, the federal government is a many-headed monster that has the greatest difficulty in coordinating its own organs: there is no way in which it can find its way around the fifty state labyrinths, let alone beyond these to the localities and communities where environmental policy actually operates.

The progress of environmental policies

It seems that everything in the field of environmental policy is beleaguered by problems of lack of knowledge, scientific uncertainty, and scarcity of data. And so it is when one comes to attempting to assess what progress has been made. It is virtually impossible to determine the effect that particular policies have had on health, amenity, and standards of living, though it is generally held that the overall effect has been a very positive one. Even the extent to which environmental quality has improved is problematic. The relevant data are sparse and scattered. The frustrated attempt by the US General Accounting Office to assess water quality (Box 13.6) is illustrative.

Nevertheless, it needs no sophisticated analysis to show that progress has been both positive and very uneven. Air quality has shown the greatest improvement, though little progress has been made with the persistent problem of ozone, and there are still 100 million people living in areas that fail to meet national standards. However, even where air quality has not improved, it is likely that it would have been very much worse had it not been for the extensive pollution controls that have been introduced over the last quarter of a century. Firm judgments are difficult because of data problems, due mainly to the lack of adequate monitoring. (It should be noted that some improvement is due to the changing industrial scene – with major decline in heavy polluting industries such as steel.)

Changes in water quality have been very varied. Much publicity has attended the cleansing of a number of rivers where the fish have returned, but such improvements are restricted to a few areas. More generally, there has been a deterioration in water quality, largely due to the lack of success (and even effort) in tackling non-point source pollution from agricultural and urban areas. On the other hand, there has been an improvement in the quality of drinking water supply and sewage disposal, though this has been far from uniform across the country, and increasing contamination is caused by the growing number of new pollutants. An overall judgment is rendered even more difficult by the crude nature of the monitoring system. It does seem, however, that such progress as has been made is frequently offset by an increase in the number of pollutants created by technological development, and even by better understanding of the processes of contamination.

BOX 13.6 THE INFORMATION GAP

GAO was not able to draw definitive, generalizable conclusions . . . because evaluating changes in the nation's rivers and streams is inherently difficult, the empirical date produced by the studies sparse, and the methodological problems reduce the usefulness of the findings. Therefore, little conclusive information is available to the Congress to use in policy debates on the nation's water quality.

Source: Rosenbaum 2007: 52

Hazardous waste policies have been faced with great technical, legal, and political problems, which would remain even if greater resources were made available. They have had the result of reducing the number of available sites for disposal, which may have been beneficial, but, in the longer run, the greater benefit will come from the trend away from land disposal to alternative methods of disposal and waste reduction. Superfund has proved extremely costly and cumbersome; and the problems it was designed to meet have proved vastly larger than initially anticipated. Indeed, even more than with other pollution problems, little is known about the nature and size of the problem, except that it is enormous and will absorb large cleanup resources for many years to come. EPA has identified more than 27,000 abandoned sites, of which 1,275 are on the National Priority List; but only 149 were cleaned up between 1980 and 1993 (Rosenbaum 2007: 59). The average cost has been about $25 million per site. Progress has been modest, to say the least.

Any assessment of environmental policies is clouded by the shortage of data and a lack of adequate measures of environmental quality. It seems clear, however, that current approaches are failing to achieve anticipated benefits. There are many reasons for this, many of which stem from the sheer complexity of the problems and the limits of our knowledge. Better policies, better administration, greater resources, more research – all these would help; but there is no quick fix.

Further reading (Chapters 12 and 13)

John Muir's Sierra Club (http://www.sierraclub.org) still exists today to lobby for conservation efforts and to provide wilderness-oriented outings for Sierra Club members. The organization has an email newsletter and online links that offer tips on how to become politically active for the environment.

Friends of the Earth (http://www.foe.org) is now an international organization. Its website lists all its campaign successes by year since 1969. The previous seven years of its annual reports are also available on internet links. Visitors will find current campaigns, online petitions and sample emails to send to companies and governments in support of environmental causes.

Earth Island Institute (https://www.earthisland.org) focuses on education, citizen activism, and creating projects to protect the environment. The institute has several of its own programs that direct their non-profit partners into action. Also on its website is a tribute to its founder, David Brower.

Earth Day has grown since it began in 1970 and its Earth Day Network (http://www.earthday.org) gives Earth Day enthusiasts a year-long opportunity to plan for the annual event. Earth Day now boasts its own television shows and blog. The website encourages visitors to learn how to conserve and to get involved with conservation politics.

NEPAnet (http://www.ceq.hss.doe.gov) is an online source of information for everything about the National Environmental Policy Act. The comprehensive website offers many internet links, such as a link to the Council on Environmental Quality (CEQ) and its studies and reports, as well as links to electronic environmental impact analyses and environment impact statements (EIS), and links to NEPA litigation and case law.

Georgia and California both have programs to reduce air pollution that have excellent websites. Georgia's Clean Air Campaign (http://www.cleanaircampaign.com) offers an air quality index email alert, promotes taking other modes of transportation rather than a single-occupied vehicle, and provides news stories on air pollution. Its Commuter Rewards program awards cash and prizes to automobile commuters who make a switch to another mode of transportation. California's Spare the Air (http://www.sparetheair.org) offers a list of Bay Area Best Workplaces for Commuters and even Ozone Movies which are animated maps that use the Air Quality Index (AQI) scale to show a daily summary of ozone concentrations.

A US Geological Survey website (http://bqs.usgs.gov/acidrain) provides a map showing acid rain deposits in the United States. Also provided are online links to presentations on acid deposition and reports ranging from the changing chemistry of lakes to a summary of the Atmospheric Deposition Program of the US Geological Survey.

The Salt Lake watershed in Hawaii has a particularly interactive website that illustrates pollution sources (http://protectingwater.com). As the website outlines the basic non-point source pollution from automobiles, residents, agriculture, urban, construction, business, and storm drains, it urges the community member to refrain from polluting.

Clean Water Action (http://www.cleanwateraction.org) is dedicated to lobbying in Washington, DC, for clean water legislation and also creating national and local programs and campaigns. The national organization has offices in approximately eighteen states, helps train canvassers to raise awareness about clean water news, and creates downloadable literature that supports its causes.

Waste is managed by several companies, and those companies helped to create the Solid Waste Association of North America (SWANA) (http://www.swana.org). Its website has information on how to educate the public about waste, innovation for different ways to use waste, and communication about the importance of proper waste management. Visitors can see SWANA's technical division programs, local chapters, and its conference offerings online.

The Global Development Research Center, a virtual organization which conducts research and education on urban planning, provides a waste management framework that is environmentally friendly (http://www.gdrc.org/uem/waste/waste.html). Taking an international perspective, the website displays examples from Japan as well as Waste Management Multilateral Environmental Agreements from Eastern Europe.

The World Bank lists Urban Development as a topic on its website (http://www.worldbank.org/urban). Among the issues listed is solid waste management, which, when clicked upon, will take the visitor to a webpage that outlines problems and possible programs for waste management solutions.

The US Department of Health and Human Services created the Agency for Toxic Substances & Disease Registry (http://www.atsdr.cdc.gov). Focused on public health, the agency strives to prevent toxic substances from reaching the public. The website offers a range of topics including case studies in environmental medicine, the congressional mandate for the agency, and internet links to toxicological profiles.

The US has its own Nuclear Regulatory Commission (http://www.nrc.gov/waste.html) that deals with nuclear waste. The website shows where low-level waste disposal and high-level waste disposal are located. It provides educational materials on a range of subjects including the transportation and storage of spent nuclear fuel.

One website that strives to be non-partisan in its discussion of radioactive waste is RadWaste.org (http://www.radwaste.org). Founded in 1996, the organization is an informational source to the public and has an impressive number of links on all nuclear-related topics such as nuclear products and services, technical reports, nuclear fusion, and industry and trade organizations for nuclear products.

The US Department of Commerce has its own Ocean and Coastal Resource Management division (http://coastalmanagement.noaa.gov). Among its offerings is a coastal management program that relies on a partnership with thirty-four coastal and Great Lake states to enhance and preserve the environment on the coast.

The Audubon Society, an organization that focuses on conservation for wildlife, offers an instructive guide on Habitat Conservation Plans (HCPs) on its website (http://web4.audubon.org/campaign/esa/hcp-guide.html). Starting with the definition of HCPs, the website offers advice on how citizens can become involved.

For an example of a comprehensive Habitat Conservation Plan, Riverside County, California, has a fact-filled website on its Multiple Species Habitat Conservation Plan (MSHCP) (http://www.rcip.org/conservation.htm). Along with internet links to the text of the Environmental

Impact Review and Statement, responses and comments, the website provides statistics of conservation for its MSHCP.

The National Mitigation Banking Association was created in 1998 to encourage mitigation banking and conservation banking as a means of compensating for adverse impacts on the nation's environment – http://www.mitigationbanking.org.

The US Environmental Protection Agency (EPA) provides online information about environmental justice and what EPA is doing about it (http://www.epa.gov/compliance/environmentaljustice). Starting with a definition of environmental justice, the website also has links to the National Environmental Justice Advisory Council, grants, and a federal interagency working group dedicated to promoting environmental justice.

The National Pollution Prevention Roundtable (http://www.p2.org), a non-profit organization, provides online visitors with case studies, current projects for pollution prevention, and information on local workshops for public education purposes. Another national organization called Pollution Prevention Resource Exchange (http://www.p2rx.org) is the result of collaboration among eight regional pollution prevention information centers. The exchange provides networking and technical assistance to individual states.

The International Emissions Trading Association (IETA) (http://www.ieta.org) works as a researcher, advocate, and marketer for emissions trading. Based on principles established by the United Nations Framework Convention on Climate Change, the IETA helps to create market-based trading systems for greenhouse gas emissions. The non-profit organization's goal is to reduce greenhouse gas emissions.

The Environmental Markets Association (http://www.emahq.org) promotes the use of market-based trading solutions for environmental concerns. The non-profit and international organization mostly focuses on emissions trading but also offers online news and links to issues of its magazine, *The Emissions Trader*.

Data and analysis tools for TRI, Toxics Release Inventory (http://www.epa.gov/tri) are available to the public on the US EPA website. Featured topics include reporting standards, training workshops, and updates to TRI reporting rules.

The non-profit Right to Know Network (http://www.rtknet.org) provides visitors with search tools for various environmentally focused government databases. The range of databases is wide and includes EPA's Toxics Release Inventory as well as the Comprehensive Environmental Response, Compensation, and Liability Information System (CERCLIS). Established in 1989 as a response to the Emergency Planning and Community Right to Know Act, the Right to Know Network is a project of the government watchdog organization OMB Watch.

The American Meteorological Society constructed a website that provides an extensive history of the Clean Air Act (http://www.ametsoc.org/sloan/cleanair/index.html). Online visitors can identify the people involved in the Act, photos of air pollution at the time of the Act, environmental events and many other historical facts.

The Clean Air Act Information Network (http://www.envinfo.com/caalead.html) was established in 1995 to provide educational and informational materials to facilitate the implementation of the Act. The Network's website has the original text as well as updates of the Act in its online document system.

It is not easy to keep up to date with advances (and reverses) in environmental affairs: events can move quickly, and there is an enormous outpouring of relevant reports. It is important to use the latest editions of standard texts. At the time of writing, the best is Rosenbaum (2007) *Environmental Politics and Policy*. Also to be recommended is Smith (2003) *The Environmental Policy Paradox*.

An informative collection of essays is Vig and Kraft (1990) *Environmental Policy in the 1990s*. For a wide discussion of environmentalism see Gottlieb (1993) *Forcing the Spring: The Transformation of the American Environmental Movement*.

There are many important postwar writings on the environment. Lewis Mumford's books of the period include *The City in History* (1961); and *Technics and Civilization* (1962). René Dubos wrote a column in *The American Scholar*, as well as *So Human an Animal* (1968); see also Piel and Segerberg (1990) *The World of René Dubos*. Also important are Aldo Leopold *Sand County Almanac* (published posthumously in 1949); Paul Ehrlich's *The Population Bomb* (1968); and Rachel Carson's *Silent Spring* (1962).

A history of environmental policy is given in Lacey (1990) *Government and Environmental Politics: Essays on Historical Development since World War II*. Two recent additions to the literature include Rosenbaum (2010) *Environmental Politics and Policy*; and Vig and Kraft (2012) *Environmental Policy: New Directions for the Twenty-first Century*. A revealing book of readings is Nash (1990) *American Environmentalism: Readings in Conservation History*. For insights into the problems of translating scientific research into policy see Unman (1993) *Keeping Pace with Science and Engineering: Case Studies in Environmental Regulation*. Assessment of risk plays a major role in environmental policy. An excellent review is provided by a range of essays in Finkel and Golding (1994) *Worst Things First? The Debate over Risk-Based National Environmental Priorities*. A fascinating, detailed case study is given in Cole (1993) *Element of Risk: The Politics of Radon*.

There is a rich literature on the subject of economic incentives. The classic statement is Dales (1968) *Pollution, Property and Prices*. A well-known work is Kneese and Schultze (1975) *Pollution, Prices, and Public Policy*. Highly recommended as thoughtful, informative, and readable is Cairncross (1992, paperback edition 1993) *Costing the Earth*.

Also useful is Kelman (1981) *What Price Incentives?* See also the articles in *Columbia Journal of Environmental Law* (1988), vol. 13.

The literature on environmental markets and emissions trading has been growing. For example, see Atkinson and Tietenberg (1991) 'Market failure in incentive-based regulations: the case of emissions trading'; Carlin (1992) 'The US experience with economic incentives to control environmental pollution'; Anderson and Lohof (1997) 'The U.S. experience with economic incentives in environmental pollution control policy'; Kosobud and Zimmerman (1997) *Market-Based Approaches to Environmental Policy: Regulatory Innovations to the Fore*; Chichilnisky and Heal (2000) *Environmental Markets: Equity and Efficiency*; Ellerman *et al.* (2000) *Markets for Clean Air: The U.S. Acid Rain Program*; Kosobud (2000) *Emissions Trading: Environmental Policy's New Approach*; Stavins (2000) *Experience with Market-Based Environmental Policy Instruments*; Ellerman *et al.* (2003) *Emissions Trading in the U.S.: Experience, Lessons and Considerations for Greenhouse Gases*; Tietenberg (2006) *Emissions Trading: Principles and Practice*; Asplund (2008) *Profiting from Clean Energy: A Complete Guide to Trading Green in Solar, Wind, Ethanol, Fuel Cell, Carbon Credit Industries, and More*; Ellerman, Convery, and de Perthuis (2010) *Pricing Carbon: The European Union Emissions Trading Scheme*; and Bergman (2011) *Cap-And-Trade Program Considerations*.

A particularly useful book on environmental law is Farber (2010) *Environmental Law in a Nutshell*. This small book provides a succinct and accessible summary. An introduction to environmental law can also be found in Kubasek and Silverman (2002) *Environmental Law* and Salzman and Thompson, Jr. (2010) *Environmental Law and Policy*. An excellent, detailed but clear account of the EIS process is given in Bass and Herson (1993) *Mastering NEPA: A Step-by-Step Approach*. For a good example of a sophisticated state environmental protection act, see Remy *et al.* (1994) *Guide to the California Environmental Quality Act*. The effectiveness of federal environmental statutes is examined by Meiners and Morriss (2000) *The Common Law and the Environment: Rethinking the Statutory Basis for Modern Environmental Law*.

For discussions of coastal zone planning and management, see Clark (1996) *Coastal Zone Management Handbook*; Klee (1999) *The Coastal Environment: Toward Integrated Coastal and Marine Sanctuary Development*; and Beatley *et al.* (2002) *An Introduction to Coastal Zone Management*.

The literature on environmental justice has increased in recent years. Representative works in this area include:

Bullard (2000) *Dumping in Dixie: Race, Class, and Environmental Quality*; Novotny (2000) *Where We Live, Work, and Play: The Environmental Justice Movement and the Struggle for a New Environmentalism*; Lester *et al.* (2001) *Environmental Injustice in the United States: Myths and Realities*; Mutz *et al.* (2002) *Justice and Natural Resources: Concepts, Strategies, and Applications*; Pellow (2002) *Garbage Wars: The Struggle for Environmental Justice in Chicago*: and Rechtschaffen, Gauna, and O'Neill (2009) *Environmental Justice: Law, Policy and Regulation*. Additional readings can be found in the Further Reading section at the end of Chapter 14.

Supplementary references for particular policy areas are:

Waste

An early popular and highly readable account is Packard (1960) *The Waste Makers*. An accessible, digestible account of the Love Canal incident and its legislative influences is given in the chapter on 'Passing Superfund' in Landy *et al.* (1994) *The Environmental Protection Agency: Asking the Wrong Questions – From Nixon to Clinton*. A detailed account of the Love Canal controversy is given in Levine (1982) *Love Canal: Science, Politics and People*. Later studies on Love Canal can be found in Gibbs (1998) *Love Canal: The Story Continues*; and Sherrow (2001) *Love Canal: Toxic Waste Tragedy*. Colten, Skinner, and Piasecki (1995) examine how disasters such as Love Canal have occurred before the formation of the Environmental Protection Agency. On packaging see Rousakis and Weintraub (1994) 'Packaging, environmentally protective municipal solid waste management, and limits to the economic premise'. A recent study of public involvement in decisions on hazardous waste siting is Rabe (1994) *Beyond NIMBY: Hazardous Waste Siting in Canada and the United States*. Other accounts of hazardous waste siting can be found in Portney (1991) *Siting Hazardous Waste Treatment Facilities: The NIMBY Syndrome*; and Munton (1996) *Hazardous Waste Siting and Democratic Choice*.

Nuclear waste

Lenssen (1991) *Nuclear Waste: The Problem that Won't Go Away*; US General Accounting Office (1993a) *Department of Energy: Cleaning up Inactive Facilities will be Difficult*; (1993b) *Much Work Remains to Accelerate Facility Cleanups*; MacFarlane and Ewing (2006) *Uncertainty Underground: Yucca Mountain and the Nation's High-Level Nuclear Waste*; Walker (2009) *The Road to Yucca Mountain: The Development of Radioactive Waste Policy in the United States*; and Alley and Alley (2012) *Too Hot to Touch: The Problem of High-Level Nuclear Waste*.

Clean air

Bryner (1993) *Blue Skies, Green Politics: The Clean Air Act of 1990*; Stensvaag (1991) *Clean Air Act: Law and Practice*; Cooper and Alley (2010) *Air Pollution Control: A Design Approach*; and Davidson and Norbeck (2011) *An Interactive History of the Clean Air Act: Scientific and Policy Perspectives*.

Water

Adler (1993) *The Clean Water Act: Twenty Years Later*; and Ryan (2012b) *The Clean Water Act Handbook*.

Endangered species and habitat conservation planning

Beatley (1994b) *Habitat Conservation Planning: Endangered Species and Urban Growth*; Stanford Environmental Law Society (2000) *The Endangered Species Act*; Burgess (2001) *Fate of the Wild: The Endangered Species Act and the Future of Biodiversity*; Czech and Krausman (2001) *The Endangered Species Act: History, Conservation Biology, and Public Policy*; Noss *et al.* (1997) *The Science of Conservation Planning: Habitat Conservation under the Endangered Species Act*; Petersen (2002) *Acting for Endangered Species: The Statutory Ark*; Duerksen and Snyder (2005) *Nature-Friendly Communities: Habitat Protection and Land Use Planning*; and Goble, Scott, and Davis (2005) *The Endangered Species Act at Thirty*.

Wetlands

Environmental Law Institute (2002) *Banks and Fees: The Status of Off-Site Wetlands Mitigation in the United States*; Kusler (1983) *Our National Wetland Heritage: Protection Guidebook*; Kusler and Kentula (1990) *Wetland Creation and Restoration: The Status of the Science*; Salvesen (1990) *Wetlands: Mitigating and Regulating Development Trends*; Mitsch and Gosselink (1993) *Wetlands*; and Kelty (2010) *Wetland Ecology: Principles and Conservation*.

Questions to discuss

1 Discuss the ways in which environmental concerns have developed since the 1960s.

2 Why is it necessary to have environmental controls over federal agencies?

3 Describe the environmental review process. What are its limitations?

4 Explain the structure of clean air controls.

5 What are the problems of water pollution, and how can they be remedied?

6 'With waste, treatment is better than disposal; and reduction is better than treatment.' Discuss.

7 What is 'Superfund'? Does it work well?

8 Compare regulatory controls and economic incentives. Do we need both?

9 Discuss the nature of risk. Why is it relevant to environmental policy?

10 Discuss the significance of vehicle emissions in environmental pollution.

11 What is the case for a greater degree of public participation in environmental policy?

12 What are Habitat Conservation Plans and how are they related to the Endangered Species Act?

13 What are wetlands and how can we mitigate their loss?

14 What are greenhouse gases and how do they affect the environment?

15 What is a cap-and-trade program for greenhouse gas emissions?

14

Transportation

The ordinary 'horseless carriage' is at present a luxury for the wealthy, and although its price will probably fall in the future, it will never, of course, come into as common use as the bicycle.

Literary Digest 1899; quoted
in Jackson 1985: 157

The centrality of transportation

Transportation is the lifeline of the economic system. It represents the essential means by which activities are linked and thus made possible. Without access, most economic activity could not take place. Transportation is thus essential, not for its own sake, but because it provides access. Since it thus serves other activities, its character is determined exogenously. Of crucial importance is the pattern of land uses – the major determinant of transport needs. The more activities are dispersed, the greater is the amount of transportation required to access them. Nevertheless, transport does not simply follow activities: its potentialities facilitate and limit the development and spread of activities. In one sense, it can be said that the history of both economic and urban development is a reflection of the history of the development of transportation.

The course of urbanization (and the disastrous attempt to arrest urban decline by improving access to the city) has been outlined in Chapter 2; here the concern is with contemporary problems and policies relating to the operation and planning of urban transportation systems. The chapter opens by summarizing some major transportation trends. This is followed by an analysis of a number of ways in which traffic might be restrained – through land use planning; by direct controls, by demand management, and by congestion charging. Finally, a brief account is given of a remarkable congressional initiative to tackle transportation problems through comprehensive state and regional planning.

Some of the issues discussed in this chapter are developed further, within different contexts, in other chapters. The role of land use planning policies in the restraint of traffic growth is dealt with in the discussion of growth management in Part 3. The chapter on development charges discusses the use of transportation impact fees and similar charges on developers.

Transportation planning

In 1956, the Interstate Highway Act established the National Interstate and Defense Highway System. Over 40,000 miles of highways were built under the act. Ninety percent of the construction costs were covered by the federal government. Two results of the new highways were the sprawling out of cities and the creation of new towns and cities near the newly constructed highways.

While the focus in earlier years was building roads and highways, planners came to realize that continuing to focus on highway construction was causing problems, such as pollution, congestion, and displacement of people and businesses. They recognized the role transportation was playing in shaping America's cities and the importance of integrating transportation planning into the general planning process of the cities. As such, attention started turning to the importance of transportation planning involving a variety of participants (local governments, private sector, the public, interest groups, etc.).

The Federal-Aid Highway Act of 1962 mandated the approval of any highway project funded by the Act, in an urban area of 50,000 or more population, based on a continuing, comprehensive urban transportation planning process that is carried out cooperatively by states and local governments – the so-called 'Three-C planning process'. The basic principles of this process, as developed by the Bureau of Public Roads, can be found in Box 14.1.

Two years later, the Urban Mass Transportation Act of 1964 was enacted. This legislation provided federal funding to states and local governments for mass transportation capital projects. It also provided the impetus for financing mass transportation research, planning, and operations. In addition to this legislation, two other items of significance emerged – the Urban Mass Transportation Administration (UMTA) was created and the Bureau of Public Roads (created in 1916) was renamed the Federal Highway Administration.

The 1970s saw continued federal promotion of transportation planning. The Urban Mass Transportation Act of 1970 required planners to develop environmental impact analyses and to hold public meetings to accommodate elderly and handicapped populations. This followed the passage of the National Environmental Policy Act (NEPA), which required an environmental impact statement (EIS) for all proposed major federal actions which might have significant environmental impacts. Environmental factors would now have to be considered when making transportation decisions.

The early 1970s witnessed the creation of a new regional entity, in urbanized areas of 50,000 or more population, which would attempt to build regional consensus/agreement on regionwide transportation systems – the Metropolitan Planning Organization (MPO). The key to an MPO was that it was planning integrated regional transportation systems, not individual transportation projects. The systems were to be based on the previously mentioned 'Three-C planning process'. According to the Association of Metropolitan Planning Organizations, there are currently over 380 MPOs in the United States. Since the 2010 US Census has identified 36 new urbanized areas, and they must be represented by an MPO, new MPOs will be formed. While some might become a part of an existing MPO, it is expected that over 400 MPOs will exist by March

BOX 14.1 TEN BASIC ELEMENTS OF A 3C PLANNING PROCESS

1 Economic factors affecting development
2 Population
3 Land use
4 Transportation facilities including those for mass transportation
5 Travel patterns
6 Terminal and transfer facilities
7 Traffic control features
8 Zoning ordinances, subdivision regulations, building codes, etc.
9 Financial resources
10 Social and community-wide value factors, such as preservation of open space, parks and recreational facilities; preservation of historical sites and buildings; environmental amenities; and aesthetics.

Source: Weiner 1997

BOX 14.2 CORE FUNCTIONS OF AN MPO

1 Establish a setting: establish and manage a fair and impartial setting for effective regional decision-making in the metropolitan area.
2 Identify and evaluate alternative transportation improvement options: use data and planning methods to generate and evaluate alternatives. Planning studies and evaluations are included in the Unified Planning Work Program or UPWP.
3 Prepare and maintain a Metropolitan Transportation Plan (MTP): develop and update a long-range transportation plan for the metropolitan area covering a planning horizon of at least twenty years that fosters (1) mobility and access for people and goods, (2) efficient system performance and preservation, and (3) good quality of life.
4 Develop a Transportation Improvement Program (TIP): develop a short-range (four-year) program of transportation improvements based on the long-range transportation plan; the TIP should be designed to achieve the area's goals, using spending, regulating, operating, management, and financial tools.
5 Involve the public: involve the general public and other affected constituencies in the four essential functions listed above.

Source: US DOT, Federal Highway Administration, and Federal Transit Administration 2007: 5–6

BOX 14.3 INTERMODAL SURFACE TRANSPORTATION EFFICIENCY ACT

The National Intermodal Transportation System shall consist of all forms of transportation in a unified, interconnected manner, including transportation systems of the future, to reduce energy consumption and air pollution while promoting economic development and supporting the Nation's preeminent position in international commerce.

2013. The core functions of an MPO can be found in Box 14.2.

There is a clear necessity for transportation planning to be on a more comprehensive basis. Transportation systems operate over wide areas: both control measures and investments can be adequately planned only over the wider region where their impacts are felt. The interconnections of transportation networks spread far beyond the locality in which they are made. These simple points present administrative, financial, and political difficulties which governments have been loath to grasp. By the end of the 1980s, however, it was clear that significant federal action was both needed

and, more surprisingly, accepted. The result was the passing in 1991 of a radical piece of legislation.

The Intermodal Surface and Transportation Efficiency Act of 1991 is usually referred to by its acronym ISTEA – commonly pronounced as 'iced tea' (which is certainly more memorable than its title). It is a remarkable piece of legislation that will either bring about a revolution in the way transportation investments are planned and implemented (as Congress intended) or it will go down in history as one of the most ambitious of congressional fantasies. Its objectives (see Box 14.3) embrace an extraordinary degree of coordinated planning. It re-emphasized the importance of metropolitan transportation planning and, in a sense, resurrected the MPO. It expanded the authority of the MPOs. The planning process is now concerned with broad issues of overall transportation and environmental efficiency rather than narrow matters of highway construction. The major elements of transportation planning in metropolitan areas under ISTEA can be found in Box 14.4. This is facilitated by the flexibility of federal funding. Many funds which were previously restricted to categorical programs can now be switched accordingly to provide the mix of projects which will best meet air quality, congestion, mobility, or other national goals.

The Act requires the preparation of *state transportation plans* and various other plans and *transportation improvement programs*. In metropolitan areas, plans are carried out jointly by the state and the *Metropolitan Planning Organizations* (MPOs) under the terms of a formal agreement. The MPO is 'the forum for cooperative transportation decision-making for an urbanized area'. In the largest metropolitan areas (those with a population of over 200,000), the MPOs are also known as *transportation management areas* (TMAs), and have additional responsibilities, particularly in connection with clean air (discussed in Chapter 12).

Curiously, but perhaps wisely, the term 'intermodal' is not defined, but it encompasses all transportation modes, including airport system plans, state rail plans, and port system plans. Public participation is a requirement of the planning process: plans are to involve all transport users. This 'proactive public involvement process' involves all 'affected public agencies, representatives of transportation agency employees, private providers of transportation, other interested parties affected', and specifically 'those traditionally underserved by existing transportation systems'.

The issues to be covered in the planning process are set out at length and in detail. They include

---

BOX 14.4 MAJOR ELEMENTS OF TRANSPORTATION PLANNING IN METROPOLITAN AREAS

1 A proactive and inclusive public involvement process;
2 Consideration of 15 specific planning factors to ensure that the transportation planning process reflects a variety of issues and considers other concerns such as land-use planning, energy conservation, and environmental management;
3 As part of plan development, major investment studies are conducted to address significant transportation problems in a corridor or subarea that might involve the use of federal funds;
4 Development and implementation of management systems including:

- Intermodal management system
- Congestion management system
- Public transit facilities management system
- Pavement management system
- Bridge management system
- Safety management system

5 Development of financial plans for implementing the transportation plan and TIP; and
6 Assurance that the transportation plan and urban plan (UP) conform to the State Implementation Plan (SIP) pursuant to the standards of the Clean Air Act Amendments of 1990 (CAAA).

Source: US DOT, no date: 7

congestion management strategies (including ridesharing, and pedestrian and bicycle facilities), the effects of transportation policy decisions on land use and development, the consistency between transportation plans and programs and land use plans, preservation of future rights-of-way, and 'the overall social, economic, energy, and environmental effects of transportation decisions'.

This remarkable initiative on the part of Congress has been prompted in part by environmental considerations. Both the Clean Air Act Amendments and ISTEA preclude the construction of new highways in areas that fail to meet federal air quality standards. In these areas (which include many major metropolitan regions), alternative (and 'specific') measures have to be proposed for reducing automobile travel (see Box 14.5).

Traffic congestion is intensely frustrating and very costly – estimated at over $40 billion a year (not counting the cost of environmental damage). It is becoming clear that the metropolitan areas simply cannot build their way out of the problem, and that insufficient relief can be obtained from traffic management measures. The peak period of congestion is lengthening, and on some metropolitan highways congestion lasts throughout the day. Innovative ways of managing road systems and of demand management have brought little relief. The result has been increased interest in exploring the validity of the theoretical advantages of congestion pricing (at least on the part of transportation experts!). Interestingly, there are indications that business leaders, who have for long opposed congestion charges, regard them as a lesser evil than the imposition of demand management schemes such as trip-reduction programs (National Research Council 1994, vol. 1: 21). It is also significant that ISTEA provides federal funding for a pilot congestion pricing program.

One of the many uncertainties facing the implementation of ISTEA is that of forging the necessary links between land use planning and transportation planning. The Act does not provide the MPOs with any new legal authority in this area; instead it lays great emphasis on a partnership of all relevant agencies to promote area-wide interests and goals. Nowhere is the political nature of the planning process more evident than on this issue. Though the new system will be greatly concerned with technical issues of great complexity, the more troublesome problems will lie in devising methods of communication, mediation, and decision-making that are acceptable to the multiplicity of agencies, authorities, and interests in a metropolitan area.

The Transportation Equity Act for the Twenty-First Century, better known as TEA-21, was enacted on June 9, 1998. It furthered the transportation initiatives promoted in ISTEA and authorized surface transportation programs in the areas of highways, highway safety, and transit for 1998–2003. The legislation was altered by the TEA-21 Restoration Act, a month later, to correct some technical inaccuracies. Ultimately, the legislation required a collaborative approach, involving a variety of stakeholders, to solving the country's transportation needs. As such, flexibility represented a key word in the legislation, in that state and local decision-makers were afforded the opportunity to consider a variety of options to address their specific transportation needs.

The legislation also emphasized the need to invest in improving the use of Intelligent Transportation Systems (ITS). For example, the Miami-Dade County Metropolitan Planning Organization received funding to enact a regional Advanced Traveler Information System (ATIS) Program to provide commuters and travelers with real-time information on existing traffic conditions. ATIS provides free information that can be accessed through a toll-free telephone number, the internet, commercial radio and television, email, and personal communication devices.

Strengthening transportation safety programs was another key focal area of TEA-21. Funding has been authorized for programs encouraging the proper use of protection devices; reducing school bus crashes; improving emergency medical services and trauma care systems; increasing pedestrian and bicyclist safety; and improving road safety.

The Safe, Accountable, Flexible, Efficient, Transportation Equity Act: A Legacy for Users (SAFETEA-LU) was signed into law by President George W. Bush on August 10, 2005. This legislation, which built on two other pieces of transportation legislation (ISTEA and TEA-21), provided $286.4

BOX 14.5 MANDATORY REDUCTIONS IN TRAFFIC

Instead of providing new capacity, those areas not in compliance with the Clean Air Act Amendments must propose specific measures for reducing automobile travel through measures such as trip reduction ordinances, employer-based transportation management, transit improvements, pricing, traffic flow improvements, and parking management. Employers with more than 100 employees in the ten metropolitan areas rated as severe or extreme 'nonattainment' are required to submit plans by 1994 that will result in reduction in the number of employees driving to work alone.

Source: National Research Council 1994: 1.20

Plate 13 Light rail transit in Salt Lake City
Photo by author

billion in guaranteed funding for highways, safety, and public transportation. It represents the largest investment in surface transportation in US history. It invests in safety, equity, innovative finance, congestion relief, mobility and productivity, efficiency, environmental stewardship, and environmental streamlining. In the August 10, 2005 signing ceremony, President Bush acknowledged the multiple benefits of the bill:

> [it] is more than a highway bill; it's a safety bill. The American people expect us to provide them with the safest possible transportation system, and this bill helps fulfill that obligation. This law makes our highways and mass transit systems safer and better, and it will help more people find work. And it accomplishes goals in a fiscally responsible way.

As previously noted, ISTEA ushered in a period of multimodal transportation planning. Relying on a single mode of transportation is no longer a viable option for the United States. Multimodal transportation consists of highways, roads, streets, bicycle paths and facilities, transit facilities, airports, rail, and even river transportation (see Plate 13). Bicycle- and car-sharing programs are starting to become commonplace throughout the US (see Plates 14 and 15). All of these

Plate 14 Pearl Street B-cycle station in San Antonio, Texas

Photo courtesy of San Antonio B-cycle

modes of transportation are designed to move people, goods, and services. A transportation system must be diverse and efficient, as well as equitable. The various modes must compete with each other.

States and local governments are now working on developing multimodal transportation systems. The Transportation Element of the City of Sumner, Washington, Comprehensive Plan has the following overall transportation goal: 'provide an efficient and safe multimodal transportation system to improve mobility for residents, employees, and visitors of Sumner while maintaining the small town quality of life within the city and supporting the economic vitality of the city' (City of Sumner, 2005: 81).

The State of Florida has authorized local governments to establish multimodal transportation districts under Chapter 163 of the Florida Growth Management Act. The Act, under Chapter 163.3180 (15)(a) of the Florida Statutes, defines a multimodal transportation district as 'an area where primary priority is placed on assuring a safe, comfortable, and attractive pedestrian environment, with convenient interconnection to transit'. The basic criteria for a multimodal transportation district can be found in Box 14.6. A reduction in automobile use and vehicle miles traveled is the ultimate goal of the district.

In 2009, the American Recovery and Reinvestment Act (ARRA) was passed by Congress, and signed into law by President Obama on February 17, 2009. This multi-purpose legislation came at a time when the US was experiencing its worst economic crisis in the last 70 years. Its goals were to create new jobs while maintaining existing jobs, stimulate the economy, and to provide accountability and transparency in government spending. To achieve these goals, some $787 billion was authorized for such items as tax cuts for families and businesses, funding for unemployment benefits, new federal contracts, grants, and loans. The amount available to areas was increased to $840 billion in 2012.

Over $48 billion of ARRA funds was to be invested in transportation infrastructure projects for items involving highways and bridges, airports, high-speed and intercity rail, and port infrastructure. Examples of specific projects funded by ARRA ranged from constructing a new river bridge, safety improvements, rehabilitating a bridge, a congestion relief project, constructing new bus and carpool lanes, new road construction, paving highways, purchasing buses, widening freeways, to constructing roadside rest areas.

The Transportation Investment Generating Economic Recovery (TIGER) Discretionary Grant Program was established in 2009 by the ARRA. This program offered federal funds for surface transportation infrastructure projects involving roads, rail, transit and port projects that would act as a catalyst for creating jobs, stimulating the economy and helping to build livable communities.

To be considered for a TIGER grant, applicants had to meet both primary and secondary selection criteria. For example, potential projects had to show that they would have significant long-term outcomes for the nation or region. Among the other items a project would have to show are that it would improve existing facilities and systems, contribute to the economic competitiveness of the US, improve the living and working environment throughout communities, contribute to sustainability goals, improve transportation safety, and create jobs. On the secondary side of selecting projects for funding, the proposed project would have to be innovative and result in a collaborative partnership with various parties. Over $1.5 billion have been allocated for discretionary projects funded by TIGER (see Plates 16 and 17).

The Moving Ahead for Progress in the 21st Century Act (MAP-21) is the latest federal legislation to fund

Plate 15 Car-sharing program

Photo by author

surface transportation programs. Signed into law by President Obama on July 6, 2012, it extended SAFETEA-LU until Fiscal Year 2014 and funded surface transportation projects in the amount of $100 billion for fiscal years 2013 and 2014. MAP-21 is designed to address multiple challenges to transportation systems, including safety, reducing traffic congestion, and protecting the environment. It funds multimodal surface transportation programs related to ports, transit, highways, bridges, rail, cycling, etc. Research, technology, training, and education programs were included in the funding. Ultimately, MAP-21 restructured highway formula programs, eliminated a number of discretionary programs while continuing some discretionary programs, created two new formula programs on construction of ferry boats and ferry terminal facilities and transportation alternatives, and created a new discretionary program focusing on Tribal Highway Priority Projects.

With the success of high-speed and intercity rail transportation in Europe and Asia, the call for such a system has garnered a great deal of attention in recent years. The Obama administration has recognized the role of rail transportation in building the nation's economy and is investing billions of dollars in economic stimulus funding for the long-term development of a national high-speed rail network that would move people and goods more efficiently. According to the Obama administration, this infusion of dollars will create jobs, revitalize manufacturing industries associated with goods to create and maintain such a network, and spur economic development opportunities throughout the US. Federal funding could be used for individual projects, corridor projects, or planning.

Areas throughout the US have received federal funding for a multitude of rail transportation projects. The types of projects receiving federal funding were those related to funding a state rail plan, installing new signaling equipment, modernizing emissions controls in locomotives, station upgrades, improving railroad tracks, completing planning studies for extending rail services, reducing delays on long distance trains, and purchasing American-made rail cars and locomotives. Recently funds have been awarded to Indiana to reduce passenger delays in the Midwest, to Virginia to develop a high-speed rail corridor in the southeastern part of the state and to construct new track and improve existing track, and to California for high-speed rail construction.

In the current period of economic problems in the US, many individuals feel the funds could be used to fund other, more needed programs in such areas as education, health, and welfare. In 2010 over $1 billion in federal funds that had been allocated to Wisconsin and Ohio were reallocated to other states to support high-speed rail projects. In Florida, Governor Charlie Crist sought federal funding for a high-speed rail project from Tampa to Orlando. Governor Rick Scott, who followed Crist, rejected the federal funding, due to the belief that the state would be on the hook for future cost overruns.

Many people tend to ignore the importance of our ports in moving people and goods from one place to another. This includes ferries, passenger ships, private recreational watercraft, and ocean-going cargo vessels. An efficient port system is a strong component of the national economy. With over 350 commercial ports providing cargo and passenger handling facilities, improving and modernizing the port infrastructure, in concert with the private sector, will help facilitate the economic development of the nation by providing jobs and by encouraging international commerce. Recently, TIGER funding has supported projects related to expanding the capacity of a port in New Jersey, renovating dock space, developing an on-site rail system, and installing a large crane at a port to help facilitate the movement of goods to other modes of transportation.

Many individuals fail to recognize that the US has had a long tradition of maritime transportation on the seas, its rivers and the Great Lakes. Maritime transportation facilitates the movement of people and goods. The ports act as transhipment points – goods arriving on one form of transportation and departing to another location via another mode of transportation. The classic example would be goods coming into a port via a container on a vessel and leaving the port for another destination by truck or rail.

According to the US Department of Transportation Maritime Administration (2007: 3), 'the marine

Plate 16 TIGER funded project

Plate 17 Cargo container ship

Photo courtesy of James Fawcett

transportation industry has become a highly sophisticated, global, intermodal transportation network that is absolutely vital to America's economy and continued prosperity'. Although the current economic downturn has hurt the marine transportation industry, the US is investing more in port infrastructure. As previously noted, MAP-21 has provided federal support for port infrastructure projects. MAP-21 also requires the development of a National Freight Strategic Plan that recognizes the important role that navigable waterways have in the movement of goods in the global economy. TIGER funds have also supported projects dealing with port infrastructure.

The arithmetic of transportation

The United States has always been concerned to have efficient transportation systems and, to a large extent, it has succeeded. There are over 231 million motor vehicles in use (a fourfold increase since 1950). Nine out of ten households have a motor vehicle, and a half have two or more. Four-fifths of passenger traffic is by private auto. Commuting by car is the most usual means of getting to work: three-quarters travel alone (now technically known as SOV travel – in a *single occupancy vehicle*), and 13 percent share the ride. Despite the amount of attention given to commuting, only a fifth of auto trips are for this purpose: shopping is almost as important, while personal and pleasure purposes each account for around a fifth of journeys.

Public transit declined with the growth of car ownership, but in more recent years has increased somewhat. Only 5 percent of work journeys are made by public transit. Two-thirds of these are by bus. Railroads are statistically insignificant in the national total of work trips, but they carry over a third of freight (measured in ton-miles). The other major freight carriers are trucks (over a quarter), and oil pipelines (almost a fifth).

BOX 14.6 BASIC CRITERIA FOR A MULTIMODAL TRANSPORTATION DISTRICT

A multimodal transportation district should be supported by community design features that provide an adequate level of multimodal mobility and accessibility within the district. Community design elements needs for establishing a multimodal transportation district include:

- Provision of a complementary mix of land uses, including residential, educational, recreational, and cultural uses
- Provision of an interconnected network of streets designed to encourage walking and bicycle use with traffic calming where desirable
- Provision of appropriate densities and intensities of land uses within walking distance of transit stops
- Provision of daily activities within walking distance of residences; public infrastructure that is safe, comfortable, and attractive for pedestrians; adjoining buildings open to the street; and parking facilities structured to avoid conflict with pedestrian, transit, automobile, and truck travel
- Provision of transit service within the designated area, or a definitive commitment to the provision of transit. This definitive commitment should be found in local planning documents and in the approved capital improvements program. For new developments, transit connectivity to the major urban area must also be included, or a definitive commitment for transit connections, again evident in both planning documents and the approved capital improvement program.

Source: State of Florida, 2003: 12

Current commuting patterns are much more complex than used to be the case when employment was concentrated in cities. Today there is more commuting from suburbs to suburbs than from suburbs to central cities. This dispersal has been made possible by the flexibility provided by auto travel. It is also highly dependent upon auto travel since (to use transportation terminology) reverse-direction and circumferential commuting poses serious difficulties for public transit.

Telecommuting

The Sacramento, California, Transportation Management Association has defined telecommuting as:

> Using telecommunications technologies to replace traditional forms of commuting. Employees work all or part of the time outside the traditional office at remote work locations, which may include the home.
>
> The work goes to the worker, rather than the worker to the work.

It essentially means getting people off the roads, thereby reducing potential congestion and pollution.

A number of areas have endorsed telecommuting. The Oregon Office of Energy promotes its use because it conserves fuel, relieves traffic congestion, and improves air quality. Other agencies and areas point to similar reasons for promoting its use.

It may be that the continued increase in commuting could be reduced by the growth of telecommuting, though this is quite uncertain. In 1992, 30 percent of the labor force worked at home for at least part of the time. Most of these were self-employed or simply working after regular hours, but a growing number are full-time employees who would otherwise be commuting. They are able to work at home because of huge advances in sophisticated telephone and computer systems. In a real sense, telecommunications services can be substituted, partly or completely, for transportation to a conventional workplace.

The extent of telecommuting, the forms it takes, and the implications it has for transport, work, and lifestyles are not clear. There are also definitional

problems as, for example, with commuting to regional telework centers (which generally appear to reduce travel, though they do not eliminate it). Some use of telecommunications may be *additional* to work undertaken at the office, or simply a more efficient means of evening and weekend office 'homework'. Telecommuting is a diffuse activity, often undertaken on an informal basis. As such, it is not well captured in current statistics; and its transportation impacts are not easily measurable. Not surprisingly, any estimate of future trends is hazardous, because of both the paucity of information on the current situation and the difficulty of prediction. But there is no doubt that it could bring about great changes in transportation, as is also possible with other dimensions of telecommunications – telebanking, teletaxes, tele-education, teleshopping, and so forth. Telecommuting is officially accepted as a congestion-reducing 'travel demand management' measure eligible for federal funding under various state and federal programs (some of which are discussed later).

A study published in 1993 by the US Department of Transportation estimated that the number of telecommuters might increase from 2 million in 1992 to between 7.5 and 15 million in 2002. On the definitions and assumptions used, this would involve between 5.2 and 10.4 percent of the labor force at the later date. The effect could be to save up to 35 billion vehicle travel miles. These and other figures are reproduced in Box 14.7. A later study showed the number of telecommuters decreased from 33.7 million in 2008 to 26.2 million in 2010 (World at Work, 2011: 3).

There is no way of knowing whether such a scenario might develop. The observable effect of reducing the amount of vehicle travel miles might be largely one of an increase in convenience, efficiency, and opportunities. There are many ways in which the possible benefits might be nullified. The beneficial effects of transportation programs to date (whether in easing congestion or in reducing air pollution) have been overtaken by increases in the number and use of motor vehicles. The most that has been achieved has been a slower rate of traffic growth. This, of course, is better than nothing, but it indicates how difficult it is to bring about significant improvements. Moreover, if

BOX 14.7 PROJECTED TELECOMMUTING AND ITS TRANSPORTATION IMPACTS

	1992	2002
Number of telecommuters (millions)	2.0	7.5–15.0
as proportion of labor force (%)	1.6	5.2–10.4
proportion working at home (%)	99.0	49.7
proportion working at telework center (%)	1	50
Average days per week telecommuting	1–2	3–4
Saving in vehicle miles traveled (billion miles)	3.7	17.6–35.1
as proportion of total passenger vehicle miles (%)	0.23	0.7–1.4
as proportion of commuting vehicle miles (%)	0.7	2.3–4.5
Saving in emissions (%)		
NO_x	0.23	1.1–2.2
HC	0.31	1.4–2.7
CO	0.36	1.7–3.4

Source: US Department of Transport 1993: viii–ix

telecommuting did have an impact on the reduction of congestion, this might simply attract more traffic to the less congested roads. This is an illustration of Downs' 'convergence' theory, which is outlined below.

Suburbanization and transportation

Historically, a city has been a node of concentrated functions with very high accessibility. Technological changes in production, energy, and transportation have dramatically reduced the advantages of the city, and a huge amount of activity has moved out to suburban locations. Even more significant has been the new growth that has located in the suburbs. At first sight, it might be expected that this would ease the urban transportation problem since there would be less traveling into city centers, and workers would have a shorter journey to their suburban homes. In fact, other factors have intervened. First, central city employment has not declined very much (in some areas it has increased slightly). Second, though many suburban residents commute to suburban job locations, their journeys are not necessarily shorter: in many cases they are much longer (whether the commute is in the same suburb or to a different one). Third, very high levels of automobile use have led to increasing congestion: suburban journeys can now be as congested as those to the city. Other factors include the design of shopping malls, office and industrial parks, and the wide range of suburban employment centers which have often been explicitly designed for automobile use: so much so that they can discourage other forms of access. This, together with abundant car parking (typically free), is a major incentive to automobile transport.

The growth of employment in the suburbs has given rise not only to circumferential commuting, but also to 'reverse' commuting – traveling from the city to the suburb. Travel distance to work has also been increased by a 'jobs–housing mismatch': the lack of affordable housing in areas close to employment centers, which compels households to seek cheaper, far-distant housing locations. Restrictive zoning plays an important role in this. Overall, the pattern of commuting is complex. As a result, dealing with congestion has become extremely difficult.

No area is immune from traffic congestion. The Texas Transportation Institute estimates that in seventy-five of the largest US cities in 2001, $69.5 billion dollars are wasted in time and fuel costs (Cambridge Systematics 2004: ES-9).

In principle, there are two major ways in which the problems can be approached – by minimizing traffic generation through land use planning measures, or by directly controlling traffic through regulatory or economic measures.

Traffic restraint through land use planning

Since transportation is a function of land use, it seems obvious that one way of effectively reducing transportation problems is by imposing tighter land use controls. However, there are several difficulties here. First, the agencies that deal with land use (mainly local governments) have little or no responsibility for transportation; and the agencies that determine transportation policies (state or regional bodies) normally have no say in the determination of land uses. There is a 'land use–transportation' disconnect among the various agencies. Second, the separatism of these agencies is reinforced by the different skills, training, and interests of local land use planners and transportation engineers. Third, as a result of the independence of local governments, action by one to restrict development (of suburban employment centers, for example) would be seized upon by other local governments in the region as an opportunity to secure development for their areas. Only a strong state or regional planning presence could deal with this type of problem (if the necessary political inclination existed). But, fourth, even if controls were imposed stringently, there are serious doubts on how effective they would be in the long run. Downs has argued that growth is impervious to local public policies: even if all local governments banned further development, migrants would still find a way in (see Box 14.8).

Although there may be little likelihood of affecting transportation by growth control measures, there is the alternative of planning land uses to minimize their transportation effects, or to make transit viable. This is not a new idea, of course. The traditional central business districts did precisely this: concentrated employment, service, retail, and other functions were served by highly developed transit systems.

The difficulty arises in implementing such a strategy when the predominant form of development is one of highly dispersed land uses. Owners of the sites selected for concentrated development might be delighted, but those that were to be 'protected' from development might see matters very differently.

Contrariwise, there can be strong objection to development, whether for NIMBY or other reasons. Thus, attempts to secure a better 'jobs–housing' balance (by building affordable housing close to employment centers) can raise the wrath of those who are already living in the area. It is instructive to note the experience of Bay Area Rapid Transit (BART) in implementing its policy of developing 'transit villages' – housing

---

BOX 14.8 LOCAL POLICIES CANNOT CONTROL GROWTH

Every US metropolitan area has at least some communities encouraging further growth. Even if none did, newcomers would continue to arrive anyway if they believed good economic opportunities were available there, as history has repeatedly proved. Such immigrants would either live on the outskirts of the metropolitan area in unincorporated places with no antigrowth policies, or they would illegally double and triple up in dwelling units within communities that had formally banned further growth. These observations lead one to conclude that growth is impervious to local public policy.

Source: Downs 1992: 33

BOX 14.9 CONTROLLING TRAFFIC BY REDUCING COMMERCIAL DEVELOPMENT – GOOD INTENTIONS IN LOS ANGELES

In 1986 the voters of Los Angeles approved a measure which reduced allowable development on most land zoned for commercial development. This particularly affected the 'strip-commercial' zones along the main streets and boulevards. The intention was to reduce the traffic congestion caused by such development.

But what was to happen to the development pressures involved? Would they simply disappear, or would they emerge elsewhere – and, if so, what would the effect then be?

Martin Wachs pointed out that much of the demand for commercial development would be redirected toward the regional centers which were exempt from the downzoning, and also to the outlying suburban centers which were beyond the jurisdiction of Los Angeles. The result could be a lengthening of journeys to work and shopping in these more distant locations.

Thus, communities which have experienced commercial downzoning in order to reduce the number of trips destined for them may well experience increases in through trips which will be in the future destined to the areas which are allowed to develop.

Regrettably, the downzoning may deprive the city of tax revenues which might be used to relieve traffic congestion through construction programs, while not relieving it of the traffic which downzoning was intended to prevent.

Source: Wachs 1990: 249

development around BART stations. Though operating under special legislation that allows the designation of station-area redevelopment districts, BART has sometimes faced considerable opposition from existing nearby residents. For example, some argue that construction of new housing can result in increased traffic congestion. More generally, BART has also had difficulty in deciding whether it should give priority to better parking or more development around its stations (Knack 1995).

This illustrates some underlying problems faced by any policy of attempting to influence patterns of behavior by land use changes: how is the choice to be made between competing desirable objectives and how effective are the plans likely to be? These problems are difficult enough with undeveloped sites; they are greatly compounded in areas where a pattern of uses is already established. There are difficulties in determining what it might be desirable to do, quite apart from the practicalities. An illustrative example is given in Box 14.9.

Transit-oriented development

Streetcars served as early focal points of development. They also helped open up suburban areas for developments. Over the years, citizens in many areas have opted to use the automobile. In fact, most individuals would claim that we have become too auto-dependent. The haphazard expansion of the cities caused a number of financial, social, and environmental costs to the cities and their inhabitants. Cities have started to encourage and guide development back to areas served by various forms of mass transit. It is hoped that by doing so, cities can increase transit ridership, generate economic development, and possibly revitalize once-vital parts of the city.

In his 1989 study *America's Suburban Centers*, Cervero pointed to the need to design these centers to encourage, or at least facilitate, commuting by transit. One way is to avoid single-use developments in favor of mixed-use developments. These are much more user-friendly than those devoted to only one use (such as offices): in the absence of other uses (such as shops, restaurants, and banks), suburban workers are forced to have their cars for use during the day for meals, banking, and other personal errands. Surveys have shown that these needs for a car during the day can be a significant factor in determining travel by automobile.

Additionally, since mixed uses peak at different times, they can give rise to economies in parking provision. This can have the further advantage of reducing the scale of a development and making walking more attractive. For such reasons, mixed-use developments can have market advantages.

However, the importance of design (as distinct from land use) should not be exaggerated. A study for the US Department of Transportation showed that transit-friendly design features were not in themselves sufficient to have any significant impact on the transit ridership. They can be a useful complement to other measures, but they are not sufficient to lure commuters out of their cars (Cervero 1994b).

Reviewing these various possibilities, it seems clear that, in the short run at least, little relief from traffic congestion is likely through land use planning measures. The alternative is to operate controls directly over roads and traffic, by management, regulation, or economic incentives. Nevertheless, high-density residential development centered on a transit facility such as a railroad station makes sense in its own terms, even if its impact on the overall transportation system is slight. At the least, it provides an alternative to the typical suburban/commuter type of development. The market success of a number of these has introduced a new term to the planning lexicon: 'transit villages' (Knack 1995).

Transportation and land use must be linked in any discussion since they feed off each other. Solving any land use problem cannot proceed without a discussion of transportation.

Increasing congestion rates and pollution rates suggest the need to locate people near transit facilities such as bus, light rail, and heavy rail stations in high-density areas. Moreover, these transit facilities should be integrated with such mixed-use activities as residential, business, entertainment, and commercial in a pedestrian-friendly setting. This 'transit-oriented development' (TOD) can serve as a model for new development or as a means to revitalize older areas of a community. The key is that a rail or bus station would serve as the center or focal point of the neighbourhood and mixed land uses would surround it. This type of development represents a true public-private venture with government offering a variety of transit services and possibly government activities and the private sector providing a host of other activities. TODs can be found throughout the country. For example, a project in San Diego, called 'Smart Corner' has a light rail station on the ground floor, five floors of offices, and nineteen floors of condominiums (see Plate 18).

A number of areas around the United States have turned to developing transit villages, Portland, Oregon being a prime example. They have become increasingly popular in areas where people cannot afford to live in the areas in which they work. Developing transit villages offers one means of increasing the amount of lower- and middle-cost housing in high-cost housing areas. The County of Arlington, Virginia, has chosen to develop commercial and residential uses around transit nodes. Open space and pedestrian walkways are also incorporated into these transit villages. The state of California passed the Transit Village Development Act of 1994, providing incentives to cities and counties to plan transit village development districts linking mixed-use developments to transit systems. Although pre-dating the California legislation, BART has encouraged high-density development around its transit stations.

The characteristics of a transit village can be found in Box 14.10. Denver's downtown 16th Street Mall is a good example of a TOD (see Plate 19). Free bus services started in the early 1980s as a means of getting people back to the downtown. With the addition of a variety of mixed uses, including shopping, residential, office, and entertainment, the area has been transformed into a thriving downtown area. In 1999, the state of New Jersey launched a transit village program that sought to create mixed-use development and investment opportunities around bus or passenger rail stations. A number of state agencies, including commerce, community affairs, environmental protection,

Plate 18 Smart Corner development in San Diego

Photo by author

Plate 19 Denver's 16th Street pedestrian mall

Photo by author

BOX 14.10 CHARACTERISTICS OF A TRANSIT VILLAGE

A city or county may prepare a transit village plan for a transit village development district that addresses the following characteristics:

a) A neighborhood centered around a transit station that is planned and designed so that residents, workers, shoppers, and others find it convenient and attractive to patronize transit.

b) A mix of housing types, including apartments, within not more than a quarter mile of the exterior boundary of the parcel on which the transit station is located.

c) Other land uses, including a retail district oriented to the transit station and civic uses, including day-care centers and libraries.

d) Pedestrian and bicycle access to the transit station, with attractively designed and landscaped pathways.

e) A rail transit system that should encourage and facilitate intermodal service, and access by modes other than single occupant vehicles.

f) Demonstrates public benefits beyond the increase in transit usage . . .

g) Sites where a density bonus of at least 25 percent may be granted pursuant to specified performance standards.

h) Other provisions that may be necessary, based on the report prepared pursuant to subdivision (b) of Section 14045.

Source: California Government Code 65460.2

and state planning, are participating in the program. As of June 2002, there were seven municipalities participating in the state transit village program: Morristown, Pleasantville, Rutherford, South Amboy, South Orange, Riverside, and Rahway.

Advocates of TODs tout a number of benefits, ranging from increased transit ridership to reducing congestion and pollution. A listing of potential benefits can be found in Box 14.11. Achieving these and other benefits will not be easy. A number of barriers or impediments might confront those involved in developing a TOD. Although federal funding encouraging TODs has increased in recent years, federal funding has historically been biased toward or favored automobiles and the expansion of highways. Another potential barrier facing some communities is that their zoning codes might discourage the development of a TOD by continuing to stress the separation of land uses. Concomitantly, some residents may oppose higher densities and mixed-use developments in their communities. Moreover, historically speaking, another

barrier may revolve around the fact that transportation decisions have been made with little or no regard to land use decisions and vice versa. Finally, simply encouraging the development of TODs is not enough. Various incentives, or the so-called 'carrots', such as tax incentives, must be available to businesses and property owners to encourage them to development a TOD.

Traffic calming

Increasing amounts of traffic in our communities have also led to heightened concerns over safety. Many individuals feel that the amount of traffic in an area affects the livability of a community. As such, many communities have initiated traffic calming programs.

A traffic calming program seeks to alter the behavior of drivers and their vehicles (see Plate 20). This means any change in street alignment or the installation of various barriers designed to reduce traffic speeds on local streets. The goal is to make residential streets

BOX 14.11 BENEFITS OF TRANSIT-ORIENTED DEVELOPMENT

1 Variety and choice in housing types, retail destinations, and office locations;
2 Catalyst and framework for revitalization and redevelopment of central urban areas into vibrant communities;
3 A structure for new growth in compact patterns, saving open space;
4 A higher activity level at transit station areas, increasing pedestrian safety through numbers;
5 Enhanced transit ridership, walking, and cycling, and reduced automobile dependence;
6 Contribution to reduced levels of congestion and improved air quality;
7 Efficient use of infrastructure due to the greater intensity of development, both in existing and new areas.

Source: Metropolitan Council 2000

Plate 20 Traffic calming device

Photo by author

less attractive and desirable than neighborhood streets by increasing the amount of travel time in residential neighborhoods. The goal can be accomplished by using such techniques as speed humps (raised sections of pavement across the road), traffic circles (raised landscape islands at the center of an intersection), chicanes (a series of two or more staggered curb extensions on alternating sides of the road), semi-diverters (curb extensions/barriers restricting movement into a street), etc.

Traffic calming programs can be found in cities across the United States. For example, the City of Seattle started a Neighborhood Traffic Control Program in 1978 that was designed to reduce accidents and speed on residential streets. One highlight of the program is that over 800 traffic circles have been installed on neighborhood streets in Seattle. Austin, Texas, recognized that traffic pressures would accompany growth and sought to protect its neighborhoods from the negative aspects of increased traffic. It enacted a Neighborhood Traffic Calming Program designed to alter the behavior of drivers. The overall goal was to provide a safe and efficient transportation system. The city of Portland, Oregon, created a Traffic Calming Program in its Office of Transportation designed to improve community safety and to preserve and enhance neighborhoods in Portland. To accomplish this mission, the Program had the following objectives:

1 To enhance neighborhood livability and sense of community by reducing excessive speeding and excessive vehicle volumes on local service streets;
2 To encourage reasonable and responsible driving behavior through education and emphasizing personal responsibility;
3 To enhance traffic safety for pedestrians, providing special attention to the safety of children in school zones;
4 To encourage alternative transportation options and the use of the arterial system for through traffic; and
5 To encourage broad citizen participation by providing service in a responsive, timely, and professional manner.

Increasing the supply of road space

Before discussing ways of restraining traffic, it is necessary to inquire whether it is not possible simply to increase the supply of road space to accommodate increased numbers of vehicles. This can be done either by new road building or by measures that increase the carrying capacity of the existing roads. The first can have spectacular results, but these are often short lived. Traffic seems to increase faster than new roads can be built. A major reason for this, of course, is the continued growth in car ownership and use (itself in part stimulated by new roads). This stems from the huge advantages of the automobile for personal mobility, and the increase in auto ownership and use. There are many issues here, including the increasing difficulties of managing *without* an auto (because of the wide dispersion of activities, and the inability of public transit to serve these); the large investment that has been made in roads; the availability of parking (provided by employers, shopping centers, etc.); the relatively low cost of auto travel (most roads can be used without direct payment).

There is, however, a limit to the extent to which the supply of road space can be continually increased. As this has become apparent, and as concern has grown about the cost and impacts of road building, increasing ingenuity has been devoted to making roads able to carry more traffic, and to reducing some of the commuting demand. Some of the techniques are discussed below, but it has to be said at the outset that none has proved particularly effective. Sooner or later any freeing of road space is taken up by increased traffic. So common has this been that it has been suggested that an underlying principle is at work.

Downs' principle of 'triple convergence'

This has been elegantly set out by Anthony Downs (1992) in his theory of 'triple convergence'. This is based on the simple fact that since every driver seeks the easiest route, the cumulative result is a convergence on that route. If it then becomes overcrowded, some drivers will switch to an alternative route that has become relatively less crowded. These switches continue until there is an equilibrium situation (which, like any human equilibrium, is not stable – conditions constantly change). On this theory, building a new road, or expanding an existing one, will have a 'triple convergence'. First, motorists will switch from other routes to the new one ('spatial convergence'); second, some motorists who avoided the peak hours will travel at the more convenient peak hour ('time convergence'); third, travelers who had used public transit will switch to driving since the new road now makes the journey faster ('modal convergence').

The eventual outcome depends upon the total amount of traffic (actual and potential) in relation to the available roads. If the increase in traffic stimulated by the new road is modest, there will be an observable benefit for all. Though peak-hour traffic may be congested, this is simply because so many drivers are traveling at the time which is most convenient to them. (There may, however, be a loss to transit passengers if the 'modal convergence' leads to a reduction in service.)

Transport demand management

Since it is so difficult to change transportation and land use systems, considerable thought has been given to the alternative of 'managing' the transportation system either by physical changes to roads or by influencing traffic behavior.

Some measures are simple, such as phasing traffic lights, changing two-way streets into one-way, controlling street parking, and carefully programming road repairs to ensure minimum disturbance to traffic. Another possibility is to increase the occupancy of autos: most peak-hour commuters travel alone ('lone rangers'). Congestion could be significantly reduced if there were more sharing. To encourage this, some areas have reserved lanes for *high occupancy vehicles* (HOV). The theory here is that the higher speeds achieved on an HOV lane will encourage drivers to change to HOV driving; that is, they will arrange to share their journey to work with others. (Definitions of HOV vary; it can be as low as two: the driver and a passenger.) HOV

lanes, of course, are of particular value to buses, and thus provide an incentive to transfer to transit (if the bus goes to a location which is convenient for the commuter). The idea is an attractive one, though the removal of a lane from general use means that the other lanes become more crowded. This naturally causes annoyance (if not fury) to lone drivers and tempts them to trespass on the HOV lane. This, in any case, is a temptation which is overcome only if there is strong enforcement (at least when they are introduced) and high fines. HOV lanes are more successful if there is an added incentive to use them, as with employer ride-share programs.

Parking policies

An apparently simple means of reducing traffic congestion is to eliminate the high tax-free subsidies granted by employers to commuters by way of free parking. Some 90 percent of American auto commuters park free at work. This significantly reduces the real cost of commuting. For instance, it has been estimated that, in Los Angeles, the effect of free parking for the average SOV commuter (i.e. a commuter in a single occupancy vehicle) is to reduce the cost from $6.07 to $1.75 a day – a reduction of 75 percent (NRC 1994, vol. 2: 518). Several studies have shown that the elimination of employer-paid parking has reduced SOV commuting significantly. Other studies have compared employees who received employer-paid parking with the groups who paid for their own parking: the share of SOV trips was much smaller for those who had to pay – ranging from 19 percent to 44 percent less. There is much evidence of a similar nature.

Free employer-provided parking not only generates SOV commuting, but is also unfair to non-auto commuters who receive no corresponding benefit. One way of rectifying this (in addition to amending the tax code to take account of the benefit) is to require employers to offer a 'parking cash-out program' that would enable employees to obtain cash in lieu of free parking. Such a scheme is in operation in California, where employers in any area designated by the Air Resources Board as a 'non-attainment area' are required to offer employees a cash allowance equivalent to the parking subsidy (see discussion in Chapter 12).

It is also to be noted that the provision of 'free' parking for employees is expensive: estimates vary around $2,000 a year in the Washington, DC, area (MacKenzie *et al.* 1992). However, though this seems persuasive in central business districts, the calculus is not so readily acceptable in the suburbs, where there is abundant cheap land.

It is necessary, however, to distinguish between central city and suburban areas. In central areas, there are public transit systems, and land for parking is expensive. The opposite is the case in the suburbs. In any case, why should individual employers try to deny their employees the convenience and low direct cost of SOV travel? Traffic congestion has to be very severe before the car-using public will accept restrictions on the basis of seemingly theoretical arguments.

One of the advantages of parking as a policy instrument is that it is very flexible. For example, charges can be varied by size of vehicle, time of arrival or departure, and duration. Targeted groups can be charged lower (or nil) rates, e.g., the disabled, local residents, and emergency staff. In non-commuter car parks, it is common to charge higher rates for longer periods of parking, but this can encourage 'reparking'. Moreover, there is a case for charging short-term parkers at a high rate since they create more travel.

It should be noted that, however effective parking controls may be in restraining trips to a local area, they do not deter through traffic. Indeed, if parking charges have the effect of reducing local congestion, there may be an increase in through traffic. The same applies with measures to prevent obstruction by autos parked in the street.

Parking measures are the most effective among the many possibilities of affecting transportation demand, but they can be more effective if they form part of a wider approach aimed at securing the benefits of cooperation, or even coordination, among different agencies. Such an approach is the essence of transport demand management (TDM) programs.

TDM programs

Transportation demand management can be implemented through voluntary arrangements, or in conjunction with a trip reduction ordinance. However, whatever the legal statutory aspect, it relies essentially on the willingness of both employers and employees to participate. When they do, it is because TDM is cheap, effective, and capable of securing tangible benefits.

At its most sophisticated, TDM operates through an organization of interested parties – employers, developers, members of business associations, landowners, and public bodies including planning and transportation agencies. There are, however, wide variations in organization, scope, and character. Some do little more than act as a source of information for employers and commuters – providing information on alternatives to SOV commuting, for instance. Some are run with no explicit budget or acknowledged cost, while others derive financial support from the private or the public sector (or both). Many have no powers of enforcement; a few have apparently draconian systems for imposing financial penalties. (The South Coast Air Quality Management District can impose a fine of up to $25,000 a day for failure to prepare a plan for reducing vehicle ridership, but this is unusual and is regarded more as an indication of the seriousness of the endeavor.) Some of the main program elements are listed in Box 14.12.

TDM programs have been shown to be worthwhile, particularly with parking controls, where they have had (relatively) the most success – though usually on a site-by-site, rather than area-wide, basis. Such measures can help if other things remain equal. Unfortunately, they seldom do. Transportation systems are composed of a myriad of elements which interact: changes in one element trigger responses in others – often a surge in new traffic to fill a 'space' created by some traffic-reduction measure. In the long run, changes in spatial structure may bring about major improvements, but congestion will remain serious in some areas while drivers continue not to meet the true costs of their use of roads. If it were to be practicable, the heart of the problem could be reached through the use of congestion charges.

BOX 14.12 TRANSPORT DEMAND MANAGEMENT

An addition to the acronyms of transportation planning is TDM – transportation demand management. This attempts a cooperative approach to 'the art of modifying travel behavior'. Program elements include:

- reduction of parking provision
- cash in lieu of free parking
- carpool matching services
- transit information center
- alternative work schedules
- parking management services
- shuttle services (e.g. to rail station)
- preferential parking for high occupancy vehicles
- transit incentives
- guaranteed emergency ride-home program
- subsidized transit fares
- subsidized vanpools
- express bus service
- alternate work hours
- home-based telecommuting
- park and ride lots
- design improvements for local pedestrians
- showers and lockers for cyclists and walkers
- secure cycle parking.

Source: Ferguson 1990

Congestion charges

No area is immune to traffic congestion. The problems of traffic congestion have reached such a point in some areas that more forceful measures need to be considered (see Box 14.13). An area might decide to institute charging for vehicles entering a specific area or areas of the city (cordon charge), to charge variable tolls on selected bridges and tolls, where the tolls are higher during peak travel hours and lower in off-peak hours, to charge all vehicles in a specific area based on the amount or level of congestion, and to charge vehicles wanting to use express toll lanes. The theoretical basis for congestion charging is essentially simple. Individual drivers are concerned only with the costs they bear. As traffic on a stretch of road increases, each additional car adds to congestion and thus imposes costs on other drivers. Congestion charging translates this cost into individual charges. Auto drivers now have an incentive to take into account the cost of congestion that they are collectively causing. Moreover, since they are forced to bear the cost if they use the congested road, some will find alternative routes or modes of transport, and traffic on the priced road will decrease.

The way in which costs arise at the margin is dramatically illustrated by the estimate of the Bay Area Economic Forum that 'a single driver entering the San Francisco area's congested roads during the peak hours can generate one hour of additional delay for all other drivers there combined'. Though this is a curiously dramatic way of illustrating the point, there can be no doubt that, above a certain level of congestion, the difference between the individual cost and the social cost is enormous. Congestion charges even this out.

Theoretical justifications for charging for the use of roads might be described as overwhelming were it not for the fact that public opposition to the idea is typically even more so. Nevertheless, there is usually far less opposition to *new* toll roads, and even less to bridge and tunnel tolls. This suggests that attitudes are a matter of perception and habit. Roads have traditionally been 'free' to the user: to introduce a charge is to take away a benefit (which, it can be argued,

is already paid for in taxes). A new facility, on the other hand, clearly requires new expenditure; and it also brings an equally clear benefit to the motorist.

In 2006, the United States Department of Transportation announced a program called the National Strategy to Reduce Congestion on America's Transportation Network, which was designed to reduce congestion on the country's transportation system. In this program, the Federal Highway Administration offered funding for metropolitan areas to partner with the Department of Transportation to help reduce congestion by using the so-called 'Four Ts' – tolling, transit, telecommuting, and technology (Federal Register, December 8, 2006: 71231). Five metropolitan areas in the United States received funding to fight

---

BOX 14.13 ROAD PRICING

Popularity of peak-period pricing

Peak-period pricing is well established and acceptable in many areas where demand fluctuates over time. Vacation prices are higher in the holiday season; air fares vary by day and hour as well as by the season; telephone calls have peak and off-peak charges; transit and parking charges differ at different times of the day. These charges do not eliminate congestion, but they limit it and divert some demand to less crowded times. Those who are able and willing to change to the less crowded times are attracted by the lower cost. Though only a small proportion may change, all benefit by the more even spread of use.

Source: NRC 1994

What is congestion pricing?

Congestion pricing – sometimes called value pricing – is a way of harnessing the power of the market to reduce the waste associated with traffic congestion. Congestion pricing works by shifting purely discretionary rush-hour highway travel to other transportation modes or to off-peak periods, taking advantage of the fact that the majority of rush-hour drivers on a typical urban highway are not commuters. By removing a fraction (even as small as 5%) of the vehicles from a congested roadway, pricing enables the system to flow much more efficiently, allowing more cars to move through the same physical space. Similar variable charges have been successfully utilized in other industries – for example, airline tickets, cell phone rates, and electricity rates. There is a consensus among economists that congestion pricing represents the single most viable and sustainable approach to reducing traffic congestion.

Source: US Department of Transportation, Federal Highway Administration 2006: 1

congestion – New York City, Miami, Minneapolis, San Francisco, and Seattle.

New York City received some $354 million to help ease traffic congestion in the city from the US Department of Transportation. In an Urban Partnership Agreement between the US Department of Transportation and the New York City Urban Partner (comprised of the New York City Department of Transportation, the New York Metropolitan Transportation Authority, and the New York State Department of Transportation), the New York City Urban Partner agreed to 1) institute a broad area pricing system in Manhattan south of 86th Street, 2) construct new transit facilities, 3) construct a series of bus rapid transit and/or bus-based corridors, 4) implement transit technologies, 5) make improvements to regional ferry service, 6) collect and analyze transportation data to support the West of Hudson regional transportation analysis, 7) construct an East River bus lane, and 8) purchase and operate additional buses to meet the mobility needs of New York City (US Department of Transportation 2007: 1). Federal funds would be used for activities one through six while the New York City Urban Partner would be responsible for funding the last two activities. Federal funding was contingent upon the New York State Legislature approving a congestion pricing or some similar pricing system within ninety days of the next legislative session and implementing it by March 31, 2009.

In 2006, the San Francisco Transportation Authority charged a toll for drivers in the city's downtown core and received federal funding to see how London's congestion pricing program might be applied in the San Francisco metropolitan area. The authority would determine if the fees should be fixed or varied by location or hour.

In 2007, five cities were selected as demonstration cities in the Urban Partnerships Congestion Initiative – Minneapolis–St Paul, Miami, Seattle, San Francisco, and New York. Some of the alternative ways of managing the increasing congestion are to increase the number of lanes (although many consider this not to be a long-term solution), variable pricing on parking, roads, and bridges during non-peak and peak hours, and charging vehicles coming into and leaving the city.

New York wanted to do the latter but was soundly rebuffed by a number of parties. San Francisco is gradually phasing-in its approaches and will begin a trial period in 2015.

Congestion charging is not unique to the United States. We should learn from other areas outside the United States. For example, in 2003, London implemented a program designed to reduce or combat the growing problem of congestion in central London. The program charged drivers a per day charge to enter central London during the period 7:00 am to 6:30 pm, Monday through Friday – the peak congestion period for the area. The funds raised by the program were used to help finance improvements in London's extensive public transit system. The program has resulted in a lessening of congestion in central London.

Stockholm has also witnessed a continuous amount of traffic growth. Congestion in the central city was commonplace and continued to get worse. In 2003, the Stockholm City Council adopted a proposal to have a trial run at congestion management in the central city. The Swedish Parliament followed with the passage of a Congestion Charges Act in 2004. The congestion management trial started in 2006, running from June to July 2006. A charge was placed on vehicles coming into and out of the central city on weekdays from 6:30 am to 6:29 pm. Higher charges would be set for peak traffic periods. There were no charges during evening hours, weekends, holidays, or the day before a public holiday. Certain types of vehicles were exempt from the charges. By introducing congestion management, the city sought to reduce the traffic flow in the central city, to improve its traffic flow, and to reduce automobile emissions. As a result of the city's experiment in 2006, the number of vehicle trips to the central part of the city has decreased, public transit ridership has increased, and the level or amount of automobile exhaust emissions has decreased.

Developing a congestion management program is one thing, but what else does an area do to actually lessen the congestion. Stockholm employed several approaches. First, it enhanced and extended public transportation by adding new buses and new routes. It increased opportunities to take the rail into the city. More park-and-ride facilities were constructed to facilitate people taking alternative modes of transportation. Ultimately, voters in Stockholm agreed to continue the congestion pricing system in the central city in 2007. The charges are still in place today.

Major objections are that congestion taxes are politically unacceptable, are difficult to administer, are unfair, and penalize low-income motorists. Certainly, there is abundant evidence about the political unpopularity of congestion charges. However, as traffic congestion worsens, attitudes to charging may change. There is already some evidence that this is happening (NRC 1994, vol. 1: 64). At the least, some pilot schemes may attract sufficient support for testing the workability and effects of charging.

It has to be acknowledged that there are many uncertainties about congestion charging. The technical problems of administering charges can be surmounted, and it seems evident that even a small switch away from priced roads would have a large beneficial effect on traffic flow. But many effects cannot even be foreseen, let alone measured. The modern metropolitan area is a highly complex urban system (perhaps better described as a multiplicity of interacting systems). Too little is understood about it for confident predictions to be made about the effects of introducing new traffic measures. Yet it is also true that there is no solution to the congestion problem without more effective means of restraint – and that congestion pricing currently seems to be one means of effectively achieving this. It is not without reason, however, that policy-makers are cautious. They have to convince a car-owning electorate that the imposition of congestion charges is not just another tax but is likely to be an effective means of improving their transportation situation.

Congestion charging and equity

Convincing a skeptical public that congestion charging will bring about tangible and widespread benefits is difficult, partly because theoretical arguments are insufficient to overcome the doubts occasioned by our degree of ignorance about the working of complex metropolitan systems. A major area of concern is whether a charging system could be designed to be sufficiently equitable. Given the inequities that exist between different socio-economic groups, complete equity is unachievable: any system will benefit the higher-income groups most – if only because, in economic terms, their time is more valuable. Of course, there are many ways in which disadvantaged groups could be compensated, but no system could offset this difference. Moreover, though compensation can be devised to benefit *groups*, there is no way in which all the *individuals* in the disadvantaged groups could be recompensed. For example, particularly vulnerable would be low-income working single mothers who can reach their employment only by driving during congested periods. Some might be able to find a substitute in improved (and subsidized) transit, but if they are unable to avoid the particular journey that involves a charge, they could suffer significant hardship.

Nevertheless, it can be argued that the inequitable effects of congestion charging have been greatly exaggerated. One reason for this is that low-income commuter travel patterns are different from those of high-income commuters. More important than income can be origin and destination patterns and the scope for substitutability by time and mode of travel. Improvements to transit and carpooling alternatives would be of specific value to low-income travelers; and, if necessary, a scheme of rebates might be possible. Kain (1994) has also suggested that those concerned about inequities have failed 'to consider the full range of urban transport technologies and the likely impacts that congestion pricing would have on the level of service provided by these alternatives'.

As with the objections which are made on equity grounds, much of this is theoretical. It is clearly important to ensure that in any scheme of congestion charges particular provision is made for vulnerable groups, and that there is careful monitoring. An explicit commitment to this is needed at the outset.

There are, of course, long-run implications of changes in the transportation system, though it is extremely difficult to identify these, let alone predict changes. Metropolitan areas have traditionally dealt with congestion by spreading out. It is difficult to predict what might happen if a major change is made

BOX 14.14 DOWNS' ADVICE TO THE WEARY COMMUTER

My advice to American drivers stuck in peak-hour traffic is not merely to get politically involved, but also to learn to enjoy congestion. Get a comfortable, air-conditioned car with a stereo radio, a tape player, a telephone, perhaps a fax machine, and commute with someone who is really attractive. Then regard the moments spent stuck in traffic simply as an addition to leisure time.

Source: Downs 1992: 164

in the complex of forces which have produced the familiar suburban pattern.

For the individual commuter, the cost of the journey to work is part of the price to be paid for the favored lifestyle provided by low-density suburbs – travel costs are lower than housing costs! Anthony Downs has expressed the point neatly in his *Stuck in Traffic* (see Box 14.14).

Transportation and public health

The relationships between transportation, land use, and public health have come under increasing scrutiny in recent years. Early transportation policy focused on automobiles and the construction of roads and highways. Zoning regulations spread out land uses and essentially required individuals to use the automobile. We were, and to some extent still are, a car-oriented society. Many people feel lost without the freedom of their automobile.

More recently, we have seen growing concerns about the reliance on automobiles in the area of public health. There is a growing body of literature that indicates how the reliance on the automobile has played a role in the sedentary lifestyles of individuals. As such, less walking has resulted in an increase of obese individuals and other health problems such as high blood pressure, coronary heart disease, and back problems.

The active living movement seeks to counter the dominance of the automobile by examining how we can design our communities so that we can improve our health. We are examining how non-motorized transportation options can improve our quality of life. People are looking for pedestrian-friendly communities so that they can walk or bike to recreational areas, shopping centers, or other areas or services. They are looking for things within walking distance of where they live. They are looking for opportunities to avoid taking trips via the automobile. The importance of how we design our communities and improve public health can be found in Box 14.15.

BOX 14.15 COMMUNITY DESIGN AND PUBLIC HEALTH

The way we design our communities appears to affect how much people will walk, how much they weigh, and their likelihood of having high blood pressure. These findings are in line with a growing body of research which shows that community design influences how people travel and how physically active they are in the course of the day. While more research is needed, urban planners, public health officials, and citizens are already looking to change communities to make it easier to get out on a bicycle or on foot. Ultimately, such long-term changes may help more Americans lead healthier and happier lives.

Source: McCann and Ewing 2003: 28

Complete Streets

Automobiles, trucks, and motorcycles are not the only users of roads and streets. Other users need to be safely accommodated on our roads and streets. To do so, many areas have completed or are in the process of creating rules and regulations designed to ensure the safety of all users. Such rules and regulations are promoting what is commonly referred to as 'Complete Streets'. While definitions of the term may vary from jurisdiction to jurisdiction, the 2012 Minnesota Statutes, Chapter 174, section 174.75, offers the following:

'Complete streets' is the planning, scoping, design, implementation, operation, and maintenance of roads in order to reasonably address the safety and accessibility needs of users of all ages and abilities. Complete streets considers the needs of the motorists, pedestrians, transit user and vehicles, bicyclists, and commercial and emergency vehicles moving along and across roads, intersections, and crossings in a manner that is sensitive to the local context and recognizes that the needs vary in urban, suburban, and rural settings.

Over half of the states are committed to and have adopted policies and regulations for Complete Streets. New York approved a state law in 2011 requiring state and local transportation agencies to consider Complete Streets as a means of making sure streets were safe and accessible to everyone regardless of age or ability. State, local, and county transportation projects that receive federal and state funding are subject to the law. In 2010, Michigan's planning enabling legislation was amended to incorporate the consideration of Complete Street policies into the required local comprehensive planning process. The legislation recognizes the importance of promoting safe and efficient travel for all legal users of roads, streets, and highways. It also requires the state transportation department to adopt a complete streets policy for itself and develop a model Complete Streets policy or policies, to be made available for use by municipalities and counties. Municipalities must also consult with the state transportation department on any project or facility on the

Complete Streets policy prior to project approval. Assistance will be provided to municipalities and counties to help develop their own complete policies. An advisory council was also created through the legislation to provide a yearly report to the governor, the state transportation commission, and the state legislature on what has transpired in regards to Complete Streets within Michigan.

Well over 250 cities have adopted Complete Streets policies. For example, Seattle adopted Ordinance 122386 in 2007 which recognized the city's policy of encouraging walking, bicycling, and transit use in Seattle's Comprehensive Plan and Transportation Strategic Plan. Complete Streets principles would now be incorporated into all appropriate plans and programs. In 2010, the Mayor of the Metropolitan Government of Nashville, Tennessee, and Davidson County recognized the importance of such a policy in Executive Order No. 40. On October 6, 2010, Mayor Karl Dean acknowledged the long-standing notion that cities were built with only cars in mind and that areas needed to give full consideration to the accommodation of the transportation needs of all users.

Recognizing that states and cities were developing Complete Streets rules and regulations, two members of the US Congress sponsored proposed legislation in the form of a Complete Streets Act of 2011. The bill, H.R. 1780, would 'ensure the safety of all users of the transportation system, including pedestrians, bicyclists, transit users, children, older individuals, and individuals with disabilities, as they travel on and across federally funded streets and highways'. It would have required each state department of transportation or MPO, within two years after the legislation had been enacted, to have a clear policy statement that federally funded transportation projects accommodate the needs of other users of the projects consistent with principles of Complete Streets. The bill did not pass. It is expected that similar legislation will appear in the future as more and more states and cities pass policies and regulations promoting Complete Streets.

Conclusions and uncertainties

There is no simple solution to the problems of congested transportation. There are too many barriers – of ignorance (of ways of reducing pollution); of cost and tax implications (of major road and transit developments); of public attitudes (which set limits to what is politically possible); of governmental machinery (which, in land matters, is essentially local and self-centered); and of understanding the sheer complexities of metropolitan areas. In the long run, changes in land use patterns may bring about a significant change in travel behavior, but this will take a very long time. Established land uses are vast compared to the incremental changes which can be brought about by new development; and there must be doubts as to whether it is possible to be sure that these changes could be effectively planned and implemented to affect travel behavior in intended ways.

Nevertheless, there are many ways in which conditions can be improved. The simplest is by action on free parking provided by employers – if there is sufficient support for this. Congestion charging could be very effective, though its promises are latent until public opinion is more favorable – which could follow from increasing congestion and pollution. Other policies which can help are ride sharing, park-and-ride connections to transit systems, speedier systems of dealing with the aftermath of road accidents, further development of TDM, planning high-density residential development at transit stations.

One final caveat is necessary. Not only is there considerable ignorance and uncertainty about many of the issues discussed in this chapter, there is also a wide variation among areas in their pattern of land uses, transportation systems, economic profile, income distribution, and many other matters. This variation may also be matched by differences in culture and attitudes. It follows that the points made in this chapter are not necessarily equally relevant in all areas, or even in those metropolitan areas that are the main centers of urban congestion.

Further reading

A discussion on 'The Transportation Planning Process: Key Issues' (updated 2007) can be found at http://www.planning.dot.gov/documents/BriefingBook/BBook.htm.

The Travel Model Improvement Program (http://tmip.fhwa.dot.gov), sponsored by the US Department of Transportation, is dedicated to improving an agency's transportation planning analysis. The Improvement Program offers seminars, training, technical assistance, research, and a peer-review program to increase the accuracy of transportation models.

The Bureau of Transportation Statistics (http://www.bts.gov) provides travel data, statistics, and current research supplemented by the National Transportation Library. Its data can be downloaded and ready-made reports analyzing the data are available online.

More transportation statistics may be found in the 2009 National Household Travel Survey (http://nhts.ornl.gov/2009/pub/stt.pdf). Last conducted in 2001–2002, the survey offers easy-to-use online analysis tools, downloadable data, and publications that analyze the data.

Telecommute Connecticut (http://www.telecommutect.com), created by the Connecticut Department of Transportation, has an informative website to help employers and employees set up a telecommuting system. The Center for Transportation Excellence (http://www.cfte.org) is a non-partisan policy research center that provides research materials, strategies, and assistance with all aspects of public transportation.

A southern California-based group called Livable Places presents a convincing case for transit villages (http://www.livableplaces.org/policy/todincentives.html). Their website has many design guidelines, case studies in California, and tips on how citizens can encourage transit villages.

The California Department of Transportation offers a Transit-Oriented Development (TOD) Searchable Database (http://transitorienteddevelopment.dot.ca.gov). Complete with locator maps and research data, the website also supplies a definition of TOD and various reports on the subject.

A more in-depth look at traffic calming is found at the Institute of Transportation Engineer's Traffic Calming for Communities (http://www.ite.org/traffic). The website offers a searchable library of traffic calming techniques with most sources available in PDF links, pictures of several common calming measures and a sample of brochures given to the public.

Anthony Downs now has his own website (http://www. anthonydowns.com). Along with promotions for his books, among them his 2004 book *Still Stuck in Traffic*, his website includes many transcripts of recent speeches and his 'provocative thoughts' on transportation and housing.

A guide to California's Parking Cash-Out Law can be found on the Air Resources Board website (http://www. arb.ca.gov/planning/tsaq/cashout/cashout.htm). Along with a guide for employers on how to apply the Parking Cash-Out Law, the website has internet links to case study research.

The Department of Transportation in Washington State has a comprehensive website on its transportation demand management (TDM) (http://www.wsdot.wa. gov/tdm). The website features a definition of TDM, shows how to reduce commute trips, and includes performance reports on the program.

Congestion Charging is a program that has been implemented in London (http://www.cclondon.com/whatis. shtml). This website provides an outline of the London program, a current schedule as well as background reports.

Regional Planning and Metropolitan Planning Organizations (MPOs) are found in almost every state. The Association of Metropolitan Planning Organizations (http://www.ampo.org) was established in 1994 as a non-profit organization to help MPOs plan their transportation investments. The website provides legislation updates, publications, technical resources, and spells out awards for planning efforts.

Recent research has connected health issues to transportation issues. The Healthy Transportation Network (http://www.healthytransportation.net) is a product of partnerships with the California Center for Physical Activity and is funded by the Federal Highway Administration's Federal Transportation Enhancement funds. The website's intent is to increase bicycle and pedestrian modes of travel.

For recent works on active transportation and urban planning, see: Frank, Engelke, and Schmid (2003) *Health and Community Design: The Impacts of the Built Environment on Physical Activity*; Frumkin, Frank, and Jackson (2004) *Urban Sprawl and Public Health*; Morris (2007) *Planning Active Communities*; and Dannenberg, Frumkin, and Jackson (2011) *Making Healthy Places: Designing and Building for Health, Well-being, and Sustainability*.

For a discussion on Complete Streets, see: National Complete Streets Coalition, http://www.smartgrowth america.org; Minnesota Complete Streets Coalition, http://www.mncompletestreets.org; Seskin (2012a) *Complete Streets: Local Policy Workbook*; and Seskin (2012b) *Complete Streets Policy Analysis*.

A non-technical, clear and interesting book on traffic congestion is Downs (1992) *Stuck in Traffic: Coping with Peak-Hour Traffic Congestion*. This is a refreshing, contentious analysis of the congestion problem, which argues that the problems are not serious enough to precipitate appropriate action – which, in any case (in Downs' view), is not likely to prove effective. Downs' earlier discussion of 'convergence' is in 'The law of peak-hour expressway convergence' (1962).

Moore and Thorsnes (1994) *The Transportation/Land Use Connection* provides a succinct outline of this subject and a summary of recent research. It also illustrates how (in the Portland Metro region) transportation and land use planning can be integrated.

Readers can get various perspectives of transportation planning from Fitch (1964) *Urban Transportation and Public Policy*; Smerk (1991) *The Federal Role in Urban Mass Transit*; and Dilger (2003) *American Transportation Policy*.

Transport Implications of Telecommuting is a review of the field carried out for the US Department of Transport and published in 1993. Three studies of telecommuting are Handy and Mokhtarian (1995) 'Planning for telecommuting'; Mitchell (1995) *City of Bits: Space, Place, and the Infobahn*; and Hanson and Giuliano (1995) *The Geography of Urban Transportation*. A report discussing barriers or impediments to telecommuting is the US General Accounting Office (2001) 'Telecommuting: overview of potential barriers facing employers'.

Cervero's *American Suburban Centers: The Land Use–Transportation Link* (1989) convincingly supports the author's contention that 'the low-density, single-use character of many suburban work centers was a root cause of the congestion problems being faced in suburbia'. Calthorpe (1993) presents a design for *The Next American Metropolis: Ecology, Community, and the American Dream*. This includes concepts of 'sustainable communities', 'pedestrian pockets', and 'transit-oriented development'.

There has been a growing literature on various aspects of 'transit-oriented development'. Representative examples include: Atash (1994) 'Redesigning suburbia for walking and transit: emerging concepts'; Belzer and Autler (2002) 'Transit-oriented development: moving from rhetoric to reality'; Bernick and Cervero (1996) *Transit-Villages in the 21st Century*; Boarnet and Compin (1999) 'Transit-oriented development in San Diego County: the incremental implementation of a planning idea'; Cervero (1994a) 'Rail transit and joint development: land market impacts in Washington, D.C. and Atlanta'; Cervero (1998) *The Transit Metropolis: A Global Inquiry*; US Federal Transit Administration (1999) *Building Livable Communities with Transit*; Kay (1997) *Asphalt Nation: How the Automobile Took Over America and How We Can Take it Back*; and Metropolitan Council (2000) 'Guidelines on smart growth: planning more livable communities with transit-oriented development'.

For views on maritime transportation, see: US Department of Transportation, Maritime Administration (2007) *The Maritime Administration and the U.S. Marine Transportation System: A Vision for the 21st Century*; and US Department of Transportation, Research and Innovative Technology Administration (2011) *America's Container Ports: Linking Markets at Home and Abroad*.

A number of good sources for literature on traffic calming exist, including: Appleyard (1981) *Livable Streets*; Ewing and Kooshian (1997) 'U.S. experience with traffic calming'; Ewing (1999) *Traffic Calming: State of the Practice*; Guzda (1998) *Slow Down, You're Going Too Fast*; Institute of Transportation Engineers Technical Council Task Force (1997) 'Guidelines for the design and application of speed humps – a recommended practice'; and Thompson (1996) 'Pedestrian road crossing safety'.

Issues relating to congestion charges are fully examined in a study commissioned by the federal government following the passage of the Intermodal Surface Transportation Efficiency Act of 1991: National Research Council (1994) *Curbing Gridlock: Peak-Period Fees to Relieve Traffic Congestion*. Volume 1 consists of the report and recommendations; volume 2 contains a set of commissioned papers. The Federal Highway Administration (2006) offers an introduction to the topic in its *Congestion Pricing: A Primer*. A comparative study of congestion charging can be found in Arnold *et al.* (2010) *Reducing Congestion and Funding Transportation Using Road Pricing in Europe and Singapore*.

A short review of 'what it really costs to drive' is given in MacKenzie *et al.* (1992) *The Going Rate*.

A review of TDM is given by Ferguson (1990) in his article 'Transportation demand management: planning, development and implementation'.

Wachs (1990) details the experience in one state where some efforts were made to regulate traffic by land use controls: 'Regulating traffic by controlling land use: the South California experience'.

Questions to discuss

1   **In what ways has transportation changed in recent decades?**

2   Do you think that telecommuting will have major impacts on transportation?

3   How far can land use planning controls be used to restrain traffic?

4   How effective are transit villages in a comprehensive transportation plan?

5   What are the characteristics of a transit-oriented development?

6   What is the purpose of traffic calming?

7   Critically discuss Downs' 'principle of triple convergence'.

8   Discuss the role of transport demand management as a solution to traffic congestion.

9   Discuss the argument that parking controls are the single most effective means of traffic restraint.

10 'If drivers paid the proper prices for their use of roads, there would be no traffic congestion problem.' Discuss.

11 How viable is comprehensive transportation planning?

12 How do cities deal with the growing problem of traffic congestion?

13 Discuss some of the key pieces of legislation that have shaped our efforts at transportation planning.

14 Does transportation influence public health?

15 What are Complete Streets and why are they important?

Housing

a decent home and a suitable living environment for every American family

Housing Act 1949

The complex of housing

Housing is of central importance in both the national economy and the individual's standard of living. It is a major land use, and its location is a crucial factor in the economy of cities, in transportation, in local economic development, and in the access to opportunities available to individuals. At the same time, its very high cost, compared with other items of household expenditure, presents some particularly difficult problems of finance. Its long life necessitates continual maintenance to prevent deterioration. The condition of individual houses can have neighborhood effects: poor maintenance can blight nearby houses. Deterioration can also result from neighborhood changes – social, economic, or physical. In addition to it providing physical shelter, an individual's position in the housing market can affect social status, capital gains (or losses), and credit availability. The ramifications of this combination of attributes make housing an extraordinarily complex matter. Its multiple dimensions include locational, architectural, physical, economic, social, medical, psychological, and financial. As a result, 'housing policy' involves very much more than the building of houses. It involves a number of policy areas that must be considered together, not independently of each other. Doing the latter may simply exacerbate existing housing problems. Moreover, the reader who has come this far will not need reminding that discrimination is a major issue

in the determination of land uses. This discrimination affects the operation of the housing market, and greatly adds to the problems of ensuring the provision of affordable housing.

Housing policy can take a number of forms. The most important has been the devising of mechanisms to facilitate home ownership. The key issue here is that the capital cost of housing is so high that few households are able to purchase a home outright: they typically require the assistance of a financial mechanism to enable them to spread payments over a long period of time. The development of various types of mortgage has enabled a large proportion of households (around two-thirds) to become home owners. (The important role of the federal government in this policy area is outlined in Chapter 2.) Tax benefits have also played a part in making home ownership cheaper and financially attractive (for example, by way of deductions for mortgage interest and capital gain benefits).

Home ownership and rental housing differ in many ways (social, economic, physical, locational), but a crucial difference is that with home ownership, the householder obtains the mortgage directly, whereas with rented housing, there is an intermediary investor. This has an obvious but important implication: if investors do not foresee a profit, they will not invest. Therefore, if alternative investments are more profitable, or if incomes are too low to enable a rental housing investor to make the expected profit, there will be a shortage of rental housing. (Rent controls can have the same effect.)

Housing market theories

In the motor car market, those who can afford to do so buy new cars; others buy used ones. As a result cars filter down to lower-income groups. Does something similar happen with houses? Not exactly, since older houses may be better than newer ones. But, if instead of age one considers quality, there are some similarities. As houses decline in quality they become cheaper and affordable by those with lower incomes. Further, as household incomes increase, better quality housing can be afforded. Thus housing of declining quality tends to filter downwards while households with rising incomes filter upwards.

These simple ideas form the basis for much theorizing about the operation of the housing market. For present purposes, the relevant issue is whether the filtering process works sufficiently well to meet the needs of lower-income households; or does something impede this neat process? Since there is a general shortage of housing for lower-income households, it is self-evident that it does not work sufficiently to deal with all needs. The reasons for this are important. First, many houses do not filter at all, since they remain occupied by higher-income households or are converted to other uses (such as commercial) or are demolished to make way for land use changes (higher-quality uses, highways, etc.). Second, houses must fall greatly in price (as do motor vehicles) to be affordable by the poor; and this fall in price may well imply a marked fall in quality (the condition of the accommodation). In short, by the time a house filters down to the lowest income level, it may be grossly inadequate, badly maintained, and have a backlog of overdue maintenance costs that cannot be afforded. Indeed, by this point, the house may well have become a public health hazard, subject to action by the local authority. Moreover, housing quality typically depends on the level of maintenance of the building and the neighborhood in which it is located. Given a stable neighborhood, a high degree of maintenance, and a continuing program for replacing outworn services and fittings, the life of a house can be infinite, and it can steadily increase in value. More generally, however, changes in neighborhood quality (itself often the cause of reduced maintenance),

obsolescence of internal fittings, changes in fashion, rising standards of heating, cooling, insulation and such-like, and a host of other factors cause values to fall. These and many other complications do not arise in the motor vehicle market.

This is a highly simplified view: theories of filtering abound; and they conflict disconcertingly. This is partly because while some analysts view it as a *process*, others view it as an *outcome*. There is no necessary connection between the two. Changes in the process may have differing outcomes; and similar outcomes may result from differing processes, depending on a multitude of variables.

However, the theory that good housing filters down to poorer households 'works' to a limited extent, though it is restricted by a host of factors. One theory rests on the differences in the size of income groups. Lower-income groups are more numerous than those of higher income, and thus will constitute a large demand for houses vacated by the latter. As a result, the fall in prices will tend to be small. But to the extent that the filtering process is successful the result may tend to be 'self-corrective': lower prices would reduce the willingness of existing owners to trade, and the supply would diminish. Whatever the validity of such theories, it is clear that filtering cannot meet the housing needs of the poor. Their very poverty makes it impossible for them to pay the costs of decent housing. Filtering stops before housing of adequate quality gets to them. Yet much argument about housing policy is centered on the efficacy of filtering.

Among the many reasons why filtering fails to meet low-income needs is that some housing is abandoned before it becomes cheap enough. At first sight, abandonment seems nonsensical: surely some income is better than no income? This is not so if the costs are higher than the income! It can cost more to demolish a building than the resultant vacant site is worth. The unfortunate result of this is a cancerous growth of decay. The tragedy of the worst inner city areas is that they are blighted by abandonment and a lack of demand.

Low-income housing

Low-income households have particular difficulty in affording market rents, and housing policy has struggled for a long time with the problems to which this gives rise (see Box 15.1). These problems are exacerbated by several special factors. First, among the many matters which affect the cost of housing are the standards imposed by government on the quality of housing: a host of regulations impose minimum standards of building for health and safety. Some of these (such as adequate sanitation) are totally accepted; others (such as high minimum house sizes) are debatable. But whether acceptable or not, they increase the cost of housing above market levels for the poor. Some analysts go further and argue, as did the report to President Bush of the Advisory Committee on Regulatory Barriers to Affordable Housing (1991), that 'millions of Americans are being priced out of buying or renting the kind of housing they otherwise could afford were it not for a web of government regulations'. This remains the case twenty-two years later. A vast number of people are still priced out of the housing market and need help. This help involves all levels of government, the private sector, and the non-profit sector.

Second, as this quotation suggests, the specific problem of providing housing for the poor merges into the general problem of affordability, which affects a wider range of income groups. There is no simple cutoff point between the poor and the not poor. Moreover,

in providing housing for the poor at a subsidized cost, there arises a basic unfairness for those of the poor (the majority) who receive neither good housing nor the financial benefits that go with it. This has been one of the reasons why programs of housing vouchers have attracted less hostility than the provision of public housing: these provide rent assistance to renters of private as well as public housing. But there is a more important reason for the opposition to public housing: quite apart from ideological issues, there is typically very strong opposition to the location of public housing. This is the classic case of NIMBY: 'public housing may be all right somewhere else, but not here'. This opposition may be racial, social, or simply a result of the fear that the construction of public housing would lead to a fall in local property values. ('We personally do not object to public housing, but others do – and this will affect house values.') The poor management and the severe problems which have arisen in some public housing projects have created an indelible image of crime-ridden, drug-infested, dangerous and decayed urban eyesores. Their unpopularity is now widespread, and more effort is being expended on transferring public housing to tenants and other owners than on expanding the supply.

There are two additional points to make here. First, despite the extent of the heated arguments on public housing, the total amount nationally is extremely small: even at its largest, the number was only about 1.5 million units. Second, though the worst projects have attracted a great deal of attention and supported the stereotype, many public housing projects are of good quality, well maintained, and popular with their tenants. Unfortunately, the 'stigma' attached to public housing has led many people to feel that all public housing projects are bad and a financial drain on the public.

Public housing and urban renewal

Public housing policies have been characterized by extraordinarily strong opposition since the very beginning. The reasons for the opposition vary: arguments about the sanctity of property rights and the limits to which government should interfere in market forces; fear of undermining individual self-reliance; concern that the private market would be jeopardized by 'unfair competition'; mistrust of the competence of government in such an area; the huge cost which a significant program would involve; and the belief that such needs as could not be met by private enterprise were best left to charity and voluntary effort. In fact, many felt that government intervention would lead to the death of private enterprise housing. Despite the appalling housing condition of the poor, these arguments held sway until tentative initiatives were made in the late 1930s. The Depression years saw both an increase in the housing problem and a worsening of social conditions in the cities. Eventually, federal legislation was passed in 1937, signaling the realization that both slum clearance and the provision of public housing were legitimate areas of public policy. The substance, however, was thin, mainly because of the continued bitter opposition of the National Association of Real Estate Boards and kindred spirits. These groups felt it was the private sector's responsibility to provide housing and to meet the housing needs of the American population. Other individuals believed the private sector had failed to provide the needed housing and that government had to step in and assist those that could not afford housing.

The opposition continued throughout the 1940s (indeed, it has never ceased and probably never will). The industry has constantly argued that the private market could meet all the nation's housing needs without the intervention of government (though 'aids to private enterprise' such as those provided by the Federal Housing Administration – discussed in Chapter 2 – were championed). The outlook for post-war housing policy was therefore bleak. A housing bill was introduced in 1945, but was killed by vociferous opposition first in 1946, and again in 1948. Truman saw the opposition as a group essentially going backwards in the nation's attempt to solve its housing woes. Philosophical differences between Truman and the Republican Party also plagued Truman's attempts at developing a national comprehensive housing strategy. After much political haggling, it eventually passed as the 1949 Housing Act – the single Fair Deal piece of legislation that Truman managed to get through Congress. The Act embraced the national goal of 'a decent home and a suitable living environment for every American family', but the means to achieve this were effectively denied by Congress. It represented one of the most often cited pieces of federal housing legislation in history.

The legislation authorized the building of 810,000 units of public housing over a period of six years, though the program was slow in starting, and it took two decades before this target was reached. Part of the reason for this was the opposition to public sector activities, which, following the passage of the Act, moved from Congress to local areas. But the task was inherently complex: areas had to be selected and designated for acquisition; sites had to be cleared; complicated negotiations were required for federal funding; and arrangements had to be concluded with private investors and developers for redevelopment. The important role given to private enterprise was part of the political price which had to be paid to secure the passage of the legislation. One aspect of this was provision for urban redevelopment by private enterprise with local government supervision and federal-local subsidies that bridged the gap between the market value of land and the (much higher) actual cost of acquisition and clearance.

The concept of linking redevelopment to the politically unpopular provision of public housing gave rise to widespread problems. This was of particular importance since redevelopment was to be 'predominantly residential'. But private developers were not interested in low-income housing (whether subsidized or not). Profits lay in other directions, particularly downtown shopping and commercial centers. These were also popular with local political and business elites. As a result, pressure built up for the rules to be altered. A major change came with the 1954 Housing Act, when the term 'urban renewal' was introduced, indicating that, in addition to redevelopment, the policy now embraced revitalization, redevelopment, conservation, and 'the renewal of cities'. The Act provided that 10 percent of project grants could be used for non-residential development. The rationale behind this 'was that there were non-residential areas around central business districts, universities, hospitals, and other institutional settings that certain city interests wished to clear and redevelop for non-residential purposes' (Weiss 1980: 267). In Mollenkopf's words (1983: 117), 'the 1954 Housing Act shifted urban renewal from a nationally directed program focusing on housing to a locally directed program which allowed downtown businesses, developers, and their political allies, who had little interest in housing, to use federal power to advance their own ends'.

The proportion allowable for non-residential purposes was increased to 20 percent in 1959 (together with the needs of dominant institutions such as hospitals and universities), and later to 35 percent. With the administrative latitude allowed, the eventual result of this was to increase the commercial part of urban renewal to one half of the total. Indeed, by manipulating definitions and procedures, it was possible to force the proportion up to two-thirds.

Despite amendments to the legislation and some notable achievements, for example in improving the physical appearance of hundreds of American cities, urban renewal became subject to increasing criticism. Above all, it failed to help the poor; indeed, it made their position worse.

While urban renewal bolstered central business districts and may even have contributed indirectly to a city's economic vitality, it also dislocated neighborhoods and often created more urban blight than it removed. Some saw it as a displacement of neighborhoods to other areas. There is no doubt that urban renewal benefited some city center businesses and upper-middle-class households who obtained desirable inner-city housing at bargain prices, but frequently the real cost was borne by the urban poor. Downs (1970: 223) estimated that households displaced by urban renewal suffered an average uncompensated loss amounting to 20 to 30 percent of one year's income. Moreover, while originally conceived as a means of increasing the supply of low-cost housing, urban renewal actually exacerbated the urban housing problem: more houses were destroyed than were replaced.

Disillusionment with urban renewal led to its decline at the end of the 1960s. As a result, it joined public housing as a cause that even its sponsors no longer supported, though, of course, its impact has lived on. But the fundamental weakness of urban renewal was that it was conceived in terms of a land use instrument that was to 'save' the declining cities. Though there were pockets of success in commercial (sometimes monumental) centers and middle-class residential areas, these typically had little or no wider effects, except of an undesirable nature such as the displacement of low-income households. Their impact on restraining the exodus to the suburbs was minimal. It is likely that a greater force for effective 'renewal' lay with the increasing number of new immigrants who, like so many before them, sought out opportunities in the cities – opportunities which urban policy had defined as problems.

As previously mentioned, the creation of public housing has always been a subject of much debate. Some individuals see it as a positive sign to help low-income people to find housing. Others view it as the federal government intruding on an area that should be left to the private sector. Still other individuals see public housing as high-rise, isolated buildings with no investment in the area surrounding the projects. People conjure up visions of the now demolished public housing projects of Cabrini Green in Chicago and Pruitt-Igoe in St Louis. People in these and other projects were, in fact, segregated from the surrounding community.

Since its creation in 1937, public housing projects have become increasingly deteriorated. There was a perceived need to transform and rehabilitate public housing projects throughout the United States. In 1989, Congress established the National Commission on Severely Distressed Public Housing to identify those projects that were severely distressed and to remedy or to eliminate the various problems associated with them by 2000. Revitalization of these developments would require investments in sites, buildings, and people. As Popkin (2010: 45) explains, 'the HOPE VI program was designed to address not only the bricks-and-mortar problems in severely distressed public housing development, but also the social and economic needs of the residents and the health of the surrounding neighborhoods'. Section 24 of the US Housing Act of 1937 as amended by Section 535 of the Quality Housing and Work responsibility Act of 1998 (P.L. 105-276) explains the purposes of the program:

(1) improving the living environment for public housing residents of severely distressed public housing projects through the demolition, rehabilitation, reconfiguration, or replacement of obsolete public housing projects (or portions thereof);

(2) revitalizing sites (including remaining public housing dwelling units) on which such public housing projects are located and contributing to the improvement of the surrounding neighborhood;

(3) providing housing that will avoid or decrease the concentration of very low-income families; and

(4) building sustainable communities

The definition of 'severely distressed public housing' is broad. According to the Quality Housing and Work Responsibility Act of 1998 (also known as the Public Housing Reform Act), Title V of Public Law 105-76, severely distressed public housing:

(A) (i) requires major redesign, reconstruction or redevelopment, or partial or total demolition, to correct serious deficiencies in the original design (including inappropriately high population density), deferred maintenance, physical deterioration or obsolescence of major systems and other deficiencies in the physical plant of the project;

(ii) is a significant contributing factor to the physical decline of and disinvestment by public and private entities in the surrounding neighborhood;

(iii) (I) is occupied predominantly by families who are very low-income families with children, are unemployed, and dependent on various forms of public assistance; or

(II) has high rates of vandalism and criminal activity (including drug-related criminal activity) in comparison to other housing in the areas;

(iv) cannot be revitalized through assistance under other programs, such as the program for capital and operating assistance for public housing under this Act, or the programs under sections 9 and 14 of the United States Housing Act of 1937 (as in effect before the effective date under section 503(a) of the Quality Housing and Work Responsibility Act of 1998), because of cost constraint and inadequacy of available amounts; and

(B) That was a project described in subparagraph (A) that has been legally vacated or demolished, but for which the Secretary has not yet provided replacement housing assistance (other than tenant-based assistance).

A number of activities are eligible to carry out the revitalization programs for severely distressed public activities. Among the eligible activities are: architectural and engineering work, demolition, rehabilitation, reasonable legal fees, moving expenses for residents displaced as a result of the revitalization, management improvements, and replacement housing.

Homeownership and Opportunity for People Everywhere (HOPE) was developed as a direct result of the commission's plan. Originally called the 'Urban Revitalization Demonstration', Congress responded to the plan in 1992 and appropriated $300 million for what would demolish 'severely distressed' or 'obsolete' public housing. It is a type of 'housing plus' program. This means that additional activities designed to help revitalize these communities can be funded through HOPE VI. Funds can be used for the following objectives:

1   Improving public housing by replacing severely distressed public housing projects, such as high-rises and barracks-style apartments, with townhouses or garden-style apartments that blend aesthetically into the surrounding community.

2   Reducing concentrations of poverty by encouraging a mix of income among public housing residents and encouraging working families to move into housing that is part of revitalized communities.

3   Providing support services, such as education and training programs, childcare services, transportation, and counseling to help public housing residents get and keep jobs.

4   Establishing and enforcing high standards of personal and community responsibility through explicit lease requirements.

5   Forging partnerships that involve public housing residents, state and local government officials, the private sector, non-profit groups and the community-at-large in planning and implementing new communities.

A number of areas and their residents have benefited from HOPE VI. The City of St Louis, Missouri, Housing Authority received funding to replace over 600 units of severely distressed public housing with some 200 new public housing units. The funding will also help develop affordable and market-rate rental units, affordable home ownership units, and market-rate home ownership units. Concomitantly, various human development services such as education, health, and workforce development services will be provided to residents. In total, federal funding is expected to leverage even more funding for projects designed to revitalize the community. The Richmond, Virginia, Redevelopment and Housing Authority received HOPE VI funding to demolish some scattered public housing and to build new homes and rental units in the community. Education, childcare, transportation, and counseling services will also be funded through the federal funding. These services are designed to help the community residents become more self-sufficient. Additional funding was obtained from various sources to assist in the revitalization of the community. The Seattle Housing Authority received $35 million in

HOPE VI funds in 1999 to replace original public housing units in the High Point community of West Seattle, built during World War II, with mixed-income housing for both renters and home owners. Plates 21 and 22 are before and after photos of the High Point transformation into a HOPE VI community.

HOPE VI, as is the case with virtually all federal programs, had its supporters and detractors. HOPE VI projects improved the lives of many people. They took people away from high-density developments and put them in low-density developments. They provided residents with needed community support services.

Other people felt HOPE VI projects reduced the amount of housing for the very poor. Some individuals felt the projects actually displaced some families. They were not provided with better housing. They were simply relocated to another area – a view also heard during the earlier urban renewal days. Social ties with friends and families were broken.

The Bush administration proposed eliminating HOPE VI in the FY 2005 and FY 2006 budgets. This was consistent with his call to reduce the federal government role in public housing. Congress resisted the call and maintained the program. Unfortunately, in 2007, Bush targeted it for elimination. Nevertheless, the FY 2006–FY 2011 HUD Strategic Plan (US Department of Housing and Urban Development 2006: 13–14) noted that HOPE VI would create 10,000 new home ownership units and leverage a total of $315 million in private sector financing between 2006 and 2011.

President Obama's latest vessel for transforming neighborhoods is the Choice Neighborhoods initiative. It intends to build on the successes of and lessons learned from HOPE VI. The discussion would now turn to housing plus the surrounding community. There could no longer be separate discussions of public housing policy and the surrounding community. They were one and the same. They had to be integrated and discussed together. As Shaun Donovan, Secretary of Housing and Urban Development, said during remarks at the national Press Club on July 14, 2009, 'and if sixteen years of HOPE VI has taught us anything, it's that building communities in a more integrated and

Plate 21 Before: High-Point Community public housing in Seattle

Photo courtesy of Seattle Housing Authority

Plate 22 After: High-Point Community mixed-income housing in Seattle

Photo courtesy of Seattle Housing Authority

inclusive way isn't separate from advancing social and economic justice and the promise of America – it's absolutely essential to it'. We would need to plan, in collaboration with various partners, for the core goals of transforming public and assisted housing, supporting positive outcomes for people in the neighborhoods, and transforming poverty-ridden neighborhoods into sustainable mixed-income neighborhoods with all of the required community services (e.g., transportation, schools, jobs, etc.).

Areas interested in pursuing federal funding under the Choice Neighborhoods program had to apply for a Planning Grant to develop a comprehensive neighborhood revitalization plan called a Transformation Plan. Having been created through a comprehensive planning process, the Transformation Plan would serve as the guiding document for areas seeking to revitalize public and assisted housing and to build a sustainable mixed-income community. Once the Transformation Plan had been developed, the area could apply for Choice Neighborhood Implementation Grant. These funds would be used to implement the Transformation Plan's tool to redevelop the affected neighborhood into a sustainable mixed-income community.

The Norwalk, Connecticut, Housing Authority received Choice Neighborhood funds in 2010 to develop a Transformation Plan for the Washington Village/South Norwalk community. Through an eighteen-month comprehensive planning process the planning document is in its final stages. In September 2012, draft strategies were identified through multiple forms of citizen and community engagement – surveys, meetings with residents, information fairs, community needs assessments, Task Force meetings, etc. The draft Transformation Plan was presented to the city in November 2012 with an anticipated submittal to HUD by March 2013.

A national non-profit organization, Preservation of Affordable Housing, was awarded a Choice Neighborhood Initiative Implementation grant in 2011 to transform the Woodlawn community of Chicago. It represented the first-ever award of this nature to a national non-profit group. The project would revitalize the area, which was a site of distressed Section 8 housing units. These distressed units were an impediment to investment in the surrounding community. Among the goals of the effort are to demolish the existing Section 8 housing units, invest in community infrastructure, improve access to educational opportunities for all people, develop a community resource center, institute a gang violence initiative, and hire local residents during the construction and operations phases of the project.

Alternatives to public housing

There continues to be a huge debate over the need for public housing. To some people, government should get out of the practice of public housing and let the private sector handle meeting the housing needs of some populations. Others will suggest that the private sector has proven that it cannot meet the need. If we have these two positions, then what can be done? How can the country meet the housing needs of the population currently residing in public housing?

The removal of the federal government from the production of public housing created a void (small though it was) which has in part been filled by state and local efforts. These can take many forms, including public-private partnerships or neighborhood non-profit bodies (Stegman and Holden 1987; Suchman *et al.* 1990). Nationally, there are bodies such as James Rouse's Enterprise Foundation and the Inner City Ventures Fund. The former assists non-profit neighborhood groups on matters such as access to capital and technical expertise.

The Enterprise Foundation was created in 1982. It was based on the idea of James Rouse and his wife. They wanted to do something to help low-income individuals and their neighborhoods. Their ideas have been realized in a number of areas throughout the United States. The Enterprise Foundation works through a vast network of organizations dedicated to building and revitalizing local neighborhoods. Its various activities include: building decent and affordable housing; providing job training skills to the poor; training childcare workers and creating neighborhood childcare centers; and working with groups to make communities safe. More recently, the Enterprise Foundation has embraced the concept of smart growth (see discussions in Chapters 10 and 11). Furthermore, in 2000, the Enterprise Foundation created a new organization called the 'Enterprise Home Ownership Partners'. Together with the US Department of Housing and Urban Development (HUD) and the City of Los Angeles, the partners were to purchase, restore, and sell homes in the developing neighborhoods of Los Angeles.

The Inner City Ventures Fund was established by the National Trust for Historic Preservation to provide financial and technical assistance to non-profit neighborhood organizations for the rehabilitation of historic buildings that would later be used as affordable housing and commercial properties. These properties must directly benefit low- and moderate-income residents in low-, moderate-, and mixed-income neighborhoods.

Some community schemes are of a partnership nature, such as two project-based corporate-community partnerships in New York City which were backed by Chemical Bank and Citibank. Such non-permanent partnerships call on financial, technical, and organizational resources to achieve a specific development objective – often the acquisition and rehabilitation of deteriorated property. In Chicago, a housing

BOX 15.2 BOSTON HOUSING PARTNERSHIP

Considered a model for the nation, the Boston Housing Partnership builds on strong state and local commitments to programs that support the provision of low income housing. The partnership was formed in 1983 under the leadership of William Edgerly, chairman of the State Street Bank and Trust Company, in response to recommendations from a group of public officials, private business interests, and neighborhood organizations. Known as Goals for Boston, the group wanted the city to devote more attention to the housing needs of its disadvantaged neighborhoods, and to address both housing abandonment in distressed areas and displacement of low and moderate income households in strong market areas.

BHP members include the city of Boston, major banks and insurance and utility companies; local universities; local businesses; community and housing development organizations; and the Massachusetts Housing Finance Agency.

Source: Suchman *et al.* 1990: 23

partnership has involved several types of organization: community, private, and local government. Financial contributions from local employers attract tax benefits as well as positive community relations. (Another housing partnership is noted in Box 15.2.)

A different type of organization is the San Francisco BRIDGE: the Bay Area Residential Investment and Development Group. It has become the largest non-profit producer of affordable housing in California. This is unlike many local organizations in that it operates over a wide area and, instead of supporting local groups, is a direct provider of housing. Over the period since its foundation in 1983, it has participated in the development of some 13,000 affordable dwellings, and has an ambitious continuing program. Over 35,000 people have been served by BRIDGE programs. It attracts funding from corporate investors, but also has obtained benefits from tax credits, and a variety of innovative techniques, tax-deductible donations, for-profit housing, density bonuses, and other support from its strong connections with the local business community.

BRIDGE can no longer be thought of as simply a producer of affordable housing. It has broadened its mission to include many more service areas. As noted in its 1999–2000 Annual Report,

While well-designed, well-maintained housing is a vital element in every community, we know

that there's often more to the picture. Dynamic, healthy communities can't exist without a solid social and economic foundation. It takes more than physical improvements to upgrade areas where incomes are low, properties are deteriorated, and services are lacking. And it is also clear that high-priced leapfrog development is not a sustainable way to meet our housing needs.

Our response is a more comprehensive approach to community-building – an effort to create infill developments that not only produce quality housing for our workforce, but also help build the strong foundation every community needs.

(BRIDGE 2000: 3–4)

There is a large number and variety of these local non-profits and it is generally agreed that they provide an acceptable and efficient way of providing additional low-income housing. Some of them have concerns and activities that extend well beyond housing: these are discussed in the next chapter.

Limited equity cooperatives offer another form of multi-family housing where a building is jointly owned by the tenants. The building is restricted or limited to people with low to moderate income levels. Tenants do not own the land. They own shares in a cooperative corporation.

While some researchers suggest that minimum

income requirements fail to make limited equity cooperatives a substitute for public housing, they do offer an alternative source of affordable housing. They remain affordable by limiting the resale pricing of the housing. Formulas devised by each cooperative determine the resale pricing.

Housing subsidies

The problem of housing affordability is a result of the level of market prices being higher than the ability to pay. The gap can be bridged either by increasing incomes or by reducing rents. Either way, a subsidy is required. However, simply increasing incomes does not, in practice, help very much since households with low incomes may (and often do) prefer to spend increased income on other necessities than housing (as the Experimental Housing Allowance program demonstrated – a finding which raises some interesting policy issues which cannot be pursued here). To target the specific problem of housing costs, financial aid has to be tied to the payment of these costs. If the subsidy is tied into the construction of affordable housing, it obviously benefits only those who obtain such housing (and do not move out of it).

Much current policy assumes that it is better to pay the subsidy direct to needy households, who then have a degree of housing choice. These matters are in practice not as simple as this, since problems remain of ensuring housing supply, of combating discrim-ination, of relating the subsidy to needs while at the same time avoiding disincentives to increase income, and so forth: housing policy issues are never easy! They are also costly (and therefore are vulnerable to public expenditure cuts). There has been considerable experience with such a system with the so-called 'Section 8' certificate and voucher programs (see Box 15.3).

Housing choice vouchers are administered locally by local public housing authorities. They receive their funding from HUD. Total annual gross income and family size are used to determine a family's eligibility for receiving a housing voucher. The income figures are published by HUD and vary by location. The figures given by the family to the public housing authority officials will be verified for accuracy to determine whether or not the family qualifies for the voucher program.

The burden is on the family to find suitable housing that meets its needs, given the income constraints imposed by the program. Any unit selected by the family must meet minimum health and safety conditions. The landlord is paid a rental subsidy by the public housing authority. The family pays the difference between what the landlord charges and what the housing voucher program pays. It must pay 30 percent of its monthly adjusted gross income for rent and utilities. The voucher pays for the remaining cost of the housing. Families can also use a voucher to purchase homes.

The housing voucher program is highly competitive. The needs outweigh the availability of vouchers.

BOX 15.3 HOUSING VOUCHERS

The housing choice voucher program is the federal government's major program for assisting very low-income families, the elderly, and the disabled to afford decent, safe, and sanitary housing in the private market. Since housing assistance is provided on behalf of the family or individual, participants are able to find their own housing, including single-family homes, townhouses and apartments. The participant is free to choose any housing that meets the requirements of the program and is not limited to units located in subsidized housing projects.

Source: Housing Choice Voucher Program Fact Sheet (Section 8), US HUD, 2001

Waiting lists are common in many areas of the United States. Public housing authorities compete with each other to receive the voucher funding from HUD. HUD evaluates and ranks the applications with respect to such criteria as number of very low-income renters with severe rent burdens, current activities by areas to promote area-wide housing opportunities for families, and number of disabled families. President Bush called for pursuing reforms to the voucher programs through regulatory and legislative levels.

The complexities of the housing problem cannot be met by any single policy. Even if the housing certificate program were fully funded, it would still be necessary to ensure that an adequate number of housing units were built, and that there were no barriers to access. In addition to the community-based provision already noted, there are other techniques which can help. These include schemes to entice builders to produce low-cost housing, and measures to remove the local barriers to such housing which are erected by municipalities.

The former includes the Low Income Housing Tax Credit (LIHTC) program, introduced in the 1986 tax reforms, and made permanent in 1993. These credits are used to raise equity for approved housing construction or rehabilitation. The developments are subject to restrictive covenants which run with the land for thirty years or more: these cover such matters as the number of housing units, guidelines relating to rents and eligible household incomes, and criteria for targeting specified needs. Individuals, partnerships, government agencies, and non-profit entities can use the LIHTC program to construct or rehabilitate affordable housing units. About three-quarters of a million rental housing units were allocated through the tax credit program between 1987 and 1994. From 1987 to 2000, over 1 million housing units were allocated through the LIHTC program. President Bush called for expanding the amount of affordable rental through the LIHTC and other funds.

The removal of regulatory barriers was the subject of a presidential commission (referred to earlier). Its NIMBY report received much publicity, and it was followed up by guidance to the states from HUD. Some states increased their pressure on local governments to ease restrictive zoning policies, and some passed legislation prohibiting discrimination against manufactured housing; but generally the report simply merged with concerns already being expressed about the need to increase the provision of affordable housing. These were given a further push by congressional action: Title I of the National Affordable Housing Act of 1990 required states to address regulatory barriers in preparing *Comprehensive Housing Affordability Strategies* (now forming part of the *Consolidated Plan* for all HUD community planning and development programs) which are a prerequisite for obtaining Community Development Block Grant funds (discussed in the following chapter). The importance attached to this by Congress is indicated by the introduction, in 1992, of a grant program for the preparation of these strategies. There can be little doubt that the issue of affordable housing has had considerable exposure, but the results of this are less easy to establish.

The barriers facing the construction of affordable housing are complex. To eliminate them requires a concerted effort on the part of many people. The federal government needs to recognize that some federal laws and policies have effects on the construction of affordable housing. States need to send signals to local governments on the importance of reducing the barriers. Outlawing certain practices, such as housing discrimination, lets everyone know that government will not tolerate it and will challenge anyone engaging in it. Local governments could do such things as revising zoning codes, development regulations, and fees. They could also provide more information to neighborhood groups that have acted as barriers to construction of lower-income housing. The reasons for their opposition could range from fear of crime, a reduction in their property values, aesthetic issues, to simply not wanting people dissimilar to themselves living close to them. Non-profit organizations can help educate residents on the importance of lifting the barriers to affordability. The private sector can help in the education process and by assisting local government in devising effective ways of reducing or eliminating the barriers to affordability (see Box 15.4).

Other options are available for jurisdictions to promote affordable housing. Affordable Housing Districts

BOX 15.4 EFFECTS OF REGULATORY REQUIREMENTS ON HOUSING CONSTRUCTION

A constricting web of regulatory requirements affects virtually every aspect of the land development and home building process, adding substantially to the cost of constructing a new home and preventing many families from becoming home owners. Imposed at the federal, state and local levels, these regulations are largely invisible to the home buyer, but nevertheless have a profound impact on affordability and home ownership. And, while some regulation is necessary, it should be sensible, appropriate and fair: it should not be imposed for the purpose of halting or limiting growth.

Source: National Association of Home Builders, n.d.: 14

BOX 15.5 FAIRFAX COUNTY ZONING ORDINANCE, AFFORDABLE DWELLING UNIT PROGRAM PURPOSE AND INTENT

The Affordable Dwelling Unit program is established to assist in the provision of affordable housing for persons of low and moderate income. The program is designed to promote a full range of housing choices and to require the construction and continued existence of dwelling units affordable to households whose income is seventy (70) percent or less of the median income for the Washington Standard Metropolitan Statistical Area. An affordable dwelling unit shall mean the rental and/or for sale dwelling unit for which the rental and/or sales price is controlled pursuant to the provisions of this Part. For all affordable dwelling unit dwellings, where the dwelling unit type for the affordable dwelling unit is different from that of the market rate units, the affordable dwelling units should be integrated within the developments to the extent feasible, based on building and development design. In developments where the affordable dwelling units are provided in a dwelling unit type which is the same as the market rate dwelling units, the affordable dwelling units should be dispersed among the market rate dwelling units.

Source: Fairfax County Zoning Ordinance, Part 8, Section 2-801 Purpose and Intent, Affordable Dwelling Unit Program

have been established to encourage the development of affordable housing. Grand Forks, North Dakota, offered a two-year deferral on special assessments and a two-year deferral on property taxes as incentives for its Affordable Housing District. In addition, some communities waive all or a portion of their local impact fees for housing developments offering affordable housing units. Communities have also designated districts solely for the use of manufactured housing units. Furthermore, Fairfax County, Virginia, adopted an Affordable Dwelling Unit Program to promote affordable housing for low- and moderate-income households. Box 15.5 shows the purpose and intent of the program. Maximum household income eligibility limits determine a household's eligibility for purchasing or renting an Affordable Dwelling Unit under the Program.

The severity of the lack of affordable housing varies by state (see Plates 23 to 26). Families in California have been suffering with this problem for years.

Plate 23 High-rise condominiums in San Diego

Photo by author

Plate 24 Single-family home in Groveland, Massachusetts
Photo by author

Plate 25 Single-family housing on Beacon Hill, Boston, Massachusetts
Photo by author

Plate 26 Single-family home in Las Vegas

Photo by author

A report by the California Budget Project (2000: 6) showed that in 1997 over 2 million renter households paid more than 30 percent of their income toward shelter and that, in 1999, only 55 percent of California households owned their own homes. The problem does not stop with low-income families. Many of the state's middle-income families cannot even qualify for a mortgage. Ultimately, the situation poses a myriad of problems for California:

> The lack of affordable housing has widespread implications for families, communities, and the vitality of the California economy. High housing costs make it difficult for businesses to attract and retain workers. The search for affordable housing is driving many metropolitan-area workers farther and farther from their jobs, creating ever-greater suburban sprawl and leading to growing traffic congestion and greater air pollution. Rising rents make it impossible for low wage workers to live in the communities where they work, forcing many to choose between a long commute and over-crowded, substandard housing. When families are forced to spend more of their earnings on shelter, they have less to spend on food, clothing, child care, and other necessities. And the lack of affordable housing contributes to the stubborn challenge of preventing homelessness and helping those who are already homeless get off the streets.
>
> (California Budget Project, 2001: 52–3)

Housing trust funds

Meeting the housing needs of the American population requires a number of different approaches. One such approach is through the creation of housing trust funds (HTFs), whose purpose is to help support, through financial resources, the development and retention of affordable housing for various populations. The populations could be very low-income, low-income, the elderly, or other identified populations. The housing needs of any area will be site-specific.

HTFs can be created by either legislation, ordinance, or resolution. It may not be a simple process. Political objections will undoubtedly surface. Individuals will wonder where the funding for the HTF will originate. Others might complain that funding is being taken away from other needed programs. Ultimately, getting an HTF created is not enough. It is a good first step. The goal is to make the HTF an ongoing entity that helps meet the housing needs of the area's identified populations.

As the previous paragraph indicates, finding the financial resources to fund the mission of the HTF may be difficult. There will be heated debates over the necessity of perhaps adding a new fee to get the needed funding. There will be discussions over the legality of using some fees. In the end, an HTF may get its funding from such sources as mortgage recording fees, transient occupancy taxes, utility user fees, general fund appropriations, filing and recording fees, building permits, hotel taxes, or proceeds from the sale of vacant lands. The sources of funding will differ by locality and state.

M. E. Brooks (2002) has indicated that there are over 275 HTFs in cities, counties, regions, and states. Moreover, Brooks (1999: 4) notes that the HTFs have spent nearly $1.5 billion building and preserving almost 200,000 units of affordable housing. The Massachusetts Affordable Housing Trust Fund was created by state legislation in 2000 for the purposes of providing resources to create or preserve affordable housing for households with incomes that are not more than 110 percent of median income throughout Massachusetts. Its funding for State Fiscal Years 2001–2005 comes from the state's General Fund.

Among the applicants eligible to apply for funding are community development corporations, local housing authorities, non-profit organizations, and private sector entities. The Nebraska Affordable Housing Trust Fund was created in 1996 as a result of the Nebraska Affordable Housing Act. Its purpose is to enhance the state's economic development through the development of affordable housing. The city of Charleston, West Virginia, created an HTF in 1998 to help meet the housing needs of low- and moderate-income families. It has provided numerous loans to rehabilitate buildings and housing for the aforementioned income groups. HTFs can be found throughout the United States.

The severity of the housing crisis for very low-, low-, and moderate-income families continues to be discussed at the highest level of government. There is currently a movement in the US Congress to create a national housing trust fund whose purpose is to create new housing opportunities for the lowest-income families in the United States. One of its proposed goals would be to construct 1.5 million units of rental housing for the lowest-income families. According to the Center for Community Change (2001: 5) 'a National Housing Trust Fund will encourage innovative housing development by allowing communities to define their own needs and design their own solutions'. A dedicated and dependable source of income is definitely needed. The National Housing Trust Fund Act of 2007 was introduced in the US House of Representatives by Massachusetts Representative Barney Frank on June 28, 2007. It was passed by the House on October 10, 2007. It was introduced in the US Senate on December 19, 2007. If history repeats itself, there is certain to be a lengthy discussion over any numerical goal over a ten-year period.

Community Land Trusts

Community Land Trusts (CLT) vary by purpose throughout the United States. A CLT represents another vehicle for developing and preserving affordable home ownership opportunities for low- and moderate-income households at or below 80 percent of the area median income. First-time home owners cannot exceed the specified household income levels. The CLT is typically a private, non-profit 501 (c) (3) membership corporation. They generally receive start-up capital from such sources as a grant from a city, community foundations, banks and other lenders, fundraising efforts, state agencies or boards, Community Development Block Grant funds, and from other sources. An example of the purposes of a CLT can be found in Box 15.6. As Box 15.6 indicates, the CLT can have purposes other than providing affordable housing.

The CLT works in the following manner. The home buyer purchases the home from the CLT at a given price and leases the ground under the home for a certain time period, generally a ninety-nine-year lease. This lease removes the value of the land from the equation. The home buyer is responsible for paying any property tax on the home.

Home buyers can receive a profit if they sell their home. However, the profit level is generally capped to make sure the housing remains affordable to qualified low- and moderate-income households. An example of how a resale, capped at 25 percent, at the Rondo Community Land Trust in St Paul, Minnesota (http://www.rondoclt.org/faqs.php, accessed January 8, 2008) would work can be found below.

Initial purchase price (market rate) of
CLT home
$165,000

Initial cost to the CLT buyer
(after buy-down grant)
$140,000

Value of the home at resale
$220,000

Increase in value of home
$220,000 − $140,000 = $80,000

Resale price to another CLT home buyer
$140,000 + (25% × $80,000) = $160,000

In this case, the home seller receives the difference between the original mortgage and the amount they currently owe, their initial down payment, plus $20,000 from appreciation – the profit. Resale formulas may vary from CLT to CLT.

A number of examples of CLTs can be found throughout the United States. Their accomplishments vary from area to area. One CLT has constructed homes on vacant inner-city lots donated to the CLT. Other CLTs have reclaimed acres in the inner neighborhood of a city and built a livable community and sold rehabilitated buildings to low-income first-time home buyers. In 2006, the Lopez Community Land Trust (p. 1) in the State of Washington announced it was partnering with 'A World Institute for Sustainable Humanity (A W.I.S.H.) to create a mixed-income, net-zero-energy, sustainable development consisting of two studio rentals and up to thirteen modest single-family homes for households of low and moderate income; and an office/resource room for the Lopez Community Land Trust's home base' (see Box 15.6).

Land banking

The number of vacant homes and properties throughout cities across the US has dramatically increased in recent years. Worsening economic conditions have led to increased unemployment rates, vacancy rates, tax delinquencies, and eventual foreclosures. These problems have severe ramifications for areas. Unemployed people need increasing amounts of public services. Properties become tax drains with no property taxes being paid. Abandoned properties become potential sites for crime and arson.

Many areas have become proactive in dealing with these problems. They are attempting to put vacant, abandoned, or foreclosed homes and properties back into productive use, thus trying to return them to community assets. One such action is the creation of land banks. These entities take vacant, tax delinquent, foreclosed homes and properties and take title to them. After obtaining the title, they can either demolish or rehabilitate them. Once this is accomplished, the houses and properties are then transferred to a new owner. Figure 15.1 shows the various functions of a land bank.

BOX 15.6 PURPOSE OF LOPEZ COMMUNITY LAND TRUST

a. Acquire and hold land in trust in order to provide for permanently affordable housing. Homes shall be built and lands shall be used in an environmentally sensitive and socially responsible manner.

b. Provide permanently affordable access to land for such purposes as quality housing, sustainable agriculture and forestry, cottage industries and co-operatives by forever removing the land from the speculative market.

c. Develop and exercise responsible and ecological practices, which preserve, protect and enhance the land's natural attributes.

d. Serve as a model in land stewardship and community development by providing information, resources and expertise.

Source: Lopez Community Land Trust, http://www.lopezclt.org (accessed January 8, 2008)

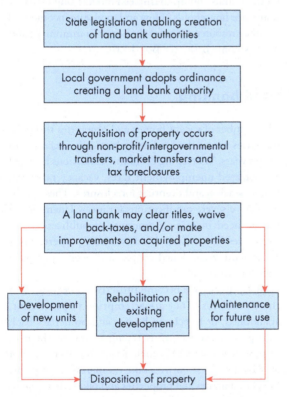

Figure 15.1 Functions of a land bank
Source: US HUD 2009

States such as Kentucky, Georgia, Michigan, and recently New York and Pennsylvania, have authorized their creation as either a non-profit corporation or within the existing local government structure in a department or agency as a tool to fight and reclaim blighted, abandoned, and tax-foreclosed properties. Recognizing the increasing quantity of abandoned housing and the problems associated with vacant properties, New York Governor Andrew Cuomo, in 2011, signed a law authorizing local governments to create land banks to turn vacant properties into community assets. In a July 29, 2011 statement, Governor Cuomo acknowledged that 'land banks will finally give local authorities the much needed ability to take community redevelopment into their own hands'. Governor Tom Corbett of Pennsylvania signed land bank legislation into law on October 24, 2012.

Accessory dwelling units

Accessory dwelling units (ADUs), also referred to as second units or granny flats when an age is attached, offer another method of providing affordable housing. An ADU is a complete, independent living unit that includes a separate kitchen, bathroom, and sleeping area. The unit must be located on a lot with a primary dwelling unit. It can be located on the second floor of the existing primary unit, above a garage, or be a freestanding addition on the lot. Local ordinances must be consulted to determine what is allowed in a specific community.

There are a number of motivations for allowing ADUs. They allow home owners to gain additional income that can be used to help them with their own mortgage payments while providing needed housing. They allow current housing to be put to a more efficient use. Moreover, they may allow individuals and families an opportunity to stay together. For example, an ADU may allow a family to care for a parent or older member of the family when that person cannot maintain or afford a larger dwelling unit. The promotion of and presence of ADUs also enable communities to increase density, a goal of many growth management programs.

Some communities oppose the creation of ADUs for a variety of reasons. One community might oppose them on the grounds of supply and demand. There may be no need for them because an adequate supply of low- and moderate-income housing is available in the community. Another community might object to allowing them because neighborhood streets are already functioning above design capacity during rush hours. This may lead to a potential increase in traffic accidents. New units, albeit small units, might exacerbate the problem. Still another community may fear that their addition may change the character of the community.

California recognizes that ADUs represent another source of affordable housing in a state where affordable housing is in short supply. It has enacted legislation for local governments to provide for the creation of ADUs (referred to as second units in the legislation). The legislation, enacted in 1982, allows local governments to designate where second units are permitted. Localities may also impose various standards on second units (e.g., minimum size, parking requirements, setback, etc.). In addition, they may also provide for a process of requiring conditional use permits for second units. Under California Government Code 65852.2(C),

> no local agency shall adopt an ordinance which totally precludes second units within single-family or multifamily zoned areas unless the ordinance contains findings acknowledging that the ordinance may limit housing opportunities of the region and further contain findings that specific adverse impacts on the public health, safety and welfare that would result from allowing second units within single-family and multifamily zoned areas justify adopting the ordinance.

The City of Newport Beach, California, Zoning Code Chapter 20.85 enacted legislation that allows granny flats while prohibiting second units. The regulations allow one or two adults over sixty years of age to reside in a detached or attached structure on a lot with a primary dwelling unit. In order for a use permit to be granted for the creation of a granny flat, the proposed unit must conform to height standards, be located on a minimum lot size, meet minimum floor area size requirements, not exceed a maximum floor area size, have an independent parking space, and have either the primary or granny flat occupied by someone having ownership in the property.

Affordable housing and growth management

Municipalities represent the interests of their electorates; and since voters object to the construction of low-cost housing, municipalities will often prevent it by one means or another. The reasons have been listed in previous chapters: they include the desire to safeguard the character of the area and its property values, to prevent increased development and the traffic it creates, and likewise to minimize tax burdens by restraining additional infrastructure needs.

There is an interesting relationship between the effects of growth management programs on the costs of housing and the effects of sprawl on the cost of housing. Restrictive land use regulations can have the effect of reducing the amount of land available for housing and lead to increasing housing prices, thus exacerbating the affordability problem. On the other hand, much can be said about the effects of sprawl on housing costs. In a study on the costs of sprawl in Pennsylvania, Clarion Associates, Inc. (2000: 6) found that 'national studies have found that at a statewide level of analysis, sprawl will raise private housing costs between 2 percent and 8 percent above what they would be under a planned growth scenario provided that the total supply of housing is high enough to meet the demand'.

The affordability crisis is not new. It has been around for many years, worse in some states and better in other states. Many people steer away from areas such as Boston, San Francisco, New York, Washington, DC, etc. because of the high costs of housing. The crisis is not simply on the low-income population. In many areas, people considered middle-income families cannot afford to pay market rents or don't have the needed income to qualify for a conventional mortgage. This affects groups such as teachers, police, fire, and other public sector workers.

Growth management programs may be designed, or used, for such purposes. Indeed, since such programs can have the effect of raising house prices, thus increasing the affordability problem, they may attract considerable opposition from supporters of affordable housing. It is generally accepted that affordability means a household paying no more than 30 percent of its annual income on housing – more than that is considered a cost-burdened household. For US federal housing purposes, the following divisions can be used: moderate-income households earn between 80 percent and 120 percent of the area-wide median income; low-income households earn between 50 percent and 80 percent of the area-wide median income; and very low-income households earn no more than 50 percent of the area-wide median income.

Debates on state growth management policies can center on this, and a number of states have broadened their land policies to encompass the provision of affordable housing (as discussed in Chapter 11).

As part of its overall statewide planning endeavors, Oregon devised a housing goal (see Box 15.7). A striking feature of this goal is that it is expressed in terms which are clearly translatable into action: this contrasts with the symbolic, empty statements which sometimes pass for 'housing policy'.

The first draft of the Oregon state goals did not address the issue of housing but, as a result of concerns about the negative impacts of growth management policies on the housing market, the specific housing goal was added. Several local governments attempted to circumvent it by changes to their charters, or by introducing conditions which prevented the building of low-cost housing (Abbott *et al.* 1994: 103). These attempts were quashed by the courts. It was held that the goal imposed an affirmative duty on local governments to facilitate the provision of housing at a price or rent which was affordable by current and *prospective* residents. There was thus an obligation to provide for households from outside the area or, to use the pertinent phrase, for 'regional needs'. There is a long legal history to this, of which a high point is the famous *Mount Laurel* case in New Jersey. It is worth looking at this in some detail.

Regional housing needs: the case of *Mount Laurel*

From the time of the *Euclid* case (1924–26) to the 1960s, the exclusionary practices of municipalities received little critical attention from the courts or from federal or state governments. The political

---

BOX 15.7 OREGON HOUSING GOALS

To provide for the housing needs of citizens of the state:
  Buildable lands for residential use shall be inventoried and plans shall encourage the availability of adequate numbers of needed housing units at price ranges and rent levels which are commensurate with the financial capabilities of Oregon households and allow for flexibility of housing location, type, and density.

independence of municipalities was largely sacrosanct. At the same time, policies in relation to both land use and housing were predominantly local in character. There was little concern for regional needs: municipalities were, in the graphic phrase of one critic, 'tight little islands'. Those needs of the exploding cities which could be met by low-density housing were welcomed by suburban municipalities, but low-income housing was nobody's responsibility – in practice, even if not in theory. There were, of course, numerous critics, but they had little political impact.

Similarly, efforts by the federal government to promote 'fair housing' and 'equal opportunity' policies, to persuade local governments to include a 'housing element' in their local plans, to introduce *area housing opportunity plans* and such-like, all proved to be generally ineffective. They never had more than a marginal effect on strong market forces and the powerful underlying social attitudes. It was against this background that the New Jersey State Supreme Court, interpreting the New Jersey State Constitution, made a frontal attack on the discriminatory practices of the local governments in that state in a set of cases collectively known as *Mount Laurel*.

Mount Laurel is a most pleasant area which has attracted large numbers of people from nearby Philadelphia and Camden. Its population doubled between 1950 and 1960 (from 2,800 to 5,200), doubled again between 1960 and 1970, and grew by another half by 1985 (to 17,600). In common with many of the 567 municipalities in New Jersey, Mount Laurel imposed minimum lot size and other restrictions. There was no doubt as to the intention, as well as the effect, of these: it was to keep out low-income families and other unwanted groups. Activist groups of academics, lawyers, and others (including Paul Davidoff's Suburban Action Institute, the National Committee Against Discrimination in Housing, and the stalwart lawyer Norman Williams) had already taken legal action against other municipalities, sometimes with success – in legal, if not in practical terms. Mount Laurel, with a blatantly exclusionary zoning policy, was an obviously important target, and in 1975 a number of public interest groups brought a case against the township. The case eventually found its way to the New Jersey Supreme Court.

The *Mount Laurel* saga was long drawn out, and the case came before the court three times. On the first occasion, the court proclaimed the doctrine that a municipality's land use regulations had to provide a 'realistic opportunity for the construction of its fair share of the present and prospective regional need for low and moderate income housing'. Much litigation followed this opinion, and there were varying interpretations of what it really meant. Indeed, far more time and effort was spent on this than on actually doing anything substantively, and it became clear that the court had not provided any effective remedy. Developers (who were keen to build for a lower-income market) were largely powerless against municipal stalling tactics. Mount Laurel's response to the court was, to put it mildly, niggardly: it rezoned twenty acres on three widely scattered plots owned by three separate individuals, two of whom were not even residential developers. Its zoning ordinance entirely prohibited the construction of apartments, townhouses, and mobile homes.

The case returned to the court, which took a most militant approach. It was determined to introduce effective means to compel municipalities to provide 'a realistic possibility' for a 'fair share' of housing opportunities for lower-income households. This it did by requiring that, where a municipality refused to fulfill 'its constitutional obligation' (of permitting low-income housing) the court itself would take over the responsibility. That this involved overriding the local zoning ordinance and various other matters concerning the actual provision of housing did not deter the court: it would if necessary employ a court-appointed master to deal with zoning approvals and 'the use of effective affirmative planning and zoning devices'. The appointment of a master was purely discretionary on the part of the court. The court acknowledged that 'the special master may also free the court from unwise direct over-involvement in the revision of the zoning ordinance, saving the court's time and eliminating the need for the court to develop an even greater expertise in the intricacies of land use regulation'.

Mount Laurel was a truly remarkable case. It was tantamount to a usurpation of municipal powers (it was certainly attacked as such by both local and state government). However, the court was at pains to stress that, though it would have preferred not to take extreme measures, it had no alternative. It was now up to the legislature to tackle the difficult political problem from which it had so far shied. The court believed it was a state's legislative function to develop public policy in this area, not a judicial function. As the court noted in its conclusion:

> The provision of decent housing for the poor is not a function of this Court. Our only role is to see to it that zoning does not prevent it, but rather provides a realistic opportunity for its construction as required by New Jersey's Constitution. The actual construction of that housing will continue to depend, in a much larger degree, on the economy, on private enterprise, and on the actions of the other branches of government at the national, state, and local level. We intend here only to make sure that if the poor remain locked into urban slums, it will not be because we failed to enforce the Constitution.

Not surprisingly, the reaction against the court's opinion was vociferous, and there was a great deal of activity aimed at both the implementation and the obstruction of the court's wishes. But the growing effectiveness of the court-imposed regime eventually forced the New Jersey Legislature to take the political action which was so clearly required and, after long and acrimonious debate, the New Jersey Fair Housing Act of 1985 was passed.

The New Jersey Fair Housing Act

New Jersey's Fair Housing Act was designed to retrieve the role which the court had usurped in relation to the provision of lower-income housing. It was intended, above all, to 'disarm' the judiciary. This achievement was made possible only by crafting the provisions of the Act in such a way that it was more acceptable (or, to be more precise, less unacceptable) than the system being operated by the court. Above all, the 'judicial monster' created by the court was replaced by a system of voluntary compliance. This, of course, implied a

considerable weakening in the system of control over exclusionary zoning. (Some idea of the depth of feelings aroused is given by the public statement of two politicians who would risk going to jail instead of allowing a court ordered builder's remedy (Kent-Smith, 1987: 929–930).

The Act established a Council on Affordable Housing (COAH) charged with carrying out 'the Mount Laurel obligation'. According to the legislation, COAH was vested with the responsibility to define housing regions, estimate low- and moderate-income housing needs, set criteria and guidelines for municipalities to determine their respective housing needs and how they would address them, and to review and approve an area's housing element. It would work with municipalities to develop plans for affordable housing. It also transferred most existing (and future) suits to the council. Transfer was dependent upon a municipality including in its zoning ordinance a 'housing element' which contained a fair share plan. If accepted by the council it is approved (certified), and the municipality is then shielded or immune from successful builders' suits – unless a builder can provide 'clear and convincing evidence' that the zoning ordinance is exclusionary. According to COAH, over 300 out of 566 municipalities are participating in its efforts.

As a part of its fair share plan, a municipality can transfer up to half of its fair share obligations to another municipality by payment through a *regional contribution agreement* (RCA). Many suburban communities have negotiated such transfers to the older urban areas of the state, such as Newark. The transferred funds can be used by the receiving municipality for new building or rehabilitation. The funds used in an RCA are raised in various ways, mostly through agreements with developers who make the payments in lieu of building affordable units in their new developments. Payments per unit were previously set at a minimum of $20,000 per housing unit. After January 1, 2001, COAH requires a minimum payment of $25,000 per unit for an RCA. For example, at its October 2, 2002 meeting, COAH approved one community paying another community $680,000 (six payments over five years) for a scattered site housing rehabilitation program (Council on Affordable Housing 2002: 1). This particular proposal was submitted prior to the new minimum figure of $25,000 per housing unit.

The Fair Housing Act tips the balance of advantage greatly in favor of a municipality (see Box 15.8). All it has to do is to satisfy the council that it is making an appropriate fair share allocation (under provisions much less demanding than under the court's system).

In 2010, Governor Chris Christie signed Executive Order No. 12, putting a ninety-day freeze on COAH and establishing a Housing Opportunity Task Force that would study the state's affordable housing laws and determine the constitutional obligations to be consistent with the Mount Laurel court decisions. Governor Christie believe the procedures in place for developing affordable housing for low- and moderate-income families were too complex and resulted in delays, unreasonable costs, and potential litigation. To him, COAH had failed in its appointed task.

On June 29, 2011, Governor Christie, under Reorganization Plan No. 001-2011, sought to eliminate COAH and transfer its functions and duties to the Department of Community Affairs. Its elimination was also consistent with the recommendations of the Housing Opportunity Task Force. An appeals panel ruled that Governor Christie lacked the power to eliminate COAH.

As of April 2012, over $150 million of unspent affordable housing funds could be found in local housing trust funds. Three months later, in July 2012, Governor Christie set a deadline for the money in the housing trust funds to be committed to approved projects. After the deadline, the state would empty the municipal trust funds. A state court ruled against Governor Christie's action by indicating any request to empty a municipal housing trust fund had to come from COAH, not the governor. The governor did not have the authority to act in this manner.

The Federal Fair Housing Act

At the federal level, the Civil Rights Act of 1968 was quickly used to outlaw racial discrimination in the sale of houses, but discrimination through zoning proved to be a much more difficult issue. Discrimination can occur because of race or color, national origin, religion, sex, familial status, and handicap. A famous 1977 US Supreme Court case is that of *Arlington Heights*. This concerned a proposal by a religious order, which owned eighty acres of land near the center of the village (a suburb of Chicago) and proposed to develop this for a federally subsidized multi-family racially integrated housing project. The prevailing land use in the village of Arlington Heights was detached single-family homes. A rezoning was necessary which, following fierce public opposition (only twenty-seven of the village's 64,000 residents were black), was denied by the local plan commission. The opposition focused its

---

BOX 15.8 NEW JERSEY REGIONAL CONTRIBUTION AGREEMENTS

The transfer of millions of dollars from suburban and rural municipalities to the more urbanized areas of the state is perhaps one of the most significant and unanticipated effects of the Fair Housing Act. This substantial affordable housing program has provided an average annual subsidy of approximately $13.5 million over a six-year period to municipalities with the greatest proportion and number of low and moderate income households in the state.

By the end of 1994, twenty-nine agreements had been signed, involving nineteen receiving and thirty-three sending municipalities. At an average of more than $19,000 per unit, 4,172 units have been or will be built or restored to a standard condition at a total cost of approximately $81 million.

Source: NJ Council on Affordable Housing Annual Report 1995

attack on the rezoning petition on two grounds. First, the area was already zoned for single-family use. It believed that many neighboring citizens had built or purchased property in the area knowing of and relying on its single-family zoning classification. Second, the proposed rezoning did not meet the village's apartment policy of using multi-family zoning as a 'buffer' between single-family development and incompatible land uses.

The case came before the courts and eventually reached the Supreme Court of the United States. There are a number of complexities (as usual) which are ignored here: the essential point is that the court held that there simply was not sufficient proof that discrimination was a motivating factor in the village's decision. In the court's words, 'official action will not be held unconstitutional simply because it results in a racially disproportionate impact . . . proof of racially discriminatory intent or purpose is required'.

As an interpretation of the equal protection clause of the Constitution, this decision was technically correct: there must be an *intent* to discriminate. But on any broader approach, it gave support to blatant discrimination. That this is so is demonstrated by a later development of the same case when it came to be dealt with under the Fair Housing Act. This legislation differs significantly from the general constitutional requirements in that it requires proof only of a racially discriminatory *effect*. It was clear to the parties concerned which way the wind was blowing; a settlement was reached out of court, and the project went ahead. This case can be compared with the earlier discussion involving *Mount Laurel*. The difference between the two cases might be the fact that the cases were decided in different courts.

Clearly, the specific provisions of the Fair Housing Act are a stronger tool to combat housing discrimination than the Constitution. Regretfully, they have not so far proved to be adequate.

Housing measures in other states

Less dramatic, though equally interesting, measures have been taken in other states. Massachusetts was one of the first when, in 1969, it passed its 'anti-snob' law, which provided a process of appeal for developers against obstructive, regulatory barriers found in municipalities. The law was designed to help address the shortage of affordable housing in Massachusetts. Communities should provide a minimum of 10 percent of their housing as affordable. Affordable is defined as a unit making 80 percent of the median income of the area. For example, the 2007 median income for a family of four in the Boston Metropolitan Statistical Area is $82,400. This figure can be broken down into $46,300 for one person, $52,950 for two people, $39,550 for three persons, and $66,150 for four persons.

A developer submits an application for a comprehensive permit to the Zoning Board of Appeals (ZBA), which can approve affordable housing developments under flexible rules if at least 25 percent of the units have long-term affordability restrictions. Local governments voiced opposition over the law and succeeded for many years in rendering it ineffective. If the proposed affordable housing development application is denied, the developer can appeal to the State Housing Appeals Committee (HAC). The Citizens' Housing and Planning Association (2008: 1) has estimated that 'as of November 2008: nearly 1,000 developments with 53,800 units have been built or are under construction using "comprehensive permits," including over 28,000 affordable units.' However, a less confrontational approach by the state (involving mediation in cases of dispute) led to changes which made the scheme less unacceptable. Other regulatory changes to the law have been made over the years allowing group homes and accessory units to be counted toward a community's 10 percent goal and requiring a thirty-day comment period for communities where an application is filed. A similar system introduced by Connecticut in 1989 provides for appeal to the courts rather than to a state administrative body. Supporting this are 'negotiated investment strategies' and 'regional fair housing compacts' aimed at increased affordable housing provision.

Another approach is to require municipalities to adopt a 'housing element' in a master plan as a prerequisite to the use of zoning powers. The New Jersey Fair Housing Act amended the Municipal Land

Use Law to this effect: 'in adopting its housing element, the municipality may provide for its fair share of low and moderate income housing by means of any technique or combination of techniques which provide a realistic opportunity for the provision of the fair share'.

Such initiatives by the states have had a limited effect. Complex unpopular laws seldom attain their objectives: there are too many ways in which they can be circumvented.

Inclusionary zoning

Housing affordability remains a serious problem for many areas around the United States. The dream of affordable housing remains out of the reach of many Americans. Unfortunately, in some areas, the dream may remain unreachable for many years.

Where a local government supports the provision of affordable housing, there still remains the problem of ensuring that it is provided. This can be difficult in high-cost areas. There are, however, as already outlined in Chapter 8, a number of incentives which can be offered to developers (sometimes with acceptance being a condition of approval). One of these is 'inclusionary zoning'.

The essential feature of inclusionary zoning is that it seeks the provision of lower-cost housing either by offering a developer a higher density in return ('incentive zoning' or 'bonusing'), or by a mandatory requirement that a certain proportion of units are affordable. Its purpose is generally to increase the provision of lower-cost housing anywhere in the municipality (to 'open up the suburbs' to lower-income households). Alternatively, it may seek to ensure the provision of housing in a central area which is predominantly commercial, thus 'bringing life back to the city after office hours' or increasing the supply of lower-cost housing in an area which is being gentrified.

Typically, however, the objective is simply to provide 'affordable' housing. This has become more prominent as the cost of housing has increased more rapidly than incomes, thus presenting increasing numbers of middle-class households with an affordability

problem which had previously been essentially confined to the poor. According to the President's 1995 National Urban Policy Report, the proportion of households owning their homes *declined* between 1980 and 1995, from 65.6 percent to 64.2 percent. Though this might appear a small drop, it is noteworthy for three reasons. First, it represents 1.4 million renters 'who would otherwise have become home owners'. Second, it was the downturn of a trend which had shown a very long-term upward movement. Third, and probably most important, there are the political implications: the 'affordability problem' was now expanding. This political perception is perhaps the most significant – more than outweighing considerations of alternative rationales for the decline (such as the possibility that a 'saturation' level for home ownership has been reached, or that changing attitudes are emerging, possibly as a result of demographic changes). Thus, the lower-income middle class has come into competition with the poor for the limited benefits of programs to deal with issues of housing affordability. There are no prizes for correctly guessing who usually wins.

A number of jurisdictions have turned to inclusionary zoning as a means of providing low- and moderate-income housing. One of the more successful programs can be found in Montgomery County, Maryland. Enacted in 1974, Montgomery County's Moderately Priced Dwelling Unit (MPDU) Ordinance requires developments of more than fifty units to include fifteen MPDUs. Some 10,000 units have been produced in Montgomery County since the program's inception.

Density bonuses also serve as an incentive to developers to construct affordable housing. A density bonus is a tool that offers developers a bonus of building over the maximum allowable residential density (specific percentages vary by jurisdiction) if they agree to construct a certain percentage of below market-rate housing units. In many cases, they are for the very low-income and low-income families. In other cases, the bonus may be used to construct housing for the elderly. California has enacted state legislation for cities and counties to provide a minimum size density bonus to serve as an incentive for the production of lower-income housing. Bonuses will vary according to the type of housing provided by the developer. Cities and counties may go above the state minimum density bonus if it will help meet their affordable housing needs.

Monterey County, California has acknowledged the existence of an inadequate supply of affordable housing. Its Board of Supervisors realized that housing costs in Monterey County had drastically increased over the years and adopted an Inclusionary Housing Ordinance in 1994 (Ordinance 3419). The ordinance called for the creation of low- and moderate-income housing through density bonuses, a charge in lieu of providing housing. The county declared that all new residential development (except for the authorized development that is allowed without inclusionary contributions) contribute to the inclusionary housing program in the amount of 15 percent of the total number of lots/units being created. In 1996, the Inclusionary Housing Ordinance was amended by Ordinance No. 03877, which 'would require that housing units either be made affordable to the Countywide standard or to affordability standards for each Planning Area of the General Plan, whichever is less, and change the affordability requirements so that they can be adjusted for household size'. The inclusionary housing program remains a top priority of the Monterey County Board of Supervisors, requiring housing developers to include a certain percentage of affordable home ownership and rental opportunities for very low, low and moderate household incomes.

Additional inclusionary housing programs can be found throughout the United States in such areas as Boulder, Colorado; Santa Monica, California; Fairfax County, Virginia; King County, Washington; and Sacramento, California.

Inclusionary housing, like any housing policy, is liable at some point to stumble over the law. Objectors will complain that the policy is unconstitutional, or that the administration is unfair. There seems to be no limit to the barriers which opponents can erect against attempts to meet the needs of minorities: there is always a chance of winning.

The current mortgage crisis

The economic and financial crises of the past decade have had lingering effects on all sectors of the economy. Companies failed and many individuals lost their jobs. Having been long considered as one of the best investments an individual or family can make, home values have plummeted in recent the years. Many people owe more on their home mortgage than their home is worth. This is what is referred to as being 'underwater'. The American dream of home ownership is under attack.

This dream of home ownership has been deeply entrenched in the US culture. When housing prices were increasing and mortgage rates low, people started borrowing too much money. After all, housing had always been a good financial investment. Unfortunately, the latest economic recession occurred. Jobs were lost. Housing prices decreased. People could not afford to make their monthly mortgage payments. Some people have tried refinancing their mortgages while others simply walked away from their homes after being turned down for refinancing or mortgage modification.

The foreclosure crisis affected countless numbers of individuals. Some feel the root of the problem was subprime lending practices. Subprime loans are targeted to individuals that don't qualify for a prime rate on a traditional loan. These individuals tend to have lower credit ratings. The lower credit rating could be due to such factors as a faulty credit history, a lack of assets, or a limited credit history. As a result, lenders saw these individuals as a risk and charged higher fees and a higher interest rate. The interest rates and terms would vary by lender.

It is generally accepted that many individuals spent more on housing that they could actually afford. However, they were drawn to the low mortgage rates offered by various lenders. Many individuals were not qualified to get traditional home mortgages and soon became the targets of predatory mortgage brokers. Individuals were charged higher interest rates because mortgage brokers deemed the loans to be risky. Many individuals were enticed to accept adjustable rate mortgages (ARM), which saw the rates fluctuate pursuant to an economic index rate. Brokers, on occasion, even failed to check the credit qualifications of individuals seeking to buy a home. Other individuals took gambles and bought homes that put them in drastic financial straits. The housing crisis was ultimately a combination of many factors. As President Obama indicated in his remarks on the home mortgage crisis on February 18, 2009:

> Our housing crisis was born of eroding home values, but it was also an erosion of common values, and in some cases, common sense. It was brought about by big banks that traded in risky mortgages in return for profits that were literally too good to be true; by lenders who knowingly took advantage of home-buyers; by homebuyers who knowingly borrowed too much from lenders; by speculators who gambled on ever-rising prices; and by leaders in our nation's capital who failed to act amidst a deepening crisis.

> So solving this crisis will require more than resources; it will require all of us to step back and take responsibility. Government has to take the responsibility for setting rules of the road that are fair and fairly enforced. Banks and lenders must be held accountable for ending the practices that got us into this crisis in the first place. And, each of us, as individuals, has to take responsibility for their own actions. That means that all of us have to learn to live within our means again and not assume that housing prices are going to go up 20, 30, 40 percent each year.

The burgeoning mortgage crisis occurred around 2006–2007. The housing bubble had burst. Home prices were decreasing and people soon found that their home values were less than what they owed on their mortgage. Some individuals sought to qualify for refinancing programs. Others walked away from their homes. Still others defaulted on their loans and the houses went into foreclosure.

The current housing mortgage crisis has been seen as the worst housing crisis since the Great Depression. There are no short cuts to solving it. Its repercussions will be felt for many years. Banks and lending institutions have failed.

Ways had to be found to assist home owners in this period of economic uncertainty. Banks, industries, and lending institutions were bailed out in an attempt to help improve the economic situation. This was considered a controversial and divisive move by many Americans. Other moves were soon to follow.

In 2008, the Housing and Economic Recovery Act established the takeover or conservatorship of two Government-Sponsored Enterprises (GSE) – the Federal National Mortgage Association (Fannie Mae) and the Federal Home Loan Mortgage Corporation (Freddie Mac) – by the Federal Housing Finance Agency. Both were designed originally to help individuals and families achieve the American dream of home ownership. A GSE is a stockholder-owned quasi-government organization designed to improve the workings and efficiency of the credit market. Fannie Mae, a privately owned corporation, was chartered in 1938 for the purpose of ensuring a reliable and affordable supply of mortgage funds in the US. Freddie Mac was chartered in 1970 as a stockholder-owned corporation with similar purposes. Both offer financial services and provide stability to the mortgage market. They compete against each other in the market. They purchase mortgages from mortgage lenders and repackage them. They can either keep the mortgages or sell them to other investors. In doing so, additional credit is made available to consumers.

As the housing crisis worsened, Fannie Mae and Freddie Mac witnessed a drop in the value of their mortgage portfolios. They experienced a decrease in their assets. They couldn't raise any capital. Placing the two into conservatorship was designed to help restore confidence in the two enterprises and to improve their financial condition. With the US Treasury Department purchasing their mortgage-backed securities, the federal government now guarantees the debt issued by Fannie Mae and Freddie Mac.

The American Recovery and Reinvestment Act of 2009 was designed to be an economic stimulus package to save existing jobs and to create new jobs. With an initial cost of some $787 billion, funds were to be used for such purposes as providing tax incentives for individuals and companies; energy infrastructure; information and communications technologies;

security technology; infrastructure investments in transportation, water, sewage facilities; and scientific research. Some $14.7 billion was dedicated to such areas as public housing; Section 8 rental assistance; rehabilitation of Native American housing; and tax credits to finance low-income housing.

In an attempt to aid or preserve home ownership, various programs were enacted by the Federal government. The Making Home Affordable Program was an Obama initiative that sought to assist home owners to avoid foreclosures by reducing monthly mortgage payment to an affordable amount and to help in the nation's economic recovery. Two initiatives were leading the program: Home Affordable Refinance Program (HARP) and the Home Affordable Modification Program (HAMP). These programs sought to rescue home owners in danger of defaulting on home loans and foreclosure by providing them options to refinance their home loans.

HARP offered loan modification assistance to home owners who were underwater or near-underwater on their mortgages. The program was designed to get back above water those home owners who couldn't qualify for private mortgage insurance because their loan-to-value ratio was more than 80 percent. The ceiling was later increased to 125 percent. If the loan-to-value ratio exceeded 125 percent, you were not eligible for HARP assistance. In order to participate in the program, additional criteria were used. For example, the home owner's mortgage had to be owned by either Fannie Mae or Freddie Mac and the home owner had to be current with their monthly mortgage payment. The home owner had to have acquired the property before May 21, 2009. The owner could not have had any previous HARP financing. Later versions of HARP relaxed some of the criteria for participation by allowing investment property owners to seek refinancing assistance and allowing those home owners not having Fannie Mae or Freddie Mac to participate in the program. Another change in the program was the elimination of the loan-to-value ceiling. The floor was now set at 80 percent, with an unlimited ceiling. As of June 2010, the national target of assisting 4–5 million home owners with refinancing their home loans has fallen short.

HAMP was the second program designed to assist home owners with FHA-insured mortgages to reduce their monthly mortgages to affordable amounts so that they could avoid foreclosure. In order to participate in the program, various eligibility requirements were established. Among the various eligibility requirements were: the home loan had to originate before January 1, 2009; a first-lien loan for owner-occupied units with an unpaid balance of no more than $729,750; and loans could be made only once through HARP. It was also required that no more than 31 percent of the gross monthly household income could be devoted to principal, interest, property taxes and home owner insurance payments. Anything over this percent indicated a financial hardship to the home owner. As a condition of participating in HAMP, mortgage providers had to enter into a contract with the federal government. The various mortgage providers would also receive financial incentive payments to modify the mortgage. This program has also fallen short of its anticipated target of assisting 4 million mortgage borrowers.

In an attempt to improve HAMP and to entice more mortgage providers to participate in the program, the federal government tripled the incentives offered to private lenders and banks that participated in the program. These incentives were now offered to Fannie Mae and Freddie Mac. They were previously unable to enjoy the incentives that private lenders and banks enjoyed. The debt-to-income ratio (DTI) was also increased so that additional home owners would be eligible for mortgage modification assistance. The DTI ratio is a percentage of the gross monthly income a family pays for housing costs and consumer debt. Since its inception in 2009, less than 1 million home owners had participated in the program – far short of the goal of assisting some 4 million home owners. The program has been extended to December 2013.

The effects of foreclosures can be felt everywhere. Many individuals went far beyond their means to purchase homes. They could not afford to make their monthly mortgage payments. However, some groups appear to have been targeted and were more susceptible to foreclosures than other groups. Discrimination became evident. Low-income and minority communities appear to have experienced more foreclosures than other communities. These communities were prime targets for subprime loans. Home owners had to pay higher costs, fees, and interest rates than the home owners in other communities. While many of their residents could qualify for traditional loans, many were steered to adjustable rate mortgages with higher fees instead of traditional fixed-rate mortgages. Women could also be added to the list of those individuals steered to the subprime market.

The Obama administration has established a number of additional programs to help rescue home owners in the struggling housing market. The Second Lien Modification Program (2MP) was established in 2009 to assist home owners with a second mortgage to their primary mortgage. In early 2010, the Hardest Hit States Fund was created to assist those states hit the hardest by the economic recession. The District of Columbia and eighteen states are participating in the program. A 2011 program, the Emergency Homeowner's Loan Program (EHLP) offers interest-free financial assistance to home owners who lose a job or suffer a loss of income in the states that do not participate in the 2010 Hardest Hit States Fund. Individuals participating in this program must have missed three monthly mortgage payments and be in danger of losing their homes. They must also show the ability to resume their payments once their income returns.

The US is still in the midst of a fragile housing market. Foreclosures are continuing in all states. However, there is hope that the repossessions are occurring at a lesser rate than in 2011. No state is immune from foreclosures. It doesn't matter which region of the country. States that have suffered the most from foreclosures include California, Florida, Texas, Michigan, Arizona, Illinois, Georgia, Nevada, and Ohio. Stockton, California, which recently filed for bankruptcy, has the distinction of being the number one city in terms of foreclosure per number of households. Las Vegas, Cleveland, Phoenix, and Detroit have also been hard hit by foreclosures.

Foreclosures do not simply affect individual families. They affect the entire neighborhood and city. Other home owners shoulder the burden of paying higher property taxes, utility fees, etc. City revenues drop and the provision of services is jeopardized. Families break up and may require additional human and social services. There is an increase in vacant properties, which might lead to more crime and vandalism. Ultimately, the aesthetics of an area change with the appearance of vacant homes. This might serve as an impediment to attracting economic investment opportunities to the community.

Conclusion

The provision of housing for lower-income groups has never been a popular policy among the electorate. As a result, the role of the federal government shifted from directly supplying housing to indirectly promoting production by subsidizing other agencies, offering tax incentives, and providing rent subsidies to tenants. Additionally, and independently, it has promoted 'fairness' in the private market.

Mount Laurel is unusual in that it assumed epic proportions, but its elements are repeated time and time again across the nation, typically with the same result – the exclusion of minorities. *Mount Laurel* does demonstrate, however, that no matter to what extremes the courts are prepared to go, they are no match for the wiles of municipalities, whose power base extends from their tiny townships to the state capital. As in other fields, the courts are 'of limited relevance', to use the term from Rosenberg's 1991 book *The Hollow Hope: Can Courts Bring About Social Change?* They are constrained by the limited nature of constitutional rights, the lack of judicial independence, and the judiciary's lack of powers of implementation. Nevertheless, as Rodney Cobb (2000: 195) has noted: 'Too often, land use law resembles a legal war rather than a set of laws designed to foster the liveability of American communities.' Many individuals and organizations share his belief.

The Fair Housing Act offers little more encouragement. Discrimination in housing has proven a far more intractable problem than the sponsors of the Fair Housing Act of 1968 anticipated. Amending legislation, passed in 1988, provides enforcement 'teeth' which the earlier Act was lacking, but it remains to be seen how strong its bite is.

Inclusionary zoning offers a different approach, but it is (at least so far) slight in its impact, even when it is implemented with Californian enthusiasm. But it does have the advantage that it adopts the coinage of American zoning: money. Thus, those concerned in dealing with it, either to promote it or to avoid as much of it as they possibly can, are talking the same language. This chapter, like others in the volume, provides testimony to the weakness of government in the face of strong socio-economic forces. Exclusionary zoning is prevalent because it is widely desired by those who have acquired, or are in process of acquiring, their share of the American Dream. Like all dreams, it is easily shattered: hence the vociferous opposition to any development which carries such a threat.

The proportion of families owning their own homes has risen since the 1995 National Urban Policy Report. At an October 16, 2001 Mortgage Bankers Association of America Conference in Toronto, Canada, Secretary of the Department of Housing and Urban Development Mel Martinez indicated that some 72 million American families own their own homes. Unfortunately, although the numbers have increased for minority families, their home ownership rates still lag behind the rates of non-minorities.

President George Bush pledged that his administration would expand some programs, eliminate others, and create new programs designed to improve housing conditions and opportunities for all families in the United States. He indicated that his top priority was to help low-income families become home owners. One means of doing so was for the development of a hybrid adjustable-rate mortgage that combines an early fixed rate with a later rate that adjusts with market conditions. He wanted to expand the use of Section 8 vouchers for home ownership and to increase the number of Section 8 rental vouchers. Reducing the barriers prohibiting the development of affordable housing was a stated priority. A 2007 Bush initiative called 'FHASecure' allowed home owners who have a history of making their mortgage payments on time under original interest rates, but who have missed payments due to their interest rates having been reset,

to refinance into the FHA's mortgage insurance program.

The Obama administration inherited an economic calamity. In addition to programs designed to stimulate the economy, the administration also faced a struggling housing market. Housing prices had plummeted. People owed more on their mortgages than their homes were worth. The amount of foreclosures was increasing monthly. Various programs have been established to help home owners in danger of being foreclosed by modifying their mortgages to affordable monthly levels. Their success is mixed. How long it will take to correct the current housing market is anyone's guess.

Many of the problems discussed in this chapter go far beyond any concept of 'housing'; and they are far too deeply embedded in the socio-economic structure to be significantly touched by current urban policies. Whether well-funded broader programs could achieve more is an open question. Unfortunately, too many are unaffected by these deeper problems to be moved to do anything about them; and even violent riots attract only temporary attention. There is, however, a glimmer of hope to be seen in an approach which is based on community involvement and development. This is discussed in the following chapter.

Further reading

The Housing and Urban Development (HUD) department has a Policy Development and Research Department that can be found online (http://www.huduser.org). Ongoing research and publications on housing issues are offered on this website.

The Regulatory Barriers Clearinghouse (http://www.huduser.org/rbc) is hosted by HUD's Policy Development and Research division as a means to 'collect, process, assemble and disseminate information on the barriers faced in the creation and maintenance of affordable housing'. The website has a searchable database on housing regulation topics as well as a listing of hundreds of solutions to regulatory barriers that were successfully used by local governments.

HUD provides an overview of its Public Housing Program online (http://www.hud.gov/renting/phprog.cfm). The website provides of definition of public housing along with application process information.

The Affordable Housing Design Advisor (http://www.designadvisor.org), sponsored by HUD, gives the what, why, and how of well-designed, affordable housing. Supplemented with pictures, the website supplies tools for good design and publishes personal accounts that tell how oppositions to affordable housing disappeared with better design.

Although there is currently not a National Housing Trust Fund, there is a National Housing Trust (http://www.nhtinc.org). The trust focuses on public policy, technical assistance, loan programs, and development of affordable housing.

An online exhibit by the New Haven Oral History Project gives a national and personal look at Urban Renewal in the 1950s. *Life in the Model City: Stories of Urban Renewal in New Haven* (http://www.yale.edu/nhohp/modelcity) illustrates in an entertaining manner many aspects of urban renewal. The website includes pictures of major project areas, stories from residents in the area, as well as the ideology, the arguments for, and the process of urban renewal.

HOPE VI, the program to eliminate extremely distressed housing, is described in detail on the US Department of Housing and Urban Development's website (http://www.hud.gov/offices/pih/programs/ph/hope6). Grant information, services, and programs for severely distressed public housing are offered on this website.

Many alternatives to public housing can be found online. The Enterprise Foundation, now known as the Enterprise Community (http://www.enterprisecommunity.org), provides information on public policy, its programs, training for developers of low-income housing, and personal stories of families with a new, low-income designated home.

An argument for accessory dwelling units (ADU) may be found in the New Urban News 2001 issue (http://www.newurbannews.com/accessory.html). The article illustrates ADU trends and enumerates their various advantages.

Maintained by the Tennessee Fair Housing Council, the National Fair Housing Advocate Online (http://www.fairhousing.com) offers legal research and information on how to access local resources for fair housing across America.

The California Coalition for Rural Housing has many programs dedicated to inclusionary zoning (http://www.calruralhousing.org/programs/inclusionary-housing). Under 'Studies and Publications' on the left side of the website, an online visitor can download a publication entitled *Inclusionary Housing in California: 30 Years of Innovation*. This is an informative work on inclusionary zoning that also includes nationwide statistics.

A useful short survey of theories of filtering is given in Bourne (1981) *The Geography of Housing*. The classic work is Grigsby (1963) *Housing Markets and Public Policy*. See also Grigsby *et al*. (1987) 'The dynamics of neighborhood change and decline'.

Inclusionary zoning is examined at length in Mallach (1984) *Inclusionary Housing Programs: Policies and Practices*. See also Merriam *et al*. (1985) *Inclusionary Zoning Moves Downtown*. See also Burchell *et al*. (1994) *Regional Housing Opportunities for Lower-Income Households: An Analysis of Affordable Housing and Regional Mobility Strategies*; Burchell *et al*. (1995) *Regional Housing Mobility Strategies in the United States*; Calavita *et al*. (1997) 'Inclusionary housing in California and New Jersey: a comparative perspective'; and Calavita and Grimes (1998) 'Inclusionary housing in California: the experience of two decades'. See also the entire edition of *New Century Housing* (Center for Housing Policy 2000) devoted to 'Inclusionary zoning: a viable solution to the affordable housing crisis?' and Netter (2000) *Inclusionary Zoning Guidelines for Cities and Towns*. A review of the Mount Laurel saga is given in Berger (1991) 'Inclusionary zoning as takings: the legacy of the *Mount Laurel* cases'. Generally on discrimination in housing, a good account is to be found in Danielson (1976) *The Politics of Exclusion*. Up-to-date discussions are to be found in Keating (1994) *The Suburban Racial Dilemma*; Yinger (1995) *Closed Doors, Opportunities Lost: The Continuing Costs of Housing Discrimination*; and Galster (1996), 'Racial discrimination and segregation', in his edited collection of essays on *Reality and Research: Social Science and US Urban Policy Since 1960*. State programs are discussed in Stegman and Holden (1987) *Nonfederal Housing Programs: How States and Localities are Responding to Federal Cutbacks in Low-Income Housing*.

President Clinton's National Urban Policy Report, *Empowerment: A New Covenant with America's Communities* (US Department of Housing and Urban Development 1995b) is a political statement outlining the urban policy agenda (and extolling its superiority over previous efforts). It discusses the challenges that faced the nation, what the current administration was doing regarding the challenges, and what remained to be done. A *States of the Cities* Report was issued in 1997, 1998, and 1999. They are published by the US Department of Housing and Urban Development.

The effect of regulation on the provision of affordable housing is dealt with at length in Advisory Commission on Regulatory Barriers to Affordable Housing (1991) *Not in My Back Yard*; see also Council of State Community Development Agencies (1994) *Making Housing Affordable: Breaking Down Regulatory Barriers: A Self-Assessment Guide for States*; Lowry and Ferguson (1992) *Development Regulation and Housing Affordability*; and Warner and Molotch (2000) *Building Rules: How Local Controls Shape Community Environments and Economies*. The California Budget Project issued two interesting reports detailing housing affordability in California, *Locked Out: California's Affordable Housing Crisis* (2000) and *Still Locked Out: New Data Confirm that California's Housing Affordability Crisis Continues* (2001). For a private sector view on housing affordability see, National Association of Home Builders (n.d.) *The Truth About Regulatory Barriers to Housing Affordability*. Listokin and Listokin (2001) prepared a two-volume report titled *Barriers to the Rehabilitation of Affordable Housing*. Volume 1 provides findings and analysis while volume 2 offers case studies.

The effects of growth controls and sprawl on housing affordability can be examined in such reports as Staley *et al.* (1999) *A Line in the Land: Urban-Growth Boundaries, Smart Growth, and Housing Affordability*; Staley and Mildner (1999) 'Urban-growth boundaries and housing affordability: lessons from Portland'; Clarion Associates, Inc. (2000) 'The costs of sprawl in Pennsylvania: executive summary'; 1000 Friends of Oregon (n.d.) 'Questions and answers about Oregon's land use program: UGBs and housing costs'; and Smart Growth Network Subgroup on Affordable Housing (2001) *Affordable Housing and Smart Growth: Making the Connection*.

The role played by public-private housing partnerships is detailed in a report of that title by Suchman *et al.* (1990). Private housing provision is discussed in a number of reports from the Urban Land Institute, including Porter (1995) *Housing for Seniors*; and Urban Land Institute (1991) *The Case for Multifamily Housing*. On manufactured housing, see Allen *et al.* (1994) *Development, Marketing, and Operation of Manufactured Home Communities*; and Suchman (1995) *Manufactured Housing: An Affordable Alternative*.

A good overall review of affordable housing policies is White (1992) *Affordable Housing: Proactive and Reactive Planning Strategies*. This also has a useful bibliography.

For works on Community Land Trusts, see Hovde and Krinsky (1996) *Balancing Acts: The Experience of Mutual Housing Associations and Community Land Trusts in Urban Neighborhoods*; Hovde *et al.* (1996) *Hands-on Housing: A Guide Through Mutual Housing Associations and Community Land Trusts for Residents and Organizations*; Packnett (2005) 'The First Homes Community Land Trust'; and Greenstein and Sungu-Eryilmaz (2007) 'Community land trusts: a solution for permanently affordable housing'.

The HOPE VI program has been widely covered in the literature. For a representative sample of the literature, see: Popkin *et al.* (2002) *HOPE VI Panel Study: Baseline Report*; Popkin *et al.* (2004) *A Decade of HOPE VI: Research Findings and Policy Challenges*; Popkin (2007) 'Testimony of Susan Popkin, Urban Institute, prepared for the hearing on S. 829 HOPE VI Improvement and Reauthorization Act'; National Housing Law Project (2002) *False Hope: A Critical Assessment of the HOPE VI Public Housing*

Redevelopment Program; Turner *et al.* (2007) *Severely Distressed Public Housing: The Costs of Inaction*; and Polikoff (2010) 'Overview: three remaining HOPE VI challenges.'

For additional information on Government-Sponsored Enterprises, the mortgage crisis, its causes, and potential solutions, see: Kosar (2007) *Government-Sponsored Enterprises (GSEs): An Institutional Overview*; Jickling (2008) *Fannie Mae and Freddie Mac in Conservatorship*; Schiller (2008) *The Subprime Solution: How Today's Global Financial Crisis Happened, and What to Do about It*; Barth (2009) *The Rise and Fall of the US Mortgage and Credit Markets: A Comprehensive Analysis of the Market Meltdown*; Wachter and Smith (2011) *The American Mortgage System: Crisis and Reform*; Weiss (2012) *Fannie Mae's and Freddie Mac's Financial Problems*.

For additional information on land banking, see the following: Strong (1979) *Land Banking*; Great Lakes Environmental Finance Center (2005) 'Best practices in land bank operation'; Fitzpatrick (2009) 'Understanding Ohio's land bank legislation'; and US Department of Housing and Urban Development, Office of Policy Development and Research (2009) 'Revitalizing foreclosed properties with land banks'.

Housing programs change with bewildering frequency, and most published material is rapidly outdated. The reader should take care to use the most recent publications.

Questions to discuss

1   What are the important features of housing? How do these affect land use and land use policies?

2   'The problem of housing affordability affects not only the poor but also many middle-income households.' Discuss.

3   What are the problems which faced the public housing program? Are these problems soluble?

4   Discuss the various ways in which the gap between market rents and affordable rents can be bridged.

5   Compare the ways in which motor vehicles and houses 'filter' down to lower-income groups.

6   Can discrimination be effectively controlled by legislation?

7   What do the *Mount Laurel* and *Arlington Heights* cases tell us about the limits of court action?

8   How can states compel or persuade local governments to permit affordable housing?

9   What role do non-profit organizations play in the provision of affordable housing?

10 How could growth controls and sprawl affect the cost of housing in an area?

11 How do Housing Trust Funds and Community Land Trusts help fill the need for affordable housing in many American cities?

12 What role does politics play in responding to the nation's housing problems?

13 How do the HOPE VI and Choice Neighborhood Initiative programs fit into the discussions of housing and community development?

14 What is a subprime mortgage and how did it contribute to the current US mortgage crisis?

15 What is land banking and why is it important?

16

Community and economic development

The days of made-in-Washington solutions, dictated by a distant government, are gone. Instead, solutions must be locally crafted, and implemented by entrepreneurial public entities, private actors, and a growing network of community-based firms and organizations.

President Clinton 1995

Changing perspectives

Public policy, like clothing, has its fashions. Services in kind and in cash alternate in popularity. Top-down and bottom-up policies change in the world of accepted program design. Discretion constantly vies with flexibility for the dominant factor in public policy. The level of government which is to be preferred as the leader in policy moves with bewildering frequency among the federal, state, and local levels. Yesterday's orthodoxy becomes today's anathema; today's accepted wisdom turns into tomorrow's target of inadequacy. Nowhere are these pendulums more clearly seen than in urban policy. In particular, the role of the federal government is always under scrutiny, if not attack: it is always too weak, or too strong; or too weak and too strong but in the wrong places.

President Johnson's policies for 'the Great Society' and the 'War on Poverty' embraced a positive role for the federal government. This 'creative federalism' was superseded by President Nixon's 'new federalism', which he claimed was to 'start power and resources flowing back from Washington to the states and communities and, more important, to the people all across America'. President Carter viewed this as a federal government 'retreat from its responsibilities, leaving state and local government with insufficient resources, interest, or leadership to accomplish all that

needed to be done'. He sought the best of all worlds with his 'new partnership' with every level of government as well as the private and community sectors. President Clinton's national urban policy is characterized by an emphasis on the empowerment of communities. Bush has been criticized on some fronts for the lack of a national urban policy. His legacy will probably read that his attention appeared to be on other areas. However, President Bush has acknowledged that federal and state programs have not eliminated our urban problems. As such, he called for enlisting the aid of what he calls the 'Armies of Compassion' – faith-based and community organizations that provide a variety of social services to residents.

Such rhetoric always exaggerates, but it does reflect shifting views on the relative importance of differing features of policy, and alternative administrative responsibilities. New administrations attempt to correct the shortcomings of their predecessors while, at the same time, strengthening the areas of proved success. In order to stimulate new endeavors to tackle old problems, and to marshal general support, the newness may be exaggerated and can be more striking in its packaging than in its content. This in part reflects the character of democratic politics, but it also masks the true uncertainties which policy-making involves. A government which stressed these uncertainties would be regarded as weak; governments have to display an appearance of conviction that their policies really will work.

The Clinton policy of community empowerment bears many features of previous approaches, but is essentially another attempt to redefine the roles of the different governmental, private, and voluntary bodies that are involved in community and economic development. The Bush idea of community empowerment seeks to stress the participation of faith-based and community organizations. It is a means of redefining the roles that various bodies have in community and economic development. It is useful to examine some of the attempts made by previous governments to tackle the difficulties that this involves.

The War on Poverty

Prior to the Kennedy and Johnson administrations, explicit federal urban policies had been largely restricted to urban renewal and public housing. Disenchantment with these spread at the same time that urban problems grew – problems which ranged from race, civil rights, poverty, and violence to state and local government finance. A response to these problems emerged as public concern developed, and as the political scene changed – with Kennedy's rediscovery of poverty in the early 1960s, and Johnson's overwhelming election victory in 1964. Kennedy started the battle of alleviating the struggles of the poor. Johnson continued the fight after the assassination of President Kennedy. The era of the 'Great Society' was at hand. To an unprecedented extent, federal policies were developed to reach 'deeply into the urban social and political fabric' (Mollenkopf 1983: 95).

As with so much in this field, the sequence of events, the policy initiatives and their impact comprise a complex and confused story. Federal programs proliferated on a bewildering scale: more than tripling in the 1960s. These covered the whole spectrum of public policy, from food stamps to regional development, from the 'War on Poverty' to health services, from education to model cities and the Community Action Program (designed to provide power to inner-city residents to improve their neighborhoods). These new

programs, each with its own budget and bureaucracy, 'generated a massive federal administrative structure and a significant transformation of federal–state–city relationships' (Frieden and Kaplan 1977: 3). Such a change demanded a greater degree of coordination among federal programs (and agencies) than had ever been required before.

Two responses emerged. First, the arguments which had been deployed for several years in favor of the establishment of a new federal department for urban policies gained the support which they had previously lacked; and the Department of Housing and Urban Development (HUD) was created in 1965. Second, a task force appointed by President Johnson to advise on the organization and responsibilities of the new department proposed that the coordinative role of HUD should be directed through a 'model cities' program toward the poverty areas of central cities. It was an ambitious program to deal with a myriad of urban problems. It incorporated programs in such areas as city planning, day-care centers, employment, drug abuse, the elderly, and housing rehabilitation programs. This started a new chapter in urban policy.

The Model Cities Program

The birth, brief life, and death of the Model Cities Program constitutes a fascinating case study in the making of public policy (see Box 16.1). Here, a few features are highlighted: a fuller story is given in Frieden and Kaplan (1977). The debate represented a classic case of how planning cannot be separated from politics (Lord 1977).

From its inception, there was debate on how many model cities there should be. One school of thought opted for a very small number; hence its original designation as 'demonstration cities' – a term which was quickly abandoned when it became associated with urban riots. Others, as indicated above, envisaged model cities being the major channel for aid to poverty-stricken urban areas. In the event, the need to obtain political support for the legislation led to an increased number of cities – eventually to 150. The idea of

---

BOX 16.1 MODEL CITIES BILL 1966

The purposes of this title are to provide additional financial and technical assistance to enable cities of all sizes [to implement] new and imaginative proposals and rebuild and revitalize large slums and blighted areas; to expand housing, job, and income opportunities; to reduce dependence on welfare payments; to improve educational facilities and programs; to combat disease and ill health; to reduce the incidence of crime and delinquency; to enhance recreational and cultural opportunities; to establish better access between homes and jobs; and generally to improve living in such areas.

'demonstration' cities was thus killed. But, more than this, Congress was not willing to see funds diverted from other programs into model cities; and so the congressional commitment became largely to a new categorical program rather than to a mechanism for reforming existing grants-in-aid.

There were also apprehensions about the concentration of power in a single agency which this implied: it was felt that this could lead to an uncontrollable, autocratic, and overpowering bureaucracy. Officials from various federal departments and agencies resented the fact that programs that they controlled would be shifted to another federal department (see also Box 1.1, p. 12).

Moreover, the redistributive features of the Model Cities Program implied that other federal agencies would be expected to divert some resources from their traditional clients. This 'went against the grain of normal agency behavior, congressional grant-in-aid policies, and ultimate reliance on established interest groups that benefited from existing programs'. Such were the considerations which doomed the original ideas underlying the Model Cities Program. Frieden and Kaplan concluded that 'if the designers of future urban policies take away any single lesson from model cities, it should be to avoid grand schemes for massive, concerted federal action'. The program, together with urban renewal and other community development programs administered by HUD, was folded into the Community Development Block Grant (CDBG) at the end of 1974.

The new federalism

While Johnson's policies embraced a positive role for federal government ('federal activism' or 'creative federalism'), Nixon (1969–74) promoted a 'new federalism' which he claimed would bring about a return to the original conception of federalism as envisaged by the Founding Fathers. He was a disciple of decreased federal involvement in state and local affairs. The main feature of this 'new federalism' was intended to be the replacement of large numbers of categorical programs (and all the controls which accompanied them) by block grants.

This was, in fact, a reaction against the federalist policies of the previous Democratic administration. One highlight of this was Senator Muskie's extensive congressional hearings launched in 1966. The senator observed that what had been created was almost 'a fourth branch of government, but one which has no direct electorate, operates from no set perspective, is under no specific control, and moves in no particular direction' (Haider 1974: 60). Virtually all the new programs were 'functionally oriented, with power, money, and decisions being vertically dispersed from program administrators in Washington to program specialists in regional offices to functional heads in state and local governments'. From the perspective of the Advisory Commission on Intergovernmental Relations (1970), this left 'cabinet ministers, governors, county commissioners, and mayors less and less informed as to what was actually taking place, and [made] effective horizontal coordination increasingly difficult'.

Given this background, there was a great deal of support for Nixon's proposals. In particular, city mayors saw them as a means of obtaining additional assistance with their fiscal problems and of enabling them to recover some of the power they had lost in the Johnson years.

Nixon's urban aid strategy had two major elements. First, and most innovative, was 'general revenue sharing', which provided federal funds on the basis of a formula encompassing population, incomes, urbanization, and tax effort. The essential policy objective was to allow localities to take spending decisions on the basis of their knowledge and understanding of local needs. Second, 'block grants' were extended by the merging of groups of categorical grants. The best known of these is the CDBG program.

Nixon's ideas were never implemented to the extent which he had envisaged, mainly because of congressional opposition and the political impact of the Watergate scandal. General revenue sharing was abolished in 1986. However, the CDBG proved so popular with local political constituencies that it survived constant financial cutbacks, though in an attenuated form. It is appropriate to examine this program more fully.

Community Development Block Grants

The three-year $8.6 billion CDBG program was signed into law by President Ford shortly after his inauguration in August 1974. The Act folded seven categorical programs administered by HUD (including urban renewal and model cities) into this single-grant program which was directly targeted on cities, particularly those showing signs of social and economic distress. It was intended to achieve a balance between providing maximum flexibility for local decisions and securing the national purpose of developing 'viable urban communities by providing decent housing and a suitable living environment, and expanding economic opportunities, principally for persons of low and moderate income'. It had three main objectives:

1 To benefit low- and moderate-income families.
2 To aid in the prevention or elimination of slum or blight.
3 To meet the community development needs having a particular urgency because existing conditions pose a serious and immediate threat to the health and welfare of the community where other financial resources are not available to meet such needs.

To help meet those objectives, activities such as the acquisition of real property, code enforcement, demolition or clearance, comprehensive planning activities, economic development activities, projects to remove architectural barriers which restrict the mobility of handicapped persons, provision of recreational facilities, rehabilitation and conservation of existing housing stock, provision of public services, and the provision of neighbourhood facilities were allowed.

This has never been an easy balance to attain. At first, there was minimal federal control: eligible local governments simply requested the allotment that was due on a predetermined formula. HUD officials checked entitlement and issued approvals. Any assessment of the value was undertaken later. Local governments took full advantage of their freedom to decide on the allocation of funds and, not surprisingly, there were a number of highly publicized cases of expenditure which *prima facie* seemed inappropriate. Tennis courts took pride of place in these indictments. For instance, Little Rock, Arkansas, used $150,000 from its CDBG to construct a tennis court in a wealthy section of the town. Chicago used $32 million for snow clearance. Other criticized schemes included golf courses, polo fields, and wave-making machines!

There is nothing surprising here: if local governments are given freedom to allocate funds as they wish, they will do precisely this. A requirement that 'maximum feasible priority' was to be given to projects benefiting low- and moderate-income families allowed a good deal of leeway. Nevertheless, grants were distributed according to a formula based on population (25 percent), housing overcrowding (25 percent), and poverty (50 percent). The formula was changed in 1977 to direct resources from high-income suburbs and urban counties to needy central cities – though not with complete success. However, HUD studies showed that 62 percent of benefits went to lower-income groups. The pattern of expenditure remained fairly constant through 1987: about a third went to housing; a fifth to public facilities and improvements, and a similar proportion to economic development. The remainder went on acquisition and clearance, administration and planning, and other activities. The pattern has changed over time, and in the 1990s housing was taking a larger share (around two-fifths). Most of the housing expenditure was on rehabilitation (Urban Institute 1995).

As is not uncommon with public policies, different sources provide different conclusions on the effectiveness of the CDBG; but it does seem that, despite an attempt at targeting needy areas, the CDBG benefits were spread widely, and became even more so after the 1974 legislation gave more discretion to cities in the allocation of funds. The increase in benefits going to wealthier areas was a result of local politicians using their discretion in favor of pleasing influential sectors of their electoral constituencies. Local discretion increased still further under the Reagan administration.

Carter's new partnership

In March 1978, Carter submitted to Congress his *National Urban Policy Report* containing proposals for 'a comprehensive national urban policy'. Reviewing previous policies, he noted that during the 1960s the federal government had taken 'a strong leadership role' in identifying and dealing with the problems of cities. This proved to be inadequate because the federal government alone had neither the resources nor the knowledge required 'to solve all urban problems'. During the 1970s, federal government 'retreated from its responsibilities', leaving state and local government with insufficient resources, interest, or leadership to accomplish all that needed to be done. The lessons had been learned: 'These experiences taught us that a successful urban policy must build a partnership that involves the leadership of the federal government and the participation of all levels of government, the private

sector, neighborhood, and voluntary organizations and individual citizens.' The 'new partnership' thus involved a positive role for the federal government, together with incentives to state and local governments, and to the private sector.

Carter's policy consisted of a large package of existing legislation and new proposals, with an emphasis on the stimulation of private investment. According to Carter, it was the private sector that would expand the economy, not the government. Consequently, attacking the economic woes of the country became a major thrust of the Carter administration. However, there was no suggestion that the migration from the northern cities should be stemmed, even if this were thought to be desirable. Local economic development was not conceived as a way of stemming powerful forces of change, but of assisting declining cities to 'a new stage of urban development' of which the main features were decentralization and the dispersal of population and economic activity.

Among Carter's specific policy initiatives was the Urban Development Action Grants (UDAG) program, which passed Congress with relative ease. The bounties of this program were distributed extensively, and it therefore proved widely popular. Grants were made to local governments which, in turn, used the funds to make loans to private developers and industrial companies. UDAG was aimed at the stimulation of private investment to create jobs in distressed communities by schemes agreed between the private and governmental sectors. Funding could be used for such activities as site acquisition, clearance and demolition, clean-up, construction, soft costs, and capital equipment. It was also aimed at stimulating housing opportunities in distressed communities. The grants were intended to create leverage on private money, particularly in distressed cities. Unfortunately, this was easier said than done since the targeting of distressed cities was not the same thing as the alleviation of distress: 'Cities that provided the best investment opportunities – where private funds were more available – were not likely to be severely distressed' (Barnekov *et al.* 1989: 79). There has been much controversy over the success (measured in differing ways) of the UDAG program. There was, however, no doubt about its popularity with developers, builders, urban chambers of commerce, and pro-development mayors. To them, the UDAG program seemed to offer benefits as profitable as those of urban renewal.

An unintended result of the policy of promoting private development (through UDAG and other programs) was increased competition among cities; this led to escalating subsidies and an increase in federal controls (thus reducing local discretion). Indeed, 'regulation gradually emerged as the key strategy for implementing the new generation of urban aid programs. The creeping growth of the new rules . . . gradually shifted power back to the federal government' (Kettl 1981: 123).

The major legacy of the Carter administration was its reorientation of policy toward the stimulation of private investment. By its last year, the Carter administration's policy had shifted away from urban issues to much broader concerns for economic growth.

This emphasis on economic development as the foundation of federal policy was embraced and increased by Reagan. His administration brought about the dramatic change of raising unfettered economic forces to the mainspring of 'policy' – a policy of 'do nothing'. This had the powerful (but highly controversial) support of the President's Commission on a National Agenda for the Eighties (which Carter had found too extreme). This report is of importance not only because of its place in the history of urban policy but also because it continues to represent some widely held views about the objectives and limitations of policies directed at urban conditions. It is therefore worth examining in some detail.

National Agenda for the 1980s: Urban America

On October 24, 1979, President Carter established an independent forum to examine the urban issues that would confront the American people. The commission would 'identify and examine the most critical public policy challenges of the 1980s. It shall examine issues related to the capacity for effective Federal governance, the role of private institutions in meeting public needs,

and underlying social and economic trends, as these issues bear on our public policy challenges in the 1980s.' According to Section 1-202 (a–e) of the order establishing the commission, the areas to be reviewed included:

1 Underlying trends or developments within our society, such as the changing structure of our economy, the persistence of inflationary forces, demands on our natural environment, and demographic shifts within our population that will shape public policy choices in the 1980s;

2 Opportunities to enhance social justice and economic well-being for all our people in the 1980s;

3 The role of private institutions, including the non-profit and voluntary sectors, in meeting basic human needs and aspirations in the future;

4 Defining the role of the public sector, and financing its responsibilities in the 1980s;

5 Impediments to building consensus, both within government – the executive branch, Congress, state and local government – and within the nation as a whole.

The essential message of *Urban America* was that unfettered market forces would benignly bring about an efficient and equitable urban settlement pattern, with the economy operating at such a high level that many 'social' problems would disappear (or at least be reduced to a level which the enhanced resources of a liberated economy could meet). Such problems as persisted should be approached directly by 'people policies' (as distinct from 'places policies'). Above all, policies which tied people to declining areas should be avoided: 'urban programs aimed solely at ameliorating poverty where it occurs may not help either the locality or the individual if the net result is to shackle distressed people to distressed places'. Such policies were inherently wasteful. By contrast, 'a federal policy presence that allows places to transform, and assists them in adjusting to difficult circumstances, can justify shifting greater explicit emphasis to helping directly those people who are suffering from the transformation process'.

The report strongly criticized the concept of a national urban policy: 'Efforts to revitalize urban areas through a national urban policy concerned primarily with the health of specific places will inevitably conflict with efforts to revitalize the larger economy.' They will therefore do more harm than good. The forces underlying urban change are 'relatively persistent and immutable', and thus are highly resistant to public policies which try to stem them or harness them to policy goals which are not consistent with wider economic development purposes (see Box 16.2).

Though the report did not enter into much detail about the translation of principles into practice, it did list 'prime candidates' that should be 'scrutinized for eventual reduction or elimination' such as economic development, community development, housing, and development planning. Also suggested was a scrutiny 'for major restructuring or elimination' of such programs as 'in-kind benefits for the poor (such as legal aid and Medicaid), the growing inventory of subsidies that indiscriminately aid the non-poor as well as the poor (such as veterans' benefits), protectionist measures for industry (trade barriers for manufacturers and price supports for farmers), and minimum wage legislation'.

Though the philosophy of *Urban America* was very much to the liking of the new president, the report was never explicitly accepted by him. Given the number of constituencies which would have been affected by the 'scrutiny' list, this is hardly surprising. But, as we shall see, President Reagan moved forcefully to develop policies which bore a strong resemblance to it.

The Reagan years

Reagan's pursuit of privatization was in the tradition of previous administrations, but he gave it a particular twist: so much so in fact that the difference became one of kind rather than of degree. The policy was of the utmost simplicity (some would say simplemindedness): free rein to private forces was the key to economic growth and thus to urban regeneration and the solution of many social problems. Government intervention was not only inadequate: it was counterproductive. Increased government action was no solution: it was part of the problem. To Reagan, it was time to curtail the expansion of government. The roadblocks created by government that slowed the economy had to be eliminated.

Reagan's first major policy statement was made in an address to Congress in February 1981. This *Program for Economic Recovery* was, as its title suggests, focused on economic matters. Most of the address dealt with general economic policy issues: proposed limitations in the growth of federal expenditure, reductions in tax rates, 'an ambitious program of reform' to reduce federal regulatory burdens, and the establishment of a

---

BOX 16.2 THE 'URBAN AMERICA' PHILOSOPHY

The federal government can best assure the well-being of the nation's people and the vitality of the communities in which they live by striving to create and maintain a vibrant national economy characterized by an attractive investment climate that is conducive to high rates of economic productivity and growth, and defined by low rates of inflation, unemployment, and dependency.

People-oriented national social policies that aim to aid people directly wherever they may live should be accorded priority over place-oriented national urban policies that attempt to aid people indirectly by aiding places directly . . . A national social policy should be based on key cornerstones, including a guaranteed job program for those who can work and a guaranteed cash assistance plan for both the 'working poor' and those who cannot work.

Source: President's Commission on a National Agenda for the Eighties 1980b: 101.

BOX 16.3 NATIONAL ECONOMIC GROWTH AS URBAN POLICY

Improving the national economy is the single most important program the federal government can take to help urban America; because our economy is predominantly an urban one, what's good for the nation's economy is good for the economies of our cities, although not all cities will benefit equally, and some may not benefit at all.

monetary policy 'to provide the financial environment consistent with a steady return to sustained growth and price stability'. Urban affairs arose only incidentally – which was precisely what was intended. There was no 'urban policy', other than cuts in programs, and an emphasis on the stimulation of national economic growth (neatly expressed by one of Reagan's senior officials in an article tendentiously entitled 'A positive urban policy for the future' – see Box 16.3). Programs which were regarded as counterproductive were reduced or completely eliminated, such as the Economic Development Administration, the Urban Development Action Grant, and subsidized housing. Much of this was, of course, along the lines proposed in the *Urban America* report. Many individuals felt that Reagan had essentially reversed and dismantled over fifty years of federal housing and community development policy.

Enterprise zones

Despite this new and growing concern to fashion economic incentives more carefully, a long-standing debate on enterprise zones continues. The introduction of legislation establishing seventy-five of these was the one and only urban policy initiative attempted by the Reagan administration. It was, as Reagan said in a March 23, 1982 Message to Congress Transmitting Proposed Enterprise Zones Legislation, an 'experimental free market-oriented program dealing with the severe problems of our Nation's economically depressed areas'. It was premised on the idea that private sector institutions (the market) can solve urban problems.

Based on an idea imported from Britain, Reagan's enterprise zone concept had three distinctive features: first, the primary aim of enterprise zones was to be the economic improvement of poor neighborhoods; second, community institutions were seen as crucial to economic development; and, third, small businesses were to be favored over large ones (Green 1991: 32). Enterprise zone benefits have mainly taken the form of reduced taxes and, since these did not explicitly appear in the federal budget, they had an obvious political attraction. However, this difference is one of appearance only: forgone revenues have the same effect as a straight subsidy. (The subsidy is given by way of non-collection of tax, rather than as a payment after the tax has been collected.) Such 'tax expenditures' are now included in the budget, and thus enterprise zones might have lost some of their attraction. Curiously, this does not seem to have dampened support for them: perhaps tax expenditures are simply less politically sensitive.

Though the Reagan enterprise zone concept was an attractive one, Congress failed to pass the necessary legislation. The reasons were partly procedural, partly technical, and partly political. Above all, one question proved difficult to answer: would enterprise zones create new jobs, or would they merely attract jobs from somewhere else? More surprisingly, the enterprise zone policy failed to be passed by the Bush administration, though this was because of a presidential veto of the legislation which contained it. Clinton recognized the merits of a zone-specific strategy and promoted its development. It was finally enacted by the Clinton administration in 1993 (see Box 16.4).

The acronym is the same but the words it stands for are changed: EZs are now *empowerment zones*. Thus, a

BOX 16.4 THE CLINTON EZ/EC INITIATIVE

The Clinton administration's EZ/EC initiative differs fundamentally from previous proposals for 'enterprise zones', which relied almost exclusively on geographically targeted tax incentives to create jobs and business opportunities in distressed communities. The EZ/EC program combines federal tax incentives with direct funding for physical improvements and social services, and requires unprecedented levels of private sector investment as well as participation by community organizations and residents. This collaborative strategic planning and co-investment exemplifies the federal government's emerging role as a catalyst for local change, and exemplifies the larger principles of President Clinton's Community Empowerment Agenda:

> We need to do more to help disadvantaged people and distressed communities . . . There are places in our country where the free enterprise system simply doesn't reach. It simply isn't working to provide jobs and opportunity . . . I believe the government must become a better partner for people in places . . . that are caught in a cycle of poverty. And I believe we have to find ways to get the private sector to assume their rightful role as a driver of economic growth.

Source: US Department of Housing and Urban Development 1995b: 45

concept espoused for a decade and broadly supported is retained. However, the concept has also been widened; and it is accompanied by *enterprise communities*. The new EZ/ECs thereby have the combined advantages of retaining established support and constituting a new initiative. The initial funding (tax waivers and block grants) amounted to $2.5 billion, with six urban EZs receiving $100 million each, three rural EZs $40 million each, and ninety-five ECs a modest $3 million each. Pressure to increase the number of areas was largely resisted (there were over 500 applications), though an additional six areas were added in new categories of *supplemental empowerment zones* and *urban enhanced enterprise communities* (see Box 16.5).

The important new features incorporated in this program are grant funding (in addition to the tax benefits), a strategic plan for the coordinated economic, human, community, and physical development of the area together with pledges of support from state, local, community, and private sources such as foundations, academic institutions, and local businesses. Community building is emphasized under the program. Residents and business owners participate in making decisions that affect their communities.

The EZ/EC program offers communities a variety of opportunities for growth and revitalization through four main principles: economic opportunity, sustainable community development, community-based partnerships, and strategic vision for change. The first principle calls for the creation of jobs for the community itself and for the region itself. Being economically self-sufficient is a main goal for all communities. The second principle calls for integrating economic development into a holistic strategy covering physical development as well as economic development. The third principle extols the need to incorporate broad-based citizen participation into the planning process of an EZ/EC. The final principle calls for the development of a strategic map or plan that describes the full comprehensive plan for the community. It details the goals and benchmarks for the program as well as a means of measuring success for each component of the overall program.

Unlike the earlier proposals, under which areas would have been selected by the federal government on the basis of statistics of distress, potential EZ/ECs were selected locally. The federal government then determined which of these had made the most persuasive bids. Those that were successful were judged to have demonstrated a commitment to a thoroughly considered strategy of local initiative. In the words of the American Association of Enterprise Zones, 'the new policy's strategic planning requirement sends the signal that recovery depends on local initiative, and holds out federal assistance as reinforcement, not as the agent of change' (Cowden 1995: 10).

There are some striking similarities between this program and Johnson's model cities initiative, with localities making plans, and Washington responding. There are also some lessons to be learned from the experiences of the earlier program, including the need to ensure that benefits are widely spread geographically (covering a good majority of the congressional districts), that adequate time is allowed for the program to get under way (resisting the natural desire for 'instant gratification' which demanded premature judgments of model cities), and that careful monitoring is undertaken to establish what works under what conditions (Hetzel 1994).

As of 2003 there were over 170 EZ/EC/RCs in urban and rural areas (US Department of Housing and Urban Development 2003: 121–123). There were also areas that have been designated Champion Communities. Although not selected for either EZ/EC designation,

these areas were rewarded for organizing and completing the strategic planning process that was required in the application process. Funds were awarded to keep their momentum going. Rural Economic Area Partnership (REAP) Zones have also been designated in two areas in North Dakota, two areas in New York, and one in Vermont. A REAP Zone is a severely economically distressed community that has unique needs that cannot be attributed solely to poverty statistics. These communities are constrained by such characteristics as geographic isolation, low density, and being historically agricultural in nature.

There are a number of notable activities that are being accomplished in the areas that have been awarded funding. Detroit, a Round 1 EZ, is stressing the need to develop working partnerships in such areas as education, job creation, health, housing, community policing, and information technology. One notable accomplishment of the Detroit EZ is renovation of a historic building with recycled materials. This building will contribute to the revitalization of the area. New Orleans, a Round 1 EC, used some of its funding to develop a project called the 'Enterprise Ice Cream Project'. This project will teach children entrepreneurial skills. They will learn how to start and operate a business. The Kentucky Highlands Rural EZ

BOX 16.5 AREAS SELECTED UNDER THE EZ/EC PROGRAMS 1995

Urban Empowerment Zones ($100m)
Atlanta, Baltimore, Chicago, Detroit, New York, and Philadelphia/Camden

Supplemental Empowerment Zones ($125m and $90m)
Los Angeles and Cleveland

Rural Empowerment Zones ($40m)
Kentucky Highlands, Mid-Delta (Mississippi), and Rio Grande Valley (South Texas)

Urban Enhanced Enterprise Communities ($25m)
Boston, Houston, Kansas City, and Oakland

Enterprise Communities ($3m)
95 areas across the country

comprises three counties in southeastern Kentucky. These areas have suffered economically due to such factors as geographical isolation, limited job skills, lack of business capital, and a lack of physical infrastructure. The Highlands program has justifiably been focusing on developing job skills and developing revolving loan programs for small farmers and business development. At the same time, constructing such facilities as a youth center swimming pool and community center are important components of the Highland program. All of the programs are using the federal funding to leverage additional funding for their various programs.

Round 2 grantees are also using their funding and leveraged funds for a variety of activities. Santa Ana, California, an EZ awardee, used some of its funding to create a partnership focusing on human service development. More specifically, a partnership was formed that provides pre-school care, childcare services, and medical screening for vision, speech, and hearing. The children serviced by the partnership come from the lowest-income families in Orange County, California. The Ogalala Sioux Tribe was a Rural EZ recipient. Its area, comprised of the entire Shannon County and parts of Jackson and Bennett County, is located in the southwest border of South Dakota. Shannon County has the dubious distinction of being the nation's poorest county. This program is encouraging business development through such activities as technical assistance and loan programs. The program also recognizes that in order to promote economic development, there is a need to develop the area's physical infrastructure, education programs, and health programs. Austin, Indiana, was awarded Rural EC funding. Located in the southeastern corner of Indiana, Austin's economy suffered over the years due to a decline of natural resource-based industries. Its program includes revolving loans to provide capital to small business for both start-up and expansion purposes. At the same time, Austin is investing in creating a revolving loan program for housing rehabilitation as well as developing a rural health clinic and skills development and training program.

Congress passed legislation in December 2000 for a third round of competition for EZ/EC designations. The competition was completed in early 2002 with nine new areas being designated as EZs. In addition, forty Renewal Communities (RC) were designated and received business tax package incentives. They did not receive direct federal funding. The Department of Agriculture created this program as a means of stimulating rural economic revitalization in the areas.

Economic development policies

Economic development policies have played a major, though variable, role in public policy. Their rationale has been a matter of wide and generally inconclusive debate, and the crucial question remains: how far, and in what ways, can public policies promote economic development?

Blakely (1989: 58) defines economic development as 'a process by which local government and/or community-based groups manage their existing resources and enter into new partnership arrangements with the private sector, or with each other, to create new jobs and stimulate economic activity in a well-defined economic zone'. The American Economic Development Council (1984: 18) defines it as 'the process of creating wealth through the mobilization of human, capital, physical, and natural resources to generate marketable goods and services'. It is a continuous, long-term process that involves developing goals and objectives and identifying various strategies to achieve them. The goals, objectives, and strategies will change over time as different needs and economic conditions arise. Ultimately, improving the economic well-being of the community represents the goal of economic development.

Multiple actors are involved in economic development. No single entity is responsible for it. The federal government is involved through various policies and programs. State governments offer policies and programs to support various aspects of economic development. Local economic development departments play a key role, as do chambers of commerce. Nonprofit organizations provide important service in promoting economic development. The key to the participation of multiple parties is to recognize that economic development represents a partnership between government, business, non-profits, and citizens. There must be communication and cooperation between those involved in the process. Failure to do this could result in a waste of resources and eventual program failure. In a period of economic uncertainty, no jurisdiction can afford to let this happen.

Economic incentives have figured significantly throughout the country's history: Alexander Hamilton received a tax exemption from New Jersey in 1791 for a manufacturing company he owned. All states now use development incentives of one kind or another, though there is increasing concern about their effectiveness, and state politicians have begun to voice doubts which echo the long-standing skepticism of economists.

It seems self-evident that firms seeking a new location will be influenced by the level of local taxes and any economic incentives offered by government. There is, however, little definitive evidence on the matter. On the contrary, there is abundant if not entirely conclusive evidence that neither local taxes nor financial incentives play a significant role in attracting economic growth. The traditional location factors (which vary according to product) are the significant ones. These include proximity to markets and materials, energy and transportation costs, labor availability and costs, the economies and diseconomies of agglomeration, and a host of more elusive qualitative factors. In fact, the number of factors that can be relevant is so large that it would take a heroic feat of economic analysis to isolate their relative importance; and many will be specific to particular places and times.

Jurisdictions have employed a variety of business incentives to promote and attract economic development. In an economic development survey jointly conducted by the International City/County Management Association and the National League of Cities (2009), the most widely used business incentives were zoning/permit assistance, infrastructure improvements, tax increment financing, tax abatements, one-stop permit issues, and grants. Relying on a single incentive is no longer an option. Using a portfolio of incentives is commonplace throughout the US.

Moreover, it is generally held that competition by incentives can be a zero-sum game, with jobs merely being shunted among different parts of the country.

This argument is persuasive at the national level, but at the state and local levels it can be argued that there is a political imperative to offer incentives, even if they simply counteract the efforts of other states (Wolman 1988). An interesting twist to the debate has been given by Bartik (1991), who has argued that the areas with high unemployment are likely to be more aggressive in the use of incentives and, thus, more jobs will be created in the neediest areas.

Whatever judgment is made on the arguments, there can be little doubt as to the need for incentives to be carefully evaluated. Quite apart from their efficacy, there is an important question as to whether the money spent on incentives would not be more effectively devoted to investment in education, housing, infrastructure, or any of many other services and amenities which make a location attractive. More accurately, local economic development policies are likely to be more effective if they encompass both the direct attraction (and retention) of business and all the other things which make a place an attractive working and living environment.

There is increasing concern about the shortcomings of economic incentives. The National Governors' Association has recommended that cost-benefit analysis be used to determine whether an incentive provides a positive return, and, if so, whether a better return could be gained from alternative investments. This may be too academic an approach, but several states do now require that incentives be examined to ensure that there is a net benefit. Others have introduced sanctions against firms that fail to produce the benefits promised. In such ways are state policies beginning to change.

Recent economic conditions have led cities to re-examine their old means of doing so. Many can no longer afford to compete with other jurisdictions offering tax breaks and other incentives to keep businesses in a community or to attract other businesses to come to the community. States are also reflecting on this issue. In a report by the Pew Center on the States (2012: 1), while continuing to rely on tax incentives, 'half the states have not taken basic steps to produce and connect policy makers with good evidence of whether these tools deliver a strong return on taxpayer dollars'.

Other changes can be seen in a more sophisticated approach to economic development (see Box 16.6), and in a concern to ensure that other state policies do not badly affect the economy (as with redevelopment schemes which bring about a physical improvement at the cost of a loss of jobs). More attention is now being given to the potential of local enterprise and to 'capacity building' in areas such as education and training. Such thinking is particularly relevant in inner-city areas where the locational advantages can be capitalized on (M. Porter 1995).

There are barriers that communities routinely face in promoting economic development. In the 2009 survey mentioned above, the top barriers to economic develop-ment identified by those jurisdictions responding to the survey were the cost of land, availability of land, and a lack of capital/funding. Box 16.7 provides a breakdown of the barriers identified by survey respondents.

It is expected to see economic development covered in a general, master, or comprehensive plan for a city or county. After all, looking at the future development of a community must include policies and programs devoted to economic development. Many states require the development of a separate economic development element. It is one of the required nine elements of a local comprehensive plan in the State of Wisconsin. According to Wisconsin State Statutes 66.1001(2)(f), the economic element is:

BOX 16.6 IOWA NEW JOBS AND INCOME ACT 1994

The general assembly finds that the public and private sectors should undertake cooperative efforts that result in improvements to the general economic climate rather than focus on subsidies for individual projects or businesses. These efforts will require a behavioral change by both the state and business, balancing short-term self-interest with the long-term common good.

Source: Quoted in Gilbert 1995: 440

BOX 16.7 BARRIERS TO ECONOMIC DEVELOPMENT (IN PERCENT)

Availability of land (52.4)
Cost of land (53.4)
Lack of building availability
  (due to space or costs) (37.3)
Inadequate infrastructure (e.g., no fiber optic
  cable, water and wastewater) (28.4)
Lack of skilled labor (17.6)
High cost of labor (8.4)
Lack of affordable, quality child care (5.9)
Limited number of major employers (34.0)
Lack of capital/funding (50.1)
Taxes (20.2)

Distance from major markets (16.2)
Lengthy permit process (12.4)
Environmental regulations (22.7)
Citizen opposition (23.0)
Lack of political support (10.8)
Declining market due to loss of population (9.0)
High cost of housing (14.9)
Poor quality of life (inadequate education, recreation,
  and arts/cultural programs) (6.3)
Traffic congestion (14.9)
Other (11.8)

Source: International City/County Management Association and National League of Cities 2009

A compilation of objectives, policies, goals, maps, and programs to promote the stabilization, retention or expansion, of the economic base and quality employment opportunities in the local governmental unit, including an analysis of the labor force and economic base of the local governmental unit. The element shall assess categories or particular types of new businesses that are desired by the local governmental unit. The element shall assess the local governmental unit's strengths and weaknesses with respect to attracting and retaining businesses and industries, and shall designate an adequate number of sites for such businesses and industries. The element shall also evaluate and promote the use of environmentally contaminated sites for commercial and industrial uses. The element shall also identify county, regional and state economic development programs that apply to the local governmental unit.

In today's period of economic uncertainty, cities are in continuous competition with other areas in the US and abroad for economic development. Businesses are going out of business. They are moving to other areas for cheaper labor and land costs. Participating in the global economy had made recruiting, sustaining, and expanding businesses more difficult than ever before.

The State of Washington's Growth Management Act added a required economic development element to the local comprehensive plan in 2002. However, as of July 2012, the requirement was postponed until adequate state funding was available. Nevertheless, a number of jurisdictions have developed economic developments within their comprehensive plans.

Economic gardening

Different times require different approaches and tactics. Relying on a single approach is extremely risky at best. Economic gardening has become an increasingly discussed topic in the area of economic development. It was pioneered in Littleton, Colorado, in the late 1980s, during an economic recession. When the area's major employer left the area, Littleton didn't sit idly by and wait for the recession to end. It became proactive. Instead of trying to attract outside businesses to come to the city, Littleton decided to grow businesses and jobs from within a city. It is something analogous to cultivating a plant. After the seed has been planted, you nurture it and then watch it grow. The benefits achieved by Littleton's economic gardening program can be seen in the following:

> Using one of the very first economic development programs focused on entrepreneurial development, Littleton's job base has grown from 15,000 to 30,000 over the last two decades and the sales tax base has grown from $6 to $20 million. The community has not recruited a single company during that period and has not spent a single penny on incentive packages.
>
> (Gibbons 2010: 5)

Various criteria are used by areas to determine which companies will be selected for assistance. One area might require that a company be a certain size while another area might require that a company be at least two years old. A start-up company would not qualify for assistance. Still another area might opt only to assist high-end technology-enabled companies.

The types of assistance offered to companies selected to receive assistance will be tailored according to their needs. A company might need information on potential markets, demographic research and analysis. It may require assistance in developing business strategies for expanding. Another company may desire assistance from a peer networking session with people that have dealt with similar issues and circumstances.

Birch (1987) recognized the role of small companies in job creation and the importance of entrepreneurial development. Entrepreneurs drive economies. It is important to support existing companies. They are vested in the community and they create jobs. Moreover, some areas cannot compete with other areas for jobs. They do not have the power to offer incentives that some businesses demand in return for moving to a particular community. Even if a community could offers incentives or inducements to a company to come to the area, there is no guarantee the company will stay in the area. Companies have a way of constantly moving to wherever they can get the best deal.

Economic gardening programs have the potential to generate job growth within a community. However, the results of such a program will take time to materialize. As Gibbons (2010: 11) notes:

> Economic gardening is not a quick fix – is not a silver bullet. It is a long term strategy. It is not a fad diet; it is a lifestyle change. It takes a while to put the infrastructure in place and to get to a scale large enough to make a difference. It also takes a while for a company to start and grow and add jobs. However, with patience and commitment it has proven to be a valuable alternative to the traditional practices of economic development.

Community development

'Community empowerment' was at the center of the Clinton urban policy. The EZ/EC program embraces this as an essential element: in the words of a White House statement, the program 'is designed to empower people and communities all across this nation by inspiring Americans to work together to create jobs and opportunity'. Other programs are similarly based, as with the new program for *community development*

banks. Resisting the claims of the existing banking system that it can well serve local communities, the program involves the establishment of new financial institutions to provide much-needed credit to poor areas which have been neglected by the traditional banks (despite the 1977 Community Reinvestment Act, which requires financial institutions to meet the credit needs of their entire communities, including low- and moderate-income neighborhoods). An effective way for these obligations to be met is through public-private partnerships. Additional support is provided by increased funding ($690 million over a five-year period) for the Community Development Block Grant. The Clinton administration estimated that this would create an additional 60,000 jobs.

The salient feature of the CDBG is its flexibility and its adaptability to local needs and initiatives. It can be used for such activities as purchasing land, rehabilitating housing, constructing public facilities, and constructing new housing for non-profit groups only. (Its adoption by six presidents is a testimony to its popularity.) It represents an extremely flexible and valuable tool for community development. At least 70 percent of CDBG funds must be used for low- and moderate-income people. The new version of the CDBG forms part of a *consolidated plan* for all HUD community planning and development programs (see Box 16.8).

BOX 16.8 CONSOLIDATED PLAN FOR COMMUNITY DEVELOPMENT

The Consolidated Plan is a creative approach to community development that encourages communities to work in collaboration to develop a comprehensive vision for action to achieve community objectives.

The Plan consolidates the planning, application and reporting requirements of HUD's programs: Community Development Block Grant, Emergency Shelter Grant, HOME Investment Partnerships, Housing Opportunities for Persons with AIDS program, as well as Comprehensive Housing Affordability Strategies.

The Plan seeks to promote a comprehensive approach to address urban problems, reduce paperwork, improve accountability to achieve results, and includes strong elements of citizen participation. A basic premise of the consolidated planning process is that local jurisdictions and citizens, not Washington, know what is best for their own communities.

Source: US HUD 1995a

The consolidated plan represented a new mechanism for citizens and communities to identify and prioritize housing, homeless, community, and economic development needs. It reduces and simplifies the steps for receiving federal funding in such block grants as the CDBG, the Emergency Shelter Grant (ESG), HOME Investment Partnerships program (HOME), and Housing Opportunities for Persons with AIDS (HOPWA). The consolidated plan consists of a three- to five-year strategic plan, annual action plans, and annual performance reports.

The ESG program provides funding to state, county, and local governments to help individuals who are homeless or in danger of becoming homeless. The funds can be used for creating homeless shelters (e.g., rehabilitating existing homeless centers, converting buildings into homeless shelters) and providing services for individuals in danger of becoming homeless.

The HOME program provides federal funding to states and local governments for a number of purposes. Funding can be used for expanding the supply of decent and affordable housing to individuals in need, with an emphasis on providing the housing for low-income and very low-income families.

The HOPWA program offers funding to both state and local governments for housing assistance and supportive services for individuals diagnosed with HIV/AIDS and their families. This includes such activities as acquisition, construction, and rehabilitation of housing units, costs for maintenance of facilities as well as supportive services needed by individuals with HIV/AIDS.

Housing is an important element in the development of community programs. It constitutes one of the most serious urban problems, but it is also a problem which communities have demonstrated an ability to tackle (though whether they can operate on a scale which will make a significant impact on housing conditions must be doubtful). Community development corporations have also operated successfully in other areas, such as starting small businesses, promoting training schemes, and providing childcare. Since they are essentially local organizations, they range widely in character, initiative, and success; and local power structures and planning offices differ in their willingness to cooperate with them. While some have little interest in them, others provide a great deal of support. Some cities are noted for their positive encouragement of community development, as with the 'equity planners' of Cleveland, Dayton, Portland, and other cities discussed in a book by Norman Krumholz and Pierre Clavel (1994). This coins the term to describe 'professional urban planners who, in their day-to-day practice, have tried to move resources, political power, and political participation away from the business elites that frequently benefit from public policy and toward the needs of low-income or working-class people of their cities'. These activist professionals are called equity planners 'because they seek greater equity among different groups as a result of their work'. Their work reflects the same philosophy as that promoted by Davidoff (whose 'advocacy planning' is discussed in Chapter 1).

Community development corporations (CDCs) are often supported by local or national foundations such as the Ford Foundation (see Box 16.9). The National Congress for Community Economic Development (NCCED), in a 1995 report *Tying It All Together*, estimated that there are over 2,000 CDCs in operation, of which about two-thirds are in urban areas. According to the NCCED, there are now more than 3,600 CDCs across the United States that offer a variety of services. They have built some 400,000 units of affordable housing and 23 million square feet of commercial and office space, and have created more than 67,000 full-time jobs. Though they have received much support from foundations and local organizations, their principal source of income is the CDBG.

Faith-based organizations

President Bush voiced the concern that government alone cannot solve the ills of our communities. He felt that 'faith-based and other community organizations', or what he referred to as 'armies of compassion', could play a crucial role in providing a variety of social services. Examples of the many 'faith-based' organizations around the United States would include such groups as Catholic charities, Habitat for Humanity, Interfaith

BOX 16.9 FORD FOUNDATION COMMUNITY DEVELOPMENT PARTNERSHIPS

By funding local community foundations that act as intermediaries in attracting and allocating funds to community development organizations, Ford's Partnership model seeks to enhance the capacity of CDCs and stimulate local support systems. Partnerships help to increase the visibility and credibility of CDCs so that they can expand their base of local support. The Partnerships augment CDC funding with vital technical assistance and training. Ford sees the Local Partnerships aiding CDC capacity by:

- brokering technical resources;
- creating local project financing mechanisms;
- sensitizing commercial financial institutions to CDC projects;
- accelerating project approval through local government;
- experimenting with means to address broader financial and social issues that impede the scale and impact of physical development activities;
- generating broader CDC support;
- disseminating the CDC model to a wider audience; and
- changing public perception of the role of local CDC initiatives.

Source: Rutgers Center for Urban Policy Research, CUPR Report 1995

Housing Development Corporation of Chicago, Jewish community services, Presbyterian services, the Salvation Army, Catholic Health Association, and the Lutheran Brotherhood of Minneapolis. These groups are funded by their congregations or by other individuals and groups. Noting their history of assisting people in times of need, these organizations know the problems and needs of the communities they serve. They were already working in the communities. To President Bush, there was no reason to prevent them from playing prominent roles in community development and from being eligible to receive federal funding. To him, we need to take advantage of their work and support them. In fact, he felt we should expand what they are doing.

On January 29, 2001, President Bush backed up his belief in faith-based and other community organizations by signing an Executive Order creating a White House Office of Faith-Based and Community Initiatives. This office would have the responsibility in the executive branch of government establishing

policies and objectives to promote and expand the use of such organizations. He also established Centers for Faith-Based and Community Initiatives in five agencies – Housing and Urban Development, Health and Human Services, Justice, Education, and Labor. These centers were to examine agency policies and programs and promote the use of faith-based and other community organizations.

One of the first major activities of the White House Office was to examine the various barriers that prevented faith-based and other community organizations from participating in federal social service programs. A report, 'Unlevel playing field: barriers to participation by faith-based and community organizations in federal social service programs' (US Department of Housing and Urban Development, Center for Faith-Based and Community Initiatives 2001) details the obstacles facing the groups. For example, the report notes that agencies may have a bias against faith-based organizations. Some agencies might be wary of being sued for providing federal funding because of the constant debate over the separation of church and state. Other agencies appear to be biased toward past grant recipients. Still other barriers preventing these groups from participating in federal social service grant programs might include a lack of awareness about the availability of some grant programs; the strict legislative requirements of some programs; and a lack of knowledge about the activities and missions of faith-based organizations on the part of the program administrator.

Many have questioned whether or not there was a domestic urban agenda under President Bush. Some feel we neglected the urban agenda at the expense of the defense agenda. We do not know his interest in urban issues except for cutting many domestic programs. His administration proposed cuts in funding for community development, for CDBG, and other programs. Nevertheless, he remained strongly supportive of faith-based community economic development.

Business Improvement Districts

An additional tool used in the United States to help revitalize community commercial districts is the Business Improvement District (BID). Local governments are able to create a BID through a state statute. MacDonald (1996: 4) defines a BID as:

> An organization of property owners in a commercial district who tax themselves to raise money for neighborhood improvement. Core functions usually include keeping sidewalks and curbs clean, removing graffiti, and patrolling the streets. Once a BID is formed, the assessment is mandatory, collected by the city like any other tax. Unlike any other taxes, however, the city returns the assessment of the BID management for use in the district.

These organizations do not rely on government dollars for their work. However, they can receive funds through various public grant programs. This goes hand in hand with the current administration's view that government should be providing these types of services.

BIDs can be found throughout the United States. For example, the Golden Triangle BID in Washington, DC covers a forty-two-block area within the city and more than 4,000 businesses as members. It was founded in 1997 as a private non-profit 501 (c) (6) corporation whose purpose is to improve the city's central business district. It provides services in the areas of public safety, capital improvements, transportation, economic development and market, and homeless services. The Oakland BID in Pittsburgh, Pennsylvania, was created in 1999 to strengthen the fifty-block commercial area. Among its work tasks are cleaning up the area and promoting the BID area as a place of business. The City of San Diego's Office of Small Business administered the city's BID program. Over fifteen districts are currently operating in the City of San Diego. Generating new businesses for the areas and helping existing businesses to prosper are major goals for the BIDs. The North Park BID, created in 1996, is also a Main Street program whose mission is 'to promote development that supports arts, culture, and entertainment, while preserving the historic integrity of North Park and to create a pedestrian-friendly destination for shopping, dining, and entertainment that supports our local businesses' (http://www.northparkmainstreet.com). Each BID member is assessed a fee for promoting and improving the business area (see Box 16.10).

Individual projects can also serve as a catalyst for community economic development. Baltimore's Inner Harbor development is the anchor for much of what is happening in the city's downtown (see Plate 27). Downtown sports stadiums in Baltimore, Denver, Indianapolis, and Pittsburgh have been driving forces in the economic development of a given city (see Plate 28).

Support for social change

There is abundant evidence that community organizations can make a significant contribution to the welfare of communities. Moreover, being local, they can aim at objectives which a locality wants, rather than being saddled with government programs which may not represent their priorities or their wishes. They work

Plate 27 Baltimore's Inner Harbor
Photo by author

Plate 28 PNC Park in Pittsburgh
Photo by author

BOX 16.10 DETERMINING BID ASSESSMENT FEES IN SAN DIEGO

The formula for determining the assessment amount is determined by the business organization that initiates the BID process, not the City. The respective group takes into account the type, size, and location of the businesses. Assessments are levied on businesses on the basis of the relative benefit from the improvement and activities to be funded. In San Diego, the fees generally range from $40 to $500 per business each year. A few of the newer BIDs have fees, ranging from $90 to $1,200 per year, with some anchor tenants paying up to $5,000 to support BID-related projects.

Source: City of San Diego Business Improvement Districts, http://www.sandiego.gov/economic-development/business-assistance/small-business/bids.shtml

for the direct provision of locally wanted facilities and services, rather than relying on the trickle-down effect of 'economic developments' such as high-rise office towers, sports stadiums, convention centers, cultural megapalaces, and other manifestations of the 'edifice complex' (Squires 1989: 289). They focus on basic needs of the poor.

Nevertheless, the inadequacy of resources available for this type of community activity, its inherent limitations in relation to stronger economic and political forces, and the deep-seated nature of problems of race, class, and poverty all point to the need for public policies as major forces for change. Some of the difficulties of devising such policies have been discussed in this chapter, and the power of the forces of discrimination has been dealt with at various places in this book. The question remains as to whether there is sufficient public understanding and support for attempting to overcome these.

American Recovery and Reinvestment Act of 2009 (ARRA)

ARRA (PL 111.5) was passed in response to the weakened and fragile economy. Its purposes could be referred to as the '3Rs': Rescue, Recovery, and Reinvestment. An economic stimulus package was contained in the law that would serve multiple functional areas. More specifically, according to Section 3 A(a), ARRA had the following purposes:

(1) To preserve and create jobs and promote economic opportunity
(2) To assist those most impacted by the recession
(3) To provide investments as needed to increase economic efficiency by spurring technological advances in science and health
(4) To invest in transportation, environmental protection, and other infrastructure that will provide long-term economic benefits
(5) To stabilize State and local government budgets, in order to minimize and avoid reductions in essential services and counterproductive state and local tax increases.

The cost of the original stimulus package was $787 billion, a figure that was later increased by almost $50 billion between 2009 and 2019.

Of the original amount, another 35 percent was to be allocated for tax relief incentives for individuals, including such items as payroll tax credits, home buyer credits for first-time home buyers, expansion of childcare tax credit, and tax credits for college tuition. Tax incentives were also offered to companies. The next largest amounts would be allocated for state and local fiscal relief, infrastructure and science, protecting vulnerable populations, health care, education and training, energy, and other areas. Examples of specific projects that were founded are given in Box 16.11.

President Obama signed ARRA on February 17, 2009 in Denver, Colorado. In his remarks, he stated it would not solve our economic problems:

BOX 16.11 EXAMPLES OF PROJECTS FUNDED UNDER ARRA 2009

Road construction projects

Job retraining programs

Rehire laid-off workers

Hire police officers

Expand solar electric facility

Aid low-performing schools

Improve small shipyard

Improve delivery of food services to the poor

Separate stormwater and sewer lines

Put young people to work

Clean-up of blighted properties

Small business loans

Upgrade housing for elderly and disabled

Purchase electric vehicles

Develop flood control project

Home weatherization program

Bring broadband network to rural areas

Establish health information center

Promote physical fitness in schools

Upgrade intercity passenger rail system

Purchase police and fire vehicles

Conduct medical research

Expand Early Head Start facility

Invest in new energy projects

Now, I don't want to pretend that today marks the end of our economic problems. Nor does it constitute all of what we're going to have to do to turn our economy around. But today does mark the beginning of the end – the beginning of what we need to do to create jobs for Americans scrambling in the wake of layoffs; the beginning of what we need to do to provide relief for families worried they won't be able to pay next month's bills; the beginning of the first steps to set our economy on a firmer foundation, paving the way to long-term growth and prosperity.

Although the bill received support from a host of individuals and groups, everyone had questions. Was the amount too small to make a dent in the economic recession? Would it be effective? Would it provide funding for wasteful and unneeded projects? Would governmental regulatory barriers impede the progress of various projects, thus delaying the creation of needed jobs? These are all valid concerns.

The issue of transparency has always surfaced whenever federal funds have been allocated. People want to see what projects are being funded and whether or not the funds are being used properly and in a timely fashion. ARRA was no different. Individual sections of ARRA discussed the need to make public cost data available to the public. ARRA also created a Recovery Accountability and Transparency Board charged with the responsibility of creating a website to foster accountability and transparency. A website was developed and then redesigned to track funds and to show how the funds, entitlements, and tax benefits were distributed and being used.

Additional opportunities

There are other opportunities for engaging in community and economic development in our cities. Idle, abandoned, or underused properties that may have some degree of contamination offer communities potential sites for community or economic development activities. Examples of these properties include gas stations, parking lots, warehouses, industrial facilities, landfills, abandoned airports or railroad yards (see Plate 29). Unfortunately, any development on these sites, better known as 'brownfields', could be hampered by concerns over potential environmental contamination.

The possibility of environmental contamination on a site causes a potential developer to become very cautious. No developer wants to be responsible for cleaning up the property. Depending upon the degree

Plate 29 Parking lot in downtown Pittsburgh

Photo by author

of contamination, the costs of such a clean-up could be in the millions of dollars. Liability issues are a constant issue of debate regarding the redevelopment of a site. One of the most publicized cases of environmental contamination can be found in the Love Canal community in Niagara Falls, New York. This environmental nightmare is covered in more detail in Chapter 13.

Not all brownfields have such a high degree of contamination as did the Love Canal site. In fact, most of them do not have such a level of contamination. Many of the sites with little or no contamination can be reclaimed and redeveloped into such beneficial uses as housing sites, park or recreation areas, or business sites. As such, a redeveloped brownfield can become

an asset to the community. It can become a link in the community revitalization chain.

There is a growing concern that blue-collar communities suffer more from the public health hazards associated with brownfields since a disproportionate number of brownfields are located in these communities. This is occasionally referred to as environmental racism. People attack this type of racism on the belief that everyone, regardless of race, ethnicity, or socioeconomic status, has a right to equal justice and the equal protection of environmental laws and regulations. This is known as environmental justice.

Many areas around the United States have benefited from a strong military presence for many years. Many have served as an integral part of the area. However, in recent years, there has been a move to reduce military overheads. This has been through the closure or realignment of numerous military bases and installations as a result of the 1988, 1991, 1993, and 1995 Base Realignment and Closure Commissions (BRAC). Examples of major base closures include: the Presidio of San Francisco; Fort Sheridan, Illinois; Norton Air Force Base, California; Fort Devens, Massachusetts; Sacramento Army Depot, California; Philadelphia Naval Shipyard; Charleston Naval Shipyard, South Carolina; Naval Training Center, San Diego, California; Fort McClellan, Alabama; and Bergstrom Air Force Base, Texas.

The BRAC process was hotly contested at virtually every corner. Congressional delegations pulled no punches in trying to save a military base or installation. It would be virtually impossible to ignore the political wheelings and dealings that took place during the four rounds of BRAC. Nevertheless, the decisions reached by the various rounds were applauded by some and cursed by others.

The BRAC decisions, although very painful to a number of areas, afforded the areas new opportunities to develop various community and economic development programs. Federal funds and technical assistance were available to areas to help plan and adjust to the closures or realignments. The process of managing the disposal of base property was the responsibility of the relevant military department. The various processes, as the local governments found out, contained numerous federal laws and regulations.

The base reuse process is still occurring in many areas. Environmental contamination problems have plagued many areas. Cleaning up the sites has taken longer than many of the areas expected. In other areas, local politics have delayed decisions on what would be done. Lawsuits have been filed by parties claiming the planning process failed to include them in the decision-making process. Nevertheless, many areas are converting bases to such uses as housing sites, business incubators, park and recreation areas, homeless facilities, government offices, homes for non-profit agencies, civilian airports, and training centers.

Further reading

The Department of Housing and Urban Development (HUD) gives an account of its current structure and its history on its website (http://www.hud.gov/about). Online visitors can view the current staff, the organization of the department, and its original mission, creation, and current strategy.

Most states use Community Development Block Grants (CDBGs), and the City of Madison, Wisconsin, has a website devoted to its program (http://www.ci.madison. wi.us/cdbg). Complete with photos of projects funded by CDBG, the website supplies internet links to the CDBG commission and lists of current and past activities.

The North Carolina Department of Commerce offers an overview of CDBGs on its website (http://www. nccommerce.com/en/CommunityServices/Community DevelopmentGrants/CommunityDevelopmentBlock Grants). It breaks down CDBGs into project categories and the purpose of those categories. The North Carolina website is easy to understand and also has internet links for more in-depth information.

The Empowerment Zones and Enterprise Community program is explained on the US Department of Health & Human Services website (http://www.acf.hhs.gov/ programs/ocs/ez-ec). Online visitors will find a general background of EZs along with current available resources.

Empowerment Zones (EZs) are so widely used they exist in almost every state. Baltimore, Maryland, has its own EZ website (http://www.ebmc.org/home/index.html) that posts a description of the EZ, best practices, resources, and news and information.

California has so many enterprise zones, the zones created their own organization called the California Association of Enterprise Zones (http://www.caez.org). Its online resources include a map of Californian enterprise zones, a list of members, zone incentives, and success stories of the zones.

Community Development Bankers have created their own association online (http://www.communitydevelopment

banks.org). Complete with an explanation of Community Development Banks, the website has links to association partners.

The Ford Foundation is still running strong and offers visitors many resources online (http://www.fordfound. org). Basic information about the foundation, such as its mission, history, and current officers, is presented next to a description of how the program works and current news and resources available.

The National Brownfields Associations (http://www. brownfieldassociation.org) is a partnership between the United States and Canada. Based in Chicago, this non-profit organization encourages redevelopment of brownfields. Online visitors can find a brownfield database and information on conferences that train and educate organizers.

The American Enterprise Institute for Public Policy Research (http://www.aei.org) provides online updates about US public policy. The website has book reviews and recommendations along with online commentary regarding public policy changes. The political and social section of the website is more tailored to urban issues and visitors can perform searches on specific issues as well.

For information on tracking ARRA funds, entitlements, and tax benefits distributed and being used, see: http:// www.recovery.gov.

There are many books on the development of urban policy. An overview is given in Robertson and Judd (1989) *The Development of American Public Policy*; and Judd and Swanstrom (1994) *City Politics: Private Power and Public Policy*. More detailed studies include Frieden and Kaplan (1977) *The Politics of Neglect: Urban Aid from Model Cities to Revenue Sharing*; Kaplan *et al*. (1970) *The Model Cities Program: The Planning Process in Atlanta, Seattle, and Dayton*; and Waste (1998) *Independent Cities: Rethinking US Urban Policy*. Gelfand's *A Nation of Cities: The Federal Government and Urban America 1933–1965* (1975) is a particularly good account of the period covered.

The literature on community development policy is rich. Among the recent books that examine various facets of community development policy are Gittell and Vidal (1998) *Community Organizing: Building Social Capital as a Development Strategy*; Ferguson and Dickens (1999) *Urban Problems and Community Development*; Green and Haines (2002), *Asset Building and Community Development*; Walker and Weinheimer (1998) *Community Development in the 1990s*; and Squires and O'Connor (2001) *Color and Money: Politics and Prospects for Community Reinvestment in Urban America*.

Non-profit organizations are examined in a number of ways in Hula and Jackson-Elmore (2000) *Nonprofits in Urban America* and Hopkins (2009) *Starting and Managing a Nonprofit Organization: A Legal Guide*.

The role of faith-based organizations in the provision of social services is examined in Vidal (2001) *The Role of Faith-Based Organizations in Community Development*. Additional information can be found in Wright (1999) *An Annotated Bibliography for Faith-Based Community Economic Development*; and Monsma (2012) *Pluralism and Freedom: Faith-Based Organizations in a Democratic Society*.

The interplay between community development and environmental justice can be found in Bullard *et al*. (2000) *Sprawl City: Race, Politics, and Planning in Atlanta*; and Cole and Foster (2000) *From the Ground Up: Environmental Racism and the Rise of the Environmental Justice Movement*.

For a discussion of the BRAC process and the effects of military base closures or realignments, see: Hix (2001) *Taking Stock of the Army's Base Realignment and Closure Selection Process*; Sorenson (1998) *Shutting Down the Cold War: The Politics of Military Base Closure*; Dardia *et al*. (1996) *The Effects of Military Base Closures on Local Communities: A Short-Term Perspective*; and Mayer (1992) *Local Officials Guide to Defense Economic Adjustment*.

Barnekov *et al*. (1989) *Privatism and Urban Policy in Britain and the United States* is more than a comparative study; it explores the implications and outcomes of the dominant cultural tradition affecting urban policy – a tradition that relies on private initiative and competition as the main agent of urban change.

Norman Krumholz, a former planning director of Cleveland, has written extensively and eloquently on 'equity planning'. See particularly Krumholz and Forester (1990) *Making Equity Planning Work: Leadership in the Public Sector*; and Krumholz and Clavel (1994) *Reinventing Cities: Equity Planners Tell Their Stories*.

These works are in the tradition of Paul Davidoff's 'advocacy planning'; Krumholz contributes to a series of articles on Davidoff in the spring 1994 issue of the *Journal of the American Planning Association* (60: 129–61). A collection of essays is edited by Squires (1989): *Unequal Partnerships: The Political Economy of Urban Development in Postwar America*.

A major text on local economic development is Blakely and Leigh (2010) *Planning Local Economic Development: Theory and Practice*. See also Blair (1995) *Local Economic Development: Analysis and Practice*; and Koven and Lyons (2010) *Economic Development: Strategies for State and Local Practice*. A detailed evaluation of the CDBG program has been undertaken by the Urban Institute and published by HUD: Urban Institute (1995) *Federal Funds, Local Choices: An Evaluation of the Community Development Block Grant Program*.

For information on Business Improvement Districts, check the following sources: J. Mitchell (1999) 'Business Improvement Districts and innovative service delivery'; Houstoun (2003) *BIDs: Business Improvement Districts*; Mitchell (2008) *Business Improvements and the Shape of American Cities*; and Feehan and Feit (2006) *Making Business Districts Work*.

Questions to discuss

1   What are the objectives of community and economic development?

2   Compare the varying approaches to economic development taken by different administrations, and outline their strengths and weaknesses.

3   Discuss how far a national urban policy is feasible.

4   Are incentives for economic development justifiable?

5   What are the ingredients for a successful local economic development policy?

6   Compare the merits of local community development and local public policy.

7   Describe the enterprise zones initiatives. Do you consider them to be effective?

8   Why is urban policy so difficult?

9   What roles do non-profit organizations play in community development?

10 How will the ARRA of 2009 help in economic recovery efforts in the US?

11 What is economic gardening?

17

Urban design and aesthetics

I think that I shall never see
A billboard lovely as a tree
Indeed, unless the billboards fall,
I'll never see a tree at all.

                    Ogden Nash

Ogden Nash may never have seen
A billboard he held dear
But neither did he see
A tree grossing 20 grand a year.
      David Flint, Turner Advertising Company

Urban design

Urban design is a term that means different things to different people. There does not appear to be any universal definition. Is it primarily concerned with architecture or planning? It covers both areas and more. It is a multidisciplinary area of inquiry. The City of Baton Rouge, Louisiana (2009: 2) indicates 'urban design is the discipline through which planning and architecture can create or renew a sense of local pride and identity' and that it 'is concerned with bringing different disciplines responsible for the components of cities into a unified vision'. Llewelyn-Davies and Alan Baxter & Associates (2000: 10) note that 'urban design is derived from but transcends related matters such as planning and transportation policy, architectural design, development economics, landscape and engineering'. It encompasses the entire urban fabric, both in horizontal and vertical dimensions – buildings, streets, public spaces, community safety, landscaping, signage, lighting, fencing, services areas,

etc. It involves physical qualities as well as visual qualities.

The City of San Diego, California (2008: UD-3) recognizes the importance of urban design through the following passage:

Urban design describes the physical features that define the character or image of a street, neighborhood, community, or the City as a whole. Urban design is the visual and sensory relationship between people and the built and natural environment. The built environment includes buildings and streets, and the natural environment includes features such as shorelines, canyons, mesas, and parks as they shape and are incorporated into the urban framework. Citywide urban design recommendations are necessary to ensure that the built environment continues to contribute to the qualities that distinguish the City of San Diego as a unique living environment.

This broad endorsement can be seen in general urban design goals of the City of San Diego (2008: UD-5, 6):

- A built environment that respects San Diego's natural environment and climate
- An improved quality of life through safe and secure neighborhoods and public spaces
- A pattern and scale of development that provides visual diversity, choice of lifestyle, opportunities for social interaction, and that respects desirable community character and context
- A City with distinctive districts, communities, neighborhoods, and village centers where people gather and interact
- Maintenance of historic resources that serve as landmarks and contribute to the City's identity
- Utilization of landscape as an important aesthetic and unifying element throughout the City.

Community character

What makes a community unique? What traits do its residents value? What are its tangible and intangible attributes and assets? These are all questions relating to an area's 'community character'.

'Community character' can be defined in various ways. It includes aesthetics and other attributes. The Georgia Department of Community Affairs (2013: 6) defines it as 'the characteristics that make a particular development unique'. Henrico County, Virginia's Vision 2026 Comprehensive Plan (2009: 55) offers the following guidance:

Community character is less quantifiable than other aspects of land use and public facilities, but it is equally important to the creation of livable communities. Community character can be defined as

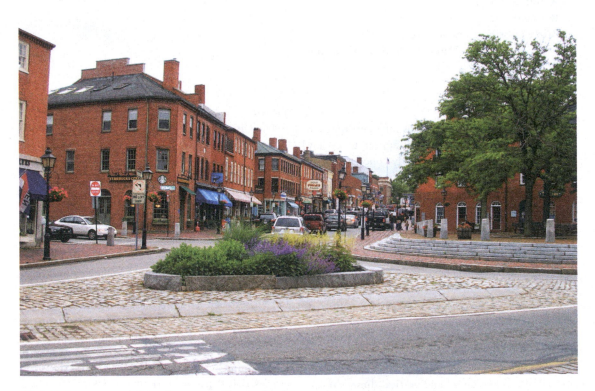

Plate 30 Downtown Newburyport, Massachusetts
Photo by author

the combined effect of the built, natural, historic and social features within a neighborhood. It encompasses the physical and social attributes that make one neighborhood different from another.

Cities throughout the US tend to address 'community character' through various plans. The City of Newburyport, Massachusetts, examines character throughout its 2001 Master Plan, which has a slogan – 'Shaping Our Future, Honoring Our Past'. With a 2011 population of approximately 17,500, Newburyport is proud of its small town heritage and character and celebrates it past (see Plate 30). It seeks to protect its historical, agricultural, cultural, recreational, and other assets for its residents and visitors. Its downtown is highly walkable and responsive to the needs of pedestrians. Its older traditional neighborhoods are cherished, but residents are concerned that increasing density threatens neighborhood character. As a result, calls for the creation of historic districts have been voiced by residents as a means of protecting the neighborhood's character.

There will be constant challenges facing an area's ability to protect its 'community character'. Newburyport is no exception. Economic development poses formidable challenges to maintaining a community's assets. Residential and commercial growth is threatening the city's agricultural heritage and natural landscape. Monitoring of development activities is of paramount importance in maintaining an area's 'community character'.

Regulating urban design and aesthetics

Despite the City Beautiful movement, aesthetic considerations always have been and probably always will be problematic in American land use planning (see Plate 31). They involve questions of preference and taste on which opinions differ, as the following examples illustrate:

The American Institute of Architects' choice of the best builder's house of 1950 was refused a mortgage by the Federal Housing Administration. Again, the Veterans Administration imposed a $1000 design penalty on an architect-designed house in Tulsa, Oklahoma, that *House and Home* had displayed on its 1954 cover. The Pruitt-Igoe public housing, which starred in a TV vehicle when HUD Secretary George Romney had it blown up, had won an architectural award in its day.

(Haar and Wolf 1989: 533)

By contrast, time tends to affect how we view things. For example, designs once despised can become popular icons: the Eiffel Tower was once described in terms of 'the grotesque mercantile imaginings of a constructor of machines'. Now it is 'the beloved signature of the Parisian skyline and an officially designated monument to boot' (Costonis 1989: 64).

The difficulties of aesthetics are great at both the practical and the philosophical levels yet, in simple terms, Americans like their neighborhoods to be pleasant and attractive, free of noxious intrusions. Fear of falling property values and unwelcome social groups plays its role here too, but there remains a real, and increasing, concern for environmental quality.

This can be seen, for example, in the increased regulation of signs and billboards; in the adoption of landscape ordinances, parking lot regulations, appearance codes, and design guidelines; and in the establishment of advisory or administrative design review boards. The following reasons are illustrative of why one city, Tuscaloosa, Alabama, regulates signage:

- To protect the City of Tuscaloosa's appearance and the quality of life of its citizens.
- To protect the public from the danger of unsafe signs, and from the degradation of the aesthetic qualities of the City.
- To preserve, protect and enhance areas of historical, architectural, cultural, aesthetic and economic value, regardless of whether they are natural or human-made.
- To establish standards and provide controls that permit reasonable use of signs and enhance the character of the City.

Plate 31 Public art in Milwaukee
Photo by author

- To support and promote the use of signs to aid the public in the identification of businesses and other activities, to assist the public in its orientation within the City, to express the history and character of the City, to promote the community's ability to attract sources of economic development and growth, and to serve other informational purposes.
- To protect the safety and efficiency of the City's transportation network by reducing the confusion and distraction to motorists, reducing collision hazards and enhancing the motorists' ability to see pedestrians, obstacles, other vehicles and traffic signs.
- To integrate sign regulations more effectively with general zoning regulations by establishing specific requirements for billboards and signs, related to

setbacks, height restrictions and spacing to allow for lighting, ventilation and preservation of views in a manner consistent with land uses in the City.
- To avoid excessive competition for large and multiple signs, so that permitted signs provide identification and direction while minimizing clutter, unsightliness, confusion, and hazardous distractions to motorists.
- To preserve the views of natural resources, green space and other open spaces from unnecessary blight and blockage caused by signage and billboards.
- To protect adjacent and nearby properties, in particular residentially zoned properties, from the impact of lighting, number, size, height, movement and location of signs and billboards.

- To enhance the impression of the City which is conveyed to tourists and visitors by controlling the location and number of signs and billboards.
- To encourage signage and other private communications which aid orientation, identify activities, express local history and character or serve other educational purposes.

(http://ci.tuscaloosa.al.us/index.aspx?NID=686)

This chapter discusses a number of these planning mechanisms. In line with historical developments, the first to be considered is the control over billboards.

Billboards

Billboards are commonly viewed as part of the urban fabric. The City of Tukwila, Washington defines a billboard in its Municipal Code, Chapter 19.08.060 as 'an off-premise, freestanding sign or visual communication device that has a sign area of at least 150 square feet in message area per face'.

Aesthetics first arose explicitly with billboards – and initially the overwhelming judicial view was that controls imposed for such reasons would not pass constitutional muster. A 1905 New Jersey case (*Passaic*) is illustrative:

> Aesthetic considerations are a matter of luxury and indulgence rather than of necessity, and it is necessity alone which justifies the exercise of the police power to take private property without compensation.

Similarly, a Denver ordinance of 1898 was held to be unconstitutional because it had specific requirements solely for billboards, including a ten feet setback from the street line. The wording of the decision became quite lyrical (see Box 17.1).

Nevertheless, a minority of courts did hold that aesthetics was a legitimate consideration in the exercise of the police power, and by the 1930s it was generally accepted that aesthetic factors could be taken into account. This involved a legal fiction, namely that while aesthetic regulations were not acceptable in themselves, they could be justified on the grounds of associated evils. A 1932 New York decision stated the view nicely: 'Beauty may not be queen but she is not an outcast beyond the pale of protection or respect. She may at least shelter herself under the wing of safety, morality or decency.' A classic statement of this view occurs in a Missouri case of 1913 (see Box 17.2).

The majority of courts today hold that the police power can be used for aesthetic purposes, whether these have the ulterior purpose of promoting some other public good such as tourism or economic development, or for 'pure' aesthetic objectives. An important factor in this change was the 1954 Supreme Court case of *Berman* v. *Parker*. In his decision, Justice Douglas delivered the following *dictum* (that is, it was a gratuitous comment, not crucial to the case in question):

> The concept of the public welfare is broad and inclusive . . . The values it represents are spiritual as well as physical, aesthetic as well as monetary. It is within the power of the legislature to determine that the community should be beautiful as well as healthy, spacious as well as clean, well balanced as well as carefully patrolled.

A later case of some notoriety concerned a Mrs Stover who, for several years, hung clotheslines of rags in the front yard of her house in Rye, New York, as a protest against the high taxes imposed by the city. Each year an additional line was festooned with a remarkable range of materials: tattered clothing, old uniforms, underwear, rags, and scarecrows. Neither the neighbors nor the city were amused, and after six years the city passed an ordinance prohibiting the erection and maintenance of clotheslines on a front or side yard abutting a street; exceptions could be granted where there were real practical difficulties in drying clothes elsewhere on the premises. Mrs Stover applied for an exemption but was refused, but she retained her clotheslines. The case (*People* v. *Stover* 1963) went to court, and it was ruled that the city was justified in preventing Mrs Stover from her unusual form of protest: a form which was 'unnecessarily offensive to the visual sensibilities of the average person'.

BOX 17.1 TASTE AND THE CONSTITUTION

The cut of the dress, the color of the garment worn, the style of the hat, the architecture of the building or its color, may be distasteful to the refined senses of some, but government can neither control nor regulate in such affairs . . . Ours is a constitutional government based upon the individuality and intelligence of the citizen, and does not seek, nor has it the power, to control him, except in those matters where the rights of others are impaired.

Source: Curran Bill Posting Co v. City of Denver 1910

BOX 17.2 THE IMMORALITY OF BILLBOARDS

Billboards endanger the public health, promote immorality, constitute hiding places and retreats for criminals and all classes of miscreants. They are also inartistic and unsightly. In cases of fire they can often cause their spread and constitute barriers against their extinction; and in cases of high wind, their temporary nature, frail structure and broad surface, render them liable to be blown down and to fall upon and injure those who may happen to be in their vicinity. The evidence shows and common observation teaches us that the ground in the rear thereof is being constantly used as privies and dumping ground for all kinds of waste and deleterious matters, and thereby creating public nuisances and jeopardizing public health; the evidence also shows that behind these obstructions the lowest form of prostitution and other acts of immorality are frequently carried on, almost under public gaze; they offer shelter and concealment for the criminal while lying in wait for his victim; and last, but not least, they obstruct the light, sunshine, and air, which are so conducive to health and comfort.

Source: St Louis Gunning Advertising Co v. St Louis 1913

Most courts now take the view that aesthetics alone is a legitimate public purpose and can be controlled by land use regulation. (The same logic is also applied to pornography.) It still remains, of course, for a municipality to ensure that the controls are properly applied.

As the *Berman* and *Stover* cases illustrate, some important court decisions on aesthetics are only indirectly concerned with signs. In the following pages, cases dealing specifically with the issues raised by signs (and billboards in particular) are discussed.

Signs can be of various kinds: directional, political, on-site business, freestanding adverts (billboards) and so on. The crucial distinction, however, is between 'informational' signs and billboards. On-premise signs

(which, of course, can be as obnoxious as the worst billboard) are generally accepted in principle, though restrictions are common on their size and number. Billboards, on the other hand, arouse a great deal of controversy – fueled by two active lobbies: one promoted by the wealthy and powerful billboard industry, and the other by Scenic America (formerly the Coalition for Scenic Beauty), dedicated to 'curb an industry that . . . has run amok'. The battle lines continue to be evident throughout the United States.

No holds are barred in the open warfare on billboards. In his legal treatise, Norman Williams (1990: 118.02) refers to the billboard lobby as 'quite intransigent in demands and quite ruthless in tactics'. He comments that 'it has been common gossip among leading planners that the billboard industry maintains (or used to maintain) a blacklist. It is certainly true that on occasion segments of the industry have intervened to try to keep a planner known to be "uncooperative" out of an important job.' Former New York Senator Thomas C. Desmond is quoted as saying that the billboard lobby 'shrewdly puts many legislators in its debt by giving them free sign space during election time, and it is savage against the legislator who dares oppose it' by favoring anti-billboard laws (Blake 1964: 11).

The billboard industry endeavors to enhance its public image by donating billboard space to candidates for political offices and to good causes such as First Lady Barbara Bush's campaign to promote family literacy, and the boosting of morale in the San Francisco Bay area following the October 1989 earthquake. These public benefits are regularly reported in *Outlook: The Newsletter of the Outdoor Advertising Association of America*.

Billboard regulation continues to make headlines across the United States. In 2002, voters in Missouri had to decide the fate of a ballot measure asking them whether they felt a statewide billboard ban was needed. Proposition A asked the voters:

Shall Missouri statutes be amended to prohibit the construction of most new outdoor advertising and to further restrict existing outdoor advertising along all National Highway System highways in Missouri; to increase the authority of any city, county, or local zoning authority to regulate outdoor advertising; and to prohibit the removal of trees and vegetation located on public rights of way except for purposes of improving aesthetic or environmental value or of eliminating safety hazards?

The measure failed to be passed by the voters in a close 51 percent to 49 percent contest.

In March 2002, voters in the City of San Francisco had to decide the fate of a ballot measure that would ban new billboard construction. The voters turned to the ballot box when they recognized code enforcement was apparently not working. Seeing the issue as essentially a quality of life issue, the voters approved the billboard ban by a wide margin.

The issue has also surfaced in Florida. In 2002, legislation was signed by Governor Jeb Bush that called for local governments to pay for billboards they wanted to remove or to enter into relocation and reconstruction agreements with the sign owner (see Box 17.3).

As expected, the governor was both praised and cursed for his decision. Supporters of the legislation praised the governor for his protection of property rights and for following the national trend of protecting property rights. Opponents claimed the governor was simply following the interests of the billboard lobby over those of the public. They also felt the legislation was undermining community beautification ordinances by placing roadblocks in front of a jurisdiction's ability and attempt to redevelop and eliminate blight and to regulate their environment.

Sign provisions in the zoning ordinance of Miami, Florida, have come under attack recently. Having been written in 1990 and amended on several occasions, Miami's ordinance allowed the city to remove billboards that were in violation of the ordinance. The ordinance required that billboards in limited commercial use areas and along certain interstates be taken down within five years. The city embarked on an aggressive course to remove the signs. Litigation quickly surfaced challenging the constitutionality of the entire Miami zoning code.

In 2002, a Special Master of the Code Enforcement Board reversed its earlier decision and now ruled that a billboard owned by a large media company was in violation of a Miami zoning ordinance. A federal judge in Miami was later ordered to decide if Miami was violating the constitutional rights of billboard companies. Specifically, the issue before the court was whether the ordinance denied the companies' freedom of speech and equal protection rights. In a related case, Florida's First District Court of Appeal ruled that restrictions on interstate highways did not constitute a prior restraint on speech. With the amount of money devoted to billboard construction in Florida, litigation is certain to continue.

BOX 17.3 2002 FLORIDA BILLBOARD LEGISLATION

70.20 Balancing of Interests – It is a policy of this state to encourage municipalities, counties, and other governmental entities and sign owners to enter into relocation and reconstruction agreements that allow governmental entities to undertake public projects and accomplish public goals without the expenditure of public funds while allowing the continued maintenance of private investment in signage as a medium of commercial and noncommercial communication.

1   Municipalities, counties, and all other governmental entities are specifically empowered to enter into relocation and reconstruction agreements on whatever terms are agreeable to the sign owner and the municipality, county, or other governmental entity involved and to provide for relocation and reconstruction of signs by agreement, ordinance, or resolution. As used in this section, a 'relocation and reconstruction agreement' means a consensual, contractual agreement between a sign owner and a municipality, county, or other governmental entity for either the reconstruction of an existing sign or the removal of a sign and construction of a new sign to substitute for the sign removed.

2   Except as otherwise provided in this section, no municipality, county, or other governmental entity may remove, or cause to be removed, any lawfully erected sign located along any portion of the interstate, federal-aid primary or other highway system, or any other road without first paying just compensation for such alteration as determined by agreement between the parties or through eminent domain proceedings. The provisions of this section shall not apply to any ordinance the validity, constitutionality, and enforceability of which the owner has by written agreement waived all rights to challenge.

Source: 2002 Florida Statutes, Chapter VI, Section 70.2

Digital billboards

As technologies have evolved, new types of billboards have become commonplace. Digital billboards can now be seen in many places throughout the US and other countries (see Plate 32). These are billboards that send messages on a rotating basis every 4–10 seconds. They are operated by a computer at an off-site location. In 2011, the Outdoor Advertising Association of America (OAAA) estimated that there are some 400,000 billboards throughout the US, about 2,400 of them being digital.

Regulations governing the use of digital billboards vary by state and jurisdiction. Some areas have banned their use. Other areas allow their use with the stipulation that the company remove a certain number of static billboards. Still other areas have implemented a temporary hold or moratorium on digital billboards.

They want to study safety and aesthetic concerns commonly associated with digital billboards. The safety concern is that the sign distracts motorists from paying attention to the road. The aesthetic concern centers on the notion that they are contributing to visual clutter.

Let us consider how one state is dealing with the issue – Nebraska. Nebraska has requirements that digital billboards or signs, on highways, be at least 5,000 feet apart. It also requires that the signs be visible from one direction and that each message must appear at least ten seconds before another message can be posted or displayed on the billboard.

The City of Lincoln, Nebraska, has its requirements detailed in its Municipal Code. Interestingly enough, the term used in Lincoln is 'electronic changeable copy sign'. Such a sign is defined in Chapter 27.69.020 as 'a sign containing a computer or digital software

Plate 32 Digital billboard in Las Vegas, Nevada
Photo by author

message or other automated or remote method of changing copy'. Additional regulations cover such issues as how many of the signs are allowed, the minimum distance between signs, outright restrictions in some districts, how many other signs have to be removed to create one, and the condition for allowing them in other districts.

The City of Omaha, Nebraska, has developed similar regulations for electronic off-premises signs. Omaha's Code of Ordinances, Chapter 55, Article XVIII, Division 2, Section 55-851 defines the sign as:

Any off-premises advertising sign which by electronic means provides for a changing, moving or otherwise animated message. Such a sign may include information of general community interest and any type of commercial messages but shall not be considered an electronic information sign. Digital and LED type billboards are examples of an electronic off-premises sign.

Electronic off-premises signs are allowed only in certain zoning districts. Regulations on setbacks and spacing of the signs can be found in the regulations as well as performance standards and replacement rations of existing signs. Section 55-855(a) dictates that 'no advertising company may have more than eight percent of its total aggregate conforming sign locations with electronic off-premises signs (in other words, number of conforming sign locations (not faces) multiplied by 0.08)'. For example, Section 55-855(e) states that in other permissible zoning districts, 'in order to erect one electronic sign in one of the aforesaid "other permissible zoning districts," at least four existing signs, which total at least four times the square footage of the electronic off-premises sign being installed, must be removed, which may include a sign being replaced'.

Salt Lake City, Utah, offers a recent example of what is happening with digital billboards. In 2011, Salt Lake City issued a ban on new electronic billboards or the conversion of existing billboards to digital along major roadways throughout the jurisdiction. In 2012, the city passed a temporary ordinance preventing any existing billboard from being converted to an electronic billboard.

The powerful billboard lobby countered this action and actions in other jurisdictions by exerting its influence in the State Legislature. It took the fight directly to the state level. A bill was introduced in the legislature to prevent municipalities in Utah from enacting or enforcing restrictions on billboard owners. Influencing a state measure that would override municipal regulations would be easier to handle or to influence than dealing with the sheer number of municipal policies and regulations. Ultimately, the bill stalled in the legislature and no action was taken. Other states, such as Arizona and Missouri, have witnessed similar tactics and passed legislation to override local regulations. In these two states, legislation to override local regulations was passed at the state level but vetoed by the governor. Arizona Governor Janice K. Brewer, in her March 18, 2012 letter to the Speaker of the Arizona House of Representatives, vetoed a bill that 'allows electronic billboards capable of changing messages to be placed in public rights-of-way along state highways, and sets standards for display transition times.' Recognizing the need to update Arizona's current outdoor advertising laws, Governor Brewer hopes that a balance or compromise between the two competing sides can be achieved. There were not enough votes to override the governors' vetoes.

Rural signs

Billboards are the art gallery of the public.
(B. L. Robbins, President, General Outdoor Advertising Company)

With rural signs, the focus of the debate is on the location of billboards in the open countryside alongside major roads and, to a lesser extent, in commercial areas. (There is relatively little controversy about the undesirability of billboards in residential areas though, as we shall see, there is a distinction to be drawn between on-site business signs and freestanding advertisements.) Billboards along highways and in rural areas have been objected to on aesthetic, safety, and other more ingenious grounds. Among the latter is the argument that the regulation of billboards takes away only that value which is created by the building of the road from which the billboard can be seen. Thus the erection of a billboard takes for private gain the value of an opportunity created by public expenditure. In New York, the state erected a screen on public land to hide a dangerously sited billboard. In 1932, the court upheld this action, claiming that no owner had a vested right for his billboard to be seen from the road.

A few states, such as Vermont, Hawaii, Maine, and Alaska, have completely banned rural billboards. It is interesting to note that each of these states prides itself in the state's natural and scenic beauty. In some states, existing billboards can be amortized without compensation, but this policy has been affected by federal legislation concerning highways. Two years after the commencement of the building of the federal interstate highway system, the Federal-Aid Highway Act of 1958 (the 'Bonus Act') provided for a voluntary program under which states could enter into an agreement with the federal government on the control of outdoor advertising within 660 feet of the edge of interstate highways. The incentive was a bonus federal grant of 0.5 percent of the construction cost of the highway project. The legislation provided for the prohibition of most off-premise signs, and some controls over on-premise signs. Later amendments exempted from control certain parts of the system: (a) areas that had been zoned or were in use for industrial or commercial purposes in September 1959, and (b) older rights of way that were incorporated into the interstate system.

Only half the states took advantage of this scheme. Three states used the power of eminent domain to eliminate non-conforming signs; seven used a combination of eminent domain and police power controls; and the remainder used police power controls alone. Six of the latter were challenged in court, but in only one case was the action declared unconstitutional: this was the highly conservative Georgia court (Floyd 1979b: 116).

The 'Bonus Act' was repealed and replaced by a more elaborate system introduced by the Highway Beautification Act of 1965 (sometimes known as the Lady Bird Johnson Act) that, in President Johnson's words, would bring about a new approach to highway planning (see Box 17.4). According to Section 131(a), the Act seeks to control outdoor advertising signs, displays, and devices adjacent to the interstate highway system and the primary system in order 'to protect the public investment in such highways, to promote the safety and recreational value of public travel, and to preserve natural beauty'. The reality bore little relation to the rhetoric. The lofty intentions of the Act were assailed by the billboard lobby and, instead of a system of effective control over roadside advertising signs – and also junkyards – a vast number of signs were in fact removed from control.

The Act was intended to make billboard control mandatory in all the states, and to extend the controls to major roads in addition to the interstates. The provisions of the Act were made mandatory (with a withdrawal of 10 percent of federal highway funds from states which did not comply), but the provisions themselves were emasculated by the efforts of the billboard lobby. Though new off-site signs are limited to commercial and industrial areas, the actual controls in these areas (which include unzoned commercial and industrial areas) are minimal. The controls are agreed between the federal government and the individual states, but there are no *national* standards: the criteria for control are based on state law and 'customary use'. On-premise signs are totally exempted from control: hence the extremely high signs that are exhibited by gas stations close to the interstates.

The biggest victory for the billboard lobby, however, was the introduction of mandatory compensation for the removal of non-conforming signs. This precluded the elimination of billboards by amortization – a favorite technique among anti-billboard communities. The provision was extended in 1978 to require compensation for the removal of billboards under any legislation (not solely under the federal Act). This constitutes a boon to owners of obsolete and abandoned signs, who can off-load them onto the states and receive compensation!

A major problem here, as in the whole of this area, is that federal funds have been very small; as a practical result of this, many states have used all their funds for acquisition of signs voluntarily surrendered by their owners. A report by the US Department of Transportation (1984: 8) on the operation of the Highway Beautification Program in Florida and Alabama notes that:

> These voluntary sales resulted in many spot purchases from areas where other signs remained.

BOX 17.4 HIGHWAY BEAUTIFICATION

In a nation of continental size, transportation is essential to the growth and prosperity of the national economy, but that economy, and the roads that serve it, are not ends in themselves. They are meant to serve the real needs of the people of this country. And those needs include the opportunity to touch nature and see beauty, as well as rising income and swifter travel. Therefore, we must make sure that the massive resources we now devote to roads also serve to improve and broaden the quality of American life.

Source: President Lyndon Johnson

Federal Highway Administration officials generally believed that the only signs acquired under the program were those that were no longer economically beneficial to their owners. The remaining non-conforming signs, presumably of value to the owners, are still visible to the traveling public, and little or no benefit can be seen from the spot purchases.

The restriction of billboards to commercial and industrial areas is a much more limited provision than appears at first sight. Many municipalities (eager for the property tax on billboards – meager though it is) have zoned large areas along interstate and other major highways as commercial. Moreover, an area can be regarded as commercial or industrial even if it is unzoned: all that is necessary is some adjacent activity that could be regarded as falling into one of these two land use categories. Floyd has described the ingenuity of some advertising companies (see Box 17.5). There are many similar stories. The problems are exacerbated by the widespread practice (whether permitted or not) of vegetation cutting undertaken to extend the economic life of signs, misunderstandings (whether intentional or not) between the states and the federal government, and weaknesses in the enforcement of violations. Underlying these specific points, however, is the general lack of political support for the program. Despite the removal of a large number of non-conforming billboards, the legislation is a failure, and is more a testimony to the resourcefulness and power of the billboard industry than to effective controls.

Urban signs

Sign controls in urban areas present trickier problems than those in rural areas, where protection of the character of the landscape is usually more clearly evident. We have seen sign regulations covering items dealing with the size, type, and location of signs. In residential areas, aesthetic issues more often relate to the 'harmony' or otherwise between new and existing developments. In commercial districts, the felt need to protect the view of a famous building, or mountain range, or vista can involve extensive controls, as can offensive satellite dishes. Some of these matters give rise to concerns about the infringement of the freedom of speech clause of the First Amendment, as witnessed by the following quote from *City of LaDue* v. *Gilleo*, 512 US 43, at 48:

> While signs are a form of expression protected by the Free Speech Clause, they pose distinctive problems that are subject to municipalities' police powers. Unlike oral speech, signs take up space and may obstruct views, distract motorists, displace alternative uses for land, and pose other problems that legitimately call for regulation. It is common ground that governments may regulate the physical characteristics of signs – just as they can, within reasonable bounds and absent censorial purpose, regulate audible expression in its capacity as noise . . . However, because regulation of a medium inevitably affects communication itself, it is not surprising that we have had occasion to review the

---

BOX 17.5 INGENUITY IN EVADING BILLBOARD CONTROLS

In Georgia one property owner erected a small shed in a rural area and put up a sign designating it as a warehouse. A large billboard was erected next to this 'warehouse' and the outdoor advertising firm then applied for a permit based on the area's being an unzoned industrial area. In South Carolina, a large national advertising company helped set up a small radio repair shop in a residence that happened to be located near Interstate 95, and then used this 'business' as justification to erect several large billboards.

Source: Floyd 1979b: 119

constitutionality of municipal ordinances prohibiting the display of certain outdoor signs.

On this, a distinction is frequently made between commercial and non-commercial free speech: commercial speech tends to receive less protection. The current situation (though by no means entirely clear) can be summarized simply. Most federal and state courts now reject free speech objections to sign ordinances; signs create visual problems that justify aesthetic controls. On-site signs advertising the business carried on at the site tend to be exempt from prohibitions, though they may be banned from certain areas for aesthetic reasons. Signs that are not subject to a blanket prohibition can, nevertheless, be subject to controls over their placement and size.

Architectural design review

Good design is an elusive quality that cannot easily be defined (see Box 17.6). Trying to define it is much like trying to define 'beauty'. It remains 'in the eye of the beholder' – and is an extremely subjective concept (see Plates 33 and 34). Yet, if it is to be regulated, definitions – or at least guidelines – are essential. If an owner cannot understand what is, or is not, permitted under an ordinance, there is a basic unfairness. The municipality has too broad a discretion, and there is a likelihood of arbitrary action. On the other hand, aesthetic matters cannot be set out in the detail possible in, for instance, a building code. The problem

is exacerbated by a lack of clarity as to what 'the underlying public purpose' actually is. A survey by Habe (1989: 199) concluded that while most communities with design control measures seemed to know why they wanted these, very few demonstrated a clear understanding of what was involved. There was little understanding of how controls could be translated into practice, how effective they might be in attaining objectives, and what their long-term implications could be.

One of the difficulties (as in many areas of public policy) is that there is typically more than one objective. Habe's survey of sixty-six American cities showed that, in addition to aesthetic considerations, each city had at least two other objectives unrelated to aesthetic concerns. These included general 'economic' and 'public welfare', protection against urban problems such as crime, slums, and traffic congestion, 'psychological well-being, ecological concern, historic/cultural concern, facilitating the functional aspect of community life, accommodating user need, and maneuvering migration'. The vagueness of many of these objectives is noteworthy, and common in this field.

A particularly frequent objective is the preservation of community character. This can, in practice, mean anything from the perpetuation of an architectural style to the exclusion of different social groups. Perhaps the most popular design control is the 'no excessive difference' rule. This is typically expressed in terms such as 'new buildings must reflect the existing character of the area', or 'be sensitive to existing architecture'.

BOX 17.6 THE ELUSIVENESS OF GOOD DESIGN

Short of requiring the builder to copy specific prototypes, it is impossible to legislate good design. No set of rules can anticipate all the situations and conflicts that will eventually surface, and there is a tendency that rules designed to prevent something bad will also prevent something good from happening. At best, we stack the odds against the worst and hope for the best. However cleverly the controls have been structured, designers have demonstrated an uncanny ability to technically meet every requirement and still evade the spirit of the underlying design objectives.

Source: Hedman and Jaszewski 1984: 136

Plate 33 Parking structure in Kansas City, Missouri

Photo by author

Plate 34 Horton Plaza

Photo by author

According to one community, harmony was defined as 'pleasant repetition of design elements to provide visual linkage, direction, orientation and connection of areas'. Often the concept is interpreted as similarity: 'cornice lines, openings and materials of new structure to be similar to those of adjacent buildings' (Concord, CA); or 'retaining and free-standing walls should be finished with brick, stone or concrete compatible with adjacent buildings' (Rochester, NY).

(Habe 1989: 202)

Habe comments that 'such overemphasis on similarity of design encourages the trend towards specificity of standards, including setting specific architectural styles, rather than encouraging innovative solutions from designers'.

'No excessive difference' seems to be generally acceptable, but 'no excessive similarity' is more problematic. However, it is inappropriate to be dogmatic on this issue since remarkably few cases involving architectural review have come before the courts. Indeed, relative absence of litigation is a feature of aesthetic controls. The reasons for this, though speculative, are interesting. A major factor is that developers prefer to have community support for (or at least to avoid community opposition to) their proposals. They are therefore generally willing to negotiate: after all, the issue at stake is 'only' one of design, not one of significant cost. And who wants to build, or live in, a dwelling to which neighbors are hostile? If a developer (or a developer's client) wants a dwelling that is unusual, the obvious path of least resistance is to choose a site occupied by, or being developed for, similar deviants. The lower the density, the easier it is to be different in peace.

The negotiation of good design is a striking feature of a number of control schemes. For instance, the Lake Forest, Illinois, ordinance provides for review by a five-member board before a building permit will be issued.

The board has not denied a permit in its twenty years of existence, choosing instead to negotiate with designers and developers over points of disagreement . . . The board's approach has been to seek improvement rather than censorship of design. The board is yet to be challenged through a lawsuit.

(Poole 1987: 305)

The information submitted to architectural and design review bodies may differ from area to area. In most cases, however, building and site plans submitted to the architectural and design review bodies contain information on site data, building location, building features, landscaping, and parking (including ingress and egress).

Architectural design controls involve particular difficulties in the large cities affected by successive property booms. San Francisco can be quoted as an illustration (see Plate 35). After a lengthy period of public controversy, the city enacted a series of design-related ordinances (an extract from which is given in Box 17.7). Whether the 'fancy tops' controls have proved effective in improving the skyline of San Francisco is debatable: they have certainly produced some very curious buildings (perhaps a nice case of beauty being in the eye of the beholder?). Seattle has similar, though less detailed, restrictions in the downtown area: 'the requirements limit building heights, establish setbacks to maintain light and air, and ensure designs that reduce wind-tunneling and retain views of Elliott Bay' (Duerksen 1986: 14).

Plate 35 San Francisco
Photo courtesy of Public Domain Photos
(http://www.pdphoto.org)

BOX 17.7 SAN FRANCISCO DESIGN ORDINANCE

- The upper portion of any tall building be tapered and treated in a manner to create a visually distinctive roof or other termination of the building facade, thereby avoiding boxy high rise buildings and a 'benching' effect of the skyline.
- New or expanded structures abutting certain streets avoid penetration of a sun-access plane so that shadows are not cast at certain times of the day on sidewalks and city parks and plazas.
- Buildings be designed so the development will not cause excessive ground level wind currents in areas of substantial pedestrian use or public seating.
- The city consider the historical and aesthetic characteristics of the area along with the impact on tourism when issuing a building permit.
- Building heights downtown be reduced from 700 to 550 feet (from about 56 to 44 stories).

Boston has produced a volume of *Design Guidelines for Neighborhood Housing*. This is part of an ambitious project 'to transform all of the city's vacant buildable lots into attractive and affordable housing'. The guidelines emphasize existing neighborhood character and also cover such matters as the site, 'the organization of the residences' (by which is meant 'public and private territory and views, security and surveillance, and construction materials and maintenance'), and the residence itself.

Washington, DC, recognizes the need to protect the past as well as enhancing its natural and built environments. In Chapter 7, Section 701.1 of its comprehensive plan, the district notes the overall urban design goal – 'to promote the protection, enhancement, and enjoyment of the natural environs and to promote a built environment that serves as a complement to the natural environment, provides visual orientation, enhances the District's aesthetic qualities, emphasizes neighborhood identities, and is functionally efficient' (see Box 17.8).

Remembering and learning from the past provides each city an opportunity to understand the forces that shaped its evolution. It also allows us to view the city's various 'personalities' that have been witnessed over the years. The City and County of Denver, Colorado, on its http://www.denvergov.org internet site, discusses the importance of an area's 'personalities':

But cities can and do have different 'personalities.' Understanding our urban design inheritance and the elements that are part of this personality is important in order to guide new development in ways that keep and clarify Denver's character. The

BOX 17.8 URBAN DESIGN OBJECTIVES OF THE DISTRICT OF COLUMBIA

1 Maintain and enhance the physical integrity and character of the District as the Nation's Capital;
2 Preserve and enhance the outstanding physical qualities of District neighborhoods;
3 Preserve and enhance the outstanding qualities of the natural park and waterfront areas; and
4 Respect the L'Enfant Plan so that it remains a positive guiding force for future development within the District.

Source: Chapter 7, Section 702.1, District of Columbia; Comprehensive Plan

challenge lies in applying these themes in the most beneficial ways for our residents and our image. Future developments such as the new airport, transit and the renewal of our neighborhoods and districts must fit into and enhance Denver's personality.

Portland, Oregon, has received much publicity (deservedly so) for its urban planning and design. Of particular interest is the incorporation of design into the urban planning process. According to Abbott's analysis, this came about in three stages:

> During the 1960s, design issues were raised piecemeal in response to specific projects and problems. During the 1970s, design goals were incorporated into general planning policies. In the 1980s, design considerations have become an accepted part of the regulatory planning system.
>
> (Abbott 1991: 1)

Though the City of Portland has its own particular character (which Abbott describes as its 'orientation to a moralistic political style which accepts the possibility of disinterested civic decisions'), some other cities are moving toward a similar use of external standards and comparisons, and toward the integration of design review with other planning goals for the area.

Mesa, Arizona's Design Review Board, advises its city council on various issues related to exterior design and landscaping guidelines for buildings, structures, and open space. It also makes decisions on design and architectural elements of development proposals. These decisions would examine such criteria as site layout, building elevation, landscaping plan, and parking arrangements. The City of Mesa's design guidelines are not meant to 'stifle creativity' or to focus solely on one property. According to Section 11-14-1 of the Mesa City Code, the guidelines are intended 'to assist the designer in achieving a quality design which will enhance the proposed development and the city and be compatible with adjacent land uses'.

It would, however, be wrong to give the impression that there is a widespread movement in this direction. Much of the United States has no design control (or certainly none that is apparent). There is considerable controversy on the reasonableness and effectiveness (and, despite evidence to the contrary, constitutionality) of design controls, except in areas with highly distinguishing features such as historic districts. One compromise is to have informal guidelines, but these are no substitute for legal sanctions (even when, as in the case of Lake Forest, they are held in abeyance). A telling case in point is the City of Philadelphia, where an unofficial height limit was set at the top of William Penn's hat on the City Hall. This limit operated from 1894 to 1984, when it was finally breached.

Poole (1987) maintains that design controls directed at preventing the construction of excessively different buildings violate the First Amendment. Kolis (1979: 304) argues that 'the general public welfare will be better served by recognizing the First Amendment rights of architects and their clients so that they may achieve great architecture'. Habe complains that in attempting to ensure legality (and also to achieve maximum efficiency) design controls tend to emphasize details (which are easier to define) and adopt the use of generalized conditions from a standard list. S. F. Williams (1977: 33) suggests establishing criteria similar to those for obscenity (for example, that the proposed design is 'blatantly offensive' to community standards). Poole has also argued that 'architectural designs sufficiently distasteful to cause measurable harm to a neighborhood occur so rarely (if ever) that regulations to prevent them amount to making mountains out of molehills'. Municipalities should 'get out of the role imposing majoritarian notions of tastefulness on the community at large. Tastefulness by a committee assures nothing more or less than mediocrity.' The final word can rest with a view from the science of economics: Hough and Kratz (1983) assert, on the basis of an hedonic price equation for office space in downtown Chicago, that 'good' new architecture passes the market test: 'tenants are willing to pay a premium to be in *new* architecturally significant office buildings, but apparently see no benefits associated with *old* office buildings that express aesthetic excellence'. In short, the market can be left to look after new buildings; for historic buildings 'those who value them must devise feasible non-market mechanisms so that their preferences for these buildings are revealed and their dollars are contributed'. Lake Forest is hardly likely to be impressed!

Landscape planning

There should be no doubt that with proper landscaping the attractiveness of an area is enhanced. With proper landscaping come aesthetic, economic, and environmental/ecological benefits. On the aesthetic side, it may improve the physical appearance of the area as well as enhancing the character of the area. It can also serve as a screen from undesirable views like dumpsters and automobile parking lots. Proper landscaping can also lead to improving property values as well as being one factor in attracting businesses to an area. Landscaping can also serve as a noise buffer, an air filter for dust, and help reduce soil erosion. The City of Wichita, Kansas, in its City Code Chapter 10.32.010, acknowledges the multiple benefits from landscaping by indicating 'properly established and maintained, landscaping can improve the livability of neighborhoods, enhance the appearance of commercial areas, increase property values, improve relationships between non-compatible uses, screen undesirable views, soften the effects of structural features, and contribute to a positive overall image of the community'.

Oklahoma City, Oklahoma, in its 2010 Municipal Code, Chapter 59, Article XI, Section 59-11100, acknowledges the many values of landscaping in achieving a multitude of goals:

A. Promote the enhancement of Oklahoma City's urban forest.
B. Promote the reestablishment of vegetation in urban areas for health, ecological and aesthetic benefits.
C. Provide new planting in concert with natural vegetation and careful grading.
D. Encourage the preservation of existing trees.
E. Establish and enhance a pleasant visual character and structure to the built environment which is sensitive to safety and aesthetic issues.
F. Promote compatibility between land uses by reducing the visual, noise and lighting impacts of specific development on users of the site and abutting properties.
G. Unify development, and enhance and define public and private places.
H. Provide an overall planting scheme that will:

(1) Reduce soil erosion, and the volume and rate of discharge of stormwater runoff.
(2) Aid in energy conservation by shading and sheltering structures from energy losses caused by weather and wind.
(3) Mitigate the loss of natural resources.
(4) Provide visual screens and buffers that mitigate the impact of conflicting land uses to preserve the appearance, character and value of existing neighborhoods.
(5) Provide shade, comfort, and seasonal color.
(6) Reduce glare, noise and heat.
(7) Provide greater perceptual clarity along major streets and roads by more consistent planting of properly sized street trees.

I. It is further recognized that good landscaping increases property values, attracts potential residents and businesses to Oklahoma City, and creates a safer, more attractive and more pleasant living and working environment for all residents and visitors of Oklahoma City.
J. These regulations are intended as minimum standards for landscape treatment. Owners and developers are encouraged to exceed this standard in seeking more creative solutions, both for the enhanced value of their land and for the collective health and enjoyment of all citizens of Oklahoma City.

For too many years, we have viewed water as an unlimited supply. We have, in essence abused and wasted our water. Today, areas throughout the US have recognized the need to use water more efficiently. This is especially true of areas with arid climates and those that have experienced drought conditions. Landscaping is no different and has become an important area of discussion in states and municipalities.

The concept of xeriscape landscaping has become

increasing popular in many areas. This type of land-scaping promotes water conservation and recommends choosing vegetation suitable for the area's soils and climate. It is based on seven water-conserving principles: planning and design; efficient irrigation systems properly designed and maintained; use of mulch; soil preparation; appropriate turf; water-efficient plant material; and appropriate maintenance. The 2012 Florida Statutes, in Title XXVIII, Chapter 13, Sections 373.185(1)(b) and (3)(a), have labeled this type of landscaping as 'Florida-friendly landscaping' and state that 'the Legislature finds that the use of Florida-friendly landscaping and other water use and pollution prevention measures to conserve or protect the state's water resources serves a compelling public interest and that the participation of homeowners' associations and local governments is essential to the state's efforts in water conservation and water quality protection and restoration'. To help residents at the local level, Section 373.185(3)(c) adds that 'a local government ordinance may not prohibit or be enforced so as to prohibit any property owner from implementing Florida-friendly landscaping on his or her land'.

Big-box retail stores and aesthetics

Big-box retail warehouses or superstores have become a common scene in many cities across the nation. These are stores ranging in size from 20,000 square feet to over 200,000 square feet. They are simple buildings and rectangular in shape. Many are prefabricated, warehouse-looking buildings. Many are stand-alone buildings. Some are membership stores while others do not require being a member to shop in them. Overall, they look sterile, are not considered attractive, and are considered eyesores by many residents. In addition to signs and billboards, they are considered examples of visual pollution.

Big-box stores appear to have a way of polarizing people and communities. Some cities actively court these stores. They offer goods for cheaper prices. They create jobs for residents. However, how long these jobs actually last is another question. Cities like the idea of

bringing in tax revenue and attracting customers from neighboring communities. Unfortunately, local businesses may fear the presence of competition from a big-box store will lead to a potential loss of jobs and possible business closures. Property owners may be concerned over potential declines in property values. The small stores cannot compete with these mega stores.

Areas are even attempting to condition the granting of building permits on a required economic impact analysis on the surrounding area. They are trying to determine the likely collateral impacts the store might have on neighboring businesses, jobs, wages, and local services like traffic.

The economics of big-box stores is only one side of the coin. The other side has to deal with the aesthetics of the building and how it fits into the surrounding area. One chain dropped plans to build a store in a historic Civil War battlefield in 2011. Today, cities are being more proactive in dealing with these stores. As McConnell (2004) notes, some cities have capped the allowable size of the store. They will no longer tolerate monstrous-sized stores. Other areas have rejected stores on negative aesthetic effects.

Preserving or maintaining community and town character has led to the development of design constraints and aesthetic restrictions on these stores. For example, cities may require certain facades on the building, impose color requirements on the building, require specific types of lighting, require them to be more pedestrian friendly, and require the presence of amenities such as landscaping, benches in front of the store, and the inclusion of public art in front of the store.

New Urbanism

There has been a growing discontent over how cities in the United States have developed. What have our development patterns helped to create — sprawling cities that have contributed to various types of environmental pollution, the destruction of open space, a growing reliance on the automobile, a lack of affordable housing, and the segregation of the rich and poor. The development of our cities and suburbs has also contributed to wasting energy and to an inefficient use of land.

A common theme among critics of our development patterns is that our cities have been designed for cars, not for people. The automobile has given individuals the ability to live increasing distances away from their work. It has contributed to a type of isolationism among individuals. Individuals have turned to living in suburban tract developments. These sterile developments have led to the separation of neighbors. Neighbors do not know each other. In essence, there is no sense of 'community or neighborhood'.

Many people have called for a reclaiming or return to traditional neighborhood patterns in recent years. They advocate reclaiming the walkable neighborhood. One of the most recent movements advocating such changes is called the New Urbanism. Other names or variations of this movement are Neotraditional Planning, Traditional Neighborhood Development, and Transit-Oriented Development. New Urbanism is clearly seen by many individuals as a reaction to sprawl and cities satisfying automobile needs at the expense of people.

Advocates of the New Urbanism, such as Andres Duany, Peter Katz, Peter Calthorpe, and others, feel the needs of the people should come before the automobile. They stress the need to incorporate ideas from the past into our development. They remember the virtues of a small, compact New England town with a traditional town center and mixed land uses. They seek a return to a place that is pedestrian-oriented and where there is a sense of community. Prime examples of areas incorporating these views and beliefs can be found in Florida (Seaside in Walton County and Celebration in Osceola County – see Plate 36); Maryland (Kentlands in Gaitherburg); and California (Laguna West in Sacramento County).

The Congress for the New Urbanism (CNU) is the leading proponent of these ideas. Founded in 1993, it acknowledges the problems of past development and calls for us to change how we currently view cities. Instead of continuing our current patterns of sprawl, we need to reinvest in our central cities, rethink our

strict separations of land uses, and restore our urban centers. In 1996, CNU members acknowledged the following:

> We advocate the restructuring of public policy and development practices to support the following principles: neighborhoods should be diverse in use and population; communities should be designed for the pedestrian and transit as well as the car; cities and towns should be shaped by physically defined and universally accessible public spaces and community institutions; urban places should be framed by architecture and landscape design that celebrate local history, climate, ecology, and building practice.
>
> (http://www.cnu.org/charter)

It has developed a series of principles to guide the development of public policy, development practice, urban planning, and design in its Charter of the New Urbanism. They range from principles at the regional level, city level, neighborhood level, block level, to the individual structure level. The Congress and its members recognize the importance of simplicity. They call for restoring existing urban centers through infill development and recognize that we cannot develop stand-alone solutions. They applaud diversity in people and in housing types. Members recognize that a city or town's history must be respected. The preservation and renewal of historic structures and landscapes is promoted. Members promote compact development with an integrated transportation system. A pedestrian-friendly environment is of paramount importance to an area. Open space is to be promoted and distributed throughout the area.

New Urbanism is not without its critics. A number of individuals feel the idea of mixed-income communities hasn't been achieved. Some claim the extra amenities included in developments such as Seaside and Celebration has caused the developments to be out of the price range of low- and moderate-income families. Others claim the developments are not diverse – they are primarily racially homogeneous communities occupied by white upper-class families.

Plate 36 Single-family house in Celebration, Osceola, Florida

Photo by author

Further reading

For more information on new urbanism, see: the Congress for New Urbanism (http://www.cnu.org).

Costonis (1989) *Icons and Aliens: Law, Aesthetics, and Environmental Change* is a lively, concise, illustrated account of the aspects of aesthetics listed in the subtitle. It nicely bridges the fields of law and design.

Duerksen (1986) *Aesthetics and Land Use Controls: Beyond Ecology and Economics* is a useful short monograph in the APA Planning Advisory Service series.

A collection of essays on a wide range of design review issues is edited by Scheer and Preiser (1994).

There are many books on different aspects of urban design, e.g., McHarg (1995) *Design with Nature*; Hedman and Jaszewski (1984) *Fundamentals of Urban Design*; Parolek, Parolek, and Crawford (2008) *Form-Based Codes: A Guide for Planners, Urban Designers, Municipalities, and Developers*; Steiner and Butler (2007) *Planning and Urban Design Standards*; Carmona, Heath, Oc, and Tiesdell (2010) *Public Places Urban Spaces: The Dimensions of Urban Design*.

The perspective of the practicing planner–politician is given by Barnett (1982) *An Introduction to Urban Design*.

A useful article is Habe (1989) 'Public design control in American communities'.

Jane Jacobs (1961) offers a penetrating view and critique of urban design in her classic *The Death and Life of Great American Cities*. This book and some of her works discussed the need for cities today to be built like early twentieth-century cities with compact development, mixed land uses, etc.

Floyd (1979a and b) provides good contemporary accounts of the passage of the billboard controls in two articles: 'Billboard control under the Highway Beautification Act', and 'Billboard control under the Highway Beautification Act – a failure of land use controls'. There is an extended discussion in (1979c) *Highway Beautification: The Environmental Movement's Greatest Failure*.

There has been an increasing amount of literature devoted to New Urbanism. Among the more notable works are: Calthorpe (1993) *The Next American Metropolis: Ecology, Community, and the American Dream*; Katz (1994) *The New Urbanism: Toward an Architecture of Community*; Calthorpe and Fulton (2001) *The Regional City: Planning for the End of Sprawl*; Duany *et al.* (2000) *Suburban Nation: The Rise of Sprawl and the Decline of the American Dream*; Hall and Porterfield (2001) *Community by Design: New Urbanism for Suburbs and Small Communities*; and Bressi (2002) *The Seaside Debates: A Critique of the New Urbanism*.

For additional information on 'community character' see: Pivo (1992) 'How do you define community character?';

City of Westminster (2009) *2009 Comprehensive Plan*; City of Newburyport (2001) *Newburyport Master Plan*; City of Norwalk (2006) *The Norwalk, Ohio Comprehensive Plan*.

Questions to discuss

1 What are the arguments for and against the regulation of aesthetic matters?

2 Do rural signs raise different issues for planners than urban signs?

3 In what circumstances (if any) do you consider that billboards should be banned?

4 Is 'preservation of community character' a legitimate matter for land use regulation?

5 Do you think that good design can be measured by market prices?

6 How do the principles of New Urbanism differ from earlier development?

7 How can local governments regulate 'big-box' stores?

8 What is 'community character' and why is it important?

9 Why are digital billboards different from traditional, static billboards?

18

Heritage and historic preservation

If it is the role of the planner concerned with land use patterns to understand them in relationship to the dynamics of the contemporary land market and its interplay with social and cultural values, then it is the task of the historic preservation planner to understand the evolution of those patterns over time and to assess the significance of remaining fragments. Historic preservation planning is one of several perspectives on, and public interests in, land.

Ames *et al.* 1989

Preservation and profit

Planning involves the resolution of conflicting claims on the use of land. This is particularly clear in the case of historic preservation since the nature of the conflict is so readily apparent. Typically, one party (often more than one) wants to preserve a historic structure for public enjoyment now and in the future. The other party (often one only) wants to use the site for a new use which produces a higher profit. The traditionalists use the language of culture and history; the redevelopers speak in terms of market trends and economic returns.

In the last century, the controversy was normally between public and private interests. This changed as it became evident that history could be molded to produce profits and (what may amount to the same thing) a good public image. For example, there was capital to be made out of a company's environmental concerns if these were manifest in the preservation of a historic building for modern use. Further profits were to be realizable from tourist attractions. Above all, changes in tax provisions transformed the attitudes of landowners and developers to preservation.

The new enthusiasm for historic preservation was not to everyone's liking. As in other fields (national parks, for instance – which in the United States are in danger of becoming theme parks or zoos), too many people seeking to enjoy 'a piece of history' can overwhelm it and destroy the very experience which is sought. Moreover, both preservationist and developer interests have become much more sophisticated than in earlier times. The step from preserving a physical structure to preserving a community is not a large one (as the New York experience with landmark preservation clearly shows). Community groups and preservation societies can be bought out by generous contributions to their good work from developers. Preservationists sit on the boards of development companies; and their development interests in turn are to be found on the managing boards of voluntary bodies.

The old lines of demarcation have become blurred. For the planner, the situation has become confused, and frequently an apparently simple clash of development and protectionist interests turns out to be something much more complex. In this chapter, a number of these issues are discussed, but the main focus is on the evolution from a simple approach to the historic preservation of landmarks toward a 'planning perspective' on cultural matters. This perspective has now merged with a concern for urban design. On this, as Abbott (1991) has pointed out, 'design considerations have become an accepted part of the regulatory planning system'. Added to this there has been such an extraordinary expansion of the field of interest of what used to be simply called 'historic preservation' that the very term is now of vintage stock.

The early days of heritage preservation

Historic preservation in the United States grew from the grass roots in an unorganized way. Its early development is the story of a large number of predominantly private endeavors to save individual structures or sites. Many of these failed, like the attempt to save the so-called 'Old Indian House' in Deerfield, Massachusetts, which was the last home in the town to escape the famous massacre of 1704 (demolished in 1848 because it had 'no intrinsic value'). Similarly, the John Hancock House in Boston was destroyed in 1863. Others had a near miss, such as Independence Hall, which the City of Philadelphia purchased for $70,000 in 1816. One of the most notable successes was Ann Pamela Cunningham's crusade to save Mount Vernon. The essentially indigenous character of the historic preservation movement in the United States was not changed by the occasional action of the federal government. This was restricted mainly to the acquisition of a small number of landmarks and individual park sites (such as Shiloh National Military Park in 1894, and Morristown Historical Park in 1933).

The national parks, of course, were already in the public domain and thus sites within these parks which needed public protection did not require acquisition. Public lands, in fact, were the scene of another development in historic preservation. This was the preservation of 'antiquities'. The Antiquities Act of 1906 provided for the designation as National Monuments of areas *in the public domain* which contained 'historic landmarks, historic and prehistoric structures, and objects of historic or scientific interest situated on federal lands'. It represents one of the early cornerstones of the American conservation effort. This was broadened in 1935 with the introduction of the Historic Sites, Buildings and Antiquities Act; this was aimed at fostering 'a national policy to preserve for the public use historic sites, buildings and objects of national significance for the inspiration and benefit of the people of the United States'. It called upon federal agencies to take account of preservation needs in their programs and plans and, for the first time, promoted the surveying and identification of historic sites throughout the country. This program became the base for the National Register of Historic Places some thirty years later.

These early endeavors in preservation were essentially concerned with history and cultural values, as distinct from architectural quality (though the line was sometimes blurred, as with Monticello, which had both historical and aesthetic features). At this time, buildings, structures, and sites were of appeal because of their associative and inspirational values. There was, however, an increasing concern for architectural values toward the end of the nineteenth century, neatly expressed in William Sumner Appleton's statement of purpose of the Society for the Preservation of New England Antiquities. He organized this in 1910 'to save for future generations structures of the seventeenth and eighteenth centuries, and the early years of the nineteenth, *which are architecturally beautiful or unique*, or have special historical significance. Such buildings once destroyed can never be replaced' (Hosmer 1965: 12). The added italics emphasize the primacy here accorded to architectural values. Of course, the historical and associative elements remain important today: in fact it is often difficult to disentangle them.

The Depression years were a lean time for historic preservation, although there were notable exceptions such as the creation of preservation commissions in Charleston in 1920, New Orleans (the Vieux Carré Commission) in 1925, and San Antonio in 1924. The World War II and early postwar period was even leaner; indeed, urban renewal and highway projects destroyed many buildings which a few years later might have been preserved. It was this very destruction which (together with the reaction to the sterility of the International Style in new architecture) acted as a catalyst to an unprecedented burst of activity in the mid-1950s. The culmination of this was the publication in 1966 by the US Conference of Mayors and the National Trust for Historic Preservation (NTHP) of a powerful, eloquent manifesto, *With Heritage So Rich*.

With Heritage So Rich and subsequent legislation

The report *With Heritage So Rich* had the advantage, which many reports lack, of appearing at precisely the right time for a positive political response. It was cogently argued, dramatically illustrated, and persuasive. It consisted of a series of essays and a concluding set of recommendations. Some of these were immediately implemented by the 1966 National Historic Preservation Act (NHPA); for example, the introduction of a National Register of Historic Places, and the establishment of an Advisory Council on Historic Preservation (ACHP) – see Box 18.1. The ACHP is vested with the authority to advise the president and Congress on historic preservation matters to promote the protection and enhancement of the nation's historic resources. In order to accomplish the above, the ACHP conducts a number of activities – see Box 18.2.

A number of pieces of federal legislation have included provisions for historic preservation. The Transportation Act of 1966 required the secretary of transportation to refuse approval for highway construction projects that would involve damaging or

---

BOX 18.1 ADVISORY COUNCIL ON HISTORIC PRESERVATION

The Advisory Council on Historic Preservation is composed of the heads of federal agencies whose departmental activities regularly affect historic properties; experts in historic preservation; a governor; a mayor; private citizens appointed by the president; and representatives of the National Trust and the Conference of State Historic Preservation Officers.

BOX 18.2 ACHP COUNCIL MISSION STATEMENT

1 Advance Federal historic preservation planning by ensuring that Federal agency policies and operating procedures adequately consider historic preservation laws and policies.
2 Oversee the Section 106 review process to ensure that it functions smoothly and effectively for the nearly 100,000 Federal actions requiring review annually.
3 Serve as a mediator in more than 1,000 individual cases annually, between project sponsors and local preservation interests to protect important historic resources from unnecessary harm.
4 Develop legally binding agreements in those cases among federal, state, and tribal officials and other affected parties to clearly set forth the treatment of historic preservation.
5 Provide essential training, guidance, and public information to make the Section 106 review process operate efficiently and with full opportunity for citizen involvement.
6 Recommend administrative and legislative improvement for protecting the nation's heritage with due recognition of other national needs and priorities.

Source: ACHP 2000: 59

demolishing historic sites unless there is 'no prudent and feasible alternative'.

A similar provision was included in the Model Cities Act 1966 in relation to urban renewal plans. Later amendments extended this policy to all federal departments. Changes in taxation were made by other legislation, such as the Tax Reform Act 1976 and the Economic Recovery Tax Act 1981, to encourage historic preservation (for example, by way of tax deductions for rehabilitation). A separate act, the National Environmental Policy Act (NEPA), included additional provisions for preserving 'important historic, cultural, and natural aspects of our national heritage'.

As this brief summary demonstrates, the fifteen years following the publication of *With Heritage So Rich* witnessed a veritable orgy of legislative activity. In the following pages, the more important features of this are discussed.

The National Register of Historic Places

The National Register of Historic Places (NRHP) is maintained by the Keeper of the National Register in the National Park Service of the Department of the Interior. It lists districts, sites, buildings, structures, and objects which are significant on a national, state, or local level in American history, architecture, archae-ology, engineering, and culture: in short, America's cultural resources (see Plate 37). Cultural resources management, a branch of archaeology, deals with the resources of past human activities and allows us to learn about our past. The concept of 'significance' is key to understanding our nation's cultural resources. Properties and objects are evaluated for inclusion into the National Register by such criteria as being associated with events that are significant in our history; being associated with the lives of famous individuals that are significant in our past; possessing distinctive characteristics of a period or the work of a master; and yielding important information in our history.

Since its creation in 1966, over 80,000 properties, such as the Alamo (see Plate 38), the Empire State Building, and the Gateway Arch, have been listed in the National Register. These are provided with a degree of protection from the harmful effects of federal action. The federal government is committed, by law, to protect these resources: agencies are required to follow a statutory process of review and consultation with the ACHP in connection with any undertaking affecting properties included in the list. Additionally, and at first sight curiously, this requirement extends to properties which, though not listed, are *eligible* for listing. (The rationale for this is summarized in Box 18.3.) Though both listed and eligible properties are subject to the review process, only listed properties are qualified to receive grant aid or tax advantages (discussed below). The process (popularly known by its

---

BOX 18.3 LISTING AND ELIGIBILITY

Before 1980, owners had no right of objection to listing. The rationale for this was:

1 The Register is a list of properties that meets an objective evaluation, which applies criteria and professional standards regardless of a current owner's opinion of the property;
2 The owner's opinion has no bearing on whether a property is historic; and
3 Inclusion in the Register does not directly restrict a private owner's use of his property in any manner.

However, as a result of political pressures, the right to object was introduced in the 1980 Amending Act. As a compromise, the concept of 'eligibility' for inclusion in the Register was added.

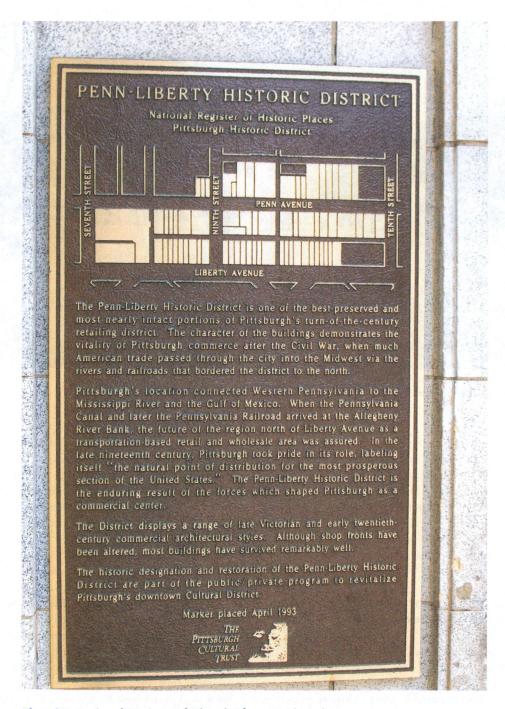

Plate 37 **National Register of Historic Places Designation**

Photo by author

Plate 38 The Alamo
Photo by author

legal reference as the 'section 106 process') is not a mere formality: all federal actions and federally funded projects are monitored or reviewed by preservationists. This usually occurs at the State Historic Preservation Office (SHPO) level. Indeed, 'review and compliance', as it is called, now occupies a dominant position in the state programs. However, as always, much depends upon the quality of the local administration.

Statewide comprehensive historic preservation planning

State governments play a significant role under the NHPA of 1966. Each state develops and administers a State Historic Preservation Plan. Each state conducts a comprehensive statewide survey on historic properties and maintains a roster of all properties. States nominate properties to the NHPA. They also help the federal

government administer financial assistance programs for property owners and carry out various public administration and training programs on all facets of historic preservation. Finally, states assist local governments to develop local historic preservation programs. This is where a great deal of activity occurs.

In 1980, the NHPA was amended to require each state to have a procedure for local governments to be certified for participating in national historic preservation programs. In this case, local government includes towns, cities, and counties. The procedure was known as a Certified Local Government (CLG). The states help accomplish the national goals and the local governments help the state accomplish the state goals. As such, a partnership between the various levels of government emerges.

For a local government to become a CLG, it must have a state-approved program for recognizing and protecting an area's historic, archaeological, and

BOX 18.4 REQUIREMENTS FOR LOCAL GOVERNMENTS TO BECOME CLGS

(a) Enforces appropriate State and local legislation for the designation and protection of historic properties;

(b) Has established an adequate and qualified historic preservation review commission by State and local legislation;

(c) Maintains a system for the survey and inventory of historic properties that furthers the purposes of subsection (b) of this section;

(d) Provides for adequate public participation in the local historic preservation program, including the process of recommending properties for nomination to the National Register; and

(e) Satisfactorily performs the responsibilities delegated to it under this subchapter.

Where there is no approved State program, a local government may be certified by the Secretary if he determines that such local government meets the requirements of subparagraphs (a) through (e); and in any such case the Secretary may make grants-in-aid to the local governments for purposes of this section.

Source: 16 USC 470 (c)(1)

---

architectural resources (see Box 18.4). The secretary of the interior and the SHPO must certify that the local government has fulfilled certain responsibilities. States provide localities with technical assistance during the application process and continue to provide them with additional types of assistance once they become certified. They are also required to set aside at least 10 percent of the funds they receive from the federal historic preservation fund to give to CLGs.

The section 106 process calls upon the federal government to consider the effects of its actions on historic properties. The 1992 Amendments to the NHPA of 1966 and the Clinton administration's desire to reinvent government led to the development of the new rules. The ACHP sought to improve and streamline how the section was being implemented. The new rules, which became effective on January 11, 2001, dealt with such issues as the roles of the SHPO, Indian tribes, and tribal historic preservation officers. It revised the role of invited signatories, gave more flexibility to involve applicants, and modified documentation standards for federal agencies.

Though historic preservation is very much a local matter, it is more than this: as with all local plans, relationships with wider plans have to be forged. (The imperative is misleading since, in practice, as has already been stressed, there are so few *plans* – as distinct from zoning provisions.) Ideally, these would include such functional elements as transportation planning, economic development planning, and environmental planning. The most promising approach is where different planning agencies integrate (or at least cooperate in) their planning processes. In the words of the Delaware Comprehensive Historic Preservation Plan:

> It is very difficult, if not impossible, to integrate complete plans that can translate the recommendations of one plan or functional area into terms relevant to another. Plans must be integrated, or information exchanged, at the points in the planning process when problems and alternative goals are defined and analyzed and decisions made.
>
> (Ames *et al.* 1989: 9)

Moreover, without coordination, historic preservation policies may conflict with land use policies. In his *Handbook on Historic Preservation Law*, Duerksen quotes the case where

preservationists have struggled to enact an ordinance to control design details or forbid demolition by private developers in a historic neighborhood, only to discover that the real threat in the area is a city zoning policy encouraging high-rise development. In short, preservationists have focused on design issues and on saving threatened buildings when the key issue is more often how landmarks and their surrounding areas will be developed according to local zoning classifications and redevelopment programs.

(Duerksen 1983: 44)

Coordination has other advantages, not the least being that it impresses courts that the municipality has a comprehensive plan, and is working to this rather than making a series of ad hoc decisions. It also facilitates the use of sophisticated zoning techniques such as incentives, bonuses, and the transfer of development rights.

Coordination is also desirable between the policies of the municipality and the state. A well-known example is Oregon's statewide planning goals, which are mandatory on municipalities. One of these goals includes the requirement that local programs shall be provided which will 'protect scenic and historic areas and natural resources for future generations, and promote healthy and visually attractive environments in harmony with the natural landscape character'. Inventories are required of historic areas, sites, structures, and objects; and cultural areas. A historic area is defined as 'lands with sites, structures, and objects that have local, regional, statewide, or national historical significance'. A cultural area is 'an area characterized by evidence of ethnic, religious or social group with distinctive traits, beliefs, and social forms' (Rohse 1987: 261). Local comprehensive plans and land use regulations are required by statute to comply with these goals.

Highways and historic preservation

The ravages of highway construction constituted one of the major reasons for the swell of public opinion against 'the federal bulldozer'. It is therefore perhaps fitting that the strongest federal provision is to be found in a transportation act. The Department of Transportation Act declares that it is a matter of national policy that a 'special effort' shall be made to preserve and enhance the natural beauty of lands crossed by transportation lines (Section 1653(f); commonly referred to (by a previous numbering) as Section 4(f)).

The scope of the provisions of the Transportation Act is much broader than that of the NHPA: it gives protection to any site considered by officials to be of historic significance – not only those listed, or eligible to be listed, on the National Register. Moreover, whereas the NHPA merely gives the ACHP opportunity for 'comment' on harmful action, the Transportation Act's more stringent provisions permit harmful use only if two conditions are met: first, that no feasible and prudent alternative exists, and second, that all possible planning is carried out to minimize harm. The courts have held that there must be 'truly unusual factors' of 'extraordinary magnitudes' for this high standard to be met.

Section 4(f) has been used in relation to a wide variety of historic sites, buildings, and objects, from the Vieux Carré (French Quarter) in New Orleans and the childhood home of Thomas Jefferson, to Hawaiian petroglyphic rocks, a truss steel bridge, Indian archaeological sites, and many others. The section also applies to privately owned historic sites as well as those in public ownership. Indeed, most of the properties listed in the National Register of Historic Places are privately owned.

Other legislation dealing with specific modes of transportation has similar provisions, e.g., the Airport and Airway Development Act 1970, the Federal-Aid Highway Act 1968, and the Urban Mass Transit Act 1976. More recently, the Transportation Equity Act for the Twenty-First Century (TEA-21), enacted in 1998, provides for funding under Section 1224 for the rehabilitation of historic covered bridges (listed or eligible for listing on the NTHP) and for educational and research programs for historic covered bridges. The legislation also includes provisions for scenic and historic highway programs, rehabilitation and operation of historic transportation facilities, and archaeological planning and research.

The National Environmental Policy Act

The National Environmental Policy Act of 1969 (NEPA) establishes a national policy of environmental protection. It requires all federal agencies to take into account the effects of their actions on the environment. This includes the built environment. The historic preservation element of this refers to the preservation of 'important historic, cultural, and natural aspects of our national heritage' and the maintenance, wherever possible, of 'an environment which supports diversity and variety of individual choice'. The legislation requires every 'major federal action' which 'significantly affects the environment' to be preceded by an environmental impact statement. This must contain a detailed analysis of the environmental impact of the proposed action, any adverse environmental effects that cannot be avoided if the proposal is implemented, and alternatives to the proposed action. (For further discussion see Chapter 12.)

There is some overlap between NHPA and NEPA (and the environmental protection acts passed by several states), and regulations have been issued by the Council on Environmental Quality in relation to co-ordination. The two Acts can be seen as reinforcing each other:

> The two laws reinforce each other and can be used effectively in tandem: if NHPA does not apply to a historic resource, NEPA might. While some courts may hold that agencies need not continue to comply with NHPA after a federal project has commenced, courts have generally agreed that NEPA does apply in such situations. If NHPA is weakened through funding cuts and revisions to the federal regulations to the ACHP, NEPA can still be used to compel agencies to consider historic properties.
> (Duerksen 1983: 305)

Nevertheless, federal acts provide no guarantee that cultural resources will be protected: the only means which guarantees protection is acquisition.

Economics of historic preservation

Taxation provisions often work against sectoral policies: they are typically designed to raise revenue, not to further public objectives. So it has been with historic preservation. Prior to 1976, tax laws actually discouraged the preservation and rehabilitation of historic properties since tax deductions were allowable for the costs of demolition. The 1976 Tax Reform Act created a number of preservation incentives: tax credits for certain rehabilitation expenditures and (as a disincentive to the demolition of historic buildings) an increase in the 'tax cost' of demolition. The Historic Preservation Tax Incentives program is administered by the Department of the Interior's National Park Service. The programs offer multiple benefits:

> The incentives have been instrumental in preserving the historic places that gave cities, towns, and rural areas their special character, and have attracted new private investment to historic cores of cities and towns. The tax incentives also generate jobs, enhance property values, create affordable housing, and augment revenues for federal, state, and local governments. Through this program, abandoned or underutilized schools, warehouses, factories, churches, retail stores, apartments, hotels, houses, and offices throughout the country have been restored to life in a manner that maintains their historic character.
> (US Department of the Interior, National Park Service 2011b: 1)

The use of preservation tax incentives increased enormously after the passing of the Economic Recovery Act of 1981, which introduced a new, and highly attractive, system of tax credits. By the mid-1980s, the program was running at an annual rate of 3,000 projects and an investment of $2 billion. Reagan's Tax Reform Act of 1986 drastically cut these incentives from 25 percent to 20 percent but, even so, some 1,000 certified rehabilitations a year (involving $900 million of investment) were undertaken in the late 1980s. Between 1976 and 1989, a total of some 21,000 historic buildings were rehabilitated with the aid of tax incentives, representing private sector investment of almost $14 billion (Blumenthal and Siler 1990: 1). A 10 percent tax credit is also available to property owners seeking to rehabilitate a non-historic, non-residential building constructed prior to 1936. In total, according to the US Department of the Interior, National Park Service (2011b: no page), program accomplishments from 1977 to 2011 resulted in the following: 38,075 historic rehabilitation projects certified; $62.94 billion in rehabilitation investment; 231,486 rehabilitated housing units; 209,913 new housing units; and 117,975 low- and moderate-income housing units.

In addition to the tax incentives program, the NHPA of 1966 provided matching grants to the states for historic preservation survey, planning, acquisition, and development. With the funding cuts made in the 1980s, little acquisition and development is being carried out, but most states continue to use grant funds for 'survey and planning' – which includes the preparation of nominations to the National Register, and developing technical preservation information.

States provide additional tax and other incentives. These vary considerably among the states, and take many forms. Many states offer tax credits that can be used in combination with federal historic credits. Maine offers a state tax credit program for projects qualifying for the federal tax credit program. New Mexico offers a program to encourage the restoration, rehabilitation, and preservation of cultural properties that are listed on, or listed as contributing to, a State Register of Cultural Properties Historic District. Missouri offers a tax credit program for non-income-producing residential properties as well as income-producing properties that have been registered as historic. When viewed together, there are, however, basically six taxation methods used to encourage historic preservation: exemption, credit or abatement for rehabilitation, special assessment for property tax, income tax deductions, sales tax relief, and tax levies.

Subsidies of this nature recognize the public interests in – and benefit from – the preservation of historic buildings. Nevertheless, private owners of historic buildings may have to carry financial burdens. These may be in the form of maintenance costs which are not covered by the income from the property, or in the form of the forgone higher profits from redevelopment. Generalization is difficult, but the constant battles to preserve buildings from demolition and redevelopment point to the frequency with which owners see redevelopment as being more profitable.

A number of old buildings have found new uses. Under what is commonly known as 'adaptive reuse', there are numerous examples of buildings which have been successfully converted into new uses. These range from adaptations of an old market place to a festival market place, an old textile mill to condominiums and offices, to adaptation of an elevated rail line to a public park (see Plates 39 to 41). A rather common sight in many cities is the conversion of old Victorian homes into shops or professional offices. Such schemes can be highly profitable as well as (and because they are) highly popular. But each site has problems and opportunities which are site specific. There are many cases where a site cannot be restored without exorbitant cost and is also of insufficient community value to warrant public subsidy. There are also sites which are more economically valuable with a rehabilitated historic building than they would be after redevelopment. As such, historic preservation efforts can provide substantial benefits to a community. As the Virginia Department of Historic Resources stated:

> Development decisions to reuse buildings benefit us greatly. Rejuvenating already developed areas conserves remaining open space, eliminating outward expansion and the need to create new and expensive infrastructure. Through state and federal tax credits, historic resources become engines for community revitalization. They preserve the integrity and vitality of the community, creating a magnet for new investment. Careful stewardship of historic resources creates communities with a strong sense of identity and place. That identity makes local heritage real and meaningful for the people who live and work there and who travel to visit.
>
> (Virginia Department of
> Historic Resources 2000: 1)

Although it is hard to quantify the value of historic preservation by simply using its dollar impact (Box

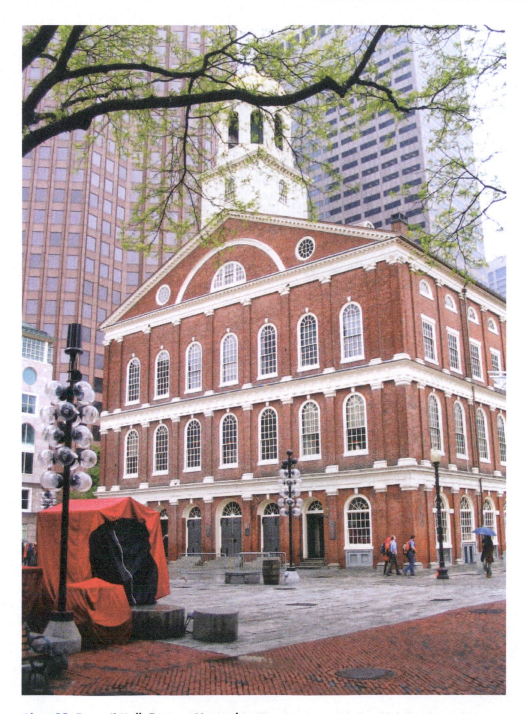

Plate 39 **Faneuil Hall, Boston, Massachusetts**
Photo by author

Plate 40 **Adaptive reuse in Lawrence, Massachusetts**

Photo by author

Plate 41 **High Line Public Park, New York**

Photo courtesy of Evan Wasserman

BOX 18.5 DIFFICULTY IN USING ONE NUMBER TO ASSESS DOLLAR IMPACT OF HISTORIC PRESERVATION

Nonetheless, the dollar impact of historic preservation on the North Carolina economy cannot be reduced to a single number. Why? Because today North Carolina is using historic resources not as an end, but as a means. A restored storefront on Main Street is historic preservation, but it is also downtown revitalization. A visitor to Baltimore represents tourism dollars, but also historic preservation dollars. A location shoot in a historic neighborhood reflects movie industry impact, but also historic preservation impact. The reinvestment in Edenton's mill village is community preservation, but at the same time historic preservation.

Source: Rypkema 1998

Plate 42 **Air rights development**
Photo by author

The sale of air rights or the transfer of development rights (TDR) affords owners of historic properties and sites in competitive real-estate markets the opportunity to sell the development rights of the property to another private entity (see Plate 42). In other words, a TDR program represents a means of separating the development potential of a site from its title. The property owner maintains title to the property but not the ability to develop it any further. This area is permanently restricted from any future development. The original property can be referred to as the 'sending area', while the development rights are sent to a 'receiving area' where more development can be handled (see Figure 18.1). Local and county governments

18.5), historic preservation is a good investment for communities. As noted in the Executive Summary of Mason's report (2005) *Economics and Historic Preservation*, 'designating a landmark or district as historic typically maintains if not boosts the value of property, and as an economic development tool, historic preservation has proved its worth. Nearly any way the effects are measured, be they direct or indirect, historic preservation tends to yield significant benefits to the economy.'

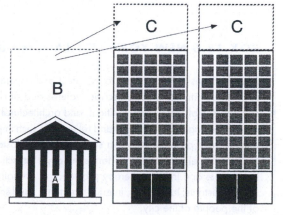

Figure 18.1 **Transfer of development rights**

throughout the United States have also used TDR programs to protect farmlands and open space and have even purchased the development rights (PDR) from landowners to protect farmlands and open space from being destroyed (see Chapter 10).

To preserve the historic building (A), the development rights above it (B) are transferred to the nearby two buildings (C) which are thereby permitted an increase in development density. The owners of the latter purchase the development rights from the owners of the historic building, who thus obtain funding for preservation.

State and local programs

To help implement the goals of the NHPA of 1966, all states have a State Historic Preservation Office (SHPO). This is a federal requirement, and the secretary of the interior has the responsibility of approving state programs that provide for the designation of an SHPO, a state historic preservation review board, and a scheme for adequate public participation in the state program. Each SHPO is required to identify and inventory historic properties in the state; nominate eligible properties to the National Register; prepare and implement a statewide historic preservation plan; serve as a liaison with federal agencies on preservation matters; and provide public information, education, and technical assistance. For example, the Virginia

Department of Historic Resources is the Virginia SHPO. Its mission is 'to foster, encourage, and support the stewardship of Virginia's significant historic, archaeological, and cultural resources'. The North Carolina SHPO 'assists private citizens, private institutions, local governments, and agencies of state and federal government in the identification, evaluation, protection, and enhancement of properties significant in North Carolina history and archaeology'. It is housed in the Division of Archives and History, North Carolina Department of Cultural Resources.

As in so many areas of land use planning, it is at the local level that most of the real action takes place.

In 1956, New York State became the first to pass legislation enabling municipalities to enact an ordinance for individual landmark buildings (as distinct from historic areas). New York City was the first to take advantage of this (see Box 18.6). A New York City Landmarks Commission was established in 1965 and empowered to designate properties of significant historic or aesthetic value. The commission can designate individual (exterior) landmarks, interior landmarks, and scenic landmarks. It can also designate historic districts. Designated properties cannot be demolished or altered without the approval of the commission. This is given only if the commission decides that the proposed work will have no effect on the protected architectural features, is otherwise consistent with the purposes of the landmarks law, or is necessary to secure a reasonable return to the owner (assessed at 6 percent).

BOX 18.6 NEW YORK CITY LANDMARKS LAW

[The purposes of the act are] to (a) effect and accomplish the protection, enhancement and perpetuation of such improvements and landscape features and of districts which represent or reflect elements of the city's cultural, social, economic, political, and architectural history; (b) safeguard the city's historic, aesthetic, and cultural heritage, as embodied and reflected in such improvements, landscape features and districts; (c) stabilize and improve property values in such districts; (d) foster civic pride in the beauty and noble accomplishments of the past; (e) protect and enhance the city's attractions to tourists and visitors and the support and stimulus to business and industry thereby provided; (f) strengthen the economy of the city; and (g) promote the use of historic districts, landmarks, interior landmarks, and scenic landmarks for the education, pleasure, and welfare of the people of the city.

Until the *Penn Central* case was settled by the Supreme Court in 1978, there was some doubt as to the constitutionality of such legislation. This case involved a proposal for the erection of a fifty-five-story office building atop the city's beaux-arts masterpiece, the Penn Central Railroad's Grand Central Terminal: a building which the city had designated as a landmark.

The New York City Landmarks Commission rejected the proposal, and the owners took the matter to court with two complaints. First, they argued, there had been, in effect, a taking of their property without just compensation. Second, by designating the terminal, the city had discriminated against the owners in requiring them to bear a financial burden which neighboring owners did not have to shoulder.

In a six-justice majority, the Supreme Court upheld the action of the Landmarks Commission. Though previous decisions had provided no clear rule for determining whether a taking has taken place, in this case it was determined that there was no taking: the owners had been left with a reasonable return, and the restrictions imposed were within the police powers of the city. It was explicitly stated that 'states and cities may enact land use restrictions or controls to enhance the quality of life by preserving the character and desirable aesthetic features of a city'.

The importance of the *Penn Central* case is underlined by Duerksen (1983: 19): '*Penn Central* made it clear that localities could forbid demolition or stop new construction for preservation purposes. Thus, a landowner who did not understand local preservation law could face serious economic consequences.' Together with the federal tax incentives introduced in 1976 (which provided landowners with significant benefits), historic preservation law suddenly emerged as a subject 'worth studying and practicing, just as environmental law had almost a decade earlier'.

The result was a major increase in historic preservation activity. Though there was some setback with the financial cuts imposed by the Reagan administration, historic preservation was clearly at the stage of becoming established as a significant land use control.

An additional note about the New York Landmarks Commission is appropriate. A major factor in its establishment was the widespread concern about the destruction of Pennsylvania Station in 1963. It was a curious twist of fate that made another railroad terminal (Grand Central Terminal) the subject of a case which confirmed the legitimacy of the commission and its functions. There have, however, been constant rumblings about the way in which these functions have been carried out. In particular, the commission has been accused of acting 'as a kind of planning commission of last resort, stepping in to prevent or slow the pace of development in circumstances in which the planning commission had failed to act'. This difficulty is to be anticipated when land use planning and historic preservation planning are administered separately. But separate administration is normal; and whether the outcome is regrettable or desirable can be very much a matter of opinion.

Any level of government (federal, state, or local) can designate an area as a historic district. By definition, a historic district can be defined as 'a geographically definable area, urban or rural, possessing a significant concentration, linkage, or continuity of sites, buildings, structures, or objects united by past events or aesthetically by plan or physical development' (36 CFR 60, revised as of July 1, 2004). A district may also comprise individual elements separated geographically, but linked by association or history.

For example, Pioneer Square, the site of Seattle's original downtown, was designated as a national historic district and as a local preservation district in 1970 – Pioneer Square Preservation District (see Box 18.7). The threat of demolition through urban renewal was a prime factor in its establishment of a historic district. The area comprises six blocks in downtown Seattle and is protected by regulations designed to protect the area's historic and architectural character. A citizen's board of nine citizens, appointed by the mayor and confirmed by the Seattle City Council, with one of the members being a resident of Pioneer Square, serves as a public review body for architectural alterations within the confines of the district, and makes recommendations to the director of the Department of Neighborhoods. The body can decide to recommend approval of the request, approve the request with conditions, or recommend denying the request.

BOX 18.7 PIONEER SQUARE PRESERVATION DISTRICT – CREATION OF DISTRICT, LEGISLATIVE FINDINGS AND PURPOSE

During the City of Seattle's relatively brief history, it has had little time in which to develop areas of consistent historical or architectural character. . . . To preserve, protect, and enhance the historic character of the Pioneer Square area and the buildings therein; to return unproductive structures to useful purposes; to attract visitors to the City; to avoid a proliferation of vehicular parking and vehicular-oriented uses; to provide regulations for existing on-street and off-street parking; to stabilize existing, and encourage a variety of new and rehabilitated housing types for all income groups; to encourage the use of transportation modes other than the private automobile; to protect existing commercial vehicle access; to improve visual and urban relationships between existing and future buildings and structures, parking spaces and public improvements within the area; and to encourage pedestrian uses, there is established as a special review district, the Pioneer Square Preservation District.

Source: Seattle Municipal Code, 23.66.100

The Vieux Carré Historic District, known as the French Quarter, in Louisiana is the site of the original French settlement in New Orleans and is the second oldest Historic Preservation District in the United States, designated in 1965 as a National Historic Landmark. It comprises an eighty-five-block area in downtown New Orleans. Fortunately, the French Quarter was spared from the major devastation that accompanied Hurricane Katrina in 2005 due to the fact that it sits on land above sea level. Some streets did receive minor flooding and some buildings were subjected to high wind damage. The French Quarter remains one of the nation's major tourist destinations.

The Vieux Carré Commission cares for the district. The nine-member volunteer commission is appointed by the mayor of New Orleans with the consent of the city council. Membership is allotted in the following fashion – three members from a list provided by the local chapter of the American Institute of Architects; one member from a list provided by the Louisiana

BOX 18.8 USE OF DESIGN GUIDELINES – INTRODUCTION

The Design Guidelines for Alexandria's regulated historic districts are a tool to assist the general public and Boards of Architectural Review in the design review process. The guidelines should be viewed as a distillation of generally accepted design approaches in the historic districts. The guidelines should not be viewed as a device that dictates a specific design response nor should the guidelines be viewed as prohibiting a particular design approach. There may be better ways to meet some design objectives that have not been reviewed by the Boards in the past. New and untried approaches to common design problems should be rejected out of hand simply because they appear to be outside the common practices outlined in the guidelines.

Source: City of Alexandria, Virginia, Design Guidelines for the Old and Historic Alexandria Districts, City of Alexandria, Department of Planning and Community Development, adopted by Board of Architectural Review, 5/25/93

Historical Society; one member from a list provided by the Louisiana State Museum Board; one member from a list provided by the Chamber of Commerce; and the remaining three members appointed at large. All members serve a four-year term.

The Old and Historic Alexandria Districts in Virginia represent excellent examples of historic districts. Established in 1946, it is the third oldest historic district designated in the United States. The areas of the historic district have expanded over the years. Nevertheless, structures must be consistent in age, architectural integrity, and signage with the properties currently in the districts.

Each district has a Board of Architectural Review which is vested with controlling development in the districts and preserving its colonial heritage. Each board has seven members of which two must be architects. Among the powers of the boards are approving a Certificate of Appropriateness for all new construction and exterior alterations and approving any permit for demolishing more than 25 square feet of a building or structure in the historic district. The boards do not rely on a specific design response in making their determinations on approving any Certificate of Appropriateness (see Box 18.8).

Neighborhood Conservation Districts

There are a number of distinctive areas worthy of protection and preservation that do not meet the historical, architectural, or cultural significance requirements of being classified as historic districts. These areas could be called Neighborhood Conservation Districts (NCD) or Heritage Districts. They are less regulatory than a registered historic district. While the purposes of an NCD vary from city to city, the following purposes can be found in the City of San Diego Unified Development Code, Article 3, Division 4, Section 35-335(a)(2):

a. To protect and strengthen desirable and unique physical features, design characteristics, and recognized identity and charm;
b. To promote and provide for economic revitalization;
c. To protect and enhance the livability of the City;

d. To reduce conflict and prevent blighting caused by incompatible and insensitive development, and to promote new compatible development;
e. To stabilize property values;
f. To provide residents and property owners with a planning tool for future development;
g. To promote and retain affordable housing;
h. To encourage and strengthen civic pride; and
i. To ensure the harmonious, orderly and efficient growth and redevelopment of the City.

While most NCDs focus on residential neighborhoods, there is nothing to preclude a commercial or mixed use from being designated an NCD. The ultimate goals of an NCD are to protect the key character of an area, keep the area's attributes from changing, and prevent incompatible land uses from entering the area. Cities are trying to maintain the integrity of the area.

Requests to designate an area as an NCD can come from a variety of parties. Nominations to have an area declared as an NCD can originate from a community resident, planning commission, a property owner, or the city council. NCDs can be found in such cities as: Austin, Texas; Boon, North Carolina; Cambridge, Massachusetts; Chapel Hill, North Carolina; Nashville, Tennessee; Portland, Oregon; and San Antonio, Texas.

Main Street™ program

In 1977, the National Trust for Historic Preservation launched a program called 'Main Street' which was designed to revitalize community commercial districts through the context of historic preservation. It is a sought-out program in many states. According to the Virginia Main Street Program, designated Main Street districts must have:

• At least 50 commercial enterprises and 70 structures or storefronts;
• At least 2/3rds of the structures are commercial (or commercially zoned) buildings and have a pedestrian scale and orientation including such elements as ground floor storefronts;

- At least 25 percent of the linear street frontage has a setback of 15' or less from the sidewalk;
- A compact size and regular pattern of sidewalks so that it can be comfortably walked by pedestrians.

(Virginia Department of Housing and Community Development, 2006: 8)

Using various preservation and economic development tools or strategies, it represents a powerful tool to revitalize older downtown historic commercial districts through improving the appearance and image of a district. It seeks to revitalize the downtown business district while retaining the district's traditional character. The importance of downtown revitalization can clearly be seen in the following quote from the Texas Main Street Program:

> There are many reasons why downtown revitalization is a crucial tool for enhancing the economic and social health of a community. In addition to being the most visible indicator of community pride and economic health, the historic downtown is also the foundation of community heritage. The historical buildings in a downtown are prime locations for the establishment of unique entrepreneurial businesses and can also be tourism attractors, all of which add to the community's sales tax collections and property values. Today, big-box development permeates the national landscape, making it even more important that communities be proactive in saving and using their historic spaces to avoid becoming featureless places.
>
> (http://www.the.state.tx.us/mainstreet/ msabout.shtml, accessed October 18, 2012)

To help communities interested in pursuing the Main Street program and to share the successes of earlier pilot projects, the National Trust for Historic Preservation, in 1980, created a National Trust Main Street Center. This Center's mission is to 'empower people, organizations, and communities to achieve on-going downtown and neighborhood district revitalization based upon the principles of self-determination, resource conservation, and incremental transformation represented through the comprehensive Main Street

Four-Point Approach®' (http://www. mainstreet.org) (see Box 18.9). The four steps are highly interconnected. The approach is designed to resuscitate and revitalize dying downtowns by putting life back into them. It is a tried and proven method of preservation-based commercial district revitalization. It is a locally community-based driven, funded, and operated revitalization program. Over 1,200 active programs exist in the United States. During the time period of 1980–2006, over $41.6 billion dollars have been reinvested in physical improvements from public and private sources; a net gain of 77,799 businesses; a 349,148 net gain in jobs, and 186,820 buildings have been rehabilitated (http://www.mainstreet.org).

The Main Street approach is used in large and small cities – this is a case of size really not being an issue! The Boston Main Street program, as noted in its 2003 Annual Report, is working with groups to create a renaissance in local communities, to revitalize local commercial areas, to renovate buildings, and to preserve the unique character of commercial areas in Boston. The North Carolina Main Street Center is working with communities in North Carolina to revitalize their downtowns by promoting economic development through the context of historic preservation. Since 1980, the program has created thousands of jobs for residents, renovated numerous buildings, bolstered existing businesses, and attracted new businesses to the downtowns. The State of Kentucky, through its Main Street program, which was started in 1979, has worked with communities throughout the state to revive dying downtowns and promote economic development that is compatible with historic preservation.

A main key to a successful Main Street program is the proactive nature of the participants involved in the project. The local community starts the program. It is a grassroots-driven program. It isn't started by an entity outside of the community or funded by an outside group. Anyone involved in or who has a stake in a community commercial district (business owner, non-profit organization, city government, or community member) can initiate the idea and run with it. Technical assistance, on-site design assistance, fundraising assistance, planning and program

BOX 18.9 MAIN STREET® FOUR POINT APPROACH

1 Organization involves getting everyone working toward the same goal and assembling the appropriate human and financial resources to implement a Main Street revitalization program.
2 Promotion sells a positive image of the commercial district and encourages consumers and investors to live, work, shop, play and invest in the Main Street district.
3 Design means getting Main Street into top physical shape.
4 Economic restructuring strengthens a community's existing economic assets while expanding and diversifying its economic base.

Source: Main Street internet site, http://www.mainstreet.org

guidance, training sessions, networking, and advocacy and leadership services can be obtained from the National Main Street Center and other operating Main Street programs, if requested.

According to the National Trust for Historic Preservation, from 1980 to December 31, 2011, the following accomplishments can be attributed for all designated Main Street communities: total reinvestment in physical improvements from public and private sources – $53.6 billion; net gain in businesses – 104,961; net gain in jobs – 448,835; number of building rehabilitations – 229,164; and $18:$1 reinvestment ratio.

Budget cuts have recently hindered designated and hope-to-be-designated Main Street communities. Some states have stopped taking applications from communities. The State of Kansas, due to a department restructuring and less federal funds, curtailed its Main Street program as of September 2012 – a program that had been in existence for 27 years. It is hoped that many of the current programs will continue, but without state-level technical assistance.

Historic preservation and tourism

One of the objectives of historic preservation is often the promotion of tourism. Sadly, success here can bring its own problems. Too many people seeking a particular experience can result in its destruction. Fitch (1990) notes that many popular places – Kyoto, New Orleans, Paris, Leningrad – are facing threats to their actual physical fabric (see Box 18.10). More generally, Americans are in danger of loving their national parks and historic sites to death. This problem is not, however, experienced in most places of historic interest. On the contrary, the economic development importance of historic resources is underlined by the use of the term 'heritage tourism'. Heritage tourism, according to the 1995 White House Conference on Travel and Tourism, can be defined as 'travel directed toward experiencing the arts, heritage, and special character of a place'. This could be an area's ethnic neighborhoods, its museums, architecture, buildings, streetscapes, etc. In Washington, DC, Judith Lanius, chairman of the Board of DC Heritage Tourism Coalition, estimates that tourism represents the number one private sector business in the city – worth $4.2 billion a year to the local economy (Smith 2000: 1).

Under the heading 'economic revitalization' the 1989 ACHP *Report to the President and Congress* summarizes three initiatives in heritage (or 'cultural') tourism. In Lockport, Illinois, the Gaylord Building rehabilitation project is the first in the National Heritage Corridor, a 120-mile historic district designated in 1984, which stretches along inland waterways from Chicago to La Salle-Peru. The canal contributed to the transformation of Chicago from a small settlement to a major transportation hub linking the East to the Midwest. This 'blend of natural and historic resources' is attracting tourist dollars to the Lockport area.

BOX 18.10 HERITAGE TOURISM – TWO PERSPECTIVES

In many famous individual monuments, tourist traffic has reached its absolute limits: at Mount Vernon, George Washington's residence, stairs and floors have had to be reinforced to carry the weight of visitors; and the abrasion of flooring surfaces is so severe that protective membranes must be replaced in a matter of weeks. Faced with the noise, confusion, and downright squalor which such overcrowding often produces it would be all too easy to reject the whole concept of mass tourism and yearn for a return to the good old days of aristocratic travel.

Source: Fitch 1990: 78

Heritage tourism is just one way in which the preservation and maintenance of historic towns and urban areas may contribute to overall economic improvement. Innovative programs initiated at all levels, public and private, illustrate how preservation efforts can support and complement economic and social developments in urban areas.

Source: ACHP 1989: 35

BOX 18.11 WHAT OTHER CITIES ARE DOING TO PROMOTE HERITAGE TOURISM

Cities across the nation are proving that heritage tourism works through investing in new tourism products and experiences, packaging, and marketing.

- Philadelphia is investing $12 million in private and public money to make heritage tourism a linchpin in its economic development planning.
- The government of Kansas City has invested $22 million in an African American jazz museum and other attractions at 18th and Vine, recognizing that African American tourists spend $30 billion a year in America.
- Los Angeles, with a tourist industry the size of the Washington area, estimates its heritage tourism initiative could bring an additional $3 billion to the region.
- Neighborhood tours sponsored by the Chicago Office of Tourism are leaving up to $250,000 a year in new, locally run craft shops.

Source: Smith 2000: 3

In the town of Port Townsend, Washington, a well-known seaport in the late 1900s, tourism tripled between 1983 and 1988 as a result of promoting Victorian-era neighborhoods and downtown façade rehabilitation. In 2000, Port Townsend was selected as one of five areas in the United States to be designated by the National Trust for Historic Preservation (NTHP) as a Great American Main Street award winner. The award is given to areas that have demonstrated excellence in revitalizing main streets and neighborhood commercial districts. In Georgia, the Antebellum Trail is being promoted (using hotel room taxes) as a tour of sites which Sherman missed on his march to the sea. Many other examples are given in the annual reports of the ACHP. Heritage tourism is growing. In fact, in a December 19, 2000 news release, the NTHP noted that more than half of the US states have established heritage or cultural tourism programs – twenty of them are less than five years old. Some of the examples of what other areas are doing in regards to heritage and cultural tourism can be found in Box 18.11.

Boston has developed a strong heritage tourism program focusing on events during the early history of the US (see Plate 43). Visitors can follow trails throughout the city. Washington, DC, has also developed a number of neighborhood heritage trails to help individuals get an understanding of the area's history, or as Cultural Tourism says, 'trace the footprints of history'. For example, following the Downtown Heritage Trail brochure allows individuals the opportunity to recreate the walk down the alley where John

Plate 43 Heritage trail at Boston Common

Photo by author

Wilkes Booth fled after shooting President Abraham Lincoln, to visit the home and office of American Red Cross founder Clara Barton, and to see the hotel where Dr Martin Luther King finished his famous 'I have a Dream' speech. Other trails enable individuals to visit the African American Civil War Memorial (the nation's sole memorial to black Civil War soldiers), John Philip Sousa's birthplace and training ground, Beaux-Arts mansions and museums, and the location of the first Toys 'R' Us store.

The widening scope of historic preservation

One of the most characteristic aspects of historic preservation today is that its domain is being constantly extended in two distinct ways. On the one hand, the *scale* of the artifact being considered as requiring preservation is being pushed upward to include very large ones (e.g., the entire island of Nantucket) as well as downward, to include very small ones (e.g., historic rooms or fragments thereof installed in art museums).

> On the other hand, the domain is being enlarged by a radical increase in the *type* of artifacts being considered worthy of preservation. Thus in addition to monumental high-style architecture – traditionally the concern of the preservationist – whole new categories of structures are now being recognized as equally meritorious: vernacular, folkloristic, and industrial structures. In a parallel fashion, the time scale of historicity is being extended to include pre-Columbian settlements at one end and Art Deco skyscrapers at the other.
>
> (Fitch 1982: 39–40)

This lengthy quotation from Fitch (1982) nearly says it all! The boundaries of 'historic preservation' are being stretched in such a way that the term is now a misnomer. One has only to peruse the volumes of *Perspectives in Vernacular Architecture* (e.g., Wells 1986; Carter and Herman 1989) to see the way in which interests have broadened. At the same time, new problems are arising, and old problems are taking on

new dimensions. Reference has been made earlier to the criticism of the New York City Landmarks Commission that it had, on occasion, exceeded its mandate. With heightened concern for 'historic' areas of the twentieth century, this may become more common. The issue is complicated by an overlapping concern to preserve low-income housing from redevelopment. In April 1990, for instance, the Landmarks Commission gave landmark status to a complex of fourteen buildings in the Yorkville section of Manhattan that were originally constructed as a privately financed experiment to provide housing for the poor. Here is a nice mixture of historic, architectural, social, and economic issues. Some idea of the flavor of the debate can be gleaned from the following quotation from the *New York Times* of April 15, 1990:

> Paul Selver of the law firm of Brown and Wood, which represents the owner, said there was 'nothing special' about the property to warrant landmarking . . .
>
> The commission, in its resolution, noted that the projects represented an attempt by a group of prominent New Yorkers 'to address the housing needs of the working poor'. Investors agreed to voluntarily limit their profits, and the apartments provided occupants with interior plumbing, more window space and more light and air than typical tenement apartments of the time.

The commission also praised the 'distinction' of the architecture, and maintained that designation would help to protect an area which represents 'an important slice of history of the Upper West Side'.

Clearly a host of different interests and values are at stake here. In such cases, the matter is settled (perhaps after recourse to the courts) by determining which interest – or interest group – is to prevail. And so we get into what Bishir (1989), in a stimulating and entertaining paper, has termed 'the politics of culture'. She ends this by stressing that, since preservationists are participants in the politics of culture, it is necessary for them to be aware of the impact on their decisions of their value system. Whether self-knowledge is sufficient is an open question.

One final (and significant) point needs to be made. It was earlier stated that 'historic and associative' elements remain as important today as they were a century ago. It is, however, important to note the emergence (particularly in the western United States) of scholarly studies of archaeology which have profoundly affected our view of material history. The traditional approaches to the field varied. As Torma (1987: 6–7) notes:

> At the onset of the national preservation program, the field was divided into at least two distinct camps – the archaeologists were on one side and the architectural historians and historians were on the other. While the orientation of the archaeologists was cultural, the orientation of the architectural historians and historians was traditional history and history of aesthetics. One group was trained in the social sciences (and some would say the sciences) and the other in the humanities. While the archaeologists looked at all aspects of the 'cultural picture' – economic base, diet, foodways, architecture and seasonal migration patterns (to name a few) – those working in the historic sites program were generally concerned with only two issues: is this structure aesthetically beautiful and/or does it have already demonstrated historic value?

The coming together of these different approaches has proved fruitful, and new perceptions of 'historic preservation' are emerging. In Stipe's words (1987: 274), the subject matter of historic preservation has become 'thoroughly democratized', and topics such as vernacular architecture, and industrial and commercial archaeology are now common and popular topics. The very term 'historic preservation' is being replaced by broader concepts of 'heritage' – a topic discussed earlier in this chapter.

Historic preservation has become a hot topic in many areas while it has remained somewhat dormant in other areas. Until some districts, structures, etc. are actually in danger of being demolished, many individuals are content with sitting back and watching events unfold. In other areas, historic preservation enthusiasts have become incensed over just the thought of losing a particular building or landmark. They are disturbed by the arrogance of some developers and seek to identify and mobilize various groups for the purpose of preserving a particular building or landmark. They recognize the importance of preserving significant examples of our past for future generations. By using federal and state tax incentives and programs, historic preservation offers a means of connecting one generation to another generation.

Further reading

In 2006, during the one hundredth anniversary of the Antiquities Act, the US Fish and Wildlife Service (USFWS) celebrated by creating a useful overview of the Landmark Act (http://www.fws.gov/historic Preservation/antiquitiesActCentennial). Its website explains how the USFWS used the Antiquities Act and it includes many links to other government websites that provide additional background for the Act.

The National Park Service's Archeology Program (http://www.nps.gov/archeology/sites/Antiquities/about.htm) gives a short history of the Antiquities Act of 1906. Along with a link to a list of national monuments, the website includes an additional reading section with more history and commentary on the Act.

The National Trust for Historic Preservation (http://preservationnation.org/), a private, non-profit organization, is dedicated to preserving historic places in America. One such project focuses on the recovery of New Orleans from Hurricane Katrina (http://www.preservationnation.org/issues/gulf-coast-recovery/overview.html#.Ufac WyLFD/) as historic monuments are being rebuilt after the hurricane. The website provides links to funding opportunities and programs for preservation in the natural disaster aftermath.

The Advisory Council on Historic Preservation (http://www.achp.gov) supplies a website highlighting its publications, reports, and guides for working within current preservation legislation.

The National Register of Historic Places can be found on the National Park Service website (http://www.nps.

gov/nr). The site provides links to the National Register database of over 80,000 listings and other information, including the procedure for listing a property.

The Certified Local Government Program (http://www.nps.gov/history/hps/clg) is coordinated through the National Park Service as a means for local governments to gain support for their historical preservation efforts. Resources on this website include an explanation of the program, funding information, Certified Local Government statistics, publications, and links to government updates on historical preservation.

The most comprehensive website offering Historic Preservation Tax Incentives is offered through the National Park Service (http://www.nps.gov/hps/tps/tax). A virtual guide to applying for tax incentives, the website also offers annual reports, publications, and online education.

PerservationDirectory.com (http://www.preservation directory.com) provides preservationists with a database of updated internet resources, and updates preservation events and practical knowledge, such as how to sell or buy a historic property.

Information on the Main Street program can be found at the following site: http://www.mainstreet.org.

The New York Landmarks Preservation Commission (http://www.nyc.gov/html/lpc/html/home/home.shtml) is still active today. Its website provides the public with a history of the commission, landmark maps, government forms, and information on public hearings.

Cultural Heritage Tourism (http://www.culturalheritage tourism.org) is an advocate for historical tourism. With its how-to website, Cultural Heritage Tourism offers marketing plans, links to funding, and educational resources for tourist sites that emphasize preservation.

Current legislation for preservation includes the Preserve America Initiative (http://www.preserveamerica.gov), which is mandated by Executive Order 13287. Many different government groups including the White House and the Advisory Council on Historic Preservation have cooperated to provide grants, awards, and information to construct Preserve America Communities.

One attempt to change the city into a living museum is Curating the City in Los Angeles (http://www.curatingthe city.org). An offshoot of the Los Angeles Conservancy, Curating the City offers the public historical tours, a continually updated memory book of Wilshire Boulevard, and other educational resources.

The best single book on historic preservation (and certainly the most enjoyable) is Costonis (1989) *Icons and Aliens: Law, Aesthetics, and Environmental Change.*

Major historical writers are Tyler, Liqibel, and Tyler (2009) *Historic Preservation: An Introduction to Its History, Principles, and Practice,* 2nd edition; Fitch (1990) *Historic Preservation: Curatorial Management of the Built World*; and Hosmer (1965) *Presence of the Past: A History of the Preservation Movement in the United States before Williamsburg*, and the same author's *Preservation Comes of Age: From Williamsburg to the National Trust, 1926–1949* (1981).

For an interesting study of heritage tourism in Atlanta, Georgia, see Newman (1999) *Southern Hospitality: Tourism and the Growth of Atlanta*. See also, Walle (1998) *Cultural Tourism: A Strategic Focus.*

For a state perspective on heritage, see: Georgia Department of Natural Resources, Historic Preservation Division (2010) *Heritage Tourism Handbook: A How-to-Guide for Georgia.*

A good and succinct account of the *Penn Central* case is to be found in Haar and Kayden (1989b) *Landmark Justice: The Influence of William J. Brennan on America's Communities.*

There is a large and burgeoning literature on the developments taking place in this field. See, for example, Upton and Vlach (1986) *Common Places: Readings in American Vernacular Architecture*; Wells (1986) *Perspectives in Vernacular Architecture*; Carter (1997) *Images of an American Land: Vernacular Architecture in the Western United*

States; and McMurry and Adams (2000) *People, Power, Places.*

For a discussion of the relationship between planning and historic preservation, see Birch and Roby (1984) 'The planner and the preservationist: an uneasy alliance'.

For a useful book on the Main Street program, see Brown, Dixon, and Gilham (2009) *Urban Design for an Urban Century: Placemaking for People.*

A useful reference book is *Landmark Yellow Pages*, edited by Dwight (1992).

For readings on economics and historic preservation, see the following: Listokin and Lahr (1997) *Economic Impacts of Historic Preservation*; Mason (1999) *Economics and Heritage Conservation*; Mason (2005) *Economics and Historic Preservation: A Guide and Review of the Literature*; Rypkema (1998) *Profiting from the Past: The Impact of Historic Preservation on the North Carolina Economy*; Rypkema (2005) *The Economics of Historic Preservation: A Community Leader's Guide*; Smith (2000) *Capital Assets: A Report on the Tourist Potential of Neighborhood Heritage and Cultural Sites in Washington, D.C.*; Throsby (2001) *Economics and Culture.*

Questions to discuss

1   Is it worth preserving buildings that have historical associations, but no architectural merit? Give examples to support your argument.

2   In what ways can historic preservation be integrated with land use planning?

3   Discuss the ways in which historic preservation policies impact on the federal government.

4   On what grounds is it justifiable to give public subsidy for the preservation of old buildings?

5   How can the impact of 'excessive tourism' on popular sites of historic value be dealt with?

6   How are states and local governments involved in historic preservation activities?

7   What is heritage tourism and how is it being practiced in the United States?

8   What are the benefits associated with preserving historic properties?

9   What is the Main Street program and has it been successful?

CONCLUSION

19

Some final questions

The search for scientific bases for confronting problems of social policy is bound to fail, because of the nature of these problems. Policy problems cannot be definitively described. Moreover, in a pluralistic society there is nothing like the indisputable public good; there is no objective definition of equity; policies that respond to social problems cannot be meaningfully correct or false . . . There are no 'solutions' in the sense of definitive and objective answers.

Rittel and Webber 1973: 155

Determining the questions

It would be satisfying for both the reader and the author if this final chapter could set out solutions to all or some of the problems raised in this book. Moreover, policy officials would obviously welcome the presence of solutions to any of the problems. It should be clear, however, from the emphasis laid on the nature of these problems that they admit of no easy solution. In fact, many of them are not 'solvable' in any real sense of the term. One might even go further and suggest that the term 'problem' is unhelpful in that it suggests that a solution is indeed possible. *Webster's Dictionary* gives helpful alternatives: 'difficult to deal with' and 'a source of perplexity, distress, or vexation'.

There remains an important distinction between a difficulty and a problem: a difficulty is a problem only if it is thought that something can be done about it. In the scientific world, 'difficulties' can be changed into problems because of advances in knowledge; and expectations of such advances stimulate optimistic expectations and further research. We live in a world of rising expectations where people feel or expect that advances in technology will resolve our problems. There is little to mirror this in the social world: with many issues, the more we study social issues, the more

complex they seem to become. Indeed, part of the difficulty facing the policy process is that public interest and support (mirrored in Congress) wanes in the face of lack of progress. As has been pointed out by Downs (1972), issues tend to rise and fall in public interest. Concern arises with the publicity given to particular events (environmental disasters, an apparent rise in crime, race riots, etc.), and a demand quickly develops for strong action to be taken to deal with the 'problem'. The importance of public awareness can be clearly witnessed with the Love Canal disaster. But, as the complexities and the resource implications become apparent, public attention wanes – often as other 'problems' claim center stage. An issue may even be transformed into its opposite, as when environmental concerns are overwhelmed by worries about unemployment: a number of areas around the country have accepted serious environmental trade-offs for jobs.

An added complication has come about as a result of the growth of pluralist politics, the demotion of the professional and the expert in public opinion, the increased mistrust of government, and the demand for greater public participation. At one time, problems were for experts to solve. There was justification for this optimism in the nineteenth century: the successes of science and the growth of the professions provided the necessary instruments to deal with the problems of the developing industrial society. The problems were – or became – capable of solution. (See the discussion in Chapter 3 on the rise of the professions in the nineteenth century and the widespread belief in both progress and efficiency: this was the era in which planning was born.) Unprecedented progress was made in engineering, sanitation, control of infectious disease, and the building of roads, bridges, schools, and hospitals. Some issues, of course – such as poverty and working-class crime – proved unamenable to the forces of progress; but these could be ascribed to moral inadequacy, or catered for by charity, or dealt with by penal measures. Hopefully, the tides of progress would eventually raise the deviants up to the new standards of cleanliness, probity, and efficiency.

Overdrawn though this analysis may be, it contrasts sharply with the contemporary scene. In retrospect, the problems that were solved in those simpler times were easy. They were 'definable, understandable and consensual' (Rittel and Webber 1973: 156). Today, there is far greater concern for equity, for debating what the important questions are, for accommodating – and even facilitating – differences of opinion and attitude, and for mediation. Questions can no longer be left to the numerous experts, since their formulation is of prior importance, and answers often involve 'unscientific' value judgments. The public demands to be heard.

Rittel and Webber (1973), in a paper which makes a major contribution to an understanding of these issues, have termed the older, easy problems 'tame' ones in contrast to the 'wicked' problems we face today. These are 'wicked' in the sense that they are incredibly complicated, multi-faceted, and elusive; to quote *Webster*'s again, 'a source of perplexity, distress, or vexation'. They are, in fact, so difficult to deal with that one is tempted to suspect that evil forces are at work to prevent solution. Indeed, 'social problems are never solved. At best they are only re-solved – over and over again' (see Box 19.1).

While this may appear very depressing, it does point to the direction in which the problems can be approached. Since there is now a major concern with the legitimacy and equity of policies, means have to be found to provide for these in the policy-making process. Encouragingly, this takes us into some familiar territory, such as public participation, mediation, conflict resolution, and intergovernmental relations.

Moreover, maintaining this book's focus on land use, urban development, and environmental issues has helped to make the discussion manageable. It needs to be stressed, however, that the intention has not been

BOX 19.1 SOLUTIONS DEFINE PROBLEMS

What would be necessary in identifying the nature of the poverty problem? Does poverty mean low income? Yes, in part. But what are the determinants of low income? Is it deficiency of the national and regional economies, or is it deficiencies of cognitive and occupational skills within the labor force? If the latter, the problem statement and the problem 'solution' must encompass the educational process. But, then, where within the educational system does the real problem lie? What then might it mean to 'improve the educational system'?

Or does the poverty problem reside in deficient physical and mental health? If so, we must add those etiologies to our information package, and search inside the health services for a plausible cause. Does it include cultural deprivation? spatial distribution? problems of ego identity? deficient political and social skills? – and so on. If we can formulate the problem by tracing it to some sorts of sources – such that we can say 'Aha! That's the locus of the difficulty' – then we have thereby also formulated a solution. To find the problem is thus the same thing as finding the solution: the problem can't be defined until the solution has been found.

Source: Rittel and Webber 1973: 161

to provide a compelling, comprehensive approach even to these selected areas. Rather, has it been intended to provide some thoughts that emerge from the previous analyses, which, it is submitted, are worth considering for the improvement of the planning process. Added confidence is given to this claim by the fact that the thoughts are not new: some (such as public involvement) have been shown to have some modest successes, though others (such as regional government) have proved to be more problematic in practice than they are convincing in theory.

Demographic shifts in our population will undoubtedly cause new problems to emerge. Some cities and regions will gain in population while other areas will suffer decreases in their population. There is currently a discussion in planning circles of 'shrinking cities' and their impact on the economy and on the residents. These changes will necessitate various new actions in terms of land uses, economic development, transportation, the environment, public facilities, social services, etc. Monitoring and responding to these demographic shifts will become a focus for planning in these areas.

Property rights – 'wise use'

For the most part, Americans are not prepared to support a land use planning system that deprives them of control over development in their local area. There continues to be a long tradition of belief in the sanctity of the rights of property. There is also a strong concern that those who are affected by government policy should be able to influence that policy; and where the effect is to impose an unreasonable burden on individuals there should be relief from that hardship either by way of exception or by payment of compensation. The mainstream of land use planning is thus essentially local, with sensitivity to local opinion and the avoidance of hardship to individuals. Zoning meets these narrow concerns neatly, and it specifically provides a safety valve with a system of variances in cases of hardship (discussed in Chapter 6). It is important to remember that the support for zoning stemmed from its essential role in the protection of property values. The broader conception of comprehensive planning

as a framework for zoning failed to obtain similar support.

However, even the restricted conception of zoning has given rise to controversy over its operation. Protecting property rights involves reducing those rights. Zoning at a low density protects the property values of owners but at the same time precludes them from maximizing potential development values. As such, zoning is not neutral. Some property owners gain value in a rezoning, while others might experience a decrease in property value. Much conflict has occurred between owners who want to preserve the status quo and those who anticipate large increases in value due to changes in the local land market.

Such controversy has burgeoned with the growth of regulatory controls over environmental quality. These controls are now very wide in scope: in addition to more sophisticated zoning controls, they include restrictions aimed at preserving historic buildings, wildlife, coastlines, and wetlands. They pertain to waste disposal, to the protection of groundwater, and to air quality. They regulate commuting, parking, pollution, and aesthetics. Every new 'protection' can involve owners in loss of value (immediate or potential). Moreover, uncertainties abound: these include the extent to which loss is acknowledged by regulatory agencies; the degree to which relief from hardship is permitted; the amount of compensation which may be payable; and the indeterminate cost and outcome of litigation. Such unknowns have contributed to a widespread anxiety, and have added to the long-standing opposition of those who are vehemently opposed in principle to government regulation. As a result, a new 'property rights' movement has developed, with claims for guaranteed compensation for diminution in property values.

Conflicts will always surface when governments attempt to regulate the use of private property. A number of attempts have been made to pass state and federal property rights legislation. Between 1991 and 1995, forty-four states considered, and nine passed, legislation providing for such compensation – typically for cases where a regulation reduces potential value by half (Tibbetts 1995: 5). Since then, a number of other states have introduced property rights legislation

(Oswald 2000). A discussion of *Kelo* v. *New London* (545 US 469), in which the US Supreme Court held that a taking of properties qualified as a 'public use' within the meaning of the 'Takings Clause' can be found in Chapter 6. The support for such (tendentiously self-styled 'wise use') legislation comes from a wide range of opinion fueled by the fear of the uncertainties of the current situation as well as by the publicity attendant on a number of court cases. At the time of writing, however, most legislative proposals (including those before Congress) had been defeated. Former President Clinton professed a belief in protecting private property with payment of just compensation to the property owner when the property was taken for a public purpose. President Bush also expressed a similar commitment. Both of them opposed various pieces of private property rights legislation on the grounds that it would weaken health and environmental standards. In addition, they opposed the proposed legislation because they felt it would hinder local land use planning efforts.

The middle ground points to the scope – and the need – for procedural remedies (akin to the zoning variance) that would deal with cases of real hardship, and for a range of legal and administrative reforms. Other issues in the debate range from the greater certainty which would be afforded by an extension of comprehensive plans, to the need to balance any system of compensation for 'wipeouts' with the 'logical corollary' of a system for capturing 'windfalls'. The latter argument is the long-established one that if owners are to be compensated for serious losses in value due to government action, they should, logically, be required to pay for major increases in the value of their land caused by government action. In fact neither policy is easy to devise and operate fairly.

Beyond localism

Though much of land use planning is essentially local in scope and intent, it is sometimes possible to marry local interests with wider concerns. Indeed, occasionally a local community can organize itself to accommodate a wider public good with profit to itself.

Such is the case, for instance, with communities that accept hazardous waste disposal sites in return for community gains. In one case these amounted to '$4.2 million in benefits that would include a new town park, trust funds for the library and the fire department, road improvements, and a scholarship fund for local youths headed to college'. Additionally, the town would receive around $1.5 million more in taxes and fees, roughly double its current tax revenues (*New York Times*, June 28, 1991).

This is by no means a unique example. The problem of finding sites for used nuclear fuel and nuclear waste has increased since the federal government mandated every state to store its waste within its own boundaries, a move necessitated by the closure to out-of-state waste of the only existing disposal sites, in South Carolina, Nevada, and Washington. Different states are following different tactics, and no doubt other bargain offers will be made. There is even a publication, *The Radioactive Exchange*, that is keeping track of efforts.

The storage of nuclear waste continues to be the subject of much debate. The US Department of Energy studied a site in southern Nevada, Yucca Mountain, to determine if it could be used as a geological repository of spent nuclear fuels and high-level radioactive waste. After seeing this and other information, President Bush notified the US Congress on February 15, 2002, that the Yucca Mountain site qualified for a construction application. His reasoning, based on what he called 'sound science', was the following:

> Currently, nuclear materials are stored in 131 above ground facilities in thirty-nine states, and 161 million Americans live within seventy-five miles of those sites. One central site provides more protection for this material than do the existing 131 sites.

Thus, presidential support was given for using Yucca Mountain as a geological repository for nuclear waste.

The state of Nevada rejected the recommendation to use Yucca Mountain as a nuclear waste repository. On June 7, 2002, Nevada challenged the US Environmental Protection Agency (EPA) radiation standard, claiming that the EPA standard 'wasn't sufficiently protective of the public health and safety, particularly Nevada's groundwater resources'.

Another state entered the debate one week later. On June 14, 2002, the Governor of South Carolina issued Executive Order 2002-14 claiming that an emergency exists and that 'the transportation of plutonium on South Carolina's roads and highways is prohibited; that any person transporting plutonium shall not enter the State of South Carolina'.

Transporting nuclear waste to the ultimate disposal site looms as a critical issue. A number of governors have opposed the Yucca Mountain site on the ground that ensuring the safety of the material en route to the site cannot be guaranteed. They simply do not want the waste going through their jurisdictions. This has become another classic example of being unable to separate politics from any decision to create a nuclear waste repository.

On July 10, 2002, the US Senate gave final congressional approval for the establishment of a nuclear waste repository inside Yucca Mountain. In 2006, the Department of Energy suggested a starting date of 2017 to open the facility and begin accepting nuclear waste. Nevada officials continue to block the project. Using Yucca Mountain to store nuclear waste continues to be a highly visible and hotly contested debate.

Although Yucca Mountain was designated in 1987 as the only site for the storage and waste disposal of used nuclear fuel and nuclear waste, opposition to its use has continued. Nevadans are against. During a campaign trip to Nevada, current President Obama wanted to find another site for the nuclear waste. The US Department of Energy has withdrawn its application to the Nuclear Regulatory Commission to use Yucca Mountain as a nuclear waste storage facility. In 2010, Obama created a Blue Ribbon Commission on America's Nuclear Future that was tasked to examine alternatives for the storage and disposal of used nuclear fuel and nuclear waste. The debate is certain to continue.

This may be a telling illustration of how a market works: increased prices will bring about additional supply. But it remains to be seen whether it circumvents the difficulties of efficiency and equity that attend zoning, and indeed whether it works at all, other than

in exceptional circumstances. And is there anyone asking whether the bargained site is desirable on wider social, economic, or environmental grounds? Surely this is more than a private matter between 'neighbors'?

A market mechanism operating in a different way is the neighborhood buyout, or 'assemblages' as they are known in the real estate world (Haar and Wolf 1989: 1094). There are a number of variations; in one, a group of home owners, realizing that commercial development is creeping up on them, get together and negotiate with the developer for the sale of the whole area. When this works, all home owners receive higher prices than they could have otherwise expected, while at the same time the deal also makes negotiations simpler for the developer. The developer may well end up paying more, but some time is saved and, of course, 'time is money'. Presumably, on completion of this market transaction the zoning designation is obligingly changed.

Examples of this nature do not take us very far, but they do illustrate a point: in deciding local futures, residents are justifiably concerned with the balance of costs and benefits. Much opposition stems from a perceived view that a development proposal involves severe costs and few or no benefits. Alternatively, the costs, which the local community are being asked to bear, are unfair. Is there a mechanism which could translate this idea of local fairness into wider social advantage? A good test case is the siting of hazardous waste sites.

Local acceptance of unwanted land uses

There are several reasons why the siting of hazardous waste facilities is problematic (discussed in Chapter 12), but there is no question that a crucially important one is public attitudes. These attitudes may be based on a legitimate fear of the implications for health or property values (which may be quite rational), and also intensified by a perception that other people's waste is being unfairly imposed on a locality. Two possible responses to this are to compensate affected localities so as to offset their costs and to attempt to distribute

BOX 19.2 PUBLIC FEARS AND MISTRUST

It is understandable that local residents and facility sponsors would fail to arrive at common understandings and agreements on siting because they have focused on different aspects of the issue – technical capabilities of a facility versus the potential harm to an individual member of the community, for example – and talked past one another. At the root of the problem is a lack of shared understanding, perceived mutual interests, and trust. Still excluded from most decisions that lead to initiating proposed new facilities, people today can and do just say 'no'. The resulting gridlock inhibits commencement of any serious improvement in toxics policy.

Source: Mazmanian and Morell 1992: 242

sites in a fair way. Of course, the definition of what constitutes 'fair' will be subject to a number of different interpretations.

Compensation for community disbenefits is a traditional economic response. In the exceptional case quoted above, it is noteworthy that the scale of the compensation was very large. An experiment in Massachusetts foundered on this very issue. As part of state-mandated negotiations between facility developers and local communities, it was found that 'virtually no reasonable amount of compensation' was sufficient to persuade local residents to accept sites (Portney 1990: 60). On the other hand, there exists some evidence to suggest that offsetting compensation at an acceptable level might be negotiable as part of a wider package that includes a reduction in the risks (Hadden 1989: 52). Since mitigation of risks is, in any case, a constructive approach, the idea merits further study.

Fair shares for all is a morally attractive policy: who can object to a scheme in which all bear the unwanted costs of hazardous waste? Mention was made in Chapter 13 of the California YIMBY program, which was to locate facilities in such a way as to ensure that each county would deal with the waste produced within its boundaries. Though accepted by all the counties, the scheme became a casualty of state politics; but it might have better luck elsewhere (Mazmanian and Morell 1992: 192–203).

These are small crumbs of comfort. Are people so blind, ignorant, and selfish that they will ignore the public necessity of finding adequate facilities for disposing of waste? The answer might be surprising. A fundamental problem is the continued widespread distrust of governments, experts, and business – indeed everyone in power. Though community and environmental groups may employ their own experts to assist in their campaigns of opposition, this practice serves (outside the courtroom) only to exacerbate the difficulties. As Harvey Brooks (1984: 48) has put it, the conflict becomes among 'noncommunicating publics that each rely on different sources and talk to different experts. Thus many public policy discussions become dialogues of the deaf' (see Box 19.2).

These basic issues of democracy have been long debated: they acquire a new urgency in the complex, interdependent, global society of today. Since apparently neither governments nor experts are to be trusted, perhaps one way forward lies with a transformation of the environmental movement into one with a positive concern for seeking acceptable and workable solutions. But, sadly, the path may be too easily diverted into yet another area of expertise.

Unwanted neighbors

If environmental issues arouse strong passions, unwanted social neighbors can create fury. Three examples are discussed here: two which offer hope (day-care facilities and group homes), and one which seems almost impossible to resolve (low-income housing). The provision of day care has given rise to neighborhood opposition on account of traffic, disturbance, and the general effect on the character of the neighborhood. The large increase in the number of working mothers, and the changed attitudes to the propriety of this, has reduced this opposition to some extent. However, a number of states have found it necessary to pass legislation to prevent municipalities from zoning against childcare facilities in residential areas.

With group homes it seems clear that either an early, earnest, and positive community consultation approach or the legal sanctions of the Fair Housing Act (or both) can be effective, though a neighborhood battle may be involved before the group home actually opens. Local fears, real or perceived, can be allayed by frank and persuasive explanations of what is involved, who will live in the home, and how it will be controlled. Of course, this becomes much easier in the case of homes for disabled children than it is for ex-prisoners or ex-drug addicts. However, even with groups such as the latter two, where there can be real neighborhood fear, good early groundwork can help reduce apprehensions.

Low-income housing is a much more difficult issue. As with group homes, a major fear is a fall in property values, but the apprehensions go further: community character, unwelcome neighbors, as well as the usual litany of tax and overcrowding impacts (ranging from schools to the fire service). Whether or not these fears are justified or simply imagined, there is no doubt about the underlying opposition to people of a different color or class.

Given the strength of the prejudice against low-income housing, and the underlying racial discrimination, there is no easy remedy. The sheer force of this opposition has to be appreciated. There are innumerable stories which demonstrate this. Historically, an important one dates from the 1960s, when the federal government was under pressure from many sources (including the Justice Department, the Equal Employment Opportunities Commission, and many officials in Washington, as well as numerous groups such as Paul Davidoff's Suburban Action Institute) to 'open up the suburbs' to low-income groups. A scheme of rent supplements was introduced

to enable low-income households to obtain access to federally subsidized housing built with FHA mortgages by bodies such as non-profit organizations. This represented a novel way of bypassing local opposition, since no formal participation by local government was necessary. In the words of one leading lobbyist, this would 'help penetrate the wall of exclusion erected by many suburban communities against the introduction of housing for low and moderate income families' (Keith 1973: 161).

In the Nixon administration, the program was coupled with *Operation Breakthrough* – a program aimed at facilitating the mass production of factory-built housing by large corporations. The 'breakthrough' was to be a major reduction in the cost of housing, from which large numbers would benefit. It was this program which federal officials stressed at the local level – studiously avoiding any mention of race, though developers were to be required to employ fair-marketing practices to maximize housing opportunities for minorities. The major emphasis, however, was on the benefits of *Operation Breakthrough* as well as generous federal grants for infrastructure and other worthy local causes. With these accompanying attractions, it was hoped that local governments would waive local zoning and building codes. It was not that easy! Though the initial reaction seemed promising, opposition soon mounted as it became clear that the program would involve subsidized housing for black people. It reached fever pitch at the Detroit suburb of Warren, where George Romney, the enthusiastic Secretary of HUD, met at first hand the fury of local people threatened by what they saw as federal enforcement of integration. The experience became known as 'the political education of George Romney'. He quickly changed his stance and argued that there was a danger of 'setting things back as a result of pushing too hard too fast' (Danielson 1976: 225).

One of the perceived fears of white suburbs is that minorities may overwhelm them. To allay this fear, while promoting a small degree of integration, various programs have been evolved. Several of these are based on the Gautreaux model. The name originates from a class action suit by Dorothy Gautreaux and other public housing residents and applicants that alleged deliberate discrimination by the Chicago Housing Authority in siting public housing developments and assigning tenants. (Virtually all public housing in Chicago was located in areas where more than a half of the residents were black.) After many years of legislation, the federal courts approved a plan to scatter public housing projects, tenants, and applicants throughout the metropolitan area.

This effectively breached the barriers defended by suburban communities. The schemes included modest housing developments planned (in scale and design) to blend into the local area. There were also 'mobility' programs that provided Section 8 certificates for housing in areas where no more than 30 percent of the population is black. Counseling was available to assist families to adjust to their new environment.

Between 1976 and 1994, such programs placed around 5,600 low-income black families, of whom over a half were located in predominantly white neighborhoods. Research undertaken by the Center for Urban Research and Policy Studies at Northwestern University has shown that these schemes can be successful: contrary to expectations, most black families are not isolated; racial harassment may be experienced by some at first, but this quickly diminishes; movers have better employment experience; and, after initial difficulty in adjusting to the higher standards, children do well in school. The most important lesson drawn from the research is that the program works because its effect on the racial composition of the neighborhood is negligible. Other areas are following the Gautreaux model (Joseph 1993).

Two other mobility programs were initiated by HUD: the Areawide Housing Opportunity Plan and the Regional Housing Mobility Program, both of which were designed to promote the voluntary cooperation of regional bodies and suburban governments in desegregating federally assisted housing. These were small in scale, and were later abandoned by the Reagan administration. However, Congress authorized another program on a demonstration basis in 1991. This *Moving to Opportunity* program's objective was assisting families to move from high to low poverty areas, where employment opportunities are much higher. Such programs offer the prospect of modest progress in a field where

progress has been rare, though their impact on the problems of the inner city seems unlikely to be significant. Time will be needed to assess the long-term effects of such programs.

The role for the states

With the advent of devolution, states have had to play a larger role in the development of affordable housing. They have their own housing programs that range from construction to mortgage lending and overseeing local zoning provisions for affordable housing. Typically they involve concerns that are wider (or resources which are greater) than those of local government. New Jersey's Fair Housing Act established a Council on Affordable Housing (see Chapter 15): this body assesses regional housing needs and municipal 'fair shares'. Oregon's state planning agency reviews local plans to ensure adequacy of planning provision for affordable housing. Illinois has a citizens' advisory body that reviews the effects of local zoning and building controls on the provision of affordable housing. Connecticut encourages regional agreements on low- and moderate-income housing (facilitated by financial incentives for infrastructure). It also has a system of appeals against local zoning decisions on housing. Several other states have a similar system, using either the courts or a state agency. Virginia requires local governments to assess the need for affordable housing as part of the comprehensive planning process.

Some of these mechanisms may be more impressive on paper than they are in fact, but they do point to an important role which state governments can play if they are prepared to accept the political responsibility – and if they can persuade their local governments to cooperate.

The inherent shortcomings of local government emerge from debates on a number of public services: housing, economic development, environmental planning, and transportation planning (see Box 19.3). Madison stated the rationale for a larger authority in magisterial terms over two centuries ago (see Box 19.4). Larger communities tend to embrace a wider range of interests, needs, and abilities, a greater concern for the

BOX 19.3 SHORTCOMINGS OF LOCAL LAND USE CONTROL

1 The absence of a comprehensive planning framework (inevitable with 'non-comprehensive jurisdictional entities');
2 The predominance of municipal self-interest and the lack of a mechanism to allocate undesirable but socially necessary land uses to optimal sites;
3 The inherent inability of local governments to address larger environmental questions;
4 The essentially negative character of local controls. Little of a positive nature (e.g. affirmative action) can be achieved.

Source: Delogu 1984

BOX 19.4 MADISON ON GOVERNMENTS AND REPRESENTATION

The smaller the society, the fewer will be the distinct parties and interests composing it; the fewer probably will be the distinct parties and interests, the more infrequently will a majority be found of the same party; and the smaller the number of individuals composing a majority, and the smaller the compass within which they are placed, the more easily they will concert and execute their plans of oppression. Extend the sphere and you will take in a greater variety of parties and interests; you make it less probable that a majority of the whole will have a common motive to invade the rights of other citizens; or if such a common motive exists, it will be more difficult for all who feel it to discover their own strength and to act in unison with each other. Besides other impediments, it may be remarked that, where there is a consciousness of unjust or dishonorable purposes, communication is always checked by distrust in proportion to the number whose concurrence is necessary.

Source: James Madison, *The Federalist*, 1788

common good, and a broader conception of the role and purpose of public policy. In short, they are less selfish. Of course, exclusionary attitudes do not wither away; but they do not flourish as they do when isolated from other influences. Instead, they are kept in rein by 'the sheer number whose concurrence is necessary'.

Unfortunately, this can be too simple a view of the matter, and increased size by itself is not necessarily enough. A large constituency may be nothing more than an assembly of small bodies, each of which works the political process to achieve its particular aims. Where this is so, mutual accommodation of interests forces out broader thinking. A large constituency may

also be so homogeneous that its character is essentially narrow minded. Such is the chemistry of democracy; but generally, the larger the authority, the more likely it is to represent a range of interests, and thus more predisposed to seek wider public policy goals.

Given the insularity of local governments, there exists a need for a compelling framework of social responsibility within which they should operate. This has two advantages. First, it provides policy objectives that have to be followed. These move to the forefront of political attention: they can develop, to use Donald Schon's (1971: 123) phrase, into 'ideas in good currency', and they become a spur to action (in much the same way as environmental protection is currently espoused by all). Second, it provides local representatives a defense against those who would have them work to a narrow interest: 'we have no alternative', the faint-hearted can plead, 'we must do what we are required to'.

Thus the rationale for policy can be a matter of high principle or of base political calculation: it makes no difference. What matters is that a public authority is operating in the broad public interest, not to the advantage of a powerful minority; and this means that it is including all needs in its civic policy.

Measures taken by states to rein the narrow-minded policies of localities include a ban on commercial rent control, regularization of local impact fees, and over-riding local prohibitions on manufactured housing and childcare facilities. (The federal government has acted in similar ways, of course, as with racial and other forms of discrimination.)

So far as land use planning is concerned, such an approach paves the way for devising a system in which zoning becomes the servant, rather than the master, of local land use planning. Here the local plans approved by a state or regional authority can assume great significance, as is the case in Oregon. Oregon provides a set of goals and procedures that have been agreed to be desirable for the future of the people of the state. These provide a policy basis for the preparation of local plans, which local governments are required to prepare and implement (see Chapter 7).

Goals, objectives, declarations, and such-like are, of course, the stuff of political discourse, and they can be totally devoid of meaning. There are many plans like this: they adorn the libraries of planning schools, and sometimes they are useful to a planning historian – as long as the concern is with plans and not with planning. Planning is a process, and like all processes must have an engine to drive it. To prepare plans, and to articulate goals, is one thing; to give them substance is another. Thus, without in any way belittling the symbolic value of declarations, it is also necessary to devise mechanisms to prompt, attract, bribe, and if necessary force, local governments to formulate and implement local plans which accord with the state goals.

The precise way in which this can be done will vary from state to state. Oregon has an effective system for consultation, agreeing plans, and enforcing and reviewing plans by means of a Land Conservation and Development Commission and the Land Use Board of Appeals (Buchsbaum and Smith 1993: 53). This is one model, but it may not work elsewhere. Each state has to evolve its own ways of devising, debating, and agreeing goals, and of implementing them.

Public participation and public confidence

A recurring theme in this book has been that the effectiveness of policies depends, in part, upon public support. This support can be proactive and enthusiastic (though often not for long), or it can be simply passive, but in either case a degree of public understanding is necessary. As discussed in the previous chapter, this has become increasingly important as the limits (and limitations) of technical knowledge have become more widely appreciated. The trust in value-free scientific solutions is now on a par with that which is placed in political judgments: neither can be accepted at face value. Some members of the public view any solution with skepticism. At the same time, the limited authority of planners has been further reduced by a greater appreciation of the unknowability of the future. Since the future cannot be predicted, the role of planning is the modest one of attempting to accommodate change efficiently, to maximize benefits, and to minimize unwanted side-effects. The role of the planner thus becomes facilitative – to assist in the public debate on planning policies, to point up the costs and benefits of alternative courses of action, and to articulate the uncertainties which have to be faced.

This is no easy task. Moral dimensions are also present which make it even more difficult. Problems of equity, of non-discrimination, of respect for differences of color and creed, of toleration for non-traditional lifestyles: these and a host of other matters that are important in human relationships cannot be ignored – at least not in the long term.

The long term is seldom at the heart of debate: the short term is too urgent. Yet the cumulative effect of policies and prejudices can be disastrous. Poorer areas become increasingly deprived as their innate abilities are overwhelmed by the sheer enormity of their problems. The social divisions of American society are increasing as those who can escape to suburban locations continue to do so. Walled communities with private policing provide a physical haven of security that may prove to be unsustainable. The social balkanization of metropolitan areas guarantees only an uncertain truce between the advantaged and the disadvantaged.

Certainly, the problems of some inner cities are viewed by suburbanites as overwhelming. The frightening description of some of these as 'no-go areas' underlines both the severity of the problems and the perceived hopelessness of public action. But much of the descriptive language obscures the truth. The *magnitude* of the problems is greatly exaggerated by their *concentration*. As Henry Cisneros (1995) has pointed out, though Hartford, Connecticut, is one of America's most distressed cities, its problems are not overwhelming when viewed on a metropolitan scale (see Box 19.5). Cisneros presented an optimistic view. Others may find it difficult to accept his view. For example, in his book *Understanding Urban Unrest*, Gale (1996) argues that effective policies for the urban poor require fundamental changes in the economic structuring of the country and a new federal agenda. Like many commentators, he sees little sign of this coming to pass.

Catastrophic events

There are some events that cause us to assess and reassess the role of planning. These events could be caused by man or a natural occurrence. Tornadoes have destroyed neighborhoods and small towns. The impact of September 11, 2001 on the redevelopment of New York has been debated and will undoubtedly change planning in New York and the entire metropolitan area. Climate change is constantly being debated. The results of climate change can have serious repercussions.

BOX 19.5 HARTFORD'S METROPOLITAN POVERTY PROBLEM IS SOLUBLE

Hartford, the capital of Connecticut, has become one of America's most distressed cities. Between 1950 and 1990, the city's population dropped 21 percent to 139,000. In 1989 city residents' average income was 53 percent of suburbanites' income, and over 27 percent of city residents were poor. Crime rates have soared, and school failure rates were so high that last year the city brought in a private management company to run the city's school system. Hartford's social agony seems unsolvable.

Yet viewed from a regional perspective, problems in the 1-million Hartford metropolitan area are not so unsurmountable: out of every 100 residents only 3 are poor and white, 2 are poor and Hispanic, and less than 2 are poor and African American. Poor whites are scattered throughout the metropolitan areas: only 12 percent live in Hartford, and only 13 percent live in poor neighborhoods. By contrast, 76 percent of poor Hispanics and 80 percent of poor African Americans live in city neighborhoods, and nearly 9 out of 10 poor minorities live in neighborhoods of concentrated poverty.

The problem is not the region's overall level of poverty – only seven out of every hundred area residents are poor – but the high concentration of minority poor in inner-city areas. Viewed in that light, greater Hartford is capable of absorbing poor minorities into the region's prosperous, middle-class society just as it already integrates poor whites into that society.

Source: Cisneros 1995:10

Rises in sea levels have been reported. Changes in the patterns and amounts of rain have been documented. The use of alternative fuels to reduce the greenhouse effect is being practiced in many parts of the United States and the world.

No area is immune from a catastrophe. An event could occur anywhere in the United States. Tornadoes routinely occur in what is referred to as 'Tornado Alley', the central states of the US. However, tornadoes are not restricted to one area. They can be seen throughout the US – the south has recently experienced severe damage.

Hurricanes have caused devastation throughout the US, most recently with Hurricane Sandy in 2012. In 2005, New Orleans and other parts of the Gulf Coast were hit by Hurricane Katrina. New Orleans is the largest city in Louisiana. Its past has been well documented. Whenever the name New Orleans is mentioned, several things come to mind – Mardi Gras, Bourbon Street, the Mississippi River, and spicy Cajun food.

New Orleans has undergone a transformation over the last thirty years or so. Industries have left the city for other locations. Jobs have been lost. According to the US Census Bureau, its population has decreased from 484,674 in 2000 to an estimated 2006 figure of 223,000. In 2000, over two-thirds of the population was African-American. Whites comprised approximately 28 percent of the city's population. The population has also been racially segregated for years.

Poverty has always plagued New Orleans. In 2000, it had the fifth highest concentration of poverty out of the 100 largest metropolitan areas in the United States. Over 80 percent of the African-American population live below the poverty level. Poverty is concentrated in 'high poverty' neighborhoods – neighborhoods where the poverty rate exceeds 40 percent. Berube and Katz (2005: 1) note that 'nearly 50,000 poor New Orleanians lived in neighborhoods where the poverty rate exceeded 40 percent'. Much of this population is African-American. Many of these individuals resided in public housing.

New Orleans is situated in a bowl, with the vast majority of the city below sea level. Moreover, the city continues to sink. Add this information to the continued loss of wetlands, and New Orleans has always been vulnerable to flooding. Drainage systems built after World War II and levees protected New Orleans for many years. To many people, New Orleans was an accident waiting to happen.

Hurricanes are no stranger to New Orleans. Residents have had to endure a number of these. However, the city was not ready for what happened in late August 2005. The city would be changed forever.

On August 29, 2005, Hurricane Katrina hit just east of the city. The storm surge caused a massive failure in the city's levee system. Some 80 percent of the city was flooded, with the poor and African-American populations experiencing the brunt of the devastation since they resided in the low-lying flood-prone areas. In fact, just over 80 percent of the city's extreme poverty census tracts were flooded.

What transpired in the days to follow was witnessed by millions of people. The devastation was extensive (see Plates 44 and 45). Over 1,700 lives were lost. Low-lying neighborhoods were destroyed. Looting of stores became commonplace. Roads were flooded or severely damaged. Pictures of people sitting on the roofs of their homes waiting to be rescued will forever be etched in people's minds.

Criticisms over how the city, state, and federal government responded to the disaster came quickly and were fierce in nature. A lack of leadership was clearly evident. While people suffered, officials blamed each other. The need for effective leadership was clearly evident.

A number of recovery efforts were launched, including the Bring Back New Orleans Commission, the New Orleans Neighborhoods Rebuilding Plan, and the Unified New Orleans Plan. An Office of Recovery Management was established to coordinate the recovery process. This office sought to coordinate the efforts of the various planning efforts. In the months that followed Katrina, debates were heard loud and clear about concentrated poverty and why the poor and African-American populations suffered disproportionately from the disaster.

There is no panacea for the problems that are facing New Orleans and its residents. Some problems have existed for years. Citizen distrust in government has

Plate 44 Devastation in New Orleans after Katrina

Photo courtesy of Fritz Wagner

Plate 45 Devastation in New Orleans after Katrina

Photo courtesy of Fritz Wagner

existed for years. Concentrated poverty is no stranger to the city. Infrastructure has been neglected for years. It will take massive amounts of funding to bring back New Orleans. Triage efforts will be needed in the short term. For the long term, it will take all levels of government, the private sector, the non-profit sector, and the citizens themselves to rebuild the city.

Rebuilding New Orleans undoubtedly covers every topic in this book. Strong and effective city planning will be needed. Neighborhoods and homes will have to be rebuilt. Land use regulations must be reviewed, revised, and ultimately enforced. Schools will need to be repaired or rebuilt. Concentrated poverty areas must become a thing of the past. Employment opportunities will be needed for the existing residents and eventual newcomers. Park and recreation areas and equipment will need to be restored and purchased. The city's infrastructure system will need to be examined and repaired. New infrastructure will have to be built. The days of deferring maintenance to a later date must be stopped. Coastal erosion issues must be addressed. Wetlands must be restored. Some of the storm surge could have been lessened by wetlands acting as natural sponges. An improved levee system must be constructed to withstand a Category 5 hurricane (winds greater than 155 miles per hour and storm surges greater than 18 feet above normal). Massive amounts of waste and debris will have to be removed. The importance of disaster and evacuation planning and hazard mitigation cannot be ignored. Cities must be prepared to deal with catastrophic events and the problems that follow such events.

Progress has been made in the rebuilding of New Orleans. It will be a work-in-progress for a number of years. Levees have been rebuilt to withstand a Category 3 hurricane; not for a Category 5, which residents wanted. Federal funds have been targeted for the rebuilding effort. Existing homes have been repaired and new housing has been built.

The issue of new housing continues to be a point of contention for many people. New housing is going up in the Ninth Ward (see Plates 46 and 47). Unfortunately, it is not affordable to many of the Ward's residents that were displaced by Katrina. They

Plate 46
**New Housing
in New Orleans**

Photo by author

Plate 47 New Housing in New Orleans
Photo by author

see it as housing to attract new residents with higher incomes to the area. What about the existing residents and their housing needs? Racial tensions continue to be visible.

In 2010, the New Orleans City Council passed an ordinance approved a new Master Plan called *Plan for the 21st Century: New Orleans 2030*. It offers a city vision for livability, opportunity, and sustainability. The Master Plan's *Land Use Plan* carries the force of law. This means that the City Council and administrative officials cannot make zoning or land use decisions that are contrary with the goals, policies, and strategies of the *Land Use Plan*. Examples of such decisions would include zoning amendments, preliminary and final approval of subdivision plans and plats, site plans, approval of planned unit development or other site-specific development plans, and variances.

Conflicts over what needs to get done and who is responsible for doing various things will occur throughout the rebuilding process. Intragovernmental and intergovernment cooperation and coordination will be needed. Rebuilding information must be distributed to all parties. It is vital that the public be involved throughout the rebuilding process. Their spirit and confidence must be restored.

In conclusion

This book has highlighted many of the problems facing US cities and states. Problems such as lack of decent and affordable housing, overburdened public services, climate change, availability of water, pollution of all types, overcrowded schools, unemployment, crime, etc., have plagued areas for years. As evidenced throughout the book, problems change over time, just as cities change. We must recognize that areas must be proactive in identifying and responding to problems. Simply sitting idly by will not reduce or eliminate the problems. All levels of government, the private sector, and the non-profit sector have roles in responding to the problems. Cooperation, communication, and coordination are of paramount importance in identifying and responding to any problem. Failure to recognize this may serve to exacerbate problems.

While the majority of students reading this book will initially be dealing with these problems in the US and other industrialized countries, it is imperative that they realize that other areas suffer from these and other planning challenges. The term 'Global North' commonly refers to wealthier and industrialized nations like the US, UK, Japan, France, Italy, Sweden, Australia, and New Zealand. Conversely, the term 'Global South' commonly refers to the less wealthy, economically and technologically advanced countries in Africa, Asia, Latin America – predominantly nations of the Southern Hemisphere. They can be also be referred to as the developing world.

The problems facing the Global South identify important challenges for the planning profession. As Watson (2009: 2259) points out, 'by 2008, for the first time in history, the majority of the world's population would be living in cities and, in future years, most of all new global population growth will be in cities in the "developing" world'. How are cities in these regions going to identify and respond to such complex issues as increasing amounts of rapid population growth, concentrated poverty, lack of water and sanitation facilities, lack of educational facilities, human and civil rights, climate change, changing spatial structures of area, population displacements, natural disasters, lack of food, building cities, etc. Planning students in the US should become aware of these daunting challenges facing planning in other parts of the world.

The title of this chapter was chosen purposely. It has been no part of the objective of the chapter (or of the book) to set out a program of reform. Other authors have embarked on this difficult task with careful thought and enthusiasm. Notable examples have included Calthorpe 1993, Cisneros 1993, Downs 1994, Rusk 1993; and many more will follow. Our intention has been to highlight a number of questions that arise in designing and tackling policies of reform. Several matters stand out in importance: freedom of information, public debate and involvement in the planning process, establishing appropriate mechanisms for dealing with different types of issue, seeking various methods of persuasion, and mediation where there is reluctance to abide by policies for the common good (and, where such methods are inadequate, designing regulatory instruments).

The overriding objective, however, has been to demonstrate the huge difficulties involved in dealing with urbanization, transportation, land use planning, discrimination, and those complexes of issues which are labeled 'urban' or simply 'social'. It is because so much effort is put into trying to solve huge insoluble matters that insufficient attention is given to the more restricted issues on which progress can be made. We must learn from the past. Some of our old ways of thinking must be rethought. New ways of examining and responding to issues must be understood so that we can effectively plan for the future.

Fortunately, there will be some who will find this approach far too clinical and unsatisfying. They will passionately strive for reform, with a disdain for the difficulties that this book has identified. Such activists have a crucial role to play in the political process.

Further reading

The US Department of Energy outlined its strategy for nuclear waste disposal at Yucca Mountain, Arizona on its website (http://www.ocrwm.doe.gov/ym_repository). Online features include technical reports, outreach programs, news updates, and facts about the project.

LGC, the Local Government Commission (http://www.lgc.org), is a non-profit organization dedicated to creating a livable community by working with the local authorities. Along with planning tips on topics such as public participation and regional planning, the LGC applies

Ahwahnee and smart growth principles to land use controls and community living.

The National Low Income Housing Coalition (http://www.nlihc.org) conducts research, and lobbies in Washington, DC, for affordable housing. Its goal is to end the affordable housing crisis and it has partnerships in almost every state. Its website gives an online visitor a summary of federal bills and updates in housing news across the country.

New Jersey's Council on Affordable Housing has an outstanding website (http://www.state.nj.us/dca/coah) that supplies internet links to almost anything related to affordable housing, including resources and research.

Another affordable housing government website that stands out is the Illinois Housing Development Authority (http://www.ihda.org). Along with updates for government programs, the website has meeting minutes and public outreach notices for local residents.

Planetizen (http://www.planetizen.com) is a planning and development network exchange that has regular correspondents as well as a blog, news posts, a job board, and comments on urban planning issues. Contributors include planners, real estate developers, government workers, architects, journalists, and others. The website offers too much to summarize and is a good source of information on current trends or, at least, will direct the visitor to an outside source of information.

For urban core enthusiasts, the International Downtown Association (IDA) (http://www.ida-downtown.org) offers a website with successful strategies and reports on making a livable and vibrant downtown area. Its resource links page is extensive and supplies internet links to organization websites as well as government publications.

The International Association for Public Participation (http://www.iap2.org) seeks to improve the relationship between the public and those that make decisions that affect the public. The Association provides an online journal and industry news for researchers and practitioners.

A similar organization with a national focus, America Speaks (http://www.americaspeaks.org), focuses on helping the public to become involved in government efforts. Along with national news, the non-profit organization offers a range of services from facilitating meetings to providing research in its 'democracy lab'.

The Brookings Institution offers an excellent site on Hurricane Katrina and the recovery efforts – The New Orleans Index: Tracking Recovery in the Region. It can be found at http://www.brookings.edu/reports/2007/08neworleansindex.aspx.

On the nature of policy 'problems' and the difficulties these pose for the policy process, see Rittel and Webber (1973) 'Dilemmas in a general theory of planning'; Kingdon (1984) *Agendas, Alternatives and Public Policies*; and Wildavsky (1987) *Speaking Truth to Power: The Art and Craft of Policy Analysis*.

Downs' 1972 paper, 'Up and down with ecology – the *issue-attention* cycle', is an interesting analysis of the fickleness of public concern.

Wills (1999) offers a historical study of distrust of government in *A Necessary Evil: A History of American Distrust of Government*.

Group homes are dealt with in Jaffe and Smith (1986) *Siting Group Homes for Developmentally Disabled Persons*. On childcare facilities, see Cibulskis and Ritzdorf (1989) *Zoning for Child Care*.

An excellent review of the issues involved in current debates on the taking issue and compensation for affected property owners in provided by Yandle (1995) *The Property Rights Rebellion: Land Use Movements in the 1990s*; and by Strong *et al.* (1996) 'Property rights and takings'. See also Tibbetts (1995) 'Everybody's taking the fifth'. A collection of papers on the legal issues involved is Callies (1996) *Takings: Land-development Conditions and Regulatory Takings after Dolan and Lucas*. Additional examinations of property rights legislation can be found in Emerson and Wise (1997) 'Statutory approaches to regulatory takings'; and Oswald (2000) 'Property rights legislation and the police power'.

A major study of class and racial prejudice is Danielson (1976) *The Politics of Exclusion*, which, though not up to date, provides an excellent perspective of the 1960s and early 1970s. The details have changed since then, but the basic themes are the same. Just how little has changed can be judged from Keating (1994) *The Suburban Racial Dilemma: Housing and Neighborhoods*. More recent examinations of this topic can be found in Yinger (1995) *Closed Doors, Opportunities Lost: The Continuing Costs of Housing Discrimination*; Coulibaly *et al.* (1998) *Segregation in Federally Subsidized Low-Income Housing in the United States*; Ellen (2000) *Sharing America's Neighborhoods: The Prospects for Stable Racial Integration*; and Rubinowitz and Rosenbaum (2000) *Crossing the Class and Color Lines: From Public Housing to White Suburbia*.

The chapter contains a quotation from Cisneros (1995) *Regionalism: The New Geography of Opportunity*, but more easily obtainable is the same author's edited collection of essays: *Interwoven Destinies* (1993). There is a vast literature on the problems of the inner city, segregation, and discrimination. A useful overview is given in Downs (1994) *New Visions for Metropolitan America*. See also Gale (1996) *Understanding Urban Unrest: From Reverend King to Rodney King*; and Darby and Anderson (1996) *Reducing Poverty in America*. A collection of papers on various aspects of urban policy is to be found in a special issue of *Urban Affairs*, May 1995.

Recent works dealing with Hurricane Katrina and New Orleans include: Birch and Wachter (2006) *Rebuilding Urban Places After Disaster: Lessons from Hurricane Katrina*; Burns and Thomas (2006) 'The failure of the nonregime: how Katrina exposed New Orleans as a regimeless city'; Hartman and Squires (2006) *There is No Such Thing as a Natural Disaster: Race, Class, and Hurricane Katrina*; Olshansky (2006) 'Planning after Hurricane Katrina'; Pastor *et al.* (2006) 'In the wake of the storm: environment, disaster, and race after Katrina'; Kales *et al.* (2007) 'Reconstruction of New Orleans after Hurricane Katrina: a research perspective'; Lukensmeyer (2007) 'Large-scale citizen engagement and the rebuilding of New Orleans: a case study'; Nelson *et al.* (2007) 'Planning, plans, and people: professional expertise, local knowledge, and governmental action in post-Katrina New Orleans'; Reible (2007) 'Hurricane Katrina: environmental hazards in the disaster area'; Olshansky *et al.* (2008) 'Planning for the rebuilding of New Orleans'; Bullard and Wright (2009) *Race, Place, and Environmental Justice After Hurricane Katrina: Struggles to Reclaim, Rebuild, and Revitalize New Orleans and the Gulf Coast*; Olshansky and Johnson (2010) *Clear as Mud: Planning for the Rebuilding of New Orleans*; and Plyer *et al.* (2011) *Resilience and Opportunity: Lessons from the U.S. Gulf Coast after Hurricane Katrina and Rita*.

Questions to discuss

1 'Since the problems are so difficult to define, wouldn't it be better simply to get on with doing something constructive, rather than escaping into clever arguments?' Discuss.

2 Why does public interest in important problems wax and wane?

3 Do you think that it is a good idea to compensate residents who have obnoxious hazardous waste facilities located close to them?

4 'Forcing local governments to allow childcare facilities is acceptable; but it is not right that locally elected representatives should be made to accept low-income housing.' Discuss.

5 What roles can states play in urban planning?

6 How far do you think that resolution of urban problems must involve the reduction of discrimination?

7 Why do you think property rights legislation is so difficult to pass at the state and national levels of government?

Main cases

Stover: People v. *Stover*: 191 N.E.2d 272 (1963)

Tahoe-Sierra Preservation Council, Inc. v. *Tahoe Regional Planning Agency*: 535 US 302 (2002)

West Montgomery County Citizens Association v. *Maryland National Capitol Park and Planning Commission*: 522 A.2d 1328 (1987)

Bibliography

1000 Friends of Oregon (n.d.) 'Questions and answers about Oregon's land use program: UGBs and housing costs'

Abbott, C. (1991) 'Urban design in Portland, Oregon, as policy and process, 1960–1989', *Planning Perspectives* 6: 1–18

Abbott, C., Howe, D., and Adler, S. (eds) (1994) *Planning the Oregon Way: A Twenty-Year Evaluation*, Corvallis, OR: Oregon State University Press

Abbott, W. W., Moe, M. E., and Hanson, M. (1993) *Public Needs and Private Dollars: A Guide to Dedications and Development Fees*, Point Arena, CA: Solano Press (*Supplement* 1995)

Ackerman, B. A. and Hassler, W. T. (1981) *Clean Coal, Dirty Air, or How the Clean Air Act Became a Multibillion Dollar Bail-Out for High-Sulphur Coal Producers and What Should Be Done About It*, New Haven, CT: Yale University Press

Adler, R. (1993) *The Clean Water Act: Twenty Years Later*, Washington, DC: Island Press

Adler, R. (1994) 'The Clean Air Act: has it worked?' *EPA Journal* 20(1/2): 10–14

Adler, S. (2012) *Oregon Plans: The Making of an Unquiet Land-Use Revolution*, Corvallis, OR: Oregon State University Press

Advisory Commission on Intergovernmental Relations (1964) *Impact of Federal Urban Development Programs on Local Government Organization and Planning*, Washington, DC: US Government Printing Office

Advisory Commission on Intergovernmental Relations (1970) *Annual Report*, Washington, DC: US Government Printing Office

Advisory Committee on Regulatory Barriers to Affordable Housing (1991) *Not in My Backyard*, Washington, DC: US Government Printing Office

Advisory Council on Historic Preservation (ACHP) (annual) *Report to the President and Congress*, Washington, DC: ACHP

Alexander, E. R. (1981) 'If planning isn't everything, maybe it's something', *Town Planning Review* 52: 131–42

Alexander, E. R. (1992) *Approaches to Planning: Introducing Current Planning Theories, Concepts and Issues* (2nd edition), Philadelphia: Gordon and Breach

Allen, G., Alley, D., and Hicks, E. (1994) *Development, Marketing, and Operation of Manufactured Home Communities*, Washington, DC: Urban Land Institute

Alley, W. M. and Alley, R. (2012) *Too Hot to Touch: The Problem of High-Level Nuclear Waste*, New York: Cambridge University Press

Altman, D. (2011) *Direct Democracy Worldwide*, New York: Cambridge University Press

Altshuler, A. A. (1965) *The City Planning Process: A Political Analysis*, Ithaca, NY: Cornell University Press (extract from this book in Stein 1995)

American Economic Development Council (1984) *Economic Development Today*, Chicago: American Economic Development Council

American Farmland Trust (1988) *Protecting Farmland through Purchase of Development Rights: The Farmers' Perspective*, Washington, DC: American Farmland Trust

American Farmland Trust (2001) 'Fact sheet: transfer of development rights', Washington, DC: American Farmland Trust

American Institute of Certified Planners (2005) *AICP Code of Ethics and Professional Conduct* (adopted March 19, 2005, effective June 1, 2005)

American Planning Association (1994) *Planning and Community Equity*, Chicago: Planners Press

American Planning Association (2002) *Growing Smart (SM) Legislative Guidelines, 2002*, Chicago: American Planning Association

American Planning Association (2007) *Policy Guide on Community and Regional Food Planning*, Chicago: American Planning Association

American Planning Association Planning Advisory Service (2012) *Incorporating Sustainability into the Comprehensive Plan: PAS Essential Info Packet EIP-33*, Chicago: APA Planning Advisory Service

Ames, D. L., Callahan, M. H., Herman, B. L., and Siders, R. J. (1989) *Delaware Comprehensive Historic Preservation Plan*, Newark, DE: Center for Historic Architecture and Engineering, University of Delaware

An, D., Gordon, P., and Richardson, H. W. (2002) 'The continuing decentralization of people and Jobs in the United States', paper presented at the 41st Annual Meeting of the Western regional Science Association, Monterey, California, February 17–20

Anderson, City of (2007) *City of Anderson General Plan: General Definitions*, Anderson, California: City of Anderson Planning Department

Anderson, L. T. (1995) *Guidelines for Preparing Urban Plans*, Chicago: Planners Press

Anderson, R. and Lohof, A. (1997) 'The U.S. experience with economic incentives to environmental pollution control policy', EE-0216A, Washington, DC: Environmental Law Institute

Annapolis, City of (2009) *Sustainable Annapolis Community Action Plan*, Annapolis, MD: City of Annapolis

Appleyard, D. (1981) *Livable Streets*, Berkeley: University of California Press

Arnold, R. *et al.* (2010) *Reducing Congestion and Funding Transportation Using Road Pricing in Europe and Singapore*, Washington, DC: Federal Highway Administration, US Department of Transportation

Arnstein, S. R. (1969) 'A ladder of citizen participation', *Journal of the American Planning Association* 35: 216–24

Asplund, R. W. (2008) *Profiting from Clean Energy: A Complete Guide to Trading in Solar, Wind, Ethanol, Fuel Cell, Carbon Credit Industries, and More*, Indianapolis, IN: Wiley

Association of New Jersey Environmental Commissions (2003) 'The master plan: smart growth, the master plan & environmental protection', Mendham, NJ: Association of New Jersey Environmental Commissions

Atash, F. (1994) 'Redesigning suburbia for walking and transit: emerging concepts', *Journal of Planning and Development* 120: 48–57

Atkinson, S. and Tietenberg, T. (1991) 'Market failure in incentive-based regulations: the case of emissions trading', *Journal of Environmental Economics and Management* 21: 17–32

Audirac, I., Shermyen, A. H., and Smith, M. T. (1990) 'Ideal urban form and visions of the good life: Florida's growth management dilemma', *Journal of the American Planning Association* 56: 470–82

Ausubel, J. H. and Sladovich, H. E. (eds) (1989) *Technology and Environment*, Washington, DC: National Academy Press

Avery, R. B., Canner, G. B., and Cook, R. E. (2005) 'New information reported under HMDA and its application in fair lending enforcement,' *Federal Reserve Bulletin* 91(Summer): 344–94

Avery, R. B., Brevoort, K. B., and Canner, G. B. (2006) 'Higher-priced home lending and the 2005 HMDA data,' *Federal Reserve Bulletin* 8 September: A123-A166

Babcock, R. F. (1966) *The Zoning Game: Municipal Practices and Policies*, Madison, WI: University of Wisconsin Press

Babcock, R. F. and Larsen, W. U. (1990) *Special Districts: The Ultimate in Neighborhood Zoning*, Cambridge, MA: Lincoln Institute of Land Policy

Babcock, R. F. and Siemon, C. L. (1985) *The Zoning Game Revisited*, Boston: Oelgeschlager, Gunn and Hain

Baden, B. M. and Coursey, D. L. (1999) *Effects of Impact Fees on the Suburban Chicago Housing Market*, Heartland Policy Study no. 93, Chicago, IL: The Heartland Institute

Bainbridge Island, City of (2004) 'Cultural element', Bainbridge Island, WA: City of Bainbridge Island

Bair, F. H. (1984) *The Zoning Board Manual*, Chicago: Planners Press

Banach, M. and Canavan, D. (1987) 'Montgomery County agricultural preservation program', in Brower and Carol (1987)

Bandy, A. (2005) *Surviving Homeowner Associations*, Victoria, Canada: Trafford Publishing

Banfield, E. C. (1959) 'Ends and means in planning', *International Social Science Journal* 11; reprinted in Faludi (1973)

Banfield, E. C. (1961) 'The political implications of metropolitan growth', *Daedalus* 90: 61–78

Banfield, E. C. (1975) 'Corruption as a feature of governmental organization', *Journal of Law and Economics* 18: 587–615

Barbour, E. (2002) 'Metropolitan growth planning in California, 1900–2000', San Francisco: Public Policy Institute of California

Barnekov, T., Boyle, R., and Rich, D. (1989) *Privatism and Urban Policy in Britain and the United States*, New York: Oxford University Press

Barnett, J. (1982) *An Introduction to Urban Design*, New York: Harper and Row

Barnett, J. (2003) *Redesigning Cities*, Chicago, IL: APA Planners Press

Barrett, C. (2002) *Everyday Ethics for Practicing Planners*, Chicago, IL: APA Planners Press

Barrett, S. and Hill, M. J. (1993) Unpublished research report quoted in Ham, C. and Hill, M. *The Policy Process in the Modern Capitalist State*, New York: Simon and Schuster

Barrett, T. (1981) 'Reaping the margins: A century of community gardening in America', *Landscape Journal* 25(2): 1–8

Barrios, S. and Barrios, D. (2004) 'Reconsidering economic development: the prospects for economic gardening', *Public Administration Quarterly* 28(1/2): 70–101

Barry, C. (2001) 'Land use regulation and residential segregation: does zoning matter?' *American Law and Economics Review* 3(2): 251–274

Barth, J. R. (2009) *The Rise and Fall of the US Mortgage and Credit Markets: A Comprehensive Analysis of the Market Meltdown*, Hoboken, NJ: John Wiley & Sons

Bartik, T. J. (1991) *Who Benefits from State and Local Economic Development Policies?*, Kalamazoo, MI: W. E. Upjohn Institute for Employment Research

Barton, S. E. and Silverman, C. J. (eds) (1994) *Common Interest Communities: Private Governments and the Public Interest*, Berkeley, CA: Institute of Governmental Studies Press

Bass, R. E. and Herson, A. I. (1993) *Mastering NEPA: A Step-by-Step Approach*, Point Arena, CA: Solano Press

Bassett, E. M. (1940) *Zoning: The Laws, Administration, and Court Decisions during the First Twenty Years* (revised edition), New York: Russell Sage Foundation

Baton Rouge, City of (2009: 2) *Urban Design Handbook*, Baton Rouge, Louisiana: City-Parish Planning Commission

Baxandall, R. and Ewen, E. (2000) *Picture Windows: How the Suburbs Happened*, New York: Basic Books

Bay Vision 2020 Commission (1991) *Bay Vision 2020: The Commission Report*, San Francisco: The Commission

Beatley, T. (1988) 'Ethical issues in the use of impact fees to finance community growth', in Nelson (1988)

Beatley, T. (1994a) *Ethical Land Use: Principles of Policy and Planning*, Baltimore: Johns Hopkins University Press

Beatley, T. (1994b) *Habitat Conservation Planning: Endangered Species and Urban Growth*, Austin: University of Texas Press

Beatley, T. (2005) *Native to Nowhere: Sustaining Home and Community in a Global Age*, Washington, DC: Island Press

Beatley, T., Brower, D. J., and Schwab, A. K. (2002) *An Introduction to Coastal Zone Management* (2nd edition), Washington, DC: Island Press

Beierle, T. C. and Cayford, J. (2002) *Democracy in Practice: Public Participation in Environmental Decisions*, Washington, DC: Resources for the Future

Bell, A. and Parchomovsky, G. (2006) 'The use of public use', *Columbia Law Review* 106: 1412–49

Bellevue, Washington, City of (2006) *Comprehensive Plan*, Bellevue, Washington: City of Bellevue

Belzer, D. and Autler, G. (2002) 'Transit-oriented development: moving from rhetoric to reality', Washington, DC: The Brookings Institution

Benjamin, L., Rubin, J. S., and Zielenbach, S. (2004) 'Community Development Financial Institutions: current issues and future prospects', *Journal of Urban Affairs* 26(2): 177–95

Ben-Zadok, E. (2005) 'Consistency, concurrency, and compact development: three faces of growth management in Florida', *Urban Studies* 42(12): 2167–90

Bergal, J., Hiles, S. S., and Koughan, F. (2007) *City Adrift: New Orleans Before and After Katrina*, Baton Rouge, LA: Louisiana State University Press

Berger, L. (1991) 'Inclusionary zoning as takings: the legacy of the *Mount Laurel* cases', *Nebraska Law Review* 70: 186–228

Bergman, M. P. (ed.) (2011) *Cap-And-Trade Program Considerations*, Hauppauge, NY: Nova Science Publishers

Berhardsen, T. (2002) *Geographic Information Systems* (3rd edition), New York: John Wiley and Sons

Bernick, M. and Cervero, R. (1996) *Transit Villages in the 21st Century*, New York: McGraw-Hill

Bermingham, T. W. (1988) Section 36: Density and Height Bonuses, Toronto: Canadian Bar Association

Berube, A. and Katz, B. (2005) *Katrina's Window: Confronting Concentrated Poverty across America*, Washington, DC: The Brookings Institution, Metropolitan Policy Program

Berube, A., Singer, A., Wilson, J. H., and Frey, W. W. (2006) *Finding Exurbia: America's Fast-growing Communities at the Metropolitan Fringe*, Washington, DC: The Brookings Institution

Big Look Task Force on Oregon Land Use Planning (2007) 'The Big Look preliminary findings and recommendations,' Salem, OR: Big Look Task Force on Oregon Land Use Planning, State of Oregon

Birch, D. L. (1987) *Job Creation in America: How Our Smallest Companies Put the Most People to Work*, New York: Free Press

Birch, E. L. and Roby, D. (1984) 'The planner and the preservationist: an uneasy alliance', *Journal of the American Planning Association* 50: 194–207

Birch, E. L. and Wachter, S. M. (2006) (eds) *Rebuilding Urban Places After Disaster: Lessons from Hurricane Katrina*, Philadelphia: University of Pennsylvania Press

Bishir, C. W. (1989) 'Yuppies, Bubbas, and the politics of culture', in Carter and Herman (1989)

Bishop, I. D. and Lange, E. (eds) (2005) *Visualization in Landscape and Environmental Planning*, London: Taylor and Francis

Blackburn, W. R. (2007) *The Sustainability Handbook: The Complete Management Guide to Achieving Social, Economic, ad Environmental Responsibility*, Washington, DC: Island Press

Blair, J. P. (1995) *Local Economic Development: Analysis and Practice*, Thousand Oaks, CA: Sage

Blake, P. (1964) *God's Own Junkyard: The Planned Deterioration of America's Landscape*, New York: Holt, Reinhart and Winston

Blakely, E. J. (1989) *Planning Local Economic Development: Theory and Practice*, Thousand Oaks, CA: Sage

Blakely, E. J. and Leigh, N. G. (2010) *Planning Local Economic Development: Theory and Practice* (4th edition), Thousand Oaks, CA: Sage

Blank, L. D. (1994) 'Seeking solutions to environmental inequity: the Environmental Justice Act', *Environmental Law* 24: 1109–36

Blumenthal, S. and Siler, B. (1990) *Tax Incentives for Rehabilitating Historic Buildings: Fiscal Year 1989 Analysis*, Washington, DC: National Trust for Historic Preservation

Boarnet, M. G. and Compin, N. S. (1999) 'Transit-oriented development in San Diego County: the incremental implementation of a planning idea', *Journal of the American Planning Association* 65: 80–90

Bogart, W. T. (2001) 'Symposium on the seventy-fifth anniversary of *Village of Euclid* v. *Ambler Realty Co.*: "trading places": the role of zoning in promoting and discouraging intrametropolitan trade', *Case Western Reserve Law Review* 51: 697–719

Bond Market Association and Temel, J. W. (2001) *The*

Fundamentals of Municipal Bonds (5th edition), Indianapolis, IN: Wiley

Booher, D. E. and Innes, J. E. (2002) 'Network power in collaborative planning', *Journal of Planning Education and Research* 21: 221–236

Boorstin, D. (1958) *The Americans: The Colonial Experience*, New York: Random House

Boorstin, D. (1965) *The Americans: The National Experience*, New York: Random House

Boorstin, D. (1973) *The Americans: The Democratic Experience*, New York: Random House

Bosselman, F. (1973) 'Can the town of Ramapo pass a law to bind the rights of the whole world?', *Florida State University Law Review* 1: 234–65

Bosselman, F. and Callies, D. (1972) *The Quiet Revolution in Land Use Control*, Washington, DC: Council on Environmental Quality

Bosselman, F., Callies, D., and Banta, J. (1973) *The Taking Issue: An Analysis of the Constitutional Limits of Land Use Control*, Washington, DC: Council on Environmental Quality

Boston, City of (1988) *Design Guidelines for Neighborhood Housing*, City of Boston Public Facilities Department

Boston Main Street Program (2003) 'Annual Report 2003', Boston: Boston Main Street Program

Boulder, City of (2007) *Social Sustainability Strategic Plan*, Boulder, CO: City of Boulder

Bourne, L. S. (1981) *The Geography of Housing*, London: Edward Arnold

Boyer, M. C. (1983) *Dreaming the Rational City: The Myth of American City Planning*, Cambridge, MA: MIT Press

Brooks, M. P. (2002) *Planning Theory for Practitioners*, Chicago, IL: APA Planners Press.

Brail, R. K. and Klosterman, R. E. (eds) (2001) *Planning Support Systems: Integrating Geographic Information Systems, Models, and Visualization Tools*, Redlands, CA: ESRI Press

Bratt, R. G., Stone, M. E., and Hartman, C. (eds) (2006) *A Right to Housing*, Philadelphia, PA: Temple University Press

Brener, K. (1999) 'Note: Belle Terre and single-family home ordinances: judicial perceptions of local government and the presumption of validity', *New York University Law Review* 74: 447–84

Brenman, M. and Sanchez, T. W. (2012) *Planning as if People Matter: Governing for Social Equity*, Washington, DC: Island Press

Bressi, T. (ed.) (1993) *Planning and Zoning New York City: Yesterday, Today, and Tomorrow*, New Brunswick: Center for Urban Policy Research, Rutgers – State University of New Jersey

Bressi, T. (ed.) (2002) *The Seaside Debates: A Critique of the New Urbanism*, New York: Rizzoli

BRIDGE (2000) '1999–2000 Annual Report', San Francisco: BRIDGE

Briffault, R. (2000) 'Localism and regionalism', *Buffalo Law Review* 48: 1–30

Brogan, H. (1986) *The Pelican History of the United States of America*, New York: Viking Penguin

Brooks, A. V. N. (1989) 'The office file box – emanations from the battlefield' [on the *Euclid* case], in Haar and Kayden (1989a)

Brooks, H. (1984) 'The resolution of technically intensive public policy disputes', *Science, Technology and Human Values* 9(1)

Brooks, H. (1988) Foreword to Graham, J. D., Green, L., and Roberts, M. J. *In Search of Safety: Chemicals and Cancer Risk*, Cambridge, MA: Harvard University Press

Brooks, M. E. (1999) *A Workbook for Creating a Housing Trust Fund*, Washington DC: Center for Community Change

Brooks, M. E. (2002) 'Housing Trust Funds Progress Report 2002: Local responses to America's housing needs', Frazier Park, CA: Center for Community Change

Brooks, M. P. (2002) *Planning Theory for Practitioners*, Chicago, IL: APA Planners Press

Brower, D. J. and Carol, D. S. (eds) (1987) *Managing Land Use Conflicts: Case Studies in Special Area Management*, Durham, NC: Duke University Press

Brown, L. J., Dixon, D., and Gilham, O. (2009) *Urban Design for an Urban Century: Placemaking for People*, Hoboken, New Jersey: John Wiley & Sons

Brueckner, J. K. and Fansler, D. A. (1983) 'The economics of urban sprawl: theory and evidence on the spatial size of cities', *The Review of Economics and Statistics* 65(3): 479–82

Bryner, G. C. (1993) *Blue Skies, Green Politics: The Clean Air Act of 1990*, Washington, DC: Congressional Quarterly Press

Buchsbaum, P. A. and Smith, L. J. (eds) (1993) *State and Regional Comprehensive Planning: Implementing New Methods for Growth Management*, Chicago: American Bar Association

Bucknall, B. (1988) *Of Deals and Distrust: The Perplexing Perils of Municipal Zoning*, Toronto: Canadian Bar Association, Unpublished paper

Budic, Z. D. (1994) 'Effectiveness of geographic information systems in local planning', *Journal of the American Planning Association* 60: 244–63

Bullard, R. D. (2000) *Dumping in Dixie: Race, Class, and Environmental Quality* (3rd edition), Boulder, CO: Westview Press

Bullard, R. D. and Wright, B. (eds) (2009) *Race, Place, and Environmental Justice After Hurricane Katrina: Struggles to Reclaim, Rebuild, and Revitalize New Orleans and the Gulf Coast*, Boulder, CO: Westview Press

Bullard, R. D., Johnson, G. S., and Torres, A. O. (eds) (2000) *Sprawl City: Race, Politics, and Planning in Atlanta*, Washington, DC: Island Press

Burchell, R. W. and Sternlieb, G. (1978) *Planning Theory in the 1980s*, New Brunswick, NJ: Center for Urban Policy Research, Rutgers University

Burchell, R. W., Listokin, D., and Pashman, A. (1994) *Regional Housing Opportunities for Lower-Income Households: An Analysis of Affordable Housing and Regional Mobility Strategies*, New Brunswick, NJ: Center for Urban Policy Research, Rutgers University

Burchell, R. W., Listokin, D., and Pashman, A. (1995) *Regional Housing Mobility Strategies in the United States*, Washington, DC: Department of Housing and Urban Development

Burchell, R. W., Listokin, D., and Galley, C. C. (2000) 'Smart growth: more than a ghost of urban policy past, less than a bold new horizon', *Housing Policy Debate* 11: 821–79

Burgess, B. B. (2001) *Fate of the Wild: The Endangered Species Act and the Future of Biodiversity*, Athens, GA: University of Georgia Press

Burns, P. and Thomas, M. O. (2006) 'The failure of the nonregime: how Katrina exposed New Orleans as a regimeless city', *Urban Affairs Review* 41(4): 517–27

Burrows, T. (ed.) (1989) *A Survey of Zoning Definitions*, Planning Advisory Service Report 421, Chicago: American Planning Association

Cairncross, F. (1992) *Costing the Earth: The Challenge for Governments, the Opportunities for Business*, Boston, MA: Harvard Business School Press

Cairncross, F. (1997) *Death of Distance*, New York: McGraw-Hill

Calavita, N. and Grimes, K. (1998) 'Inclusionary housing in California: the experience of two decades', *Journal of the American Planning Association* 64(2): 150–69

Calavita, N., Grimes, K., and Mallach, A. (1997) 'Inclusionary housing in California and New Jersey: a comparative analysis', *Housing Policy Debate* 8: 109–42

California Budget Project (2000) *Locked Out: California's Affordable Housing Crisis*, Sacramento: California Budget Project

California Budget Project (2001) *Still Locked Out: New Data Confirm California's Affordability Crisis Continues*, Sacramento: California Budget Project

California Resources Agency, State Water Resources Control Board (2002) *Addressing the Need to Protect California's Watershed: Working with Local Partnerships*, Report to the Legislature (as required by AB 2117, Chapter 735, Statutes of 2000), Sacramento, CA: California Resources Board

Callies, D. L. (1980) 'The quiet revolution revisited', *Journal of the American Planning Association* 46: 135–44

Callies, D. L. (1984) *Regulating Paradise: Land Use Controls in Hawaii*, Honolulu: University of Hawaii Press

Callies, D. L. (1994) *Preserving Paradise: Why Regulation Won't Work*, Honolulu: University of Hawaii Press

Callies, D. L. (ed.) (1996) *Takings: Land-Development Conditions and Regulatory Takings after Dolan and Lucas*, Chicago: American Bar Association

Callies, D. L. and Grant, M. (1991) 'Paying for growth and planning gain: an Anglo-American comparison of development conditions, impact fees, and development agreements', *Urban Lawyer* 23: 221; reprinted in Freilich and Bushek (1995)

Callies, D. L. and Tappendorf, J. A. (2001) 'Symposium on the seventy-fifth anniversary of *Village of Euclid* v. *Ambler Realty Co.*: unconstitutional land development conditions and the development agreement solution: bargaining for public facilities after Nollan and Dolan', *Case Western Reserve Law Review* 51: 663–96

Callies, D. L., Freilich, R. H., and Roberts, T. E. (1994) *Land Use: Cases and Materials* (2nd edition), St Paul, MN: West

Calthorpe, P. (1993) *The Next American Metropolis: Ecology, Community, and the American Dream*, Princeton, NJ: Princeton Architectural Press

Calthorpe, P. and Fulton, W. (2001) *The Regional City: Planning for the End of Sprawl*, Washington, DC: Island Press

Cambridge Systematics with the Texas Transportation Institute (2004) *Traffic Congestion and Reliability: Linking Solutions to Problems*, report prepared for the Federal Highway Administration, Cambridge, MA: Cambridge Systematics

Campbell, J. B. (2007) *Introduction to Remote Sensing* (4th edition), New York: Guilford Publications

Campbell, S. (1996) 'Green cities, growing cities, just cities?: Urban planning and the contradictions of sustainable development', *Journal of the American Planning Association* 62(3): 296–312

Campbell, S. and Fainstein, S. (eds) (1996) *Readings in Planning Theory*, Cambridge, MA: Blackwell (see also companion volume on urban theory edited by Fainstein and Campbell)

Carlin, A. (1973) 'The Grand Canyon controversy; or how reclamation justifies the unjustifiable', in Enthoven, A. and Freeman, A. M. (eds) *Pollution, Resources, and the Environment*, New York: Norton

Carlin, A. (1992) 'The US experience with economic incentives to control environmental pollution', EE-0216, Washington, DC: US Environmental Protection Agency, Office of Policy Planning and Evaluation

Carlino, G. and Chatterjee, S. (1999) *Postwar Trends in Metropolitan Employment Growth: Decentralization and Deconcentration*, Working Paper No. 99–10, Philadelphia: Federal Reserve Bank of Philadelphia

Carlson, T. and Dierwechter, Y. (2007) 'Effects of urban growth boundaries and residential development in Pierce County, Washington', *Professional Geographer* 59(2): 209–20

Carmona, M. and Tiesdell, S. (2007) *Urban Design Reader*, Oxford, UK: Architectural

Carmona, M, Heath, T., Oc, T., and Tiesdell, S. (2010) *Public Places Urban Spaces: The Dimensions of Urban Design* (2nd edition) Oxford, UK: Architectural Press

Carson, R. (1962) *Silent Spring*, New York: Houghton Mifflin (Penguin Edition 1965, reprinted 1991) (extract from this in Stein 1995)

Carter, T. (ed.) (1997) *Images of an American Land: Vernacular Architecture in the Western United States*, Albuquerque: University of New Mexico Press

Carter, T. and Herman, B. L. (1989) *Perspectives in Vernacular Architecture III*, Columbia, MI: University of Missouri Press

Casey, J. J. (1998) 'Bridging the great divide: SEWRPC, politics, and regional cooperation', *Marquette Law Review* 81: 505–760

Castle Coalition (2007) '50 State Report Card: tracking eminent domain reform since *Kelo*', Arlington, VA: Castle Coalition

Catlin, R. A. (1997) *Land Use Planning, Environmental Protection, and Growth Management: The Florida Experience*, Chelsea, MI: Ann Arbor Press

Caves, R. W. (1992) *Land Use Planning: The Ballot Box Revolution*, Newbury Park, CA: Sage

Caves, R. W. and Walshok, W. G. (1999) 'Adopting innovations in information technology', *Cities* 16: 3–1

Cedar Miller, S. (2003) *Central Park, an American Masterpiece: A Comprehensive History of the Nation's First Urban Park*, New York: Harry N. Abrams

Center for Community Change (2001) 'Home sweet home: why America needs a national housing trust fund', Frazier Park, CA: Center for Community Change

Center for Housing Policy (2000) 'Inclusionary zoning: a viable solution to the affordable housing crisis?', *New Century Housing* 1: 1–37 (entire issue)

Cervero, R. (1989) *America's Suburban Centers: The Land Use–Transportation Link*, Boston: Unwin Hyman

Cervero, R. B. (1994a) 'Rail transit and joint development: land market impacts in Washington, D. C. and Atlanta', *Journal of the American Planning Association* 60: 83–94

Cervero, R. (1994b) 'Transit-based housing in California: evidence on ridership impacts', *Transport Policy* 1: 174–83

Cervero, R. (1998) *The Transit Metropolis: A Global Inquiry*, Washington, DC: Island Press

Champion, A. G. (ed.) (1989) *Counterurbanization: The Changing Pace and Nature of Population Deconcentration*, London: Edward Arnold

Chapin, T., Connerly, C., and Higgins, H. (eds) (2007) *Growth Management in Florida: Planning for Paradise*, London: Ashgate

Charlemont, Town of (2003) Charlemont Master Plan. Charlemont, MA: Town of Charlemont

Checkoway, B. (ed.) (1994) 'Paul Davidoff and advocacy planning in retrospect', *Journal of the American Planning Association* 60: 139–61

Cheney, C. H. (1920) 'Zoning in practice', *National Municipal Review* 9: 31–43

Chichilnisky, G. and Heal, G. (eds) (2000) *Environmental Markets: Equity and Efficiency*, New York: Columbia University Press

Chifos, C. (2007) 'The sustainable communities experiment in the United States: insights from three federal-level initiatives,' *Journal of Planning Education and Research* 26(4): 435–49

Chinitz, B. (1990) 'Growth management: good for the town, bad for the nation?', *Journal of the American Planning Association* 56: 3–8

Chused, R. H. (2001) 'Symposium on the seventy-fifth anniversary of *Village of Euclid* v. *Ambler Realty Co.*: Euclid's historical imagery', *Case Western Reserve Law Review* 51: 597–616

Cibulskis, A. and Ritzdorf, M. (1989) *Zoning for Child Care* (Planning Advisory Service Report 422), Chicago: American Planning Association

Cisneros, H. G. (ed.) (1993) *Interwoven Destinies: Cities and the Nation*, New York: Norton

Cisneros, H. G. (1995) *Regionalism: The New Geography of Opportunity*, Washington, DC: US Department of Housing and Urban Development

Citizens' Housing and Planning Association (2008) *40B Housing Production Update*, Boston, MA: Citizens' Housing and Planning Association

Clarion Associates, Inc. (2000) 'The costs of sprawl in Pennsylvania: executive summary', Philadelphia, PA: 10,000 Friends of Pennsylvania

Clark, J. K., McChesney, R., Munroe, D. K., and Irwin, E. G. (2005) 'Spatial characteristics of exurban settlements', paper presented at the 52nd Annual North American Meeting of the Regional Science Association, Las Vegas, Nevada, November 12

Clark, J. R. (1996) *Coastal Zone Management Handbook*, Boca Raton, FL: Lewis Publishers

Clarke, K. C. (2000) *Getting Started with Geographic Information Systems* (3rd edition), New York: Prentice Hall College Division

Clavel, P., Forester, J., and Goldsmith, W. (eds) (1980) *Urban and Regional Planning in an Age of Austerity*, Oxford: Pergamon

Clawson, M. (1971) *Suburban Land Conversion in the United States: An Economic and Governmental Process*, Washington, DC: Resources for the Future

Cobb, R. L. (2000) 'Land use law: marred by public agency abuse', *Washington University Journal of Urban Law and Policy* 3: 195–214

Cochise County, Arizona (2004) 'Subdivision Regulations'

Cohen, C. E. (2006) 'Eminent domain after *Kelo v. City of New London*: an argument for banning economic development takings', *Harvard Journal of Law and Public Policy* 29(2): 492–568

Cohen, J. R. (2002) 'Maryland's "smart growth": using incentives to combat sprawl', in Squires, G. (ed.) *Urban Sprawl: Causes, Consequences and Policy Responses*, Washington, DC: Urban Institute

Cohen, R. E. (1992) *Washington at Work: Back Rooms and Clean Air*, New York: Macmillan

Cole, L. A. (1993) *Element of Risk: The Politics of Radon*, New York: Oxford University Press

Cole, L. W. and Foster, S. R. (2000) *From the Ground Up: Environmentalism, Racism and the Rise of the Environmental Justice Movement*, New York: New York University Press

Cole, R. L. (1974) *Citizen Participation and the Urban Policy Process*, Lexington, MA: Lexington Books

Colten, C. E., Skinner, P. N., and Piasecki, B. (1995) *The Road to Love Canal: Managing Industrial Waste before EPA*, Austin, TX: University of Texas Press

Columbia Journal of Environmental Law (1988), vol. 13.

Commoner, B. (1971) *The Closing Circle: Nature, Man and Technology*, New York: Knopf

Commoner, B. (1990) *Making Peace with the Planet*, New York: Pantheon

Commoner, B. (1994) 'Pollution prevention: putting comparative risk assessment in its place', in Finkel and Golding (1994)

Condon, P. M. (2007) *Design Charrettes for Sustainable Communities*, Washington, DC: Island Press

Congress for the New Urbanism (2004) *Codifying New Urbanism*, Chicago, IL: APA Planning Advisory Service

Cooper, C. D. and Alley, F. C. (2010) *Air Pollution Control: A Design Approach* (4th edition), Long Gorve, IL: Waveland Press

Corey, S. H. and Boehm, L. K. (2010) *The American Urban Reader: History and Theory*, New York: Routledge

Correll, M. R., Lillydahl, J. H., and Singell, L. D. (1978) 'The effects of greenbelts on residential property values: some values on the political economy of open space', *Land Economics* 54(2): 207–17

Cortese, A. D. (1999) *Education for Sustainability*, Boston, MA: Second Nature.

Costonis, J. J. (1989) *Icons and Aliens: Law, Aesthetics, and Environmental Change*, Urbana, IL: University of Illinois Press

Coulibaly, M., Green, R. D., and James, D. M. (1998) *Segregation in Federally Subsidized Low-Income Housing in the United States*, Westport, CT: Praeger

Council on Affordable Housing (2002) 'Springfield Township/Beverly City RCA approved', October, Trenton, New Jersey: Council on Affordable Housing

Council on Environmental Quality (2006) *Conserving America's Wetlands 2006: Two Years of Progress Implementing the President's Goal*, Washington, DC: Council on Environmental Quality

Council of State Community Development Agencies

(1994) *Making Housing Affordable: Breaking Down Regulatory Barriers: A Self-Assessment Guide for States*, Washington, DC: US Department of Housing and Urban Development

Cowardin, L. M., Carter, V., Golet, F. C., and LaRoe, E. T. (1979) *Classification of Wetlands and Deepwater Habitats of the United States*, Washington, DC: US Department of the Interior, Fish and Wildlife Service

Cowden, R. (1995) 'Power to the zones: HUD offers a new twist on an old standby', *Planning* 61 (February): 8–10

Creighton, J. L. (2005) *The Public Participation Handbook: Making Better Decisions Through Citizen Involvement*, Hoboken, NJ: John Wiley & Sons

Cunningham, J. V. (1972) 'Citizen participation in public affairs', *Public Administration Review* 32: 589–602

Curtin, D. J. Jr. and Merritt, R. E. (2002) *Subdivision Map Act Manual: A Desk Reference Covering California's Subdivision Law and Processes*, Point Arena, CA: Solano Press

Curtin, D. J. Jr. and Talbert, C. T. (eds) (2005) *Curtin's California Land Use and Planning Law* (25th edition), Point Arena, CA: Solano Press

Czech, B. and Krausman, P. R. (2001) *The Endangered Species Act: History, Conservation Biology, and Public Policy*, Baltimore: Johns Hopkins University Press

Dales, J. H. (1968) *Pollution, Property and Prices: An Essay in Policy-Making and Economics*, Toronto: University of Toronto Press

Dandekar, H. C. (2003) *The Planner's Use of Information* (2nd edition), Chicago, IL: APA Planners Press

Daniels, T. L. (1991) 'The purchase of development rights: preserving agricultural land and open space', *Journal of the American Planning Association* 57: 421–31

Daniels, T. L. (1999) *When City and Country Collide: Managing Growth in the Metropolitan Fringe*, Washington, DC: Island Press

Daniels, T. and Daniels, K. (2003) *The Environmental Planning Handbook*, Chicago, IL: APA Planners Press

Danielson, M. N. (1976) *The Politics of Exclusion*, New York: Columbia University Press

Dannenberg, A. L., Frumkin, H., and Jackson, R. J. (eds) (2011) *Making Healthy Places: Designing and Building for Health, Well-Being, and Sustainability*, Washington, DC: Island Press

Darby, M. R. and Anderson, J. E. (eds) (1996) *Reducing Poverty in America*, Thousand Oaks, CA: Sage

Dardia, M., McCarthy, K. F., Malkin, J., and Vernez, G. (1996) *The Effects of Military Base Closures on Local Communities: A Short-Term Perspective*, Santa Monica, CA: The Rand Corporation

Davenport, T. E. (2002) *Watershed Project Management Guide*, Boca Raton, FL: CRC Press

Davidoff, P. (1965) 'Advocacy and pluralism in planning', *Journal of the American Institute of Planners* 31: 331–8; reprinted in Faludi (1973), Stein (1995), and Campbell and Fainstein (1996)

Davidoff, P. (1975) 'Working toward redistributive justice', *Journal of the American Planning Association* 41: 317–18

Davidson, J. M. and Norbeck, J. M. (2011) *An Interactive History of the Clean Air Act: Scientific and Policy Perspectives*, Waltham, MA: Elsevier

Dear, M. J. and Wolch, J. R. (1987) *Landscapes of Despair: From Deinstitutionalization to Homelessness*, Princeton, NJ: Princeton University Press

DeGrove, J. M. (1984) *Land, Growth and Politics*, Chicago: Planners Press

DeGrove, J. M. (2005) *Planning Policy and Politics*, Cambridge, MA: Lincoln Institute of Land Policy

DeGrove, J. M. and Miness, D. A. (1992) *The New Frontier for Land Policy: Planning and Growth Management in the States*, Cambridge, MA: Lincoln Institute of Land Policy

Delogu, O. E. (1984) 'Local land use control: an idea whose time is passed', *Maine Law Review* 36: 261–310

DeMers, M. N. (2000) *Fundamentals of Geographic Information Systems*, New York: John Wiley

DePlace, E. and Kaleda, C. (2007) 'Two years of Measure 37: Oregon's property wrongs', Seattle, WA: Sightline Institute

Dernbach, J. C. (2002) *Stumbling Toward Sustainability*, Washington, DC: Island Press

Dernbach, J. C. (2009) (ed.) *Agenda for Sustainable America*, Washington, DC: Environmental Law Institute

Derthick, M. (1972) *New Towns In-Town: Why a Federal Program Failed*, Washington, DC: Urban Institute

Dilger, R. J. (1982) *The Sunbelt/Snowbelt Controversy: The War Over Federal Funds*. New York: New York University Press

Dilger, R. J. (2003) *American Transportation Policy*, Westport, CT: Praeger

Dilger, R. J. and Beth, R. S. (2011) 'Unfunded Mandates Reform Act: history, impact, and issues', Washington, DC: Congressional Research Service

Dowall, D. E. and Landis, J. D. (1982) 'Land-use controls and housing costs: an examination of San Francisco Bay Area communities', *Journal of the American Real Estate and Urban Economics Association* 10(1): 67–93

Downs, A. (1962) 'The law of peak-hour expressway convergence', *Traffic Quarterly* 16: 393–409

Downs, A. (1967) *Inside Bureaucracy*, Boston: Little, Brown

Downs, A. (1970) *Urban Problems and Prospects*, Chicago: Markham

Downs, A. (1972) 'Up and down with ecology – the *issue-attention* cycle', *The Public Interest* 28 (Summer): 38–50

Downs, A. (1991) 'The Advisory Commission on Regulatory Barriers to Affordable Housing: its behavior and accomplishments', *Housing Policy Debate* 4(2): 1095–137

Downs, A. (1992) *Stuck in Traffic: Coping with Peak-Hour Traffic Congestion*, Washington, DC: Brookings Institution, and Cambridge, MA: Lincoln Institute of Land Policy

Downs, A. (1994) *New Visions for Metropolitan America*, Washington, DC: Brookings Institution, and Cambridge, MA: Lincoln Institute of Land Policy

Downs, A. (2002) 'Have housing prices risen faster in Portland than elsewhere', *Housing Policy Debate* 13(1): 7–31

Downs, A. (2004) *Growth Management and Affordable Housing: Do They Conflict?*, Washington, DC: Brookings Institution

Downs, A. (2004) *Still Stuck in Traffic*, Washington, DC: Brookings Institution

Dresch, M. and Sheffrin, S. M. (1997) *Who Pays for Development Fees and Exactions?* San Francisco: Public Policy Institute of California

Duany, A. and Brain, D. (2005) 'Regulating as if humans matter: the transect and post-suburban planning', in Ben-Joseph, E. and Szold, T. (eds), *Regulating Place: Standards and the Shaping of Urban American*, New York: Routledge

Duany, A. and Talen, E. (2002) 'Transect planning', *Journal of the American Planning Association* 68(3): 245–66

Duany, A., Plater-Zyberk, E., and Speck, J. (2000) *Suburban Nation: The Rise of Sprawl and the Decline of the American Dream*, New York: North Point Press

Duany, A., Speck, J., and Lydon, M. (2009) *The Smart Growth Manual*, New York: McGraw-Hill

Dubos, R. (1968) *So Human an Animal*, New York: Scribner

Duerksen, C. J. (ed.) (1983) *A Handbook on Historic Preservation Law*, Washington, DC: Conservation Foundation and National Center for Preservation Law

Duerksen, C. J. (ed.) (1986) *Aesthetics and Land Use Controls: Beyond Ecology and Economics* (Planning Advisory Service Report 399), Chicago: American Planning Association

Duerksen, C. J. and Snyder, C. (2005) *Nature-Friendly Communities: Habitat Protection and Land Use Planning*, Washington, DC: Island Press

Duerksen, C. J., Dale, C. G., and Elliott, D. L. (2009) *The Citizen's Guide to Planning* (4th edition), Chicago: American Planning Association Planners Press

Dunbar, P. M. and Dunbar, M. W. (2004) *The Homeowners Association Manual* (5th edition), Sarasota, FL: Pineapple Press

Dunham-Jones, E. and Williamson, J. (2011) *Retrofitting Suburbia, Updated Edition: Urban Design Solutions for Redesigning Suburbs*, Hoboken, NJ: John Wiley & Sons

Durchslag, M. R. (2001) 'Symposium on the seventy-fifth anniversary of *Village of Euclid* v. *Ambler Realty Co.*: *Village of Euclid* v. *Ambler Realty Co.*, seventy-five years later: this is not your father's zoning ordinance', *Case Western Reserve Law Review* 51: 645–61

Dwight, P. (ed.) (1992) *Landmark Yellow Pages: Where to Find All the Names, Addresses, Facts, and Figures You Need*, Washington, DC: Preservation Press

Echevarria, J. D. (2007) 'Property values and Oregon Measure 37: Exposing the false premise of regulation's harm to landowners', Washington, DC: Georgetown University Environmental Law and Policy Institute

Ehrlich, P. (1968) *The Population Bomb*, New York: Ballantine

Eisinger, P. (1988) *The Rise of the Entrepreneurial State: State and Local Economic Development Policy in the United States*, Racine, WI: University of Wisconsin Press

Elazar, D. J. (1984) *American Federalism: A View from the States*, New York: Harper and Row

Ellen, I. G. (2000) *Sharing America's Neighborhoods: The Prospects for Stable Racial Integration*, Cambridge, MA: Harvard University Press

Ellerman, A. D., Bailey, E. M., Schmalensee, J., Joskow, P. L., and Montero, J. (2000) *Markets for Clean Air: The U.S. Acid Rain Program*, Cambridge, UK: Cambridge University Press

Ellerman, A. D., Joskow, P. L., and Harrison, Jr., D. (2003) *Emissions Trading in the U.S.: Experience, Lessons and Considerations for Greenhouse Gases*, Arlington, VA: Pew Center on Global Climate Change

Ellerman, A. D., Convery, F. J., and de Perthuis, C. (2010) *Pricing Carbon: The European Union Emissions Trading Scheme*, New York: Cambridge University Press

Ellickson, R. C. (1977) 'Suburban growth controls: an economic and legal analysis', *Yale Law Journal* 86: 385–511

Elliott, J. (2012) *An Introduction to Sustainable Development*, NY: Routledge

Elliott, M. L. P. (1984) 'Improving community acceptance of hazardous waste facilities through alternative systems of mitigating and managing risk', *Hazardous Waste* 1: 397–410

El Paso, City of (2008) *Livable City Sustainability Plan*, El Paso, TX: City of El Paso

Emerson, K. and Wise, C. R. (1997) 'Statutory approaches to regulatory takings: state property rights legislation issues and implications for public administration', *Public Administration Review* 57: 411–22

Environmental Law Institute (2002) *Banks and Fees: The Status of Off-Site Wetland Mitigation in the United States*, Washington, DC: Environmental Law Institute

Erie, S. P. (1988) *Rainbow's End: Irish-Americans and the Dilemmas of Urban Machine Politics, 1840–1985*, Berkeley, CA: University of California Press

Etzioni, A. (1967) 'Mixed-scanning: a "third" approach to decision-making', *Public Administration Review* December 1967; reprinted in Faludi 1973

Etzioni, A. (1986) 'Mixed scanning revisited', *Public Administration Review* January/February, 46: 8–14

Ewing, R. H. (1999) *Traffic Calming: State of the Practice*, Washington, DC: Institute of Transportation Engineers

Ewing, R. H. and Cervero, R. (2001) 'Travel and the built environment', *Transportation Research Record* 1780: 87–114

Ewing, R. H. and Kooshian, C. (1997) 'U.S. experience with traffic calming', *Institute for Transportation Engineers Journal* 67: 28–37

Faga, B. (2006) *Designing Public Consensus*, Hoboken, NJ: John Wiley & Sons

Fainstein, S. and Campbell, S. (eds) (2011) *Readings in Urban Theory* (3rd edition), Maldon, MA: Wiley-Blackwell

Fainstein, S. S., Fainstein, N. I., Smith, M. P., Hill, R. C., and Judd, D. R. (1986) *Restructuring the City: The Political Economy of Urban Development* (2nd edition), New York: Longman

Faludi, A. (1973) *A Reader in Planning Theory*, Oxford: Pergamon

Fantone, D. M. (2011) 'Federal mandates: few rules trigger Unfunded Mandates Reform Act,' Testimony before the Subcommittee on Technology, Information Policy, Intergovernmental Relations, and Procurement Reform, Committee on Oversight and Government Reform, House of Representatives, Washington, DC: Government Accounting Office, 15 February

Farber, D. A. (2010) *Environmental Law in a Nutshell* (8th edition), St Paul, MN: West

Farley, R. (ed.) (1995) *State of the Union: America in the 1990s, Vol. 2: Social Trends*, New York: Russell Sage Foundation

Feagin, J. R. (1989) 'Arenas of conflict: zoning and land use reform in critical economic perspective', in Haar and Kayden (1989a)

Federal Highway Administration (2006) *Congestion Pricing: A Primer*, Washington, DC: Federal Highway Administration, US Department of Transportation

Federal Register (1995) 'Federal guidelines for the establishment, use and operation of mitigation banks' (November 28): 58606

Federal Register (2006) 'Application for urban partnership agreements as part of congestion agreement' (December 8): 71231

Federal Reserve Bank of San Francisco (2005) *Community Development Investment Review* 1(1): entire issue

Fee, J. (2006) 'Eminent domain and the sanctity of home', *Notre Dame Law Review* 81: 783–819

Feehan, D. and Feit, M. D. (eds) (2006) *Making Business Districts Work*, Binghamton, NY: Haworth Press

Feldstein, S. G. and Fabozzi, F. J. (2008) *The Handbook of Municipal Bonds*, Indianapolis, IN: Wiley

Ferguson, E. (1990) 'Transportation demand management: planning, development and implementation', *Journal of the American Planning Association* 56: 442–56

Ferguson, R. F. and Dickens, W. T. (eds) (1999) *Urban Problems and Community Development*, Washington, DC: Brookings Institution Press

Findley, R. W. and Farber, D. A. (2000) *Environmental Law in a Nutshell* (5th edition), St. Paul, MN: West Group

Finkel, A. M. and Golding, D. (eds) (1994) *Worst Things First? The Debate over Risk-Based National Environmental Priorities*, Washington, DC: Resources for the Future

Fiorino, D. J. (1995) *Making Environmental Policy*, Berkeley, CA: University of California Press

Fiscelli, C. (2005) 'New approaches to affordable housing: overview of the housing affordability problem', Policy Update 20, Los Angeles, CA: Reason Foundation

Fischel, W. A. (1982) 'The urbanization of agricultural land: a review of the National Agricultural Lands Study', *Land Economics* 58: 236–59

Fischel, W. A. (1990) *Do Growth Controls Matter? A*

Review of Empirical Evidence, Cambridge, MA: Lincoln Institute of Land Policy

Fischel, W. A. (1991), 'Growth management reconsidered: good for the town, bad for the nation?' *Journal of the American Planning Association*, 57(3): 341–4

Fischer, F. (2000) *Citizens, Experts, and the Environment: The Politics of Local Knowledge*, Durham, NC: Duke University Press

Fischer, M. L. (1985) 'California's coastal program: larger-than-local interests built into local plans', *Journal of the American Planning Association* 51: 312–21

Fisher, R., Ury, W., and Patton, B. (1991) *Getting to Yes* (2nd edition), New York: Penguin

Fishman, R. (1987) *Bourgeois Utopias: The Rise and Fall of Suburbia*, New York: Basic Books

Fitch, J. M. (1982) *Historic Preservation: Curatorial Management of the Built World*, New York: McGraw-Hill

Fitch, J. M. (1990) *Historic Preservation: Curatorial Management of the Built World*, Charlottesville, VA: University of Virginia Press

Fitch, L. C. (1964) *Urban Transportation and Public Policy*, San Francisco: Chandler Publishing Company

Fitzgerald, J. (2010) *Emerald Cities: Urban Sustainability and Economic Development*, NY: Oxford University Press

Fitzpatrick, IV, T. J. (2009) 'Understanding Ohio's land bank legislation', Policy Discussion Papers: Special Series from the Program on Consumer Finance, Cleveland, OH: Federal Reserve Bank of Cleveland

Flack, T. A. (1986) '*Euclid* v. *Ambler*: a retrospective', *Journal of the American Planning Association* 52: 326–37

Fletcher, W. A. (2006) 'Keynote Address: Kelo, Lingle, and San Remo Hotel: takings law now belongs to the states', *Santa Clara Law Review* 46: 767–79

Florida State Comprehensive Plan Committee (1987) *The Keys to Florida's Future: Winning in a Competitive World*. Tallahassee, FL: State Comprehensive Plan Committee

Florida, State of, Department of Community Affairs (2000) 'Growth management programs: a comparison of selected states', Tallahassee, FL: Department of Community Affairs

Florida, State of, Department of Transportation (2003) 'Multimodal transportation districts and areawide quality of service handbook', Tallahassee, FL: Department of Transportation

Floyd, C. F. (1979a) 'Billboard control under the Highway Beautification Act', *Real Estate Appraiser and Analyst* July–August: 19–26

Floyd, C. F. (1979b) 'Billboard control under the Highway Beautification Act – a failure of land use controls', *Journal of the American Institute of Planners* 45: 115–26

Floyd, C. F. (1979c) *Highway Beautification: The Environmental Movement's Greatest Failure*, Boulder, CO: Westview

Ford Foundation (1949) *Report of the Ford Foundation on Policy and Program*. Detroit: Ford Foundation.

Foreign Affairs and International Trade Canada (2013) *Sustainable Development*, Ottawa: Foreign Affairs and International Trade Canada

Foresman, T. (ed.) (1998) *The History of Geographic Information Systems*, New York: Prentice Hall

Forester, J. (1999) *The Deliberative Practitioner: Encouraging Participatory Planning Processes*, Cambridge, MA: MIT Press

Form-Based Codes Institute (2008) 'Definition of a form-based code', Blacksburg, VA: Form-Based Codes Institute

Fossedal, G. A. (2005) *Direct Democracy in Switzerland*, Edison, NJ: Transaction Publishers

Foster, V. and Hahn, R. W. (1995) 'Designing more efficient markets: lessons from Los Angeles smog control', *The Journal of Law and Economics* 38: 19–48

Fox, J. (2003) *The Legend of Proposition 13*, Philadelphia, PA: Xlibris Corporation

France, R. L. (2005a) *Facilitating Watershed Management: Fostering Awareness and Stewardship*, Lanham, MD: Rowman and Littlefield

France, R. L. (2005b) *Introduction to Watershed Development: Understanding and Managing the Impacts of Sprawl*, Lanham, MD: Rowman and Littlefield

Frank, L. D. and Engelke, P. (n.d.) 'How land use

and transportation systems impact public health: a literature review on the relationship between physical activity and built form', Active Community Environments Initiative Working Paper #1

Frank, L. D., Engelke, P., and Schmid, T. (2003) *Health and Community Design*, Washington, DC: Island Press

Frece, J. W. (2009) *Sprawl and Politics: The Inside Story of Smart Growth*, Albany, NY: SUNY Press

Freeman, A. M. (1990) 'Water pollution policy', in Portney (1990)

Freilich, R. H. and Bushek, D. W. (1995) *Exactions, Impact Fees and Dedications: Shaping Land-Use Development and Funding Infrastructure*, Chicago: American Bar Association

Freilich, R., Sitkowski, R., and Mennillo, S. (2010) *From Sprawl to Sustainability: Smart Growth, New Urbanism, Green Development and Renewable Energy*, Chicago: American Bar Association Section on State and Local Government Law

Frey, W. H. (1989) 'United States: counter-urbanization and metropolis depopulation', in Champion (1989)

Frey, W. H. (1994a) 'The new urban revival in the United States', in Paddison *et al.* (eds), *International Perspectives in Urban Studies 2*, London: Jessica Kingsley

Frey, W. H. (1994b) 'Minority suburbanization and continued "white flight" in US metropolitan areas: assessing findings from the 1990 Census', *Research in Community Sociology* 4: 15–42

Frey, W. H. (1995) 'The new geography of population shifts: trends toward balkanization', in Farley (1995)

Frey, W. H. and Fielding, E. L. (1994) *New Dynamics of Urban-Suburban Change: Immigration, Restructuring and Racial Separation*, Ann Arbor, MI: Population Studies Center, University of Michigan

Frey, W. H. and Speare, A. (1992) 'The revival of metropolitan growth in the US: an assessment of findings from the 1990 census', *Population and Development Review* 18: 129–46

Frieden, B. J. (1979) *The Environmental Protection Hustle*, Cambridge, MA: MIT Press

Frieden, B. J. and Kaplan, M. (1977) *The Politics of Neglect: Urban Aid from Model Cities to Revenue*

Sharing (2nd edition), Cambridge, MA: MIT Press

Frieden, B. J. and Sagalyn, L. B. (1989) *Downtown Inc. – How America Rebuilds Cities*, Cambridge, MA: MIT Press

Friedmann, J. (1987) *Planning in the Public Domain: From Knowledge to Action*, Princeton, NJ: Princeton University Press; extract from this book in Stein (1995)

Frumkin, H., Frank, L., and Jackson, R. (2004) *Urban Sprawl and Public Health*, Washington, DC: Island Press

Fulton, W. (1990) 'The trouble with slow-growth politics: it wins elections, but the subdivisions keep going up', *Governing* 3(7): 27–33

Fulton, W. (1991) *Guide to California Planning*, Point Arena, CA: Solano Press

Fulton, W. and Shigley, P. (2012) *Guide to California Planning* (4th edition), Thousand Oaks, CA: Solano Press

Gale, D. E. (1996) *Understanding Urban Unrest: From Reverend King to Rodney King*, Thousand Oaks, CA: Sage

Gallagher, J. (2010) *Reimagining Detroit: Opportunities for Redefining an American City*, Detroit, MI: Wayne State University Press

Galster, G. (ed.) (1996) *Reality and Research: Social Science and US Urban Policy Since 1960*, Washington, DC: Urban Institute Press

Garmise, S. (2006) *People and the Competitive Advantage of Place*, Armonk, NY: M. E. Sharpe

Garner, J. F. and Callies, D. L. (1972) 'Planning law in England and Wales and in the United States', *Anglo-American Law Review* 1: 292–334

Garreau, J. (1991) *Edge City: Life on the New Frontier*, New York: Doubleday

Gelfand, M. I. (1975) *A Nation of Cities: The Federal Government and Urban America, 1933–1965*, New York: Oxford University Press

Georgia, State of, Department of Community Affairs (2013) *Discovering and Planning Your Community Character: A Guidebook for Citizens and Local Planners*, Atlanta: Department of Community Affairs

Georgia, State of, Department of Natural Resources, Historic Preservation Division (2010) *Heritage*

Tourism Handbook: A How-to-Guide for Georgia, Atlanta, Georgia: Department of Natural Resources, Historic Preservation Division

Gerckens, L. C. (1988) 'Historical development of American planning', in So and Getzels (1988)

Getzels, J. and Jaffe, M. (1988) *Zoning Bonuses in Central Cities* (Planning Advisory Service, Report 410), Chicago: American Planning Association

Gibbons, C. (2010) 'Economic gardening,' *IEDC Economic Development Journal* 9(3): 5–11

Gibbs, L. M. (1998) *Love Canal: The Story Continues,* Gabriola Island, British Columbia: New Society Publishing

Gilbert, J. L. (1995) 'Selling the city without selling out: new legislation on development incentives emphasize accountability', *Urban Lawyer* 27: 427–93

Giles, S. L. and Blakely, E. J. (2001) *Fundamentals of Economic Development Finance,* Thousand Oaks, CA: Sage Publishing

Gillette, C. P. (2005) 'Kelo and the local political process', *Hofstra Law Review* 34: 13–22

Gilroy, L. C. (2006) 'Statewide regulatory takings reform: Exporting Oregon's Measure 37 to other states', Policy Study 343, Los Angeles, CA: Reason Public Policy Institute

Gittell, R. J. and Vidal, A. (1998) *Community Organizing: Building Social Capital as a Development Strategy,* Thousand Oaks, CA: Sage Publications

Glaab, C. N. and Brown, A. T. (1983) *A History of Urban America* (3rd edition), New York: Macmillan

Glaeser, E. L. and Kahn, M. E. (2001) 'Decentralized employment and the transformation of the American City,' NBER Working Paper 8117, Cambridge, MA: National Bureau of Economic Research

Glendenning, P. N. (2007) 'Smart growth 10 years later', Opening speech at Smart Growth at 10 Conference in Annapolis, MD, College Park, MD: National Center for Smart Growth Research and Education, University of Maryland

Goble, D. D., Scott, J. M., and Davis, F. W. (eds) (2005) *The Endangered Species Act at Thirty,* Vol. 1, Washington, DC: Island Press

Godschalk, D. R., Brower, D. J., McBennett, L. D., Vestal, B. A., and Herr, D. C. (1979) *Constitutional Issues of Growth Management* (revised edition), Chicago: Planners Press

Goebel, T. (2002) *Direct Democracy in America: A Government by the People,* Chapel Hill, NC: University of North Carolina Press

Goldberger, P. (1981) *The Skyscraper,* New York: Knopf

Goldman, S. and Jahnige, T. P. (1985) *The Federal Courts as a Political System,* New York: Harper and Row

Gordon, R. J. and Gordon, L. (1990) 'Neighborhood responses to stigmatized urban facilities', *Journal of Urban Affairs* 12: 437–47

Gori, P. L., Jeer, S., and Schwab, J. (2005) *Landslide Hazards and Planning,* Chicago, IL: APA Planning Advisory Service

Gottlieb, R. (1993) *Forcing the Spring: The Transformation of the American Environmental Movement,* Washington, DC: Island Press

Gottman, J. (1961) *Megalopolis: The Urbanized Northeaster Seaboard of the United States,* New York: Twentieth Century Fund

Goulder, L. H., Perry, I. W. H., Williams III, R. C., and Burtraw, D. (1999) 'The cost-effectiveness of alternative instruments for environmental protection in a second best setting', *Journal of Public Economics* 72(3): 329–60

Governor's Council on Vermont's Future (1988) *Report of the Governor's Commission on Vermont's Future: Guidelines for Growth,* Montpelier, VT: State of Vermont

Governor's Office of Planning and Research (2003) 'State of California General Plan Guidelines', Sacramento, CA: State of California, Governor's Office of Planning and Research

Graham, S. (ed.) (2004) *The Cybercities Reader,* London: Taylor and Francis

Graham, S. and Marvin, S. (1996) *Telecommunications and the City: Electronic Spaces, Urban Places,* London: Routledge

Graham, S. and Marvin, S. (2001) *Splintering Urbanism: Networked Infrastructures, Technological Mobilities and the Urban Condition,* London: Routledge

Gramlich, E. M. (2004) 'Subprime mortgage lending: benefits, costs and challenges,' remarks to the Financial Services Roundtable Annual Housing Policy Meeting, Chicago, IL, May 21

Gray, R. C. (2007) 'Ten years of smart growth: a nod to policies past and a prospective glimpse into the future', *Cityscape: A Journal of Policy Development and Research* 9(1): 109–30

Great Lakes Environment Finance Center (2005) 'Best practices in land bank operation', Cleveland, OH: Great Lakes Environment Finance Center, Maxine Goodman Levin College of Urban Affairs, Cleveland State University

Green, G. P. and Haines, A. (2002) *Asset Building and Community Development*, Thousand Oaks, CA: Sage Publications

Green, R. E. (ed.) (1991) *Enterprise Zones: New Directions in Economic Development*, Newbury Park, CA: Sage

Green Leigh, N. and Hoelzel, N. (2012) 'Smart growth's blind side', *Journal of the American Planning Association* 78(1): 87–103

Greenstein, R. and Sungu-Eryilmaz (2007) 'Community land trusts: a solution for permanently affordable housing', *Land Lines*, Cambridge, MA: Lincoln Institute of Land Policy

Grigsby, W. G. (1963) *Housing Markets and Public Policy*, Philadelphia: University of Pennsylvania Press

Grigsby, W. G., Baratz, M., Galster, G., and Maclennan, D. (1987) 'The dynamics of neighborhood change and decline', *Progress in Planning* 28(1): 1–76

Grodach, C. (2011) 'Barriers to sustainable economic development: the Dallas-Fort Worth experience,' *Cities*, 28(4): 300–9

Gruen, C. (1985) 'The Economics of Requiring Office Space Development to Contribute to the Production of Housing,' in Porter, D. R. (1985) *Downtown Linkages*, Washington, DC: Urban Land Institute

Guzda, M. K. (1998) *Slow Down, You're Moving Too Fast*, Washington, DC: Public Technology, Inc.

Haar, C. M. and Kayden, J. S. (1989a) *Zoning and the American Dream: Promises Still to Keep*, Chicago: Planners Press

Haar, C. M. and Kayden, J. S. (1989b) *Landmark Justice: The Influence of William J. Brennan on America's Communities*, Washington, DC: Preservation Press

Haar, C. M. and Wolf, M. A. (1989) *Land Use Planning: A Case Book on the Use, Misuse, and Re-use of Urban Land* (4th edition), Boston: Little, Brown

Habe, R. (1989) 'Public design control in American communities', *Town Planning Review* 60: 195–219

Hadden, S. G. (1989) *A Citizen's Right to Know: Risk Communication and Public Policy*, Boulder, CO: Westview

Hagerstown, City of (2006) 'City of Hagerstown Capital Improvement Program 2006/2011', Hagerstown, MD: City of Hagerstown

Hahn, R. W. (1994) 'United States environmental policy: past, present and future', *Natural Resources Journal* 34: 305–48

Haider, D. H. (1974) *When Governments Come to Washington: Governors, Mayors, and Intergovernmental Lobbying*, New York: Free Press

Hall, K. B. and Porterfield, G. A. (2001) *Community by Design: New Urbanism for Suburbs and Small Communities*, New York: McGraw-Hill

Hall, P. (1988) *Cities of Tomorrow: An Intellectual History of Urban Planning and Design in the Twentieth Century*, New York: Blackwell

Hall, P. (1992) *Urban and Regional Planning* (3rd edition), New York: Routledge, Chapman and Hall

Hambleton, R. and Thomas, H. (1995) *Urban Policy Evaluation: Challenge and Change*, London: Paul Chapman

Hanak, E. (2005) *Water for Growth: California's New Frontier*, San Francisco, CA: California Public Policy Institute

Handy, S. L. and Mokhtarian, P. L. (1995) 'Planning for telecommuting: measurement and policy issues', *Journal of the American Planning Association* 61: 99–111

Handy, S., Paterson, R. G., and Butler, K. (2003) *Planning for Street Connectivity*, Chicago, IL: APA Planning Advisory Service

Hanson, S. and Giuliano, G. (eds.) (1995) *The Geography of Urban Transportation* (3rd edition), New York: Guilford Publications

Harr, C. M. (1977) *Land-use Planning: A Casebook on the Use, Misuse, and Re-use of Urban Land*. Boston: Little, Brown and Company

Harrigan, J. J. (1993) *Political Change in the Metropolis* (5th edition), New York: HarperCollins

Hartman, C. (1994) 'On poverty and racism, we have

had little to say', *Journal of the American Planning Association* 60: 158–9

Hartman, C. W. and Squires, G. D. (2006) *There is No Such Thing as a Natural Disaster: Race, Class, and Hurricane Katrina*, New York: Routledge

Harvey, D. (2005) *A Brief History of Neoliberalism*, New York: Oxford University Press

Harvey, F. (2008) *A Primer of GIS: Fundamental Geographic and Cartographic Concepts*, New York: Guilford Publications

Hawaii, State of, Department of Business, Economic Development and Tourism (1999) *The State of Hawaii Data Book 1999*, Honolulu: Hawaii Department of Business, Economic Development and Tourism

Hawaii, State of, Office of the Auditor (2007) *Hawai'i 2050 Sustainability Task Force: Draft Hawaii 2050 Sustainability Plan: Charting a Course for Hawai'i's Sustainable Future*, Honolulu, Hawaii: Office of the Auditor

Hawaii, State of (2008) *Hawai'i 2050 Sustainability Plan: Charting a Course for Hawai'i Sustainable Future*, Honolulu, Hawaii: State of Hawaii

Hayden, D. (2003) *Building Suburbia: Green Fields and Urban Growth, 1820–2000*, New York: Vintage

Healy, R. G. and Rosenberg, J. S. (1979) *Land Use and the States*, Baltimore: Johns Hopkins University Press

Heathcote, I. W. (2009) *Integrated Watershed Management: Principles and Practices* (2nd edition), Hoboken, NJ: John Wiley & Sons

Hecht, B. L. (2006) *Developing Affordable Housing: A Practical Guide for Nonprofit Organizations* (3rd edition), San Francisco: Jossey-Bass

Hedman, R. and Jaszewski, A. (1984) *Fundamentals of Urban Design*, Chicago: Planners Press

Hellman, P. C. (2006) *Designing Greenways*, Washington, DC: Island Press

Hendler, S. (ed.) (1995) *Planning Ethics: A Reader in Planning Theory, Practice and Education*, Piscataway, NJ: Center for Urban Policy Research, Rutgers University

Henrico County, Virginia (2009) *Henrico County Vision 2026 Comprehensive Plan*, Henrico County: Planning Department

Hetzel, O. (1994) 'Some historical lessons for imple-menting the Clinton administration's empower-ment zones and enterprise communities program: experiences from the model cities program', *Urban Lawyer* 26: 63–81

Heyman, I. M. and Gilhool, T. K. (1964) 'The con-stitutionality of imposing increased community costs on new subdivision residents through sub-division exactions', *Yale Law Journal* 73: 1119–57

Heyward, I., Cornelius, S., and Carver, S. (1999) *An Introduction to Geographical Information Systems*, Boston: Addison-Wesley Longman

Hibbard, M., Seltzer, E., Weber, B., and Emshoff, B. (2011) *Toward One Oregon: Rural-Urban Interdependence and the Evolution of a State*, Corvallis, OR: Oregon State University Press

Hiemstra, H. and Bushwick, N. (eds) (1989) *Plowing the Urban Fringe: An Assessment of Alternative Approaches to Farmland Preservation*, Miami, FL: Florida Atlantic University and Florida International University Joint Center for Environmental and Urban Problems

Hill, R. C. (1986) 'Crisis in the Motor City: the politics of economic development in Detroit', in Fainstein *et al.* (1986)

Hiscock, K. (1995) 'Groundwater pollution and protection', in T. O'Riordan (ed.), *Environmental Science for Environmental Management*, Harlow: Longman

Hix, W. M. (2001) *Taking Stock of the Army's Base Realignment and Closure Selection Process*, Santa Monica, CA: The Rand Corporation

Hoch, C., Dalton, L. C., and So, F. S. (eds) (2000) *The Practice of Local Government Planning* (3rd edition), Washington, DC: International City Manager/Management Association

Hofstadter, R. (1948) *The American Political Tradition*, New York: Vintage

Hollander, J. B. (2011) *Sunburnt Cities: The Great Recession, Depopulation and Urban Planning in the American Sunbelt*, New York: Routledge

Hollander, J. B. and Nemeth, J. (2011) 'The bounds of smart decline: a foundational theory for planning shrinking cities', *Housing Policy Debate* 21(3): 349–67

Hopkins, B. (2009) *Starting and Managing a Nonprofit*

Organization: A Legal Guide, Hoboken, NJ: John Wiley and Sons, Inc.

Hornstein, D. T. (1994) 'Paradigms, process, and politics: risk and regulatory design', in Finkel and Golding (1994)

Hosmer, C. F. (1965) *Presence of the Past: A History of the Preservation Movement in the United States before Williamsburg*, New York: Putnam

Hosmer, C. F. (1981) *Preservation Comes of Age: From Williamsburg to the National Trust, 1926–1949*, Charlottesville, VA: University Press of Virginia

Hou, J., Johnson, J. M. and Lawson, L. J. (2009) *Greening Cities Growing Cities: Learning from Seattle's Urban Community Gardens*, Seattle: University of Washington Press

Hough, D. E. and Kratz, C. G. (1983) 'Can "good" architecture meet the market test?', *Journal of Urban Economics* 14: 40–54

Houstoun, L. (2003) *BIDs: Business Improvement Districts* (2nd edition), Washington, DC: Urban Land Insitute and the International Downtown Association

Hovde, S. and Krinsky, J. (1996) *Balancing Acts: The Experience of Mutual Housing Associations and Community Land Trusts in Urban Neighborhoods*, New York: Community Service Society of New York

Hovde, S., Krinsky, J., and Mottola, M. (1996) *Hands-on Housing: A Guide Through Mutual Housing Associations and Community Land Trusts for Residents and Organizations*, New York: Community Service Society of New York

Howe, F. C. (1913) *European Cities at Work*, New York: C. Scribner's Sons

Howe, E. (1994) *Acting on Ethics in City Planning*, New Brunswick, NJ: Center for Urban Policy Research, Rutgers University

Howland, M. and Sohn, J. (2007) 'Will Maryland's Priority Funding Areas Initiative Contain Urban Sprawl', *Land Use Policy* 24(1): 175–86

Hubbard, T. K. and Hubbard, H. V. (1929) *Our Cities of Today and Tomorrow: A Survey of Planning and Zoning Progress in the United States*, Cambridge, MA: Harvard University Press

Hula, R. C. and Jackson-Elmore, C. (eds) (2000) *Nonprofits in Urban America*, Westport, Connecticut: Quorum Books

Huxhold, W. E. (1991) *An Introduction to Urban Geographic Information Systems*, New York: Oxford University Press

Huxhold, W. E. and Levinsohn, A. G. (1995) *Managing Geographic Information Projects*, New York: Oxford University Press

Huxhold, W. E., Fowler, E. M., and Parr, B. (2004) *ArcGIS and the Digital City: A Hands-on Approach for Local Government*, Redlands, CA: ESRI Press

ICMA (International City/County Management Association) (2008) *Call 311: Connecting Citizens to Local Government Final Report*, Washington, DC: International City/County Management Association

Ihlanfeldt, K. R. (1995) 'The importance of the central city to the regional and national economy: a review of the arguments and empirical evidence', *Cityscape: A Journal of Policy Development and Research* 1(2): 125–50

Immergluck, D. and Wiles, M. (1999) *Two Steps Back: The Dual Mortgage Market, Predatory Lending, and the Undoing of Community Development*, Chicago, IL: The Woodstock Institute

Inam, A. (2005) *Planning for the Unplanned*, London, UK: Routledge, Taylor & Francis

Ingram, G. K, Carbonell, A., Hong, Y., and Flint, A. (2009) *Smart Growth Policies: An Evaluation of Programs and Outcomes*, Cambridge, MA: Lincoln Institute of Land Policy

Ingram, H. M. and Mann, D. E. (1984) 'Preserving the Clean Water Act: the appearance of environmental victory', in Vig, N. J. and International City Management Association (1994) *Transportation Planning under ISTEA*, Washington, DC: ICMA

Inhaber, H. (1998) *Slaying the NIMBY Dragon*, New Brunswick, NJ: Transaction Publishers

Innes, J. E. (1996) 'Planning through consensus building: a new view of the comprehensive planning ideal', *Journal of the American Planning Association*, 62: 460–72

Innes, J. E. (2004) 'Consensus building: clarification for the critics', *Planning Theory* 3(1): 5–20

Innes, J. E. and Booher, D. E. (2003) 'The impact of collaborative planning on governance capacity', Working Paper 2003–03, Berkeley, CA: Institute of Urban and Regional Development

Innes, J. E., Gruber, J., Thompson, R., and Neuman, M. (1994) *Coordinating Growth and Environmental Management through Consensus Building*, Berkeley, CA: California Policy Seminar, University of California

Institute of Transportation Engineers (2005) *Improving the Pedestrian Environment Through Innovative Transportation Design*, Washington, DC: Institute of Transportation Engineers

Institute of Transportation Technical Council Engineering Task Force (1997) 'Guidelines for the design and application of speed humps – a recommended practice', RP-023A, Washington, DC: Institute for Transportation Engineers

Internal Revenue Service (2010) *New Markets Tax Credit*, Washington: Internal Revenue Service

International City/County Management Association and National League of Cities (2009) *Economic Development 2009 Survey Summary*, Washington: International City/County Management Association and National League of Cities

International Joint Commission (on the Great Lakes) (1993) *A Strategy for Virtual Elimination of Persistent Toxic Substances*, Windsor, Ontario: IJC

International Joint Commission (on the Great Lakes) (1994) *Seventh Biennial Report on Great Lakes Water Quality*, Washington, DC: IJC

Ischida, T. and Isbister, K. (eds) (2000) *Digital Cities: Technologies, Experiences and Future Perspectives*, New York: Springer-Verlag

Jackson, J. (2009) 'How risky are sustainable real estate projects? An evaluation of LEED and ENERGY STAR development options', *Journal of Sustainable Real Estate* 1(1): 91–106

Jackson, K. T. (1985) *Crabgrass Frontier: The Suburbanization of the United States*, New York: Oxford University Press

Jacobs, J. (1961) *The Death and Life of Great American Cities*, New York: Random House

Jacobson, J. W., Burchard, S. N., and Carling, P. J. (eds) (1992) *Community Living for People with Developmental and Psychiatric Disabilities*, Washington, DC: Johns Hopkins University Press

Jaeger, W. K. (2006) 'The effects of land use regulations on property values', *Environmental Law* 36: 105–30

Jaffe, M. and Smith, T. P. (1986) *Siting Group Homes for Developmentally Disabled Persons* (Planning Advisory Service Report 397), Chicago: American Planning Association

Jefferson Parrish, Louisiana (2004) 'Envision 2020 Jefferson Comprehensive Plan background slide-show', Jefferson Parrish, Louisiana: Jefferson Parrish Planning Department, slide 33

Jefferson, T. (1892–99) *The Writings of Thomas Jefferson* (ed. P. L. Ford), New York: G. Putnam's Sons

Jickling, M. (2008) *Fannie Mae and Freddie Mac in Conservatorship*, RS22950, Washington, DC: Congressional Research Service, 15 September

Johnson, C. L. and Man, J. Y (eds) (2001) *Tax Increment Financing and Economic Development: Uses, Structures, and Impact*, Albany: State University of New York Press

Johnson, L., Samant, L. D., and Frew, S. (2005) *Planning for the Unexpected*, Chicago, IL: APA Planning Advisory Service

Johnston, R. A. (1994) 'The evolution of multimodal transportation systems for economic efficiency and other impacts', UTCC #272, Berkeley, CA: University of California Berkeley University of California Transportation Center

Joseph, L. B. (ed.) (1993) *Affordable Housing and Public Policy: Strategies for Metropolitan Chicago*, Chicago: Center for Urban Research and Policy Studies, Northwestern University

Journal of the American Planning Association (1992) Issue largely devoted to growth management, *Journal of the American Planning Association* 58(4) (Autumn)

Judd, D. R. (1988) *The Politics of American Cities: Private Power and Public Policy* (3rd edition), Glenview, IL: Scott, Foresman

Judd, D. R. and Swanstrom, T. (1994) *City Politics: Private Power and Public Policy*, New York: HarperCollins

Juergensmeyer, J. C. and Roberts, T. E. (1998) *Land Use Planning and Control Law*, St Paul, MN: West Group

Kain, J. (1994) 'Impacts of congestion pricing on transit and carpool demand and supply', in National Research Council, Transportation Research Board, Committee for the Study on Urban Transportation

Congestion Pricing, *Curbing Gridlock: Peak Period Fees to Relieve Traffic Congestion*, Vol. 2, Washington DC: National Academy Press

Kales, R. W., Colton, C. E., Lacka, S., and Leatherman, S. P. (2007) 'Reconstruction of New Orleans after Hurricane Katrina: a research perspective', *Cityscape* 9(3): 5–22

Kalinosky, L. (2001) 'Smart growth for neighborhoods: affordable housing and the regional vision', Washington, DC: National Neighborhood Coalition

Kaplan, H., Gans, S. P., and Kahn, H. M. (1970) *The Model Cities Program: The Planning Process in Atlanta, Seattle, and Dayton*, New York: Praeger

Kaplan, J. and Valls, A. (2007) 'Housing discrimination as a basis for Black reparations', *Public Affairs Quarterly* 21(3): 255–72

Katz, P. (1994) *The New Urbanism: Toward an Architecture of Community*, New York: McGraw-Hill

Katz, P. (2008) 'Eight advantages of a form-based code', Blacksburg, VA: Form-Based Codes Institute

Katznelson, I. (2005) *When Affirmative Action Was White: An Untold History of Racial Inequality in Twentieth Century America*, NY: W.W. Norton

Kaufman, J. and Bailkey, M. (2000) 'Farming inside cities: entrepreneurial urban agriculture in the United States', Cambridge, MA: Lincoln Institute of Land Policy Working Paper

Kay, J. H. (1997) *Asphalt Nation: How the Automobile Took Over America, and How We Can Take It Back*, New York: Crown Publishers

Kayden, J. S. (1990) 'Zoning for dollars: new rules for an old game? Comments on the *Municipal Art Society* and *Nollan* cases', in Lassar (1990)

Kayden, J. S., the New York City Department of City Planning, and the Municipal Art Society of New York (2000) *Privately Owned Public Space: The New York City Experience*, New York: John Wiley & Sons

Keating, W. D. (1994) *The Suburban Racial Dilemma: Housing and Neighborhoods*, Philadelphia: Temple University Press

Keith, N. S. (1973) *Politics and the Housing Crisis Since 1930*, New York: Universe Books

Kelly, B. M. (1993) *Expanding the American Dream: Building and Rebuilding Levittown*, Albany, NY: State University of New York Press

Kelly, E. D. (2009) *Community Planning: An Introduction to the Comprehensive Plan* (2nd edition), Washington, DC: Island Press

Kelman, S. (1981) *What Price Incentives? Economics and the Environment*, Boston: Auburn House

Kelty, P. A. (2010) *Wetland Ecology: Principles and Conservation*, New York: Cambridge University Press

Kemp, D. D. (1998) *The Environment Dictionary*, London: Routledge

Kendall, D., Dowling, T., and Bradley, J. (2006) *The Good News About Takings*, Chicago, IL: APA Planners Press

Kendig, L. H. and Keast, B. C. (2010) *Community Character: Principles for Design and Planning*, Washington, DC: Island Press

Kendig, L. H. and Keast, B. C. (2011) *A Guide to Planning for Community Character*, Washington, DC: Island Press

Kennedy, J. (1961) 'Special Messsage to the Congress on Housing and Community Development', March 9

Kent, T. J. (1964) *The Urban General Plan*, Chicago: American Planning Association (reprint 1990)

Kent-Smith, H. L. (1987) 'The Council on Affordable Housing and the Mount Laurel Doctrine: Will the Council Succeed?' *Rutgers Law Journal* 18: 929–960

Kettl, D. F. (1981) 'Regulating the cities', *Publius* 11: 111–25

Kingdon, J. W. (1984) *Agendas, Alternatives, and Public Policies*, New York: HarperCollins

Kinkead, E. (1990) *Central Park, 1857–1995: The Birth, Decline, and Renewal of a National Treasure*, New York: Norton

Klee, G. A. (1999) *The Coastal Environment: Toward Integrated Coastal and Marine Sanctuary Development*, Upper Saddle River, NJ: Prentice Hall

Klein, W. R. (2000) 'Building consensus,' Chapter 17 in Hoch, C. J., Dalton, L. C., and So, F. S. *The Practice of Local Government Planning* (3rd edition), Washington, DC: International City/County Management Association, pp. 421–38

Kmiec, D. W. (2007) 'Eminent domain Post-Kelo: introduction to the 2006 Templeton Lecture: hitting home – the Supreme Court earns public notice opining on public use', *University of Pennsylvania Journal of Constitutional Law* 9: 501–43

Knack, R. E. (1995) 'BART's village vision', *Planning* 61(1): 18–21

Knack, R., Meck, S., and Stollman, I. (1996) 'The real story behind the standard planning and zoning acts of the 1920's', *Land Use Law*, February: 3–9

Knapp, G. J. (1985) 'The price effects of urban growth boundaries in metropolitan Portland, Oregon', *Land Economics* 61: 26–35

Knapp, G. and Frece, J. (2007) 'Smart growth in Maryland: looking forward and looking back', *Idaho Law Review* 43: 1–30

Knapp, G.-J. and Lewis, R. (2007) 'State agency spending under Maryland's Smart Growth Areas Act: who's tracking, who's spending, how much, and where?', College Park, MD: National Center for Smart Growth Research and Education, University of Maryland

Knapp, G. and Nelson, A. C. (1992) *The Regulated Landscape: Lessons on State Land Use Planning from Oregon*, Cambridge, MA: Lincoln Institute of Land Policy

Kneese, A. and Schultze, C. L. (1975) *Pollution, Prices, and Public Policy*, Washington, DC: Brookings Institution

Koenig, J. (1990) 'Down to the wire in Florida: concurrency is the byword in the Nation's most elaborate statewide growth management scheme', *Planning* 56(10): 4–11

Kolis, A. B. (1979) 'Architectural expression: police power and the first amendment', *Urban Law Annual* 16: 272–304

Koll-Schretzenmayr, M., Kelner, M., and Nussbaumer, G. (eds) (2003) *The Real and Virtual Worlds of Spatial Planning*, New York: Springer-Verlag

Korngold, G. (2001) 'Symposium on the seventy-fifth anniversary of *Village of Euclid* v. *Ambler Realty Co.*: the emergence of private land use controls in large-scale subdivisions: the companion story to *Village of Euclid* v. *Ambler Realty Co.*', *Case Western Reserve Law Review* 51: 617–43

Kosar, K. R. (2007) *Government-Sponsored Enterprises*, RS21663, Washington, DC: Congressional Research Service, April 23

Kosobud, R. F. (ed.) (2000) *Emissions Trading: Environmental Policy's New Approach*, New York: John Wiley & Sons

Kosobud, R. F. and Zimmerman, J. M. (eds) (1997) *Market-Based Approaches to Environmental Policy: Regulatory Innovations to the Fore*, New York: Van Nostrand Reinhold

Koven, S. And Lyons, T. (2010) *Economic Development: Strategies for State and Local Practice*, Washington, DC: International City/County Management Association

Kowsky, F. R. (1998) *Country, Park and City: The Architecture and Life of Calvert Vaux*, New York: Oxford University Press

Krumholz, N. (1994) 'Advocacy Planning: Can it Move the Center.' *Journal of the American Planning Association*, 60(2): 150–151

Krumholz, N. and Clavel, P. (1994) *Reinventing Cities: Equity Planners Tell Their Stories*, Philadelphia: Temple University Press

Krumholz, N. and Forester, J. (1990) *Making Equity Planning Work: Leadership in the Public Sector*, Philadelphia: Temple University Press; extract from this in Stein 2004

Krygier, J. and Wood, D. (2005) *Making Maps: A Visual Guide to Map Design for GIS*, New York: Guilford Publications

Kubasek, N. and Silverman, G. (2002) *Environmental Law* (4th edition), Englewood Cliffs, NJ: Prentice Hall College Division

Kulick, P. J. (2000) 'Comment: rolling the dice: determining public use in order to effectuate a "public-private taking" a proposal to redefine "public use"', *Detroit College of Law at Michigan State University Law Review* (Fall): 639–89

Kushner, J. A. (2004) *The Post-Automobile City*, Durham, NC: Carolina Academic Press

Kusler, J. S. (1983) *Our National Wetland Heritage: Protection Guidebook*, Washington, DC: Environmental Law Institute

Kusler, J. A. and Kentula, M. E. (eds) (1990) *Wetland Creation and Restoration: The Status of the Science*, Washington, DC: Island Press

Lacey, M. J. (ed.) (1990) *Government and Environmental Politics: Essays on Historical Development since World War II*, Lanham, MD: University Press of America

LaCour-Little, M. (1999) 'Discrimination in mortgage lending: A critical review of the literature', *Journal of Real Estate Literature* 7: 15–49

Laguerre, M. S. (2006) *Digital City: The American Metropolis and Information Technology*, New York: Palgrave Macmillan

Landis, J. D. (1992) 'Do growth controls work? A new assessment', *Journal of the American Planning Association* 58: 489–508

Landy, M. K., Roberts, M. J., and Thomas, S. R. (1994) *The Environmental Protection Agency: Asking the Wrong Questions – From Nixon to Clinton* (expanded edition), New York: Oxford University Press

Lang, R. E. and Dhavale, D. (2005) 'Beyond megalopoli: exploring America's new "megapolitan" geography', Census Report 05: 01, Alexandria, VA: Metropolitan Institute Census Report Series, Virginia Tech

Lapping, M. B. and Leutwiler, N. R. (1987) 'Agriculture in conflict: right-to-farm laws and the peri-urban milieu for farming', in W. Lockeretz (ed.), *Sustaining Agriculture Near Cities*, Ankenny, IA: Soil and Water Conservation Society

La Porte, City of, (n.d.) 'Comprehensive Plan update – introduction', La Porte, TX: City of La Porte

Lash, J. (1994) 'Integrating science, values, and democracy through comparative risk assessment', in Finkel and Golding (1994)

Lassar, T. J. (1989) *Carrots and Sticks: New Zoning Downtown*, Washington, DC: Urban Land Institute

Lassar, T. J. (1990) *City Deal Making*, Washington, DC: Urban Land Institute

Lavelle, M. and Loyle, M. (1992) 'Unequal protection – the racial divide in environmental law', *National Law Journal*, September 21

Lawson, L. (2004) 'The planner in the garden: a historical view into the relationship between planning and community gardens', *Journal of Planning History* 3(2): 151–76

Lawson, L. (2005) *City Bountiful: A Century of Community Gardening in America*, Berkeley, CA: University of California Press

Lee, E. G. III (2001) 'Symposium on the seventy-fifth anniversary of *Village of Euclid* v. *Ambler Realty Co.*: introduction', *Case Western Reserve Law Review* 51: 593–5

LeGates, R. T. (2005) *Thinking Globally, Acting Regionally: GIS and Data Visualization for Social Science and Public Policy Research*, Redlands, CA: Esri Press

Lennertz, B. and Lutzenhiser, A. (2006) *The Charrette Handbook*, Chicago, IL: APA Planners Press

Lenssen, P. (1991) *Nuclear Waste: The Problem that Won't Go Away*, Washington, DC: Worldwatch Foundation

Leopold, A. (1949) *Sand County Almanac*, New York: Oxford University Press

Lester, J. P., Allen, D. W., and Hill, K. M. (2001) *Environmental Injustice in the United States: Myths and Realities*, Boulder, CO: Westview Press

Levine, A. (1982) *Love Canal: Science, Politics and People*, Lexington, MA: Lexington Books

Levine, J. (2006) *Zoned Out*, Washington, DC: Resources for the Future Press

Levy, F., Meltsner, A., and Wildavsky, A. (1973) *Urban Outcomes*, Berkeley, CA: University of California Press

Lewis, R. and Knapp, G. J. (2012) 'Targeting spending for land conservation: an evaluation of Maryland's Rural Legacy Program', *Journal of the American Planning Association* 78(1): 34–52

Lincoln, City of (2003) 'Lincoln City – Lancaster County Comprehensive Plan', Lincoln, Nebraska

Lincoln Metropolitan Planning Organization (2011) *LPlan 2040: Lincoln-Lancaster County Comprehensive Plan*, Lincoln, NE: Lincoln Metropolitan Planning Organization

Lindblom, C. E. (1959) 'The science of "muddling through"', *Public Administration Review* (Spring 1959; reprinted in Faludi (1973), and in Campbell and Fainstein (1996)

Lindsey, G., Todd, J. A., and Hayter, S. J. (2003) *A Handbook for Planning and Conducting Charrettes for High-Performance Projects*, Golden, CO: National Renewable Energy Laboratory

Lipset, S. and Schneider, W. (1983) *The Confidence Gap: Business, Labor, and Government in the Public Mind*, New York: Free Press

Listokin, D. and Lahr, M. (1997) *Economic Impacts of Historic Preservation*, New Brunswick, NJ: Center for Urban Policy Research, Rutgers University

Listokin, D. and Listokin, B. (2001) *Barriers to the Rehabilitation of Affordable Housing*, Vols 1 and 2, Washington, DC: US Department of Housing and Urban Development, Office of Policy Development and Research

Litman, T. (2003) 'Integrating public health objectives in transportation decision-making', *American Journal of Health Promotion* 18(1): 103–8

Llewelyn-Davies and Alan Baxter & Associates (2000) *Urban Design Compendium*, London, UK: Llewelyn-Davies and Alan Baxter & Associates

Local Government Commission (2001) 'Building livable communities: a policymaker's guide to infill development,' Sacramento, CA: Local Government Commission

Lockeretz, W. (ed.) (1987) *Sustaining Agriculture Near Cities*, Ankenny, IA: Soil and Water Conservation Society

Lopez Community Land Trust (2006) 'Sustainable community homes', Lopez Island, WA: Lopez Community Land Trust

Lord, T. F. (1977) *Decent Housing: A Promise to Keep*, Cambridge, MA: Schenkman Publishing

Los Angeles, City of (1999) 'City of Los Angeles General Plan – transportation element', Los Angeles, CA: City of Los Angeles Department of City Planning

Los Angeles, County of (2007) 'Los Angeles County Draft Preliminary General Plan 2007 – Technical Appendix 2', Los Angeles, CA: County of Los Angeles, Department of City Planning

Lovell, C. H. *et al.* (1979) 'Federal and state mandating on local governments: report to the National Science Foundation', Riverside, CA: University of California

Lowry, I. S. and Ferguson, B. W. (1992) *Development Regulation and Housing Affordability*, Washington, DC: Urban Land Institute

Lubove, R. (1962) *The Progressives and the Slums: Tenement House Reform in New York City 1890–1917*, Pittsburgh: University of Pittsburgh Press

Lucy, W. H. and Phillips, D. L. (2000) *Confronting Suburban Decline: Strategic Planning for Metropolitan Renewal*, Washington, DC: Island Press

Lukensmeyer, C. J. (2007) 'Large-scale citizen engagement and the rebuilding of New Orleans: a case study', *Cityscape* 6(3): 3–15

Lukensmeyer, C. J. and Torres, L. H. (2006) 'Public deliberation: a manager's guide to citizen engagement', Armonk, NY: IBM Center for the Business of Government Collaboration Series

McArdle, N. (1999) *Outward Bound: The Decentralization of Population and Employment*, Boston: Joint Center for Housing Studies, Harvard University

McAvoy, G. E. (1999) *Controlling Technocracy: Citizen Rationality and the NIMBY Syndrome*, Washington, DC: Georgetown University Press

McCann, B. A. and Ewing, R. (2003) 'Measuring the health effects of sprawl: a national analysis of physical activity, obesity and chronic disease', Washington, DC: Smart Growth America Surface Transportation Policy Project

McCarthy, D. J., Jr. (1995) *Local Government Law*, St Paul, MN: West Publishing

McConnell, A. S. (2004) 'Making Wal-Mart pretty: trademarks and aesthetics restrictions on big-box retailers', *Duke Law Journal* 53: 1537–67

McCormack, A. (1946) 'A law clerk's recollections' [on the *Euclid* case], *Columbia Law Review* 46: 710–18

MacDonald, H. (1996) 'Why Business Improvement Districts work', *Civic Bulletin* 4: 1–3

MacFarlane, A. and Ewing, R. (eds) (2006) *Uncertainty Underground: Yucca Mountain and the Nation's High-Level Nuclear Waste*, Cambridge, MA: MIT

McGuigan, J. and Downey, J. W. (eds) (1999) *Technocities: The Culture and Political Economy of the Digital Revolution*, Thousand Oaks, CA: Sage Publications

McHarg, I. (1995) *Design with Nature*, Hoboken, NJ: John Wiley & Sons

McKenzie, E. (1994) *Privatopia: Homeowner Associations and the Rise of Residential Private Government*, New Haven, CT: Yale University Press

MacKenzie, J. J., Dower, R. C., and Chen, D. T. (1992) *The Going Rate: What it Really Costs to Drive*, Washington, DC: World Resources Institute

McLean, B. (1997) 'Evolution of marketable permits: the U.S. experience with sulfur dioxide allowance trading', *International Journal of Environment and Pollution* 107(5): 19–36

McMurry, S. and Adams, A. (eds) (2000) *People, Power, Places*, Knoxville: University of Tennessee Press.

Mahoney, J. D. (2005) 'Kelo's legacy: eminent domain and the future of property right', *Supreme Court Review* 2005: 103–33

Maine, State of, State Planning Office (2006) 'An evaluation of the Growth Management Act and its implementation: in response to resolve 2004, Chapter 73, Joint Standing Committee on Natural Resources', Augusta, ME: Maine State Planning Office

Maine, State of, State Planning Office (2007) 'Four-year Growth Management Program evaluation: report to the Joint Standing Committee on Natural Resources', Augusta, ME: Maine State Planning Office

Mallach, A. (1984) *Inclusionary Housing Programs: Policies and Practices*, New Brunswick, NJ: Center for Urban Policy Research, Rutgers University

Mandelbaum, S. J., Mazza, L., and Burchell, R. W. (1996) *Explorations in Planning Theory*, New Brunswick, NJ: Center for Urban Policy Research, Rutgers University

Mandelker, D. R. (1962) *Green Belts and Urban Growth*, Madison, WI: University of Wisconsin Press

Mandelker, D. R. (1993) *Land Use Law* (3rd edition), Charlottesville, VA: Michie

Mandelker, D. R. (1997) *Land Use Law* (4th edition), Charlottesville, VA: Lexis Law Publishing

Mandelker, D. R. (ed.) (2005) *Planning Reform in the New Century*, Chicago, IL: APA Planners Press

Mandelker, D. R. and Cunningham, R. A. (1990) *Planning and Control of Land Development: Cases and Materials* (3rd edition), Charlottesville, VA: Michie

Margerum, R. D. (2002) 'Collaborative planning: building consensus and building a distinct model for practice', *Journal of Planning Education and Research* 21: 237–53

Margerum, R. D. (2011) *Beyond Consensus: Improving Collaborative Planning and Management*, Cambridge, MA: Massachusetts Institute of Technology

Marin, M. C. (2007) 'Impacts of urban growth boundary versus exclusive farm use zoning on agricultural land uses', *Urban Affairs Review* 43(2): 199–220

Markusen, A. R., Hall, P., Campbell, S., and Deitrick, S. (1991) *The Rise of the Gunbelt: The Military Remapping of Industrial America*, New York: Oxford University Press

Marsh, B. C. (1909) *An Introduction to City Planning: Democracy's Challenge to the American City*, New York: Marsh

Marsh, W. (2010) *Landscape Planning: Environmental Applications*, Hoboken, NJ: John Wiley & Sons

Marshall, A. (2000) *How Cities Work: Suburbs, Sprawl, and the Roads Not Taken*, Austin: University of Texas Press

Maryland Department of Planning (2011) *PlanMaryland: A Sustainable Growth Plan for the 21st Century*, Baltimore, Maryland: Maryland Department of Planning

Maryland Office of Planning (1995) 'Urban Growth Boundaries', Publication #95-09, Baltimore, MD: Maryland Office of Planning

Mason, R. (ed.) (1999) *Economics and Heritage Conservation*, Los Angeles: Getty Conservation Institute

Mason, R. (2005) *Economics and Historic Preservation: A Guide and Review of the Literature*, Washington, DC: Brooking Institution Metropolitan Policy Program

Massey, D. S. and Denton, N. A. (1993) *American Apartheid: Segregation and the Making of the Underclass*, Cambridge, MA: Harvard University Press

Mayer, C. and Pence, K. (2009) 'Subprime mortgages: what, where, and to whom?', in Glaser, E. L. and Quigley, J. M. (eds) *Housing Markets and the Economy: Risk, Regulation, and Policy*, Cambridge, MA: Lincoln Institute of Land Policy, pp. 149–96

Mayer, V. M. (1992) *Local Officials Guide to Defense Economic Adjustment*, Washington, DC: National League of Cities

Mazmanian, D. and Morell, D. (1992) *Beyond Superfailure: America's Toxics Policy for the 1990s*, Boulder, CO: Westview

Meadows, D. H., Meadows, D. L., Randers, J., and Behrens III, W. W. (1972) *The Limits of Growth: A Report to the Club of Rome*, New York: Universe Books

Meck, S. (1996) 'Model planning and zoning enabling legislation: a short history', in *Modernizing State Planning Statutes: The Growing Smart Working*

Papers, Vol. 1, pp. 1–18, by American Planning Association, Chicago, IL: American Planning Association

Meck, S., Retzlaff, R., and Schwab, J. (2003) *Regional Approaches to Affordable Housing*, PAS Report Nos 513/514, Chicago, IL: American Planning Association

Meiners, R. E. and Morriss, A. P. (eds) (2000) *The Common Law and the Environment: Rethinking the Statutory Basis for Modern Environmental Law*, Lanham, MD: Rowan and Littlefield

Meisal, A. (2010) *LEED Materials: A Resource Guide to Green Buildings*, NY: Princeton Architectural Press

Merriam, D., Brower, D. J., and Tegeler, P. D. (1985) *Inclusionary Zoning Moves Downtown*, Chicago: Planners Press

Merrill, T. W. (1986) 'The economics of public use', *Cornell Law Review* 72: 61–116

Mesa, City of (2006) 'City of Mesa Capital Improvement Plan for Fiscal Years 2006–2011', Mesa, AZ: City of Mesa

Meshenberg, M. J. (1976) *The Language of Zoning: A Glossary of Words and Phrases* (Planning Advisory Service Report 322), Chicago: American Planning Association

Metro (1992) 'Portland Metropolitan Services District Charter', Portland, OR: Metro

Metropolitan Council (2000) 'Guidelines on smart growth: planning more livable communities with transit-oriented development', St Paul, MN: Metropolitan Council

Metzger, J. T. (1996) 'The theory and practice of equity planning: an annotated bibliography', *Journal of Planning Literature* 11(1): 112–26

Meyerson, M. (1956) 'Building the middle-range bridge for comprehensive planning', *Journal of the American Institute of Planners* 22: 58–64

Meyerson, M. and Banfield, E. C. (1955) *Politics, Planning and the Public Interest: The Case of Public Housing in Chicago*, Glencoe, IL: Free Press

Meystedt, R. S. (2011) '*Stop the Beach Renourishment*: why judicial takings may have meant taking a little too much', *Missouri Environmental Law and Policy Review* 18(2): 378–96

Mieszkowski, P. and Mills, E. S. (1993) 'The causes of metropolitan suburbanization', *Journal of Economic Perspectives* 7(3): 135–47

Mikelbank, B. A. (2004) 'A typology of U.S. surburban places', *Housing Policy Debate* 15(4): 935–64

Miles, M. E., Haney, R. L., and Berens, G. (1996) *Real Estate Development: Principles and Process* (2nd edition), Washington, DC: Urban Land Institute

Miller, Z. L. and Melvin, P. M. (1987) *The Urbanization of Modern America* (2nd edition), New York: Harcourt Brace Jovanovich

Mills, E. S. and Hamilton, B. W. (1994) *Urban Economics* (5th edition), New York: HarperCollins

Mitchell, A. (1997/1998) *Zeroing In: Geographic Information Systems at Work in the Community*, Redlands, CA: Environmental Systems Research Institute

Mitchell, J. (1999) 'Business Improvement Districts and innovative service delivery', Arlington, VA: The PricewaterhouseCoopers Endowment for The Business of Government

Mitchell, J. (2008) *Business Improvement Districts and the Shape of American Cities*, Albany, NY: SUNY Press

Mitchell, W. J. (1995) *City of Bits: Space, Place, and the Infobahn*, Cambridge, MA: Massachusetts Institute of Technology

Mitchell, W. J. (1999) *e-topia*, Cambridge, MA: MIT Press

Mitchell, W. J. (2004) *Me ++: The Cyborg Self and the Networked City*, Cambridge, MA: MIT Press

Mitsch, W. J. and Gosselink, J. D. (1993) *Wetlands*, New York: Van Nostrand Reinhold

Mogulof, M. B. (1971) 'Regional planning, clearance and evaluation: a look at the A-95 process', *Journal of the American Planning Association* 37: 418–22

Mollenkopf, J. H. (1983) *The Contested City*, Princeton, NJ: Princeton University Press

Monkkonen, E. (1988) *America Becomes Urban: The Development of U.S. Cities and Towns, 1780–1980*, Berkeley: University of California Press

Monmonier, M. (1996) *How to Lie With Maps* (2nd edition), Chicago: University of Chicago Press

Monsma, S. V. (2012) *Pluralism and Freedom: Faith-Based Organizations in a Democratic Society*, Lanham, MD: Rowman & Littlefield

Moore, C. (2001) 'Smart growth and the Clean

Air Act', Washington, DC: Northeast-Midwest Institute

Moore, C. G. and Siskin, C. (1985) *PUDs in Practice*, Washington, DC: Urban Land Institute

Moore, T. and Thorsnes, P. (1994) *The Transportation/Land Use Connection* (Planning Advisory Service Report 448/449), Chicago: American Planning Association

Moraga, City of (2006) *City of Moraga Municipal Code*, 8.100.040

Morgan, M. G. (1993) 'Risk analysis and management', *Scientific American* 169: 2–41

Morris, M. (ed.) (2007) *Planning Active Communities*, Chicago, IL: APA Planning Advisory Service

Moss, M. L. (1987) 'Telecommunications, world cities, and urban policy', *Urban Studies* 24: 534–46

Moynihan, D. P. (1969) *Maximum Feasible Misunderstanding: Community Action in the War on Poverty*, New York: Free Press

Mumford, L. (1961) *The City in History: Its Origins, Its Transformations, and Its Prospects*, New York: Harcourt Brace (Penguin edition 1991)

Mumford, L. (1962) *Technics and Civilization*, New York: Harcourt, Brace and World

Municipal Research Services Center of Washington (1997) 'Infill development: strategies for shaping livable neighborhoods,' Report No. 38, Seattle, WA: Municipal Research Services Center of Washington

Munton, D. (ed.) (1996) *Hazardous Waste Siting and Democratic Choice*, Washington, DC: Georgetown University Press

Muse, D. E., Driscoll, R. L., and Green, R. L. (1995) 'A municipal landfill Superfund response cost and contribution action – the city's response', *Urban Lawyer* 27: 129–61

Mutz, K. M., Bryner, G. C., and Kenney, D. S. (eds) (2002) *Justice and Natural Resources: Concepts, Strategies, and Applications*, Washington, DC: Island Press

Nash, R. F. (1990) *American Environmentalism: Readings in Conservation History* (3rd edition), New York: McGraw-Hill

National Association of Home Builders (n.d.) *The Truth About Regulatory Barriers to Housing Affordability*, Washington, DC: National Association of Home Builders

National Association of Home Builders (2004) 'Housing policy for the 21st century', Washington, DC: National Association of Home Builders

National Commission on the Environment (1993) *Choosing a Sustainable Future: The Report of the National Commission on the Environment*, Washington, DC: Island Press

National Congress for Community Economic Development (1995) *Tying It All Together: The Comprehensive Achievements of Community-Based Development Organizations*, Washington, DC: NCCED

National Housing Law Project (2002) *False HOPE: A Critical Assessment of the HOPE VI Public Housing Redevelopment Program*, Oakland, CA: National Housing Law Project

National Research Council (NRC), Transportation Research Board (1993) *Moving Urban America*, Washington, DC: National Academy Press

National Research Council (NRC), Transportation Research Board (1994) *Curbing Gridlock: Peak-Period Fees to Relieve Traffic Congestion* (2 vols), Washington, DC: National Academy Press

Nelson, A. C. (1988) *Development Impact Fees: Policy Rationale, Practice, Theory and Issues*, Chicago: Planners Press

Nelson, A. C. (1992a) 'Characterizing exurbia', *Journal of Planning Literature* 6(4): 350–68

Nelson, A. C. (1992b) 'Preserving prime farmland in the face of urbanization: lessons from Oregon', *Journal of the American Planning Association* 58: 467–88

Nelson, A. C. (1994) 'Development impact fees: the next generation', *Urban Lawyer* 26: 541–62

Nelson, A. C. and Dawkins, C. J. (2004) *Urban Containment in the United States*, Chicago, IL: APA Planning Advisory Service

Nelson, A. C. and Duncan, J. B. (1995) *Growth Management Principles and Practices*, Chicago: Planners Press, American Planning Association

Nelson, A. C., Pendall, R., Dawkins, C. J., and Knapp, G. J. (2002) 'The link between growth management and housing affordability: the academic evidence', discussion paper, Washington, DC: Brookings Institution Center on Urban and Metropolitan Policy

Nelson, M., Ehrenfeucht, R., and Laska, S. (2007) 'Planning, plans, and people: professional expertise, local knowledge, and governmental action in post-Katrina New Orleans', *Cityscape* 9(3): 23–52

Nelson, R. R. (1977) *The Moon and the Ghetto*, New York: Norton

Netter, E. M. (2000) *Inclusionary Zoning Guidelines for Cities and Towns*, Boston: Massachusetts Housing Partnership Fund

Netusil, N. (2005) 'The effect of environmental zoning and amenities on property values: Portland, Oregon', *Land Economics* 81(2): 227–46

Netzer, D. (ed.) (2003) *Property Tax, Land Use and Land Use Regulation*, Northampton, MA: Edward Elgar Publishing

Neuman, M. (1991) 'Utopia, dystopia, diaspora', *Journal of the American Planning Association* 57(3): 344–348

Newburyport, City of (2001) *Newburyport Master Plan*, Newburyport, Massachsetts: Newburyport, City of

New Jersey Pinelands Commission (2006) 'Annual report 2006', Lisbon, NJ: New Jersey Pinelands Commission

Newman, H. K. (1999) *Southern Hospitality: Tourism and the Growth of Atlanta*, Tuscaloosa: University of Alabama Press

New York, City of (1982) *Midtown Zoning*, New York: New York City Planning Commission

New York, City of (2011) *PlaNYC*, New York: City of New York

Nicholas, H. G. (1986) *The Nature of American Politics*, Oxford: Oxford University Press

Nicholas, J. C. and Steiner, R. L. (2000) 'Growth management and smart growth in Florida', *Wake Forest Law Review*, 35: 645–70

Nicholas, J. C., Nelson, A. C., and Juergensmeyer, J. C. (1991) *A Practitioner's Guide to Development Impact Fees*, Chicago: Planners Press

Nolon, J. R. and Salkin, P. E. (2006) *Land Use in a Nutshell*, St Paul, MN: Thomson/West Group

Norwalk, City of (2006) *The Norwalk, Ohio Comprehensive Plan*, Norwalk, Ohio: City of Norwalk

Noss, R. F., O'Connell, M. A., and Murphy, D. D. (1997) *The Science of Conservation Planning: Habitat Conservation under the Endangered Species Act*, Washington, DC: Island Press

Novek, E. (2001) 'You wouldn't fit here: the subtle and blatant forms of communication that keep segregation going', *NLI Online*, http://www.nli.org/online/issues/Novek.html, March/April, accessed July 27

Novotny, P. (2000) *Where We Live, Work, and Play: The Environmental Justice Movement and the Struggle for a New Environmentalism*, Westport, CT: Praeger

Obermeyer, N. J. and Pinto, J. K. (2007) *Managing Geographic Information Systems* (2nd edition), New York: Guilford Publications

O'Connell, L. (2009) 'The impact of local supporters of smart growth policy adoption', *Journal of the American Planning Association* 75(3): 281–91

Office of the Federal Environmental Executive (2007) 'FACT SHEET: Executive Order 13423 Strengthening Federal Environmental, Energy, and Transportation Management,' Washington, DC: Office of the Federal Environmental Executive

Ohmae, K. (1995) *The End of the Nation State: The Rise of Regional Economies*, New York: Free Press

Oliver, G. (1992) '1000 friends are watching: checking out the record of Oregon's pace-setting public interest group', *Planning* 58(11): 9–13

Olmsted, F. L. (1910) 'An Introductory Address on City Planning', *Proceedings of the Second National Conference on City Planning and the Problems of Congestion*, Rochester, NY, May 2, Cambridge, MA: The University Press, pp. 15–32.

O'Looney, J. (1995) *Economic Development and Environmental Control: Balancing Business and Community in the Age of NIMBYs and LULUs*, Westport, CT: Quorum Books

O'Looney, J. (2000) *Beyond Maps: GIS and Decision Making in Local Government*, Redlands, CA: Environmental Systems Research Institute

Olshansky, R. B. (2006) 'Planning after Hurricane Katrina,' *Journal of the American Planning Association* 72(2): 147–52

Olshansky, R. B. and Johnson, L. A. (2010) *Clear as Mud: Planning for the Rebuilding of New Orleans*, Chicago: APA Press

Olshansky, R. B. *et al.* (2008) 'Planning for the rebuilding of New Orleans,' *Journal of the American Planning Association* 74(3): 273–87

Onsrud, H. J. and Rushton, G. (eds) (1995) *Sharing Geographic Information*, New Brunswick, NJ: Center for Urban Policy Research, Rutgers, State University of New Jersey

Oregon, State of, Secretary of State (2004) 'Voters' guide for November 2, 2004 General Election, volume 1 – state measures', Salem, OR: State of Oregon, Secretary of State

Oregon, State of, Secretary of State (2007) 'Voters' pamphlet for November 6, 2007 Special Election', Salem, OR: State of Oregon, Secretary of State

Orfield, M. and Luce, Jr., T. F. (2010) *Region: Planning the Future of the Twin Cities*, Minneapolis, MN: University of Minnesota Press

O'Riordan, T. (ed.) (1995) *Environmental Science for Environmental Management*, Harlow: Longman

Orman, L. (1984) 'Ballot-box planning: the boom in electoral land use control', Public Affairs Report 25, Berkeley: Bulletin of the Institute of Government, University of California

Oswald, L. J. (2000) 'Property rights legislation and the police power', *American Business Law Journal* 37: 527–62

Packard, V. (1960) *The Waste Makers*, New York: David McKay

Packnett, D. (2005) 'The First Homes Community Land Trust', Working Paper, Cambridge, MA: Lincoln Institute of Land Policy

Palen, J. J. (1995) *The Suburbs*, New York: McGraw-Hill

Pappas, D. (2001) 'A new approach to a familiar problem: the New Markets Tax Credit', *Journal of Affordable Housing and Community Development Law* 10(4): 323–52

Parolek, D. G., Parolek, K., and Crawford, P. C. (2008) *Form-Based Codes: A Guide for Planners, Urban Designers, Municipalities, and Developers*, Indianapolis: IN: Wiley

Pastor, M., Bullard, R., Boyce, J., Fothergill, A., Morello-Frosch, R., and Wright, B. (2006) 'In the wake of the storm: environment, disaster, and race after Katrina', New York: Russell Sage Foundation

Paul, E. F. (1987) *Property Rights and Eminent Domain*, New Brunswick, NJ: Transaction Books

Peirce, N. R., Johnson, C. W., and Hall, J. S. (1993) *Citistates: How Urban America Can Prosper in a Competitive World*, Washington, DC: Seven Locks Press

Peiser, R. (1990) 'Who plans America? Planners or developers?', *Journal of the American Planning Association* 56: 496–503

Pelham, Town of (2005) 'Town of Pelham Capital Improvements Plan 2006–2012', Pelham, NH: Town of Pelham

Pellow, D. N. (2002) *Garbage Wars: The Struggle for Environmental Justice in Chicago*, Cambridge, MA: MIT Press

Penalver, E. M. (2006) 'Property metaphors and *Kelo* v. *New London*', *Fordham Law Review* 74(6): 2971–6

Penalver, E. M. and Strahilevitz, L. (2012) 'Judicial takings or due process', *Cornell Law Review* 97(2): 305–68

Perry, M. J. and Machun, P. J. (2001) *Population Change and Distribution, 1990–2000*, Washington, DC: US Census Bureau

Peterson, J. A. (2003) *The Birth of City Planning in the United States, 1840–1917*, Baltimore, MD: Johns Hopkins University Press

Peterson, S. C. (2002) *Acting for Endangered Species: The Statutory Ark*, Lawrence: University Press of Kansas

Pew Center on the States (2012) *Evidence Counts: Evaluating State Tax Incentives for Jobs and Growth*, Washington and Philadelphia: The Pew Charitable Trusts

Philadelphia City Planning Commission (2011) 'Citywide Vision – Philadelphia 2035', Philadelphia: Philadelphia City Planning Commission

Philbrick, F. S. (1938) 'Changing concepts of property in law', *University of Pennsylvania Law Review* 86: 691–732

Phoenix, Arizona, City of (2008) *General Plan Amendment*, Phoenix, AZ: Planning and Development Department

Pickering, K. T. and Owen, L. A. (1994) *An Introduction to Global Environmental Issues*, New York: Routledge

Piel, G. and Segerberg, O. (eds) (1990) *The World of René Dubos*, New York: Holt

Pincetl, S. S. (1999) *Transforming California: A Political History of Land Use and Development*, Baltimore: Johns Hopkins University Press

Pinchot, G. (1910) *The Fight for Conservation*, New York: Doubleday

Pivo, G. (1992) 'How do you define community character?' *Small Town* 23(3): 4–17

Planner's Collaborative (2001) 'Wilmington Master Plan 2001', Boston, MA: Planner's Collaborative

Planning Advisory Services (1998) *Principles of Smart Development*, N. 479, Chicago: American Planning Association

Plantinga, A. and Jaeger, W. (2007) 'How have land use regulations affected property values in Oregon?', Corvallis, OR: Oregon State University

Platt, R. H. (2004) *Land Use and Society* (revised edition), Washington, DC: Island Press

Plyer, A. *et al.* (eds) (2011) *Resilience and Opportunity: Lessons from the U.S. Gulf Coast after Katrina and Rita*, Washington, DC: Brookings Institution Press

Policicchio, J. (2011) '*Stop the Beach Renourishment, Inc. v. Florida Department of Environmental Protection*', *Harvard Environmental Law Review* 35(2): 541–3

Polikoff, A. (2010) 'Overview: three remaining HOPE VI challenges', *Housing Policy Debate* 20(1): 147–51

Poole, S. E. (1987) 'Architectural appearance review regulations and the First Amendment: the good, the bad, and the consensus ugly', *Urban Lawyer* 19: 287–344

Popkin, S. J. (2007) 'Testimony of Susan Popkin, Urban Institute, prepared for the hearing on S. 829 HOPE VI Improvement and Reauthorization Act', before the US Senate Committee on Banking, Housing and Urban Affairs, Subcommittee on Housing, Transportation, and Community Development, June 20.

Popkin, S. J. (2010) 'A glass half empty? New evidence from the HOPE VI panel study', *Housing Policy Debate* 20(1): 42–63

Popkin, S. J. *et al.* (2002) *HOPE VI Panel Study: Baseline Report*, Washington, DC: The Urban Institute

Popkin, S. J. *et al.* (2004) *A Decade of HOPE VI: Research Findings and Policy Challenges*, Washington, DC: The Urban Institute

Porter, D. R. (1986) *Growth Management: Keeping on Target?* Washington, DC: Urban Land Institute

Porter, D. R. (ed.) (1992) *State and Regional Initiatives for Managing Development: Policy Issues and Practical Concerns*, Washington, DC: Urban Land Institute

Porter, D. R. (ed.) (1995) *Housing for Seniors: Developing Successful Projects*, Washington, DC: Urban Land Institute

Porter, D. R. (1996) *Performance Standards for Growth Management*, Chicago: American Planning Association

Porter, D. R. (1997) *Managing Growth in America's Communities*, Washington, DC: Island Press

Porter, D. R. (2002) *Making Smart Growth Work*, Washington, DC: Urban Land Institute

Porter, D. R. (2003) 'Tennessee's Growth Policy Act: purposes, implementation, and effects on development', Washington, DC: Growth Management Institute

Porter, M. (1995) 'The competitive advantage of the inner city', *Harvard Business Review*, May–June: 55–71

Portney, K. E. (1991) *Siting Hazardous Waste Treatment Facilities: The NIMBY Syndrome*, New York: Auburn House

Portney, K. E. (2003) *Taking Sustainable Cities Seriously*, Cambridge, MA: MIT Press

Portney, P. R. (ed.) (1990) *Public Policies for Environmental Protection*, Washington, DC: Resources for the Future

President's Commission on a National Agenda for the Eighties (1980a) *A National Agenda for the Eighties*, Washington, DC: US Government Printing Office

President's Commission on a National Agenda for the Eighties (1980b) *Urban America*, Washington, DC: US Government Printing Office

President's Council on Sustainable Development (1993) Executive Order 12852, June 29, Washington DC: President's Council for Sustainable Development

President's Council on Sustainable Development (1996) *Sustainable America: A New Consensus for Prosperity, Opportunity, and a Healthy Environment for the Future*, Washington, DC: President's Council on Sustainable Development

President's Council on Sustainable Development (1997a) *Building on Consensus: A Progress Report on Sustainable America*, Washington, DC: President's Council on Sustainable Development

President's Council on Sustainable Development (1997b) *The Road to Sustainable Development: A Snapshot of Activities in the United States of America*, Washington, DC: President's Council on Sustainable Development

President's Council on Sustainable Development (1999) *Towards a Sustainable America: Advancing Prosperity, Opportunity, and a Healthy Environment for the 21st Century*, Washington, DC: President's Council on Sustainable Development

Pressman, J. L. and Wildavsky, A. B. (1984) *Implementation: How Great Expectations in Washington Are Dashed in Oakland* (3rd edition), Berkeley, CA: University of California Press

Preutz, R. (2003) *Beyond Takings and Givings*, Burbank, CA: Arje Press

Pritchett, W. E. (2006) 'Beyond Kelo: thinking urban development in the 21st century', *Georgia State University Law Review* 22: 895–933

Rabe, B. G. (1994) *Beyond NIMBY: Hazardous Waste Siting in Canada and the United States*, Washington, DC: Brookings Institution

Randolph, J. (2003) *Environmental Land Use Planning and Management*, Washington, DC: Island Press

Rapp, D. (1988) *How the US Got into Agriculture: and Why It Can't Get Out*, Washington, DC: Congressional Quarterly

Rechtschaffen, C., Gauna, E., and O'Neill, C. A. (2009) *Environmental Justice: Law, Policy and Regulation*, Durham, NC: Carolina Academic Press

Reible, D. (2007) 'Hurricane Katrina: environmental hazards in the disaster area', *Cityscape* 9(3): 53–68

Reichl, A. J. (1999) *Reconstructing Times Square: Politics and Culture in Urban Development*, Lawrence: University Press of Kansas

Remy, M. H., Thomas, T. A., and Moose, J. G. (1994) *Guide to the California Environmental Quality Act*, Point Arena, CA: Solano Press

Reps, J. W. (1965) *The Making of Urban America: A History of City Planning in the United States*, Princeton, NJ: Princeton University Press

Retzlaff, R. C. (2009) 'The use of LEED in planning and development regulation: an exploratory analysis,' *Journal of Planning Education and Research* 29(1): 67–77

Reynolds, G. (2007) *Federal Housing Subsidies: To Rent or To Own?* No. 6, Washington, DC: The Urban Institute

Rice, D. and Sard, B. (2007) 'Cuts in Federal Housing Assistance are undermining community plans to end homelessness', Washington, DC: Center on Budget and Policy Priorities

Riggins, R. E., Jones, E. B., Singh, R., and Rechard, P. E. (eds) (1990) 'Watershed planning and analysis in action', Reston, VA: American Society of Civil Engineers

Riis, J. A. (1890) *How the Other Half Lives: Studies Among the Tenements of New York*, New York: C. Scribner's Sons

Reynarsson, B. (1999) 'The planning of Reykjavik, Iceland: three ideological waves – a historical overview', *Planning Perspectives* 14: 49–67

Rittel, H. W. J. and Webber, M. M. (1973) 'Dilemmas in a general theory of planning', *Policy Sciences* 4: 155–69

Robertson, D. B. and Judd, D. R. (1989) *The Development of American Public Policy: The Structure of Policy Restraint*, Glenview, IL: Scott Foresman

Robinson, C. M. (1903) *Modern Civic Art, or the City Made Beautiful*, New York: Putnam

Robinson, C. M. (1907) *The Improvement of Towns and Cities*, New York: Putnam

Rohse, M. (1987) *Land-Use Planning in Oregon: A No-Nonsense Handbook in Plain English*, Corvallis, OR: Oregon State University Press

Rosenbaum, W. A. (2007) *Environmental Politics and Policy*, Washington, DC: CQ Press

Rosenbaum, W. A. (2010) *Environmental Politics and Policy* (8th edition), Washington: CQ Press

Rosenberg, G. N. (1991) *The Hollow Hope: Can Courts Bring About Social Change?* Chicago: University of Chicago Press

Rosenweig, R. and Blackmar, E. (1998) *The Park and the People: A History of Central Park*, Ithaca, NY: Cornell University Press

Ross, B. H., Levine, M. A., and Stedman, M. S. (1991) *Urban Politics: Power in Metropolitan America* (4th edition), Itasca, IL: Peacock

Rousakis, J. and Weintraub, B. A. (1994) 'Packaging, environmentally protective municipal solid waste management, and limits to the economic premise', *Ecology Law Quarterly* 21: 947–1005

Rubinowitz, L. S. and Rosenbaum, J. E. (2000) *Crossing the Class and Color Lines: From Public Housing to White Suburbia*, Chicago: University of Chicago Press

Rusk, D. (1993) *Cities Without Suburbs*, Washington, DC: Woodrow Wilson Center

Rutlow, E. (2006) '*Kelo* v. *City of New London*', *Harvard Environmental Law Review* 30(1): 261–79

Ryan, B. (2012a) *Design After Decline: How America Rebuilds Shrinking Cities*, Philadelphia, PA: University of Pennsylvania Press

Ryan, M. A. (2012b) *The Clean Water Act Handbook* (3rd edition), Chicago: American Bar Association

Rybczynski, W. (1999) *A Clearing in the Distance: Frederick Law Olmsted and America in the Nineteenth Century*, New York: Scribner

Rypkema, D. (1998) *Profiting from the Past: The Impact of Historic Preservation on the North Carolina Economy*, Raleigh: Preservation North Carolina

Rypkema, D. (2005) *The Economics of Historic Preservation: A Community Leader's Guide*, Washington, DC: National Trust for Historic Preservation

Sabatier, P., Focht, W., Lubell, M., Trachterberg, Z., Vedlitz, A., and Matlock, M. (2005) *Swimming Upstream: Collaborative Approaches to Watershed Management*, Cambridge, MA: MIT Press

Saco, Maine, City of (2003) *Zoning Ordinance*

Saha, D. and Paterson, R. G. (2008) 'Local government efforts to promote the "Three Es" of sustainable development: survey in medium to large cities in the United States', *Journal of Planning Education and Research*, 28(1): 21–37

Saleem, O. (1994) 'Overcoming environmental discrimination: the need for a disparate impact test and improved notice requirements on facility siting decisions', *Columbia Journal of Environmental Law* 19: 211–47

Salkin, P. E. and Freilich, R. H. (eds) (2000) *Hot Topics in Land Use Law: From the Comprehensive Plan to Del Monte Dunes*, Chicago: Section of State and Local Government Law, American Bar Association

Salsich, P. W., Jr. (2000) 'Neighborhoods: grassroots consensus building and collaborative planning',

Washington University Journal of Law and Policy, 3: 709–40

Salsich, P. W., Jr. and Tryniecki, T. J. (1998) *Land Use Regulation: A Legal Analysis and Practical Application of Land Use Law*, Chicago: Real Property, Probate and Trust Law, American Bar Association

Salvesen, D. (1990) *Wetlands: Mitigating and Regulating Development Trends*, Washington, DC: Urban Land Institute

Salzman, J. and Thompson, B. H., Jr. (2010) *Environmental Law and Policy* (3rd edition), New York: Thomson Reuters/Foundation Press

San Clemente, City of (2003) 'Nuclear safety element', San Clemente, CA: City of San Clemente

San Diego, City of (2008) *City of San Diego General Plan – Urban Design Element*, San Diego, CA: City of San Diego

Santos, A., McGickin, N., Natamoto, H. Y., Gray, D., and Liss, S. (2009) Summary of Travel Trends: 2009 National Household Travel, Washington, DC: US Department of Transportation, Federal Highway Administration

Saunders, W. S. (ed.) (2005) *Sprawl and Suburbia*, Minneapolis, MN: University of Minnesota Press

Savas, E. S. (1983) 'A positive urban policy for the future', *Urban Affairs Quarterly* 18: 447–53

Sax, J. L. (2006) 'Federalism issues following Kelo v. City of New London: Kelo: a case rightfully decided', *University of Hawaii Law Review* 28: 365–71

Scheer, B. C. and Preiser, W. (eds) (1994) *Design Review: Challenging Urban Aesthetic Control*, New York: Chapman and Hall

Schiffman, I. (1990) *Alternative Techniques for Managing Growth*, Berkeley, CA: Institute of Governmental Studies

Schiller, R. J. (2008) *The Subprime Solution: How Today's Global Financial Crisis Happened, and What to Do about It*, Princeton, NJ: Princeton University Press

Schmelzkopf, K. (1995) 'Urban community gardens as contested spaces', *Geographical Review* 85(3): 264–80

Schnidman, F., Smiley, M., and Woodbury, E. G. (1990) *Retention of Land for Agriculture: Policy, Practice and Potential in New England*, Cambridge, MA: Lincoln Institute of Land Policy

Schon, D. A. (1971) *Beyond the Stable State*, New York: Norton

Schon, D. A., Sanyal, B., and Mitchell, W. J. (eds) (1999) *High Technology and Low-Income Communities: Prospects for the Positive Use of Advanced Information Technologies*, Cambridge, MA: MIT Press

Schwab, J. (1994) 'Environmental LULUs: is there an equitable solution?', in American Planning Association (1994)

Schwab, J. and Meck, S. (2005) *Planning for Wildfires*, Chicago, IL: APA Planning Advisory Service

Schwab, J. *et al.* (1998) *Planning for Post-Disaster Recovery and Reconstruction*, Chicago, IL: APA Planning Advisory Service

Scott, M. (1969) *American City Planning since 1890*, Berkeley, CA: University of California Press

Scott Hanson, C. and Scott Hanson, K. (2005) *The Cohousing Handbook*, Gabriola Island, Canada: New Society Publishers

Seattle Department of Neighborhoods, 'P-Patch community gardens: *Growing Communities*', http://www.seattle.gov/Neighborhoods/ppatch/, accessed July 6, 2012

Seidman, K. F. (2005) *Economic Development Finance*, Thousand Oaks, CA: Sage Publications

Selmi, D. P. and Kushner, J. A. (1999) *Land Use Regulation: Cases and Materials*, Gaithersburg, MD: Aspen Law and Business

Seltzer, E. and Carbonell, A. (2011) *Regional Planning in America: Practice and Prospect*, Cambridge, MA: Lincoln Institute of Land Policy

Servon, L. (2002) *Bridging the Digital Divide: Technology, Community, and Public Policy*, Boston: Blackwell Publishers

Seskin, S. (2012a) *Complete Streets: Local Policy Workbook*, Washington, DC: Smart Growth America

Seskin, S. (2012b) *Complete Streets Policy Analysis 2012*, Washington, DC: Smart Growth America

Shabecoff, P. (1993) *A Fierce Green Fire: The American Environmental Movement*, New York: Hill and Wang

Shalala, D. (1978) 'A Pilgrim's Progress: Moving toward a National Urban Policy', address at Trinity College, Hartford, Connecticut, 20 April: quoted in Dilger (1982: 139)

Sheller, M., Urry, J., McCabe, P. T., and Bust, P. D. (eds) (2006) *Mobile Technologies of the City*, London: Taylor and Francis

Sherrow, V. (2001) *Love Canal: Toxic Waste Tragedy*, Berkeley Heights, NJ: Enslow Publishers

Shires, M., Ellwood, J. W., and Sprague, M. (1998) *Has Proposition 13 Delivered? The Changing Tax Burden in California*, San Francisco: Public Policy Institute of California

Sies, M. C. and Silver, C. (eds) (1996) *Planning the Twentieth-Century American City*, Baltimore, MD: Johns Hopkins University Press

Silver, C. (1991) 'The racial origins of zoning: southern cities from 1910–1940', *Planning Perspectives* 6: 189–205

Simon, v. Needham, 42 N. E. 2d 516 (1942)

Sims, R. (2001) 'News release: Executive Sims' growth plan adopted; measure supports smart growth initiatives', February 12

Sinclair, U. (1906) *The Jungle*, New York: Doubleday

Smart Growth Network Subgroup on Affordable Housing (2001) *Affordable Housing and Smart Growth: Making the Connection*, Washington, DC: US Environmental Protection Agency and the Smart Growth Network

Smerk, G. M. (1991) *The Federal Role in Urban Mass Transit*, Bloomington, IN: Indiana University Press

Smith, K. S. (2000) *Capital Assets: A Report on the Tourist Potential of Neighborhood Heritage and Cultural Sites in Washington, D.C.* (2nd edition), Washington, DC: DC Heritage Tourism Coalition

Smith, M. M. and Hevener, C. C. (2010) 'Subprime lending over time: the role of race', Discussion Paper, Philadelphia, PA: Federal Reserve Bank of Philadelphia

Smith, R. E. and Ferryman, K. (2006) 'Saying goodBye: relocating senior citizens in the HOPE VI panel study', Brief No. 10, Washington, DC: The Urban Institute Center for Metropolitan Housing and Communities

Smith, Z. A. (2003) *The Environmental Policy Paradox* (4th edition), Englewood Cliffs, NJ: Prentice Hall

Snow, D. (ed.) (1992) *Inside the Environmental Movement*, Washington, DC: Island Press

Snyder, T. and Stegman, M. A. (1986) *Paying for Growth:*

Using Development Fees to Finance Infrastructure, Washington, DC: Urban Land Institute

So, F. S. and Getzels, J. (eds) (1988) *The Practice of Local Government Planning* (2nd edition), Washington, DC: International City Management Association

Solove, D. J. (2006) *Digital Person: Technology and Privacy in the Information Age*, New York: New York University Press

Sorenson, D. S. S. (1998) *Shutting Down the Cold War: The Politics of Military Base Closure*, New York: St Martin's Press

Soule, D. C. (ed.) (2006) *Urban Sprawl*, Westport, CT: Greenwood Press

South Pasadena, City of (1998) 'Historic preservation element' South Pasadena, CA: City of South Pasadena

Spann, E. K. (1996) *Designing Modern America: The Regional Planning Association of America and Its Members*, Columbus: Ohio State University Press

Spectorsky, A. C. (1955) *The Exurbanites*, New York: J. B. Lippincott Co.

Sprankling, J. G. (2007) 'Perspective on Kelo v. City of New London: introduction: the impact of Kelo v. City of New London on eminent domain', *McGeorge Law Review* 38: 369–70

Squires, G. D. (ed.) (1989) *Unequal Partnerships: The Political Economy of Urban Redevelopment in Postwar America*, New Brunswick, NJ: Rutgers University Press

Squires, G. D. and O'Connor, S. (2001) *Color and Money: Politics and Prospects for Community Reinvestment in Urban America*, Albany: State University of New York Press

Squires, J. F. (1992) 'Growth management redux: Vermont's Act 250 and Act 200', in Porter (1992)

Staley, S. R. and Mildner, G. C. S. (1999) 'Urban-growth boundaries and housing affordability: Lessons from Portland', Policy Brief No. 11, Los Angeles, CA: Reason Public Policy Institute

Staley, S. R., Edgens, J. G., and Mildner, G. C. S. (1999) *A Line in the Land: Urban-Growth Boundaries, Smart Growth, and Housing Affordability*, Policy Study No. 263, Los Angeles: Reason Public Policy Institute

Stanfield, R. L. (1991) 'Strains in the family', *National Journal* 23(39): 2316–33

Stanford Environmental Law Society (2000) *The Endangered Species Act*, Palo Alto, CA: Stanford Environmental Law Society

Stanilov, K., Pailthorp, M., Carlson, D., and Pivo, G. (1993) *A Literature Review of Community Impacts and Costs of Urban Sprawl*, Washington, DC: National Trust for Historic Preservation

Starr, R. (1998) 'How to fix New York's heavy-handed zoning laws', *Civic Bulletin* No. 12, New York: The Manhattan Institute

Stavins, R. (2000) *Experience with Market-Based Environmental Policy Instruments*, RFF Discussion Paper 00–09, Washington, DC: Resources for the Future

Steffens, L. (1904) *The Shame of Our Cities*, New York: Farrar, Strauss, and Giroux

Stegman, M. A. (1999) *State and Local Affordable Housing Programs*, Washington, DC: Urban Land Institute

Stegman, M. A. and Holden, J. D. (1987) *Nonfederal Housing Programs: How States and Localities are Responding to Federal Cutbacks in Low-Income Housing*, Washington, DC: Urban Land Institute

Stein, J. M. (ed.) (1993) *Growth Management: The Planning Challenge of the 1990s*, Newbury Park, CA: Sage

Stein, J. M. (ed.) (1995) *Classical Readings in Real Estate and Development*, Washington, DC: Urban Land Institute

Stein, J. M. (ed.) (2004) *Classic Readings in Urban Planning* (2nd edition), Chicago, IL: APA Planners Press

Steiner, F. (2000) *The Living Landscape: An Ecological Approach to Landscape Planning*, New York: McGraw-Hill

Steiner, F. and Butler, K. (2007) *Planning and Urban Design Standards*, Hoboken, New Jersey: John Wiley & Sons

Steinman, L. D. (1988) *The Impact of Zoning on Group Homes for the Mentally Disabled: A National Survey*, Chicago, IL: Section of Urban, State, and Local Govt Law, American Bar Association

Stensvaag, J. M. (1991) *Clean Air Act: Law and Practice*, New York: Aspen Publishers

Stern, A. (1982) 'History of air pollution legislation

in the United States', *Journal of the Air Pollution Control Association* 32: 44–61

Stipe, R. (1987) *The American Mosaic: Preserving a National Heritage*, Washington, DC: US Committee, International Council on Monuments and Sites

Stocker, F. D. (ed.) (1991) *Proposition 13: A Ten Year Retrospective*, Cambridge, MA: Lincoln Institute of Land Policy

Strong, A. (1979) *Land Banking*, Baltimore: Johns Hopkins University Press

Strong, A. L., Mandelker, D. R., and Kelly, E. K. (1996) 'Property rights and takings', *Journal of the American Planning Association* 62: 5–16

Suchman, D. R. (1995) *Manufactured Housing: An Affordable Alternative*, Washington, DC: Urban Land Institute

Suchman, D. R. (2002) *Developing Successful Infill Housing*, Washington, DC: Urban Land Institute

Suchman, D. R., Scott, D., and Giles, S. L. (1990) *Public–Private Housing Partnerships*, Washington, DC: Urban Land Institute

Sugrue, T. J. (2011) 'A dream still deferred', *New York Times* Sunday Opinion, March 27: 11

Sumner, City of (2005) *Transportation Element – City of Sumner Comprehensive Plan*, Sumner, WA: City of Sumner

Susskind, L. and Cruikshank, J. (1987) *Breaking the Impasse: Consensual Approaches to Resolving Public Disputes*, New York: Basic Books

Sutton, P. C., Cova, T. J., and Elvidge, C. D. (2006) 'Mapping exurbia in the conterminous United States using nighttime satellite imagery', *Geocarto International* 21(2): 39–45

Szold, T. S. and Carbonell, A. (2002) *Smart Growth: Form and Consequences*, Cambridge, MA: Lincoln Institute of Land Policy

Talen, E. (2005) *New Urbanism & American Planning*, London: Routledge

Taylor, F. W. (1911) *The Principles of Scientific Management*, New York: Harper & Brothers

Taylor, N. (1998) *Planning Theory since 1945*, Thousand Oaks, CA: Sage Publications

Teaford, J. C. (1990) *The Rough Road to Renaissance – Urban Revitalization in America*, Baltimore: Johns Hopkins University Press

Teaford, J. C. (1993) *The Twentieth-Century American City* (2nd edition), Baltimore: Johns Hopkins University Press

Tempe, City of (2008) *City of Tempe General Plan 2030*, Tempe, AZ: Development Services Department

Thomas, J. M. and Ritzdorf, M. (1997) *Urban Planning and the African-American Community*, Thousand Oaks, CA: Sage Publications

Thompson, L. (1996) 'Pedestrian road crossing safety', Council of Planning Librarians Bibliography 331/332/333, *Journal of Planning Literature* 11: 263–300

Throsby, D. (2001) *Economics and Culture*, New York: Cambridge University Press

Tibbetts (1995) 'Everybody's taking the fifth', *Planning* 61(1): 4–9 (January)

Tietenberg, T. H. (2006) *Emissions Trading: Principles and Practice*, Washington, DC: Resources for the Future

Tocqueville, A. de (1848) *Democracy in America* (1969 edition edited by J. P. Mayer), London: Fontana Press

Toll, S. I. (1969) *Zoned American*, New York: Grossman

Tomioka, S. and Tomioka, E. M. (1984) *Planned Unit Developments: Design and Regional Impact*, New York: Wiley

Tomlinson, R. (2007) *Thinking about GIS: Geographic Information System Planning for Managers*, Redlands, CA: ESRI Press

Torma, C. (1987) 'Assessing the work to date', in *Proceedings of the Workshop on Historic Mining Resources*, Pierre, SD: South Dakota State Historical Preservation Society

Toronto City (1988) *Section 36 Guidelines: Further Report on Guidelines for Bonusing Pursuant to Section 36 of the Planning Act*, Toronto: City of Toronto Planning and Development Department

Tribe, L. H. (1985) *God Save this Honorable Court: How the Choice of Supreme Court Justices Shapes Our History*, New York: New American Library

Tschinkel, V. J. (1989) 'The rise and fall of environmental expertise', in Ausubel and Sladovich (1989)

Turner, M. A. and Skidmore, F. (1999) 'Mortgage lending discrimination: a review of existing evidence', Washington, DC: The Urban Institute

Turner, M. A. *et al.* (2007) *Severely Distressed Public Housing: The Costs of Inaction*, Washington, DC: The Urban Institute

Tyler, N., Liqibel, T. J., and Tyler, I. R. (2009) *Historic Preservation: An Introduction to Its History, Principles, and Practice* (2nd edition), New York: W.W. Norton & Company

UN (1972) *Report of the United Nations Conference on the Human Environment*, NY: UN

UN (1992) *Report of the United Nations Conference on Environment and Development*, NY: UN

UN (2001) *Johannesburg Summit 2002: World Summit on Sustainable Development*, NY: UN

UN (2002) *UN Report of the World Summit on Sustainable Development*, NY: UN

UN (2012) *Report of the United Nations Conference on Sustainable Development*, NY: UN

United Church of Christ, Commission on Racial Justice (1987) *Toxic Waste and Race in the United States*, New York: United Church of Christ

United States Census Bureau (2012) *Statistical Abstract of the United States*, Washington, DC: US Department of Commerce

Unman, M. F. (ed.) (1993) *Keeping Pace with Science and Engineering: Case Studies in Environmental Regulation* (National Academy of Engineering), Washington, DC: National Academy Press

Upton, D. and Vlach, J. M. (eds) (1986) *Common Places: Readings in American Vernacular Architecture*, Athens, GA: University of Georgia Press

Urban Institute (1995) *Federal Funds, Local Choices: An Evaluation of the Community Development Block Grant Program*, Washington, DC: US Department of Housing and Urban Development

Urban Land Institute (1991) *The Case for Multifamily Housing*, Washington, DC: Urban Land Institute

Urban Land Institute and National Building Museum (2005) *Affordable Housing*, Washington, DC: Urban Land Institute and National Building Museum

Urban Mobility Corporation (1995) 'The Dulles Greenway', *Innovation Briefs*, 6–5 (October)

US Army Corps of Engineers, US Environmental Protection Agency, US Fish and Wildlife Service, and National Oceanic and Atmospheric Administration (2000) 'Federal guidance on the use of in-lieu-fee arrangements for compensatory mitigation under Section 404 of the Clean Water Act and Section 10 of the Rivers and Harbors Act', Washington DC: US Army Corps of Engineers, US EPA, US Fish and Wildlife Service, and National Oceanic and Atmospheric Administration

US Census Bureau (2002) *2002 Census of Governments – Government Organization*, Volume 1, Number 1, Washington, DC: Department of Commerce

US Census Bureau (2007) *Statistical Abstract of the US*, Washington, DC: Department of Commerce

US Conference of Mayors (1966) *With Heritage So Rich: A Report of a Special Committee on Historic Preservation*, New York: Random House

US Congressional Budget Office (1994) *Cleaning up the Department of Energy's Nuclear Weapons Complex*, Washington, DC: CBO

US Department of Agriculture (1981) *National Agricultural Lands Study*, Washington, DC: GPO

US Department of Housing and Urban Development (1994) *Impact Fees and the Role of the State: Guidance for Drafting Legislation*, Washington, DC: HUD

US Department of Housing and Urban Development (1995a) 'Consolidation plan for community development', Washington, DC: HUD

US Department of Housing and Urban Development (1995b) *Empowerment: A New Covenant with America's Communities: President Clinton's National Urban Policy Report*, Washington, DC: HUD

US Department of Housing and Urban Development (1997) *State of the Cities – 1997*, Washington, DC: US Department of Housing and Urban Development

US Department of Housing and Urban Development (1998) *State of the Cities – 1998*, Washington, DC: US Department of Housing and Urban Development

US Department of Housing and Urban Development (1999) *State of the Cities – 1999*, Washington, DC: US Department of Housing and Urban Development

US Department of Housing and Urban Development (2003) *Tax Incentive Guide for Businesses in the Renewal Communities, Empowerment Zones, and Enterprise Communities FY 2003*, Washington DC: US HUD

US Department of Housing and Urban Development (2006) 'HUD Strategic Plan: FY 2006 – FY 2011', Washington, DC: US Department of Housing and Urban Development

US Department of Housing and Urban Development, Center for Faith-Based and Community Initiatives (2001) 'Unlevel playing field: barriers to participation by faith-based and community organizations in federal social service programs', Washington, DC: US Department of Housing and Urban Development

US Department of Housing and Urban Development, Office of Policy Development and Research (2009) 'Revitalizing foreclosed properties with land banks', Report prepared by Sage Computing, Inc., Washington: US Department of Housing and Urban Development, Office of Policy Development and Research

US Department of the Interior, National Park Service (2011a) *Federal Tax Incentives for Rehabilitating Historic Buildings Annual Report for Fiscal Year 2011*, Washington, DC: DOI

US Department of the Interior, National Park Service (2011b) *Federal Tax Incentives for Rehabilitating Historic Buildings Statistical Report and Analysis for Fiscal Year 2011*, Washington, DC: DOI

US Department of Transportation (n.d.) 'A guide to metropolitan transportation under ISTEA – How the pieces fit together', Washington, DC: DOT

US Department of Transportation (1984) *Report on Highway Beautification Program*, Washington, DC: DOT

US Department of Transportation (1993) *Transport Implications of Telecommuting*, Washington, DC: DOT

US Department of Transportation (2007) 'Urban Partnership Agreement by and between U.S. Department of Transportation and the New York City Urban Partner', Washington, DC: DOT

US Department of Transportation, Federal Highway Administration, Office of Transport Management (2006) 'Congestion pricing: a primer', FHWA-HOP-07-074, Washington, DC: US DOT, Federal Highway Administration, Office of Transport Management

US Department of Transportation, Federal Highway Adminstration, and Federal Transit Administration (updated 2007), 'The transportation planning process: key issues', FHWA-HEP-07-039, Washington, DC: US Department of Transportation, Federal Transit Administration

US Department of Transportation, Maritime Administration (2007) *The Maritime Administration and the U.S. Marine Transportation System: A Vision for the 21st Century*, Washington, DC: US Department of Transportation, Maritime Administration

US Department of Transportation, Research and Innovative Technology Administration (2011) *America's Container Ports: Linking Markets at Home and Abroad*, Washington, DC: US Department of Transportation, Research and Innovative Technology Administration

US Environmental Protection Agency (1987) *Unfinished Business*, Washington, DC: EPA

US Environmental Protection Agency (1992) *Environmental Equity – Reducing Risk for All Communities*, Washington, DC: EPA

US Environmental Protection Agency (1993) *Technical and Economic Capacity of States and Public Water Systems to Implement Drinking Water Regulations: Report to Congress*, Washington, DC: EPA

US Environmental Protection Agency (1994a) *Quality of Our Nation's Rivers 1992*, Washington, DC: EPA

US Environmental Protection Agency (1994b) *President Clinton's Clean Water Initiative*, Washington, DC: EPA

US Environmental Protection Agency (1994c) *Clean Water: A Memorial Day Perspective*, Washington, DC: EPA

US Environmental Protection Agency (1994d) *Safe Drinking Water Act Reauthorization Review*, Washington, DC: EPA

US Environmental Protection Agency (1995) *America's Wetlands: Our Vital Link Between Land and Water*, Washington, DC: EPA

US Environmental Protection Agency (2005a) *Draft Handbook for Developing Watershed Plans to Restore and Protect Our Waters*, EPA 841-B-05-005, Washington, DC: EPA Office of Water, Nonpoint Source Control Board

US Environmental Protection Agency (2005b) 'Review of the National Ambient Air Quality Standards for Particulate Matter: policy assessment of scientific and technical information', EPA-452/R-05-005a, Research Triangle Park, NC: EPA

US Environmental Protection Agency (2006) *Proposed Wetlands Conservation Rule*, Washington, DC: EPA

US Federal Transit Administration (1999) *Building Livable Communities with Transit*, Washington, DC: US Federal Transit Administration

US General Accounting Office (1983) *Siting of Hazardous Waste Landfills and their Correlation with Racial and Economic Status of Surrounding Communities*, Washington, DC: GAO

US General Accounting Office (1986) *The Nation's Water: Key Unanswered Questions about the Quality of Rivers and Streams*, Washington, DC: GAO

US General Accounting Office (1993a) *Department of Energy: Cleaning up Inactive Facilities will be Difficult*, Washington, DC: GAO

US General Accounting Office (1993b) *Much Work Remains to Accelerate Facility Cleanups*, Washington, DC: GAO

US General Accounting Office (2001) 'Telecommuting: overview of potential barriers facing employers', Washington, DC: US Governmental Accounting Office

US Office of Technology Assessment (1989) *Facing America's Trash: What Next for Municipal Solid Waste?* Washington, DC: OTA

Varady, D. P. (1998) *New Directions in Urban Public Housing*, New Brunswick, New Jersey: Center for Urban Policy Research

Varady, D. P. and Raffel, J. (1995) *Selling Cities: Attracting Homebuyers through Schools and Housing Programs*, Albany, NY: State University of New York Press

Vermont, State of, Department of Housing and Community Affairs (2007) 'Growth centers planning manual for Vermont communities', Montpelier, VT: Department of Housing and Community Affairs

Vidal, A. (2001) *The Role of Faith-Based Organizations in Community Development*, Washington, DC: US Department of Housing and Urban Development, Office of Policy Development and Research

Vig, N. J. and Kraft, M. E. (1990) *Environmental Policy in the 1990s*, Washington, DC: Congressional Quarterly Press

Vig, N. J. and Kraft, M. E. (eds) (2012) *Environmental Policy: New Directions for the Twenty-First Century* (8th edition), Thousand Oaks, CA: CQ Press

Virginia, State of, Department of Historic Resources (2000) 'Virginia's sustainable future: solutions through historic preservation', *Preservation in Progress Newsletter* 4(4): 1

Virginia, State of, Department of Housing and Community Development, Virginia Main Street Program (2006) *Program Guidelines*, Richmond, Virginia: Virginia Department of Housing and Community development, Virginia Main Street Program

Wachs, M. (ed.) (1985) *Ethics in Planning*, New Brunswick, NJ: Center for Urban Policy Research

Wachs, M. (1990) 'Regulating traffic by controlling land use: the South California experience', *Transportation* 16: 249–50

Wachter, S. M. and Smith, M. M. (eds) (2011) *The American Mortgage System: Crisis and Reform*, Philadelphia, PA: University of Pennsylvania Press

Waldon, R. (2006) *Planner and Politics*, Chicago, IL: APA Planners Press

Walker, C. J. and Weinheimer, M. (1998) *Community Development in the 1990s*, Washington, DC: Urban Institute

Walker, J. S. (2009) *The Road to Yucca Mountain: The Development of Radioactive Waste Policy in the United States*, Berkeley, CA: University of California Press

Walle, A. H. (1998) *Cultural Tourism: A Strategic Focus*, Boulder, CO: Westview Press

Wallis, A. D. (1991) *Wheel Estate: The Rise and Decline of Mobile Homes*, New York: Oxford University Press

Walters, D. (2007) *Designing Community: Charrettes, Masterplans and Form-based Codes*, Amsterdam: Elsevier

Waltman, J. L. and Holland, K. M. (eds) (1988) *The Political Role of Law Courts in Modern Democracies*, New York: St Martin's Press

Warner, K. and Molotch, H. (2000) *Building Rules:*

How Local Controls Shape Community Environments and Economies, Boulder, CO: Westview Press

Warner, S. B. (1968) *The Private City: Philadelphia in Three Periods of Its Growth*, Philadelphia: University of Pennsylvania Press

Warner, S. B. (1972) *The Urban Wilderness: A History of the American City*, New York: Harper and Row (reprinted 1995, Berkeley, CA: University of California Press)

Warner, S. B. (1978) *Streetcar Suburbs: The Process of Growth in Boston 1870–1900* (2nd edition), Cambridge, MA: Harvard University Press

Wascott, Town of (2003) 'Town of Wascott Comprehensive Planning Process: 'Public Participation Plan'', Wascott, WI: Town of Wascott Planning Committee

Waste, R. J. (1998) *Independent Cities: Rethinking US Urban Policy*, New York: Oxford University Press

Watson, D. W., Plattus, A., and Shibley, R. (2003) *Time-Saver Standards for Urban Design*, Columbus, OH: McGraw-Hill

Watson, V. (2009) 'Seeing from the South: refocusing urban planning on the globe's central urban issues,' *Urban Studies* 46(11): 2259–2274

Weaver, C. L. and Babcock, R. F. (1979) *City Zoning: The Once and Future Frontier*, Chicago: Planners Press

Weber, R. and Goddeeris, L. (2007) *Tax Increment Financing: Process and Planning Issues*, Cambridge, MA: Lincoln Institute of Land Policy

Webster's New Universal Unabridged Dictionary (1989) New York: Barnes and Noble

Weiner, E. (1997) *Urban Transportation Planning in the United States: An Historical Overview* (5th edition), Washington, DC: US Department of Transportation

Weiss, M. A. (1980) 'The origins and legacy of urban renewal', in Clavel *et al.* (1980)

Weiss, M. A. (1987) *The Rise of the Community Builders: The American Real Estate Industry and Urban Land Planning*, New York: Columbia University Press

Weiss, N. E. (2012) *Fannie Mae's and Freddie Mac's Financial Problems*, 7-5700, Washington, DC: Congressional Research Service, August 10

Weitz, J. (1999) *Sprawl Busting: State Programs to Guide Growth*, Chicago: American Planning Association

Wells, C. (ed.) (1986) *Perspectives in Vernacular Architecture*, Annapolis, MD: Vernacular Architecture Forum

Werner, D. (2001) 'Note: the public use clause, common sense and takings', *Boston University Public Interest Law Journal* 10: 335–59

Westminster, City of (2009) *2009 Comprehensive Plan*, Westminster, Maryland: City of Westminster

Wheeler, J. O., Warf, B., and Aoyama, Y. (eds) (1999) *Cities in the Telecommunications Age: The Fracturing of Geographies*, London: Taylor and Francis

Wheeler, S. M. (2004) *Planning for Sustainability: Creating Livable, Equitable and Ecological Communities*, NY: Routledge

Wheeler, S. M. (2008) 'Smart infill: creating more livable communities in the Bay Area – a guide for Bay Area leaders', San Francisco, CA: Greenbelt Alliance

Wheeler, S. M. and Beatley, T. (2009) *Sustainable Urban Development Reader* (2nd edition), NY: Routledge

White, M. S. (1992) *Affordable Housing: Proactive and Reactive Planning Strategies* (Planning Advisory Service, Report 441), Chicago: American Planning Association

White, S. B., Bingham, R. D., and Hill, E. W. (eds) (2003) *Financing Economic Development in the 21st Century*, Armonk, NY: M. E. Sharpe

White House Office on Environmental Policy (1993) *Protecting America's Wetlands: A Fair, Flexible, and Effective Approach*, Washington, DC: White House Office on Environmental Policy

White House Office of Faith-Based and Community Initiatives (2001) *Unlevel Playing Field: Barriers to Participation by Faith-Based and Community Organizations in Federal Social Service Programs*, Washington, DC: White House Office of Faith-Based and Community Initiatives

Whyte, W. H. (1988) *City: Rediscovering the Center*, New York: Doubleday

Wickersham, J. H. (1994) 'The Quiet Revolution continues: the emerging new model for state growth management statutes', *Harvard Environmental Law Review* 18: 489–548

Wildavsky, A. (1987) *Speaking Truth to Power: The Art and Craft of Policy Analysis* (2nd edition), New Brunswick, NJ: Transaction

Williams, N. (1990) *American Land Planning Law*, Deerfield, IL: Callaghan

Williams, S. F. (1977) 'Subjectivity, expression, and privacy: problems of aesthetic regulation', *Minnesota Law Review* 62: 1–58

Wills, G. (1999) *A Necessary Evil: A History of American Distrust of Government*, New York: Simon & Schuster

Wolf, M. A. (2008) *The Zoning of America: Euclid v. Ambler*, Lawrence, KS: University Press of Kansas

Wolman, H. (1988) 'Local economic development policy: what explains the divergence between policy analysis and political behavior?', *Journal of Urban Affairs* 10: 19–28

WorldatWork (2011) Telework2011: A WorldatWork Special Report, Scottsdale, AZ: WorldatWork

World Commission on Environment and Development (1987) *Our Common Future*, New York: Oxford University Press

Wright, E. (1999) *An Annotated Bibliography for Faith-Based Community Economic Development*, Washington, DC: National Congress for Community Economic Development

Wright, R. R. and Gitelman, M. (1982) *Land Use: Cases and Materials* (3rd edition), St Paul, MN: West Group

Wright, R. R. and Gitelman, M. (2000) *Land Use in a Nutshell*, Belmont, CA: West/Wadsworth

Wurtzebach, C. H. and Miles, M. E. (1991) *Modern Real Estate* (4th edition), New York: Wiley

Wylie, J. (1989) *Poletown: Community Betrayed*, Urbana, IL: University of Illinois Press

Yandle, B. (ed.) (1995) *The Property Rights Rebellion: Land Use Movements in the 1990s*, Lanham, MD: Rowman and Littlefield

Yinger, J. (1995) *Closed Doors, Opportunities Lost: The Continuing Costs of Housing Discrimination*, New York: Russell Sage Foundation

Zimmerman, R. and Horan, T. (eds) (2004) *Digital Infrastructures: Enabling Civil and Environmental Systems through Information Technology*, London: Taylor and Francis

Ziph, R. (1995) *How Municipal Bonds Work*, New York: New York Institute of Finance

Zovanyi, G. (1998) *Growth Management for a Sustainable Future: Ecological Sustainability as the New Growth Management Focus for the Twenty-First Century*, Westport, CT: Praeger

Index